Y0-BUA-643

Video Dialtone Technology

Other McGraw-Hill Books of Interest

BARTLETT • *Cable Communications*

CHETTY • *Satellite Component Data Handbook*

GORALSKI • *Introduction to ATM Networking*

GRINBERG • *Computer/Telecom Integration*

KESSLER • *ISDN, 2/e*

KUMAR • *Broadband Networks*

MCDYSAN/SPOHN • *ATM*

MINOLI • *1st, 2nd, and Next Generation LANs*

MINOLI • *Imaging in Corporate Environments*

MINOLI/VITELLA • *ATM and Cell Relay Service for Corporate Environments*

RUSSELL • *Signaling System #7*

SZUPROWICZ • *Multimedia Networking*

WINCH • *Telecommunication Transmission Systems*

To order or receive additional information on these or any other McGraw-Hill titles, in the United States please call 1-800-822-8158. In other countries, contact your local McGraw-Hill representative.

BC15XXA

Video Dialtone Technology

Digital Video over ADSL, HFC, FTTC, and ATM

Daniel Minoli
DVI Communications, Inc.
and
Stevens Institute of Technology

McGraw-Hill, Inc.

New York San Francisco Washington, D.C. Auckland Bogotá
Caracas Lisbon London Madrid Mexico City Milan
Montreal New Delhi San Juan Singapore
Sydney Tokyo Toronto

Library of Congress Cataloging-in-Publication Data

Minoli, Daniel, date.
Video dialtone technology : digital video over ADSL, HFC, FTTC, and ATM / Daniel Minoli.
p. cm.
Includes bibliographical references and index.
ISBN 0-07-042724-0
1. Digital video. 2. Cable television. 3. Data transmission systems. I. Title. II. Title: Video dial tone technology.
TK6680.5.M56 1995
621.388—dc20 94-49661
CIP

Copyright © 1995 by McGraw-Hill, Inc. All rights reserved. Printed in the United States of America. Except as permitted under the United States Copyright Act of 1976, no part of this publication may be reproduced or distributed in any form or by any means, or stored in a data base or retrieval system, without the prior written permission of the publisher.

2 3 4 5 6 7 8 9 0 DOC/DOC 9 0 0 9 8 7 6 5

ISBN 0-07-042724-0

The sponsoring editor for this book was Marjorie Spencer and the production supervisor was Pamela A. Pelton. It was set in Century Schoolbook by North Market Street Graphics.

Printed and bound by R. R. Donnelley & Sons Company.

McGraw-Hill books are available at special quantity discounts to use as premiums and sales promotions, or for use in corporate training programs. For more information, please write to the Director of Special Sales, McGraw-Hill, Inc., 11 West 19 Street, New York, NY 10011. Or contact your local bookstore.

Information contained in this work has been obtained by McGraw-Hill, Inc., from sources believed to be reliable. However, neither McGraw-Hill nor its authors guarantee the accuracy or completeness of any information published herein and neither McGraw-Hill nor its authors shall be responsible for any errors, omissions, or damages arising out of use of this information. This work is published with the understanding that McGraw-Hill and its authors are supplying information but are not attempting to render engineering or other professional services. If such services are required, the assistance of an appropriate professional should be sought.

For Gino, Angela,
Anna,
Emmanuelle, Emile, and Gabrielle

". . . acquistate ogni speranza voi che uscite . . ."
—an Escherism on Dante's Inferno

Contents

Preface

"A new day is dawning," FCC Commissioner Susan Ness said in July 1994. "No longer will telephone companies simply provide telephone services and Cable companies merely provide video programming services . . ."

There is now feverish activity in the U.S. telecommunications and cable TV industries, as well as within the realm of other interested parties, to deploy what has been metaphorically called the Information Superhighway. The global economic scene is changing and is evolving at an expedited pace to an information-intensive economy. Video is entering the workplace of many corporations and institutions, whether in the form of desk-to-desk video conferencing, group-based videoconferencing, imaging, distance learning and training (locally based and/or remotely based), production, or other type of video delivery. New domicile-distribution video systems are being contemplated for introduction during the next couple of years. These systems are distinguished from their predecessors in a number of fundamental aspects, which are the topic of this text:

1. New players (e.g., Regional Bell Operating Companies) are entering the cable TV market, increasing competitive pressures on existing providers and possibly leading to some consolidation.
2. The FCC has re-regulated the cable TV industry and has directed that competition be enhanced. New entrants and new technologies will be seen.
3. The new entrants are planning to take a quantum leap in technology as they approach the market. Not only will video delivery be based on fiber systems (at least for a portion of the network) and the video signal be digital, but the video signal will be compressed using sophisticated new algorithms, it will be delivered over switched networks, and it will be brought to the residential user employing broadband asynchronous transfer mode (ATM) techniques part of

the way or even all the way to a powerful setup box on top of (or inside) the TV set. Broadband techniques imply the availability of hundreds of channels (say, around 500).

4. The TV set will be transformed from a fairly passive device to a device that looks more like (or in fact uses) the Windows PC desktop. The user will interact through such an interface to order movies, use VCR-like functions, play video games, sustain interactive sessions for access to library information, purchase goods from point-and-click menus, obtain individualized news programming, take a course, and undertake other tasks, including computing.

Hence, in the view of the technology providers, TV as we know it is about to take a high-tech turbocharge. New delivery systems, based on electronically enhanced twisted pair, enhanced coaxial cable, fiber, hybrid fiber-coax, radio, and direct broadcast satellite, will employ digital transmission techniques, likely to include ATM (to support switching). The signal will be compressed using Motion Picture Expert Group 1 (MPEG-1) or, more likely, MPEG-2 techniques. Setup boxes supporting from 10 to 100 MIPS are contemplated for the home. GUI-like interfaces to the user (including object-based approaches) will be supported.

This timely text examines the many issues involved in this technological enhancement of video delivery, including technical considerations, regulatory considerations, market considerations, and service-development considerations.

The first chapter examines the age of digital video; Chapter 2 examines the service drivers and the markets. Regulatory aspects are addressed in Chapter 3. The next few chapters form the core of the technical discussion: Chapter 4 provides some technical preliminaries on TV technology; Chapter 5 describes typical video dialtone platforms being contemplated by carriers; Chapter 6 looks at digital video compression; Chapter 7 looks at digital video servers; Chapter 8 examines ATM and ATM switches. The next set of chapters covers the evolving distribution systems: Chapter 9 covers terrestrial/wireline distribution systems based on asymmetric digital subscriber line (ADSL); Chapter 10 discusses hybrid fiber coax; Chapter 11 discusses Fiber To The Curb (FTTC) and related methods; Chapter 12 looks at a number of other distribution systems, including direct broadcast satellite (DBS). Chapter 13 covers the home-based setup box. The last two chapters (Chapters 14 and 15) look at the key players and the competitors: telephone carriers moving to video (Chapter 14) and Cable TV providers moving to telephony while upgrading their cable plant (Chapter 15). Other aspects of competition are also covered in this chapter.

This text is aimed at technology developers, carriers, planners, manufacturers, students, and end users. We hope that the timely nature of the information will be of value to these constituencies. Many of these concepts are only one or two years old.

Acknowledgment

The author would like to warmly thank Mr. Benedict Occhiogrosso, DVI Communications, for his support of this effort.

Dan Minoli

Acronyms

AAL	ATM adaptation layer
AALM	AAL management
AAP	alternate access providers
AAU	audio access unit
ADSL	asymmetric digital subscriber line
AE	application entity
AIS	alarm indication signal
ALI	application level interface
AM-FDM	amplitude modulated-frequency division multiplexed
AM-VSB	amplitude modulation-vestigial sideband
AMS	audiovisual multimedia service
APU	audio presentation unit
ATIS	Alliance for Telecommunications Industry Solutions
ATM	asynchronous transfer mode
ATU-C	ADSL-3 terminal unit at the CO
ATU-R	ADSL-3 terminal unit at the remote (site)
B-ICI	Broadband InterCarrier Interface
B-ISDN	Broadband Integrated Services Digital Network
B-ISUP	Broadband ISDN User Part
BCH	Bose-Chaudhuri-Hocquenghem
BER	bit error-rate
BONT	broadband optical network termination
BOPS	billion operations per second
BSS	broadband switching system
CAP	carrierless amplitude/phase
CASE	common application service elements
CATV	Community Antenna Television

CBR	constant bit rate
CD-DA	compact disk digital audio
CD-I	compact disk interactive
CD-R	CD recordable
CDMA	code division multiple access
CDV	cell delay variation
CIE	Commission Internationale de L'Eclairage
CLR	cell loss ratio
CMC	connection management computer
CNM	customer network management
CO	central office
CPAAL5	Common Part of AAL Type 5
CPCS	Common Part CS
CPE	customer premises equipment
CRC	cylic redundancy check
CRS	cell relay service
CRT	cathode ray tube
CS	convergence sublayer
CSA	carrier serving area
CSO	composite second order
CSPS	constrained system parameter stream
CTB	composite triple beat
CTTC	Coax To The Curb
CTTH	Coax To The Home
DBS	direct broadcast satellite
DC	distribution center
DCT	discrete cosine transform
DDM	direction division multiplexing
DLC	digital loop carrier
DMT	discrete multitone
DNC	digital node controller
DP	demarcation point
DRAM	dynamic random access memory
DSL	digital subscriber line
DSM	digital storage media
DSP	digital signal processing
DSU	digital service units
DTS	decoding time stamp

DXI	Data Exchange Interface
EDFA	erbium-doped fiber amplifiers
EISA	Extended Industry Standard Architecture
ESCR	elementary stream clock reference
ESP	Enhanced Service Providers
FDDI	Fiber Distributed Data Interface
FDM	frequency division multiplexing
FITL	Fiber In The Loop
FM-FDM	frequency modulated-frequency division multiplexed
FNOI	Further Notice of Inquiry
FNPRM	Further Notice of Proposed Rule Making
FR&O	First Report and Order
FTTC	Fiber To The Curb
FTTH	Fiber To The Home
FTTN	Fiber To The Node
GCM	generalized control model
HDT	host digital terminal
HDTV	high-definition TV
HE	headend
HEC	header error control
HFC	hybrid fiber coaxial
I-TV	interactive TV
IM	intensity modulation
IMVOD	impulse VOD
ISDN	Integrated Services Digital Network
ISP	Information Services Providers
ISUP	ISDN User Part
ITFS	Instructional Television Fixed Service
ITU-T	International Telecommunications Union-Telecommunication
IVOD	interactive VOD
IVSN	Video Services Network
IVS	Interactive Video Services
IW	information warehouse
JPEG	Joint Photographic Experts Group
LCM	logical control modules
LEA	line extender amplifier
LEC	local exchange carriers

LOS	line of sight
MDS	Multipoint Distribution Service
MFJ	modified final judgment
MIB	management information base
MIPS	million instructions per second
MMDS	multichannel MDS
MO	magneto-optic
MPEG	Motion Picture Expert Group
MPEG-1	Motion Picture Expert Group 1
MPEG-2	Motion Picture Expert Group 2
MPG	multiplayer games
NAN	Neighborhood Area Network
NAS	network access signaling
NII	National Information Infrastructure
NNI	Network Node Interface
NOI	Notice of Inquiry
NPRM	Notice of Proposed Rule Making
NREN	National Research and Education Network
NS	network signaling
NTSC	National Television System Committee
NVOD	near real-time video on demand
O-QPSK	offset quadrature phase shift keying
OAM	operation and maintenance
OAM&P	operations, administration, management, and provisioning
ODN	Optical Distribution Network
OLT	optical line termination
ONU	optical network unit
OSIRM	Open Systems Interconnection Reference Model
PCI	protocol control information
PCR	program clock references
PCS	Personal Communication Services
PDN	passive distribution network
PDU	payload data unit
PDU	protocol data unit
PES	packetized elementary stream
PID	packet identifier
PIN	personal identification number

PLV	production level video
PMD	physical medium dependent
POS	passive optical splitter
POS	point of sale
PSI	program-specific information
PSTN	Public Switched Telephone Network
PTM	packet transfer mode
PTS	presentation time stamp
PU	presentation unit
PVC	permanent virtual connection
PVD	point of video delivery
QAM	quadrature amplitude modulation
QOS	quality of service
QPSK	quaternary phase shift keying
RAID	redundant arrays of inexpensive disks
RARP	Reverse Address Resolution Protocol
RBOCs	Regional Bell Operating Companies
RISC	Reduced Instruction Set Computing
RRD	revised resistance design
RTS	residual time stamp
SAP	service access point
SAR	segmentation and reassembly sublayer
SASE	specific application service element
SCSI	Small Computer System Interface
SDH	synchronous digital hierarchy
SDT	structured data transfer
SDU	service data unit
SFNOI	Second Further Notice of Inquiry
SIF	source input format
SMATV	Satellite Master Antenna Television
SMDS	Switched Multimegabit Data Service
SNMP	Simple Network Management Protocol
SOA	semiconductor optical amplifier
SONET	synchronous optical network
SPE	synchronous payload envelope
SRM	self-routing modules
SRTS	synchronous residual time stamp
SSCS	service-specific cs

STC	system time clock
STD	system target decoder
STM	synchronous transfer mode
STM	synchronous transport module
STS-1	synchronous transport signal-level 1
STT	set-top terminal
STV	Subscription Television
SVC	switched virtual connection
TA	terminal adapter
TA-BA	trunk-bridger amplifier
TC	transmission convergence
TCM	time compression multiplexing
TDM	time division multiplexing
TDMA	time division multiple access
TN	transport network
TS	time stamp
TVRO	television receive-only
UDP	User Datagram Protocol
UDT	unstructured data transfer
UNI	User-Network Interface
UNISON-1	Unidirectional Synchronous Optical Network
VAS	video application signaling
VAU	video access unit
VBR	variable bit rate
VC	virtual channel
VCC	virtual channel connection
VCI	virtual channel identifier
VCL	virtual channel link
VDT	video dialtone
VIP	video information provider
VIP-NI	VIP network interface
VIU	video information user
VIU-NI	VIU network interface
VIW	VIP-video information warehouse
VMTP	VDT Message Transfer Part
VMTP-T	VDT Messages Transfer Part-Translator
VOD	video on demand
VP	virtual path

VPC	virtual path connection
VPI	virtual path identifier
VPL	virtual path link
VPU	video presentation unit
VSB	vestigial sideband
VSCCP	VDT Signaling Connection Control Part
VSCP	VDT Session Control Part
WAN	wide area network
WDM	wavelength division multiplexing
WORM	write once read many

Video Dialtone Technology

Chapter

1

The Age of Digital Video

1.1 Overview

There is now feverish activity in the U.S. telecommunications and cable TV industries, as well as within the realm of other interested parties, such as Information Services Providers (ISP), to deploy what has been called the *Information Superhighway.* The global economic scene is changing and is evolving, at an expedited pace, to an information-intensive economy. Digital business techniques, based on powerful point-of-access processors capable of executing from 10 to 100 million instructions per second (MIPS), are permeating many, if not most, economic activities from education, medicine, and entertainment to corporate functions and underlying work paradigms. Economic planners palpate the strong correlation between economic viability of 21st-century nations and the broadband telecommunication infrastructure these nations now start to put in place. The analogy with the introduction of railroads in the 19th century and highways and airports in the 20th century is compelling.

A key driver of the Information Superhighway is video. Video is entering the workplace of many corporations and institutions, whether in the form of desk-to-desk videoconferencing, group-based videoconferencing, imaging, video production or development, distance learning and training (locally based and/or remotely based), or other type of video delivery. As the reading habits of most of the population follow a downward spiral, video and visually based work, business, and education practices become almost an imperative for economic survival. New domicile-distribution video systems*, including what have been called

* Unless noted otherwise, in this text the terms *video delivery* and *video distribution* refer to the delivery of video to the residential customer (as contrasted to, for example, backbone video transport).

video dialtone (VDT*) systems, are being contemplated for introduction in the United States during the next couple of years. These systems are distinguished from their predecessors in a number of fundamental ways, and are the topic of this text:

1. New players (e.g., the Regional Bell Operating Companies, or RBOCs) are entering the cable TV market, increasing competitive pressures on existing providers and possibly leading to some industry consolidation.
2. The Federal Communications Commission (FCC) has re-regulated the cable TV industry and has directed that competition be enhanced. New entrants and new technologies will be seen.
3. The new entrants are planning to take a quantum leap in technology as they approach the market.
   a. video delivery will be based on fiber systems (at least for a portion of the network)
   b. the video signal will be digital
   c. the video signal will be compressed using sophisticated new algorithms
   d. the video signal will be delivered over switched networks
   e. the video signal will be brought to the residential user employing broadband asynchronous transfer mode (ATM) techniques, part of the way or even all the way, to a powerful setup† box on top of (or inside) the TV set. Broadband techniques imply the availability of hundreds of channels (say, around 500) when channels are talked about and delivered in the traditional sense; with switching capabilities, even a system with enough capacity to simultaneously deliver only 12 channels can actually give the consumer access to thousands of program sources.
4. The TV set will be transformed from a fairly passive device to a device that looks more like (or in fact could use) the Windows PC desktop or a similar derivative. The user will interact through such interface to order movies, use VCR-like functions, play video games, sustain interactive sessions for access to library information, pur-

---

* At the grammatical level, the terms *VDT* and *video dialtone* should be used as modifiers, i.e., VDT systems, VDT services, VDT platforms, and VDT signaling. Sometimes we use VDT by itself to refer to the entire field, including the technology, the infrastructure, the blueprint, and the principles.

† The term *set-top* box is also used in the literature. We prefer the term *setup* since this is consistent with the goal of delivering video over ATM and using this device to set up connections through (principally) the SETUP signaling message of Q.2931. Set-top has a parochial localized topological flavor and is not representative of the function. In some cases, the box could actually be under the TV; one would not use the term set-bottom, set-side, etc.

chase goods from point-and-click menus, obtain individualized news programming, take a course, and undertake other tasks, including computing (e.g., checkbook maintenance).

In the view of the technology providers, TV as we know it is about to take a high-tech turbocharge. Observers see *profound* technological changes coming to television in general, and video delivery in particular.[1] The industry is being pushed into new technologies by both fear and greed, as current providers "fear threats to the current business and feel greed in seeing new opportunities . . . for substantial revenues."[2] This realignment of the industry has been called *convergence.* New delivery systems, based on electronically enhanced twisted pair providing 1.5 Mbps, enhanced coaxial cable supporting one gigaHertz of bandwidth, fiber, hybrid fiber-coax, radio, or direct broadcast satellite delivering 150 channels, will employ digital transmission techniques, likely to include end-to-end or edge-to-edge ATM services to support switching functions. The signal will be compressed using Motion Picture Expert Group 1 (MPEG-1) or, more likely, MPEG-2 techniques. Setup boxes supporting from 10 to 100 MIPS of computing power are being contemplated for the home. GUI-like interfaces to the user (including object-based approaches) will be supported. Two-way interactivity will be routine. Numerous carriers, traditional cable TV companies, equipment providers, and sponsors have initiated trials of this new "TV-2000" (to quote a 1994 CNN documentary on this topic) concept.

There are (melo)dramatic characterizations of the impending changes to which this text addresses itself, such as the following:[3]

> Telecommunication carriers are experiencing cataclysmic changes in business circumstances during this last decade of the 20th century. Carriers using media such as radio, cable, and satellite are invading the kingdom of the conventional carriers. The battle has begun, and only those carriers who can provide the highest levels of performance, reliability, and flexibility will be the winners. Customers are ascertaining how this melee will develop. They are judging the carrier based on existing and potential services. Only sophisticated, truly open-ended systems will have any chance of gaining customer acceptance. Many problems must be overcome . . . but if carriers hesitate . . . they will be in a hopeless situation in the 21st century.

Given such stakes, this text examines the many issues involved in this technological enhancement of video delivery in general, and video dialtone in particular, focusing on a whole gamut of Interactive Video Services (IVSs). The text includes technical considerations, regulatory considerations, market considerations, and service-development considerations. Initially, it appeared that the RBOCs' road to residential video distribution was going to be through the acquisition of existing cable TV companies, in well-publicized multibillion dollar deals such as Bell Atlantic/TCI and Southwestern Bell Telephone Company*/Cox Enterprises.† Now it appears that they have taken the road of developing their own system by deploying broadband systems (that is, broadband ATM interoffice facilities, broadband ATM switching, and broadband local loop technologies such as Asymmetric Digital Subscriber Line, Hybrid Fiber Coax, Fiber To The Node, Fiber To The Curb, and Fiber To The Home), thereby constructing their own infrastructure. Among other things, this text examines issues related to the deployment of such new infrastructure supporting residential (also, business) video requirements.

The first chapter examines, in encapsulated form, the entire field of digital video and digital video distribution, in order to provide the reader a sense of this emerging technology. The chapters that follow take various aspects introduced in Chap. 1 and provide more detailed coverage. Chapter 2 examines the service drivers and the markets. Regulatory aspects are addressed in Chap. 3. The next few chapters form the core of the technical discussion: Chap. 4 provides some technical preliminaries on TV and related technology; Chap. 5 describes typical video dialtone platforms being contemplated by carriers; Chap. 6 looks at digital video compression; Chap. 7 looks at digital video servers; Chap. 8 examines ATM and ATM switches. The next set of chapters cover the evolving distribution systems: Chap. 9 covers terrestrial-wireline distribution systems based on Asymmetric Digital Subscriber Line (ADSL); Chap. 10 discusses Hybrid Fiber Coax (HFC) and related methods such as Fiber To The Node (FTTN); Chap. 11 discusses Fiber To The Curb (FTTC) and Fiber To The Home (FTTH); Chap. 12 looks at a number of other distribution systems, including Direct Broadcast Satellite (DBS). Chapter 13 covers the home-based setup box. The last two chapters (Chaps. 14 and 15) look at the key players and the competitors: telephone carriers moving to video through the deployment of broadband platforms (Chap. 14) and cable TV providers moving to telephony while upgrading at the same time their cable plant (Chap. 15); other aspects of competition are also covered in this chapter.

There are many technical aspects of this developing technology that are still at the research stage. Two areas of particular importance are

---

* Now known as SBC Communications Inc.

† Other recent corporate alliances and/or overtures include: U S WEST/Time Warner; NYNEX/Time Warner; NYNEX/Viacom; BellSouth/Prime Management; BellSouth/QVC; Southwestern/Hauser; Sprint/TCI/Comcast/Cox; U S WEST/TCI; AT&T/Time Warner; Bell Atlantic/FutureVision; Cablevision/Ameritech; TCI/Jones Intercable; and GTE/various cable TV companies.

signaling and control and carrying compressed digital video over ATM networks. Both aspects are treated in detail in Chap. 5 and Chap. 8, respectively.

This text can be read at two levels. At the first level, it can be read in terms of opportunities, technologies, and challenges related to the RBOCs' entry into the video distribution market under the auspices of VDT. Second, it can be read as an assessment of the latest technologies to support enhanced delivery of video, regardless of who does it or under what banner. Some technical material applies principally to cable TV systems; other technical material applies principally to VDT systems; a lot of the material applies to both systems. The book aims at being clinical and neutral. It aims at presenting not a single side of the debate—for example, who is best suited to bring video to the consumer, if such a best party could, in fact, exist—but, rather, in taking all sides of the debate. However, the book is on VDT and, therefore, examines the topic from that specific perspective, rather than from a completely undirected perspective.

1.2 The Environment and the Players

Major new applications for video are expected in the next few years. Wider choice of programming should become a reality as both existing providers and new providers add new delivery systems or expand existing ones. Interactive TV (I-TV), including teleshopping, distance learning, and video games, may become commonplace in the next few years. Video telephony and business videoconferencing are expected to see increased penetration. Desktop multimedia programs on PCs will bring the value of video, PC-based video, and interactivity to the forefront of people's minds. People can now see their favorite soap opera or TV news in a small window in their PCs at the office, right inside a word-processing or spreadsheet application.* This section looks at the current environment for video delivery over terrestrial links as well as the emerging trends, enabling some of these services, as well as others, to become a reality.

---

* In 1994, Intel and Turner Broadcasting started a service for the delivery of *CNN Headline News* to a 2- by 3-in window on a PC (the goal was for a nationwide service by the end of 1994). A satellite dish at a company site picks up the news broadcast; the signal is processed by a PC (80486- or Pentium-based, plus new Intel software). The signal is processed to digitize and compress it. The signal is then sent over a local area network to all PCs in the organization. The user's PC employs Intel software to decompress the digital video signal (a sound card needs to be added). Each PC can also store up to 30 minutes of video on the hard disk for delayed display (the delayed display is either linear or can be based on an indexed access pointing to the headlines' content).[4]

1.2.1 Video delivery

There are at least two major network possibilities for the guided*† delivery of video: the cable TV company approach and the telephone company video dialtone approach.

Figures 1.1*a* and *b* depicts various aspects of a traditional video distribution system. In almost all instances, the video distribution is analog. Traditionally, the cable systems used a trunk-and-branch,‡ one-to-many broadcast tree topology; the return bandwidth has been small or none at all. The cable TV company is likely to employ a coaxial-based system which carries analog TV signals. Over 60 percent of U.S. households have the physical coaxial cable installed. Even when the cable TV companies replace some of the coaxial system with fiber, analog techniques are still employed in most instances; additionally, the signal is reconverted to a coaxial-based delivery mechanism (see Fig. 1.2). Occasionally, video has been carried in digital form by common carriers, for example, between TV studio sites. This type of video operates at DS3 (45 Mbps) rates, using a small amount of compression (additional discussion on compression follows later).§

Cable TV companies have their roots in the broadcast of analog video through a coaxial medium, even though, as noted, an increasing amount of fiber with analog modulation is being used. Their principal concern at this point is to achieve higher capacity (this being dictated by the fact that they broadcast all the channels into the medium rather than just supplying those that the consumer wants at any point in time). The increased capacity is in support of enhanced entertainment services. Additionally, there is interest in adding other consumer services such as I-TV, video on demand (VOD), home shopping, viewer polling, information on demand, and possibly telephony (some of the interactive services could actually be achieved using the public

---

* We use the term *guided* in the electromagnetic sense of delivering a signal over a physical medium such as copper, fiber, or coaxial cable. The term *unguided* refers to the delivery of signals over the air, either in the radio domain or in the infrared (laser) domain. The terms *wireline* and *wireless*, respectively, can also be used.

† Unguided systems such as satellite, "wireless" cable, and cellular cable, are discussed in Chap. 12.

‡ Also called *tree-and-branch*.

§ Some distance learning networks that support two-way video with "continuous presence" (i.e., round-robin or on-demand view of all classrooms, including quad-split screens) have used 45-Mbps switched digital video since the early 1990s (other distance learning systems use analog TV). For example, Northern Telecom's Integrated Community Network (ICN) product used in the Mississippi FiberNet 2000 network, in the Vision Caroline network, and in other educational networks, employs digital codecs at 45 Mbps for use over a switched DS3 network—compressed video at 6 Mbps for use over ATM network should follow in the future for these types of applications.

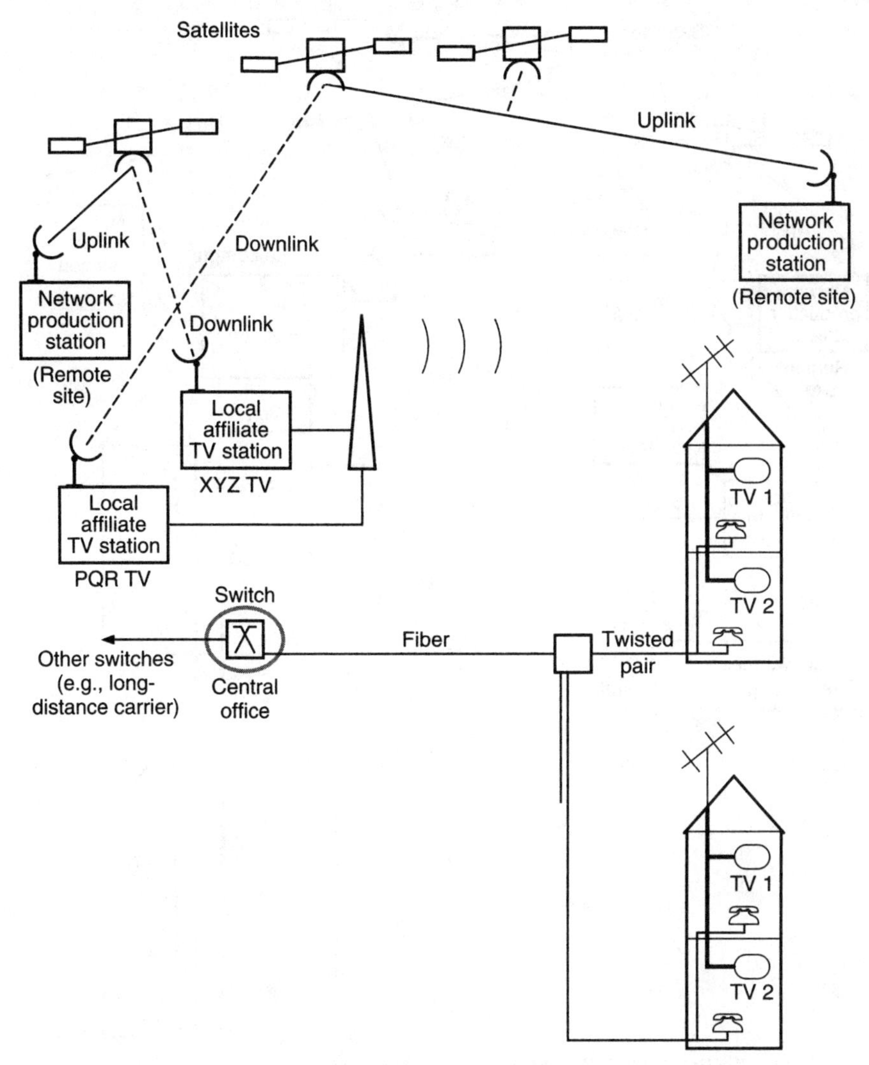

Figure 1.1*a* Traditional TV distribution without cable service.

switched telephone network (PSTN) for the return path, namely, employing the user's telephone line). Cable companies, such as TCI and others, are reportedly aiming at providing both video and voice-data services over a single coaxial cable; there is also interest in providing data services to the business user.[5,6] The problem faced by the cable TV industry in pursuing these goals is that the existing cable plant is inflexible, is not open, and does not easily interconnect with the rest of the information world (that is to say, beyond the video programming that is offered). The upgrade of this collection of about

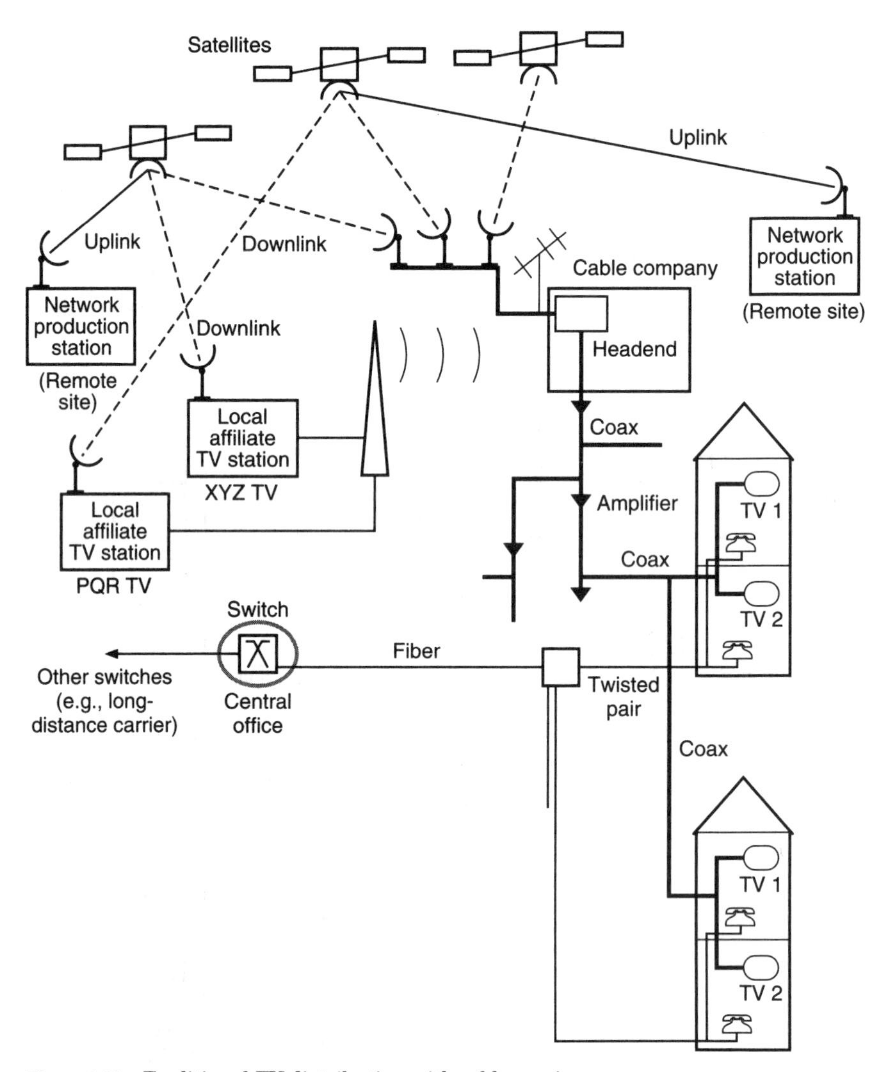

Figure 1.1*b* Traditional TV distribution with cable service.

11,200 U.S. networks to a switched, high-capacity,* open, standard-based system will be challenging.

The public switched telephone network made available by the RBOCs (along with the interexchange carriers) used to be analog, that

* Although one might consider 450 MHz to be high (supporting from 450 Mbps to 3 Gbps if Quadrature Amplitude Modulation or Vestigial SideBand techniques could be used for the entire spectrum, even though higher bandwidth is unlikely because of the noise on the cable), the fact that the channels are broadcast and are in analog form (each requiring 6 MHz), in reality limits the system to a few dozen channels.

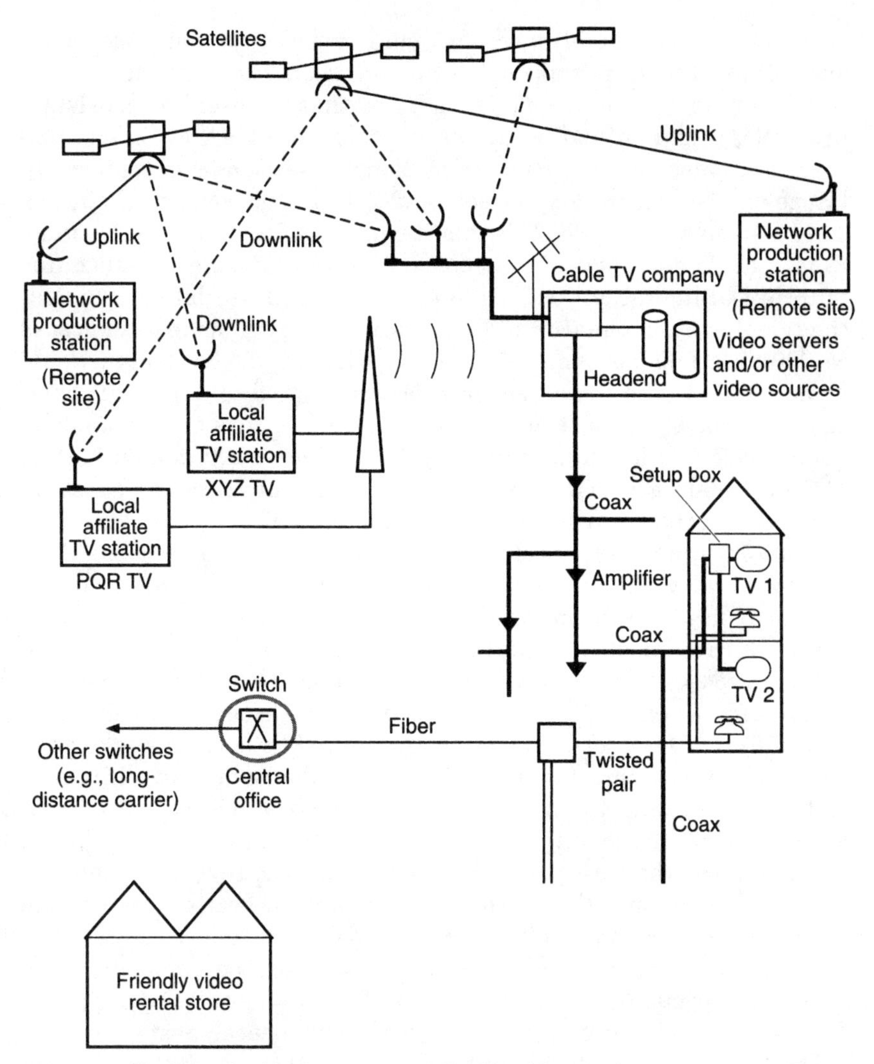

Figure 1.2 Traditional cable distribution with other video services, e.g., Pay Per View.

is, it employed analog distinct copper loops, analog switching, and analog interoffice transmission systems. Now, most of the switching systems are digital and, in the local loop, transmission occurs over either digital T1 feeder copper-based systems or over fiber-based feeder systems. However, at this juncture the PSTN does not support video distribution to consumers for technical and regulatory reasons.

Many hold that telephone companies are technically advantaged exactly where the cable TV companies are technically disadvantaged. The telephone companies' network is switched and is two-way. The network is effectively a multimedia network in the sense that voice, video, and data can be simultaneously sent to a user. The network provides nearly any-to-any connectivity (assuming that the user had consistent protocols, sent information at the appropriate speed, etc.). The network is robust and can support millions of sessions simultaneously. Usage can be tracked very closely and detailed bills can be produced. The problem with the PSTN is that it currently does not have adequate capacity into a domicile. This capacity bottleneck is not so much due to technical limitations as it is to economics and regulatory issues. If the telephone companies could install fiber to each home (at a cost of $1,000 to $2,000 per home) and recover the cost, the bottleneck would be eliminated. The issue, however, is also related to universal coverage: it is one thing to rewire a few thousand or tens of thousands of homes; it is another thing to rewire 10 million or 100 million homes. Some estimate the cost to rewire every home, business, and institutional site in the United States to be $400 billion.*[4]

Table 1.1 compares aspects of the telephone company and cable TV networks today and in the near future. A system with a high ratio of downstream-to-upstream bandwidth supports information *delivery;* here, the receiver of the information is simply a consumer. A system with a ratio closer to the value 1 (either on a fixed basis or on a demand basis) is more amenable to true *communication* (in the PSTN, the ratio is usually 1). Designers of next-generation public networks take such issue under consideration, in their effort to deploy a network that is sufficiently flexible to support functions such as videoconferencing, work-at-home, collaborative work, and distributed computing. In the video context, the ratio need not always be close to 1, although one wants it to be greater than zero. The question, however, is whether one should deploy a common infrastructure to deliver video and the other services, or whether multiple networks may also be employed. This question impacts the final architecture of the system(s) to be deployed. The key candidate architectures are briefly introduced next.

The FTTN† approach, likely to be used in the near future, follows today's telephony model of using a specialized multiplexer (called the Digital Loop Carrier, or DLC, system) at the end of the feeder plant, somewhere in the neighborhood. From that point, electronic-free loops are used (except for ISDN and ADSL line electronics). In this configuration, the telephony DLC is replaced with a node that is suited for

---

* Although that figure looks high, it is an investment of only about 8 percent of the $5 trillion of the yearly gross domestic output.

† As a collection, the various fiber-based architectures for the local loop are known as Fiber In The Loop (FITL).

TABLE 1.1 Simplified Comparison of Telephone Company Networks and Cable Company Networks

Telephone company today	Telephone company near future	Cable company today*	Cable company near future
Available to nearly everyone	Available to nearly everyone	Used by about 60% of U.S. homes	Unknown penetration
Backbone: very high percentage of digital fiber (SONET-based)	Backbone: very high percentage of digital fiber (ATM/SONET-based)	Backbone: satellite, coax, and small amounts of analog fiber	Backbone: satellite, coax, analog and digital fiber
Access (residential): large amount of fiber in the feeder plant; twisted pair wire (analog or digital, i.e., ISDN) in the distribution plant (i.e., the "last mile")	Access (residential): large amount of ATM/SONET in the feeder plant; a number of systems in the future: twisted pair with electronics, providing 1.544 Mbps; fiber to the curb (ATM/SONET); fiber/coax (ATM/SONET) providing analog and digital TV; fiber to the home (more in the future). Two-way communication is the rule	Access (residential): analog coaxial cable	Access (residential): high-capacity coaxial systems operating at 1 GHz (new coaxial amplifiers needed); some digitally compressed video over analog carrier; telephone service over coax (two-way communication)
Network structure: low-speed switched, symmetrical, interactive, star-based	Network structure: high-speed switched, symmetrical, interactive, star- and/or ring-based	Network structure: medium-speed unswitched, asymmetrical†, not interactive (distributive), tree-based	Network structure: medium-speed switched and unswitched, asymmetrical, interactive, tree- and star-based

* This table does not focus on "wireless" cable operators and DBS providers.

† It is estimated that only 5 percent of all cable systems can send Ethernet-level data (10 Mbps) upstream; about 25 percent of the cable networks can send 0.5 Mbps upstream.[6]

video delivery. Typically from 100 to 500 (sometimes 2000) homes are supported. The node can support a variety of functions such as:

- Terminating the central office (CO)-originated fiber and transferring the signal to a coaxial or ADSL-based twisted pair cable
- Accepting an analog broadcast TV feed and digitizing and compressing it to deliver the signal to the user
- Splitting the signal from the CO into a video component for delivery over a coaxial cable and into the telephony component for delivery over traditional twisted pair cable
- Supporting some signaling and control functions, including generation of menus
- Injecting specialized commercials for targeted advertisement

The FTTC approach puts the fiber termination closer to the consumer; typically, this arrangement supports 4, 8, or 16 homes.

The FTTH approach puts the fiber termination equipment right into the home, thereby supporting a single consumer. At this time, this solution is fairly expensive due to the cost of the optoelectronics. FTTH requires consumer-provided powering; this impacts telephone service in case of a power outage (in FTTC arrangements, the photoelectronics are powered by the carrier, thereby assuring, when using two cables into the home, telephone service in case of a power interruption).

There are internal operational advantages in a single network. However, recent economic history has shown that nothing that is a single-provider system (e.g., a central government-run bank, a single national airline, a single railroad system, a single road to connect city A to city B) is ultimately efficient from a macroeconomic point of view. We extrapolate this argument to the case of a single communications network. For example, it is desirable for a physical highway to have numerous entry and exit ramps. If a traveler misses an exit or the exit is blocked by an accident, the traveler can, without major inconvenience, move up to the next exit. The same goes for accessing the highway from a local street.

Studies show that, in most instances, the consumer does not really care if the desired services come from a single wire in the wall into a single device (or multiple devices); from multiple wires in the wall into a single device; or (as may be the case today) from multiple wires in the wall into multiple devices in a home office (e.g., fax machine, answering machine, or telephone set). What matters most is having access to the desired capabilities and cost-effectiveness. This is a simple concept but is often unappreciated by the carriers. In the business context, the issue is even more serious. A *Fortune* 1500 company would never want to put all of its voice, data, video, and image (fax) communication on a single wire because any failure of that wire would incapacitate the entire company.

The deployment of broadband (as well as narrowband) services in the local loop is now a major focus for both telephone companies and cable television operators. Some of the general requirements of a network for delivery of video are shown in Table 1.2, which also provides a qualitative assessment of cable and carrier networks.[5]

There is also keen interest on the part of the telephone companies as well as cable TV operators in deploying Personal Communication Services (PCS).* Although this topic is not covered in this text, there are cases in which there is some overlap; some mention of this will be made

* There are over 148 million cars in the United States, 15 million trucks, 18,500 buses, 2000 trains, 275,000 civilian planes, and 3190 commercial planes.

TABLE 1.2 General Requirements of a Network for Delivery of Video

Requirement	Cable TV networks today	Carrier networks today
Affordability	Fair	Good
Availability	Fair	Good
Bandwidth	Good	Marginal
Bandwidth on demand	Marginal	Marginal
Ease of use	Good	Good
Flexible bill and billing	Fair	Good
Information content	Passing	Marginal
Interactivity (two-way communication)	Poor	Good
Openness	Marginal	Fair
Security	Poor	Fair

as appropriate (for more information on this topic see Refs. 7 and 8). Three early video dialtone-PCS trials were undertaken in Manhattan (NYNEX), West Hartford (SNET), and Boston (PCS, Cablevision).[9] Recently, Comcast Corp., Continental Cablevision Inc., Cox Cable Communications, Tele-Communications Inc., and Time Warner Entertainment announced that they had agreed in principle to form a joint venture to develop new communications services using digital video, fiber-optic, and wireless technologies. The joint venture wants to establish national service organizations and local service providers, and to seek additional participation and investment by cable operators.[10]

In conclusion, although initially fairly distinct, video and information are converging into a single medium that might be called *vimation* (*vi*deo infor*mation*). The video-information providers now include telephone carriers (local and long distance), cable service distributors, content providers (on-line studios, publishers, newspapers, studios), computer equipment manufacturers, and even the Internet community.

1.2.2 The Information Superhighway

Some* features and services developed under video dialtone and the new cable TV systems are in support of accessing the Information Superhighway,† which is basically the vision of the blueprint for the next-generation public switched network. Before narrowing the discussion to video for the rest of the text, this section briefly covers some aspects of this topic.‡

---

* The regulatory aspects of the superhighway, properly known as the National Information Infrastructure (NII), are discussed in more detail in Chap. 3.

† Entering 1995, some consider the "information superhighway" metaphor as inadequate and inappropriate. We use the term sparingly in this textbook.

‡ Now there is also talk about a Global Information Infrastructure (GII); this topic is not treated in the book.

Fiber-optic technology now nearing market introduction operates at about 10 Gbps (at OC-192 speed in the Synchronous Optical Network, or SONET, hierarchy) per fiber using wavelength division multiplexing (WDM) techniques. Planners would like to extend the benefit of such bandwidth to as many users and segments of the economy as possible. Hence, proponents see the need for hundreds of billions of dollars, to come from both public and private funds, to build the superhighway during the next 10 years. Besides the United States, countries such as Canada, Germany, and Japan have similar projects underway. There is now a goal to provide broadband connectivity to every Japanese home by the year 2015*[11,12]. Given the economic consequences at stake, some in the United States are lobbying to achieve the same goal by the year 2000. Such a goal is obviously very ambitious.

The superhighway is being viewed as a "network of networks," with "shopping" malls† of information providers. The information will be available based on both client-server and peer-peer methods. In the view of some, the superhighway already exists in the form of communication systems based on fiber, coaxial cable, microwave links, satellite links, and even twisted-pair copper that span most of the industrialized world. What is needed are better "on- and off-ramps" to the superhighway.[5] Links from businesses, community institutions (e.g., hospitals, schools, courts, libraries), and residences to the high-speed backbone are required to complete the superhighway. Such a ubiquitous, seamlessly interconnected communication apparatus would allow businesses, suppliers, customers, employees, agencies, and other parties to communicate effectively, expeditiously, and in a bandwidth- and feature-rich manner. In addition to traditional applications, such as transaction processing and electronic mail, the superhighway will stimulate applications that hereto have been on the sideline, such as (but not limited to) videoconferencing, networked multimedia, image archival, retrieval and distribution, compound document sharing (documents including text, voice, and video) for computer-supported cooperative work applications. Even lower-speed applications, such as telecommuting and telemetry, will be facilitated by such infrastructure.

There are disagreements, however, at the detailed level, as to what such a superhighway encompasses. Some providers of telecommunica-

* By 1995, the Japanese consumers are slated to begin receiving video at 1.5 Mbps and by 1998, they should have access to 155-Mbps channels supporting multiple video programs. The 45-trillion-yen investment in the Superhighway equates with $430 billion (22 percent of the budget is for FTTH upgrades; 22 percent for digital telecommunication infrastructure and switches; 18 percent for software; 11 percent for R&D; and 27 percent for operational costs).

† Also called by some Information Warehouses.

tion services and products simply label the next phase of their product line as being the promised superhighway. Others (e.g., cable service distributors) see the superhighway as a commercial distribution vehicle for video and perhaps other services (e.g., I-TV, audio, games, and data bank access); their vision is tilted to a fairly asymmetric superhighway with lots of bandwidth (information) delivered to the consumer, but little information in the upstream direction. One might call this a video client-server paradigm. Yet others see the superhighway more based on the Minitel model, where almost anyone can become an information provider. This is more of a video peer-to-peer paradigm, being, in a sense, a more symmetric system. In this scenario, the superhighway is less commercial and/or entertainment-oriented.

Telephone networks are viewed as meeting many of the desirable criteria for information superhighways, but the copper-based loops providing 3 kHz of analog bandwidth, or, at most 144 kbps under the Integrated Services Digital Network (ISDN), present a bottleneck in the "ramps" to the superhighway. Interestingly, fiber providing adequate bandwidth is already available in close proximity to the intended user; however, the last mile is the current problem. A number of techniques are emerging to facilitate the delivery of the bandwidth necessary to support video to the home over telephone carrier networks. One key design consideration of telephone carriers is the ubiquitous availability of services, video on demand and telephony, over an integrated access. These carriers are, however, defining video dialtone services in phases (e.g., Phase 1, Phase 2), where Phase 1 services will be available, say by 1996, while Phase 2 services might be available by 1999.

Even traditional cable TV systems, although (generally) supporting from 36 to 60 channels, are bandwidth-limited in terms of adding new programming, because each program is present all the time over each cable, rather than being switched in when needed. Additionally, existing cable systems do not support two-way communication that is so important for the evolving new applications, including home-based distance learning, video games, point-and-click interactive shopping, video on demand, and telephone-data communication service. Cable companies have been adversely impacted by poor consumer perception because of service limitations, degradation, and interruptions, as well as cost issues.*

---

* This text focuses on video distribution—in particular, residential video distribution—rather than on the more generic issues related to superhighways. However, it should be briefly noted that there is a third contender for constructing the superhighway. This is the Internet. The traditional limitations of the Internet were the fact that it does not support the real-time delivery of video (it can deliver store-and-forward video), it is not "easy" to use, it can be difficult to access as well as being expensive (with weak billing capabilities). Some of these limitations are now being addressed with commercial Internet service. See Chap. 2.

1.3 Emerging Local Distribution of Video

The past decade has seen major technological improvements in terms of both transmission capacity over fiber and computing power. Just a few years ago a mainframe supported 0.5 MIPS; now a $5000 workstation supports 100 MIPS. An Intel 80386 chip had approximately 300,000 transistors, while a Pentium chip has about 4 million transistors. High-power chips are making video compression as well as broadband ATM switching possible, both of which will be playing a critical role in the emerging video distribution systems.

Figure 1.3 depicts a next-generation cable TV network as contemplated by the cable TV operators. Figure 1.4 depicts a video distribution system, as contemplated by the telephone companies. (Notice that both architectures employ an FTTN approach.) Figure 1.5 shows the scenario that has been called the two-wire alternative; here both the local telephone company and the cable TV company provide video ser-

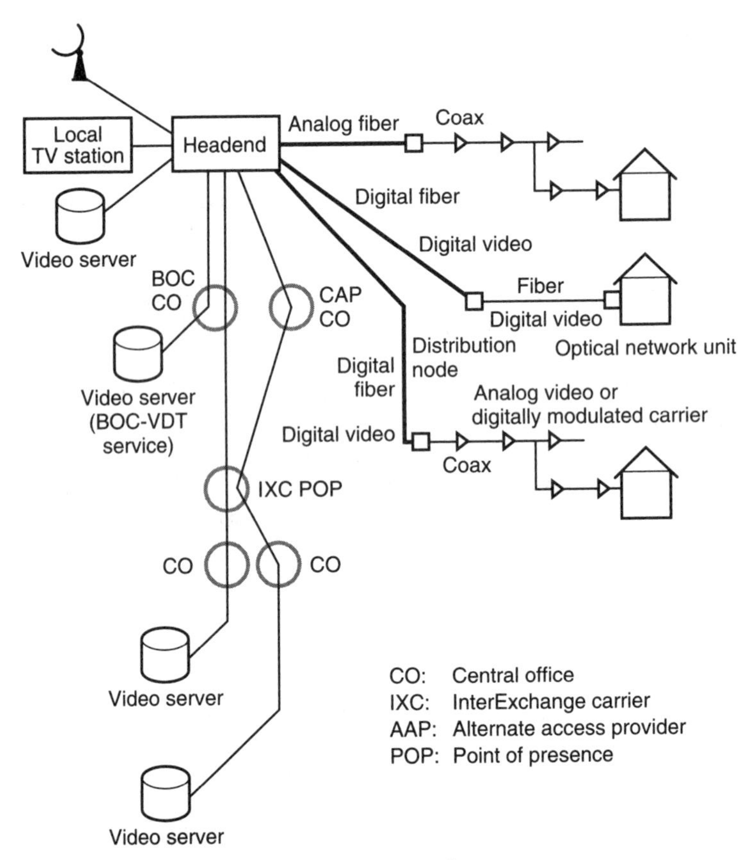

Figure 1.3 Next-generation cable TV network.

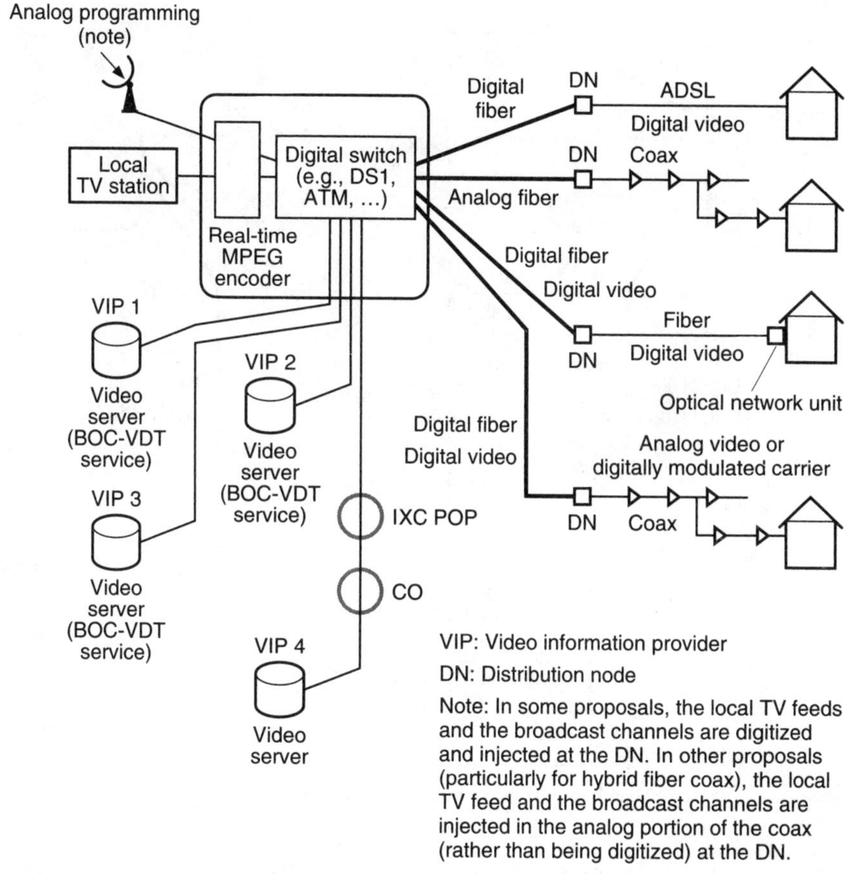

Figure 1.4 Proposed VDT distribution mechanisms. Note: In the future, satellite programming (from traditional broadcasters) will also be digital.

vices—both of their wires pass each home and the consumer decides which service to accept. This is a scenario that is favored by many regulators and consumer groups.

1.3.1 Technological options for physical distribution

The key guided delivery technologies, briefly introduced previously, that are of relevance to VDT (and cable TV) and that are examined in this book are:

- Existing coax/system
- Enhanced coax

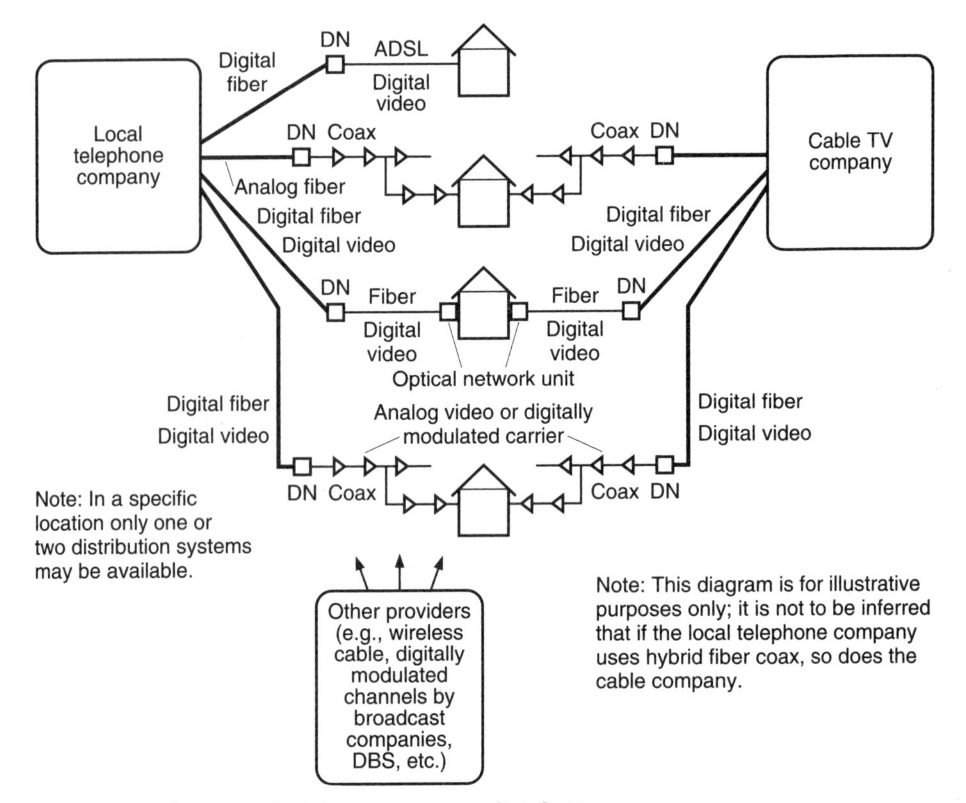

Figure 1.5 Two-wire television programming distribution.

- Extended coax with digitally encoded and compressed channels (in the upper portion of the bandwidth, say beyond 450 MHz)
- Hybrid fiber coax
- ADSL-1 and ADSL-3
- Fiber To The Node, Fiber To The Curb, Fiber To The Home

The key unguided delivery technologies are:

- DBS
- Wireless cable (e.g., Multipoint Distribution Service)
- Cellular cable (CellularVision)
- Digitally encoded channels in the broadcast TV spectrum (similar to traditional Subscription TV, or STV).

The deployment of Broadband Integrated Services Digital Network (B-ISDN) in the next few years will force an upgrade of the PSTN, enabling it to support many new video services. New FITL systems are near parity in cost with twisted pair for the benchmark of delivering telephone service (however, the FITL cost of delivering video is not comparable to the FITL cost of delivering telephone service only). One of the issues associated with single-wire fiber-based access in the FTTH mode and potentially impacting the choice being made by carriers is the powering of the telephone set to retain service in the case of a power failure (the photoelectronics in the home as well as the setup box need AC power). The FTTN and FTTC approaches mitigate this problem since the pedestal electronics (at the curb or at the node) are powered by the carrier—as long as the telephone component (in the setup box) has a passive bypass switch or uses a separate wire.

Some of the new video delivery mechanisms are also driven by the introduction of High Definition TV (HDTV) in the United States. An FCC-sanctioned HDTV standard is expected by 1995; HDTV service is set to start in 1998 (by the year 2008, the National Television System Committee, or NTSC, should be phased out). Figure 1.6 depicts a possible timetable of developments of advanced video distribution systems in the United States at the technical as well as regulatory level.

These physical local distribution technologies will be discussed in detail in Chaps. 9 to 12.

1.3.2 Cable TV companies' migration

In spite of the historical limitations, cable TV companies are now spearheading greatly improved HFC networks that support more bandwidth (currently up to 1 GHz, more later), use compressed digital

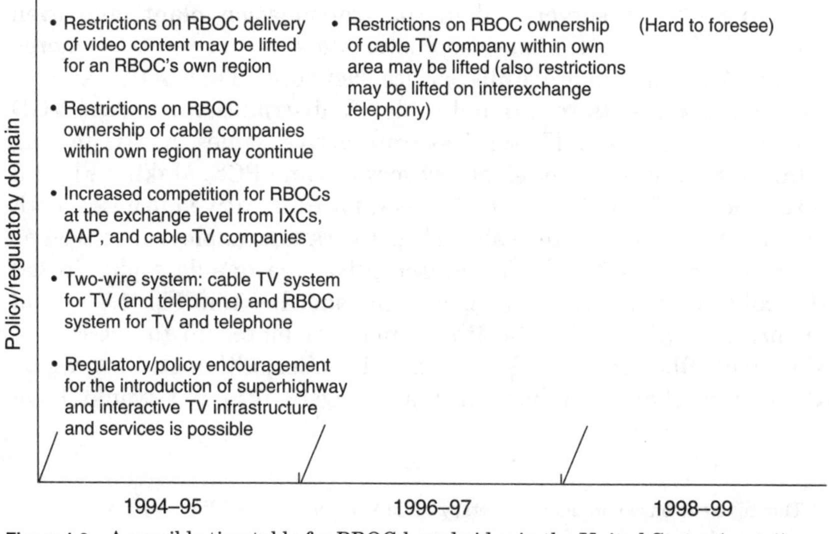

Figure 1.6 A possible timetable for RBOC-based video in the United States (part 1).

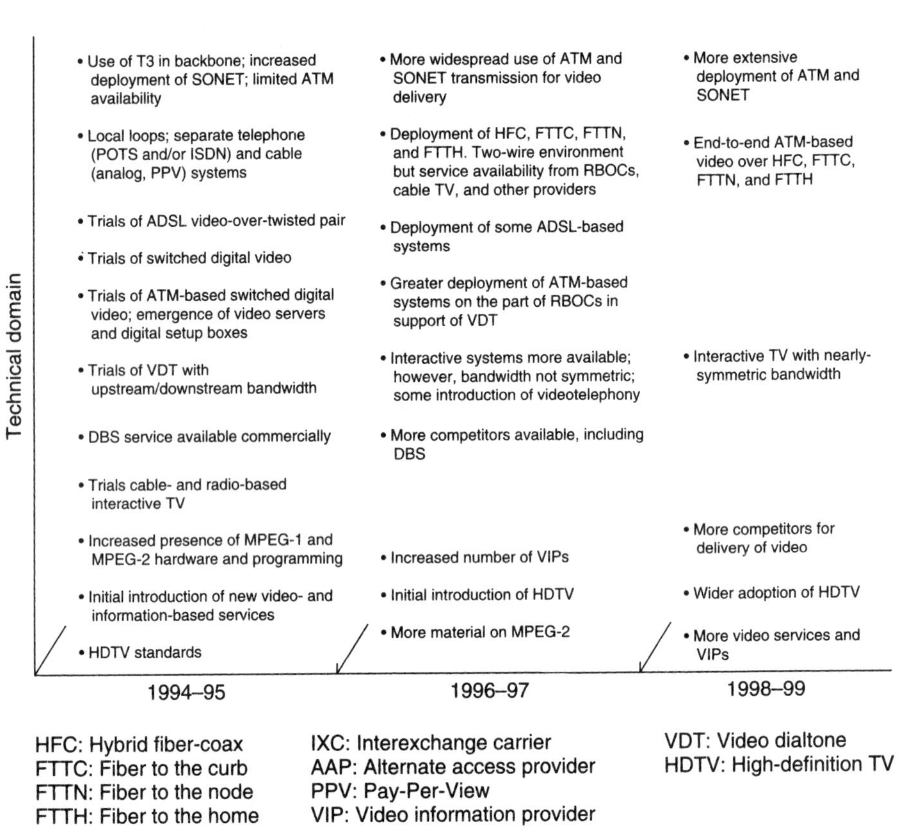

Figure 1.6 A possible timetable for RBOC-based video in the United States (part 2).

video in both the server and in the transmission plant, and even employ ATM. One of the more sophisticated such networks to emerge is Time Warner Cable's Full Service Network. This network, now under trial, supports traditional cable TV distribution services, VOD with instant access, I-TV services, interactive games, access to long distance telephony, video telephony services and PCS. Making Fig. 1.3 more concrete, Figs. 1.7*a* and *b** depict, for illustrative purposes, both a reasonably sophisticated cable TV network now deployed, as well as the Full-Service Network. A representative, reasonably sophisticated HFC cable TV network of today (e.g., the one shown in Fig. 1.7) would support 450 MHz of bandwidth (sometimes more, up to 1 GHz); it would have fiber spans no longer than 10 miles, although that can be extended further with hub sites and regeneration; it would have

* This figure is based on nonproprietary Time Warner Cable RFP materials.

approximately five total miles of coaxial cable per node, with a maximum of three amplifiers in cascade; each node would serve about 500 homes; and it would typically use the band 3 to 33 MHz for the upstream channel to support some level of interactivity.

1.3.3 Video dialtone systems—more than transport

In the United States, the next-generation video distribution apparatus (transport, control, information warehouses), at least from the RBOCs' perspective, is coming about under the framework of video dialtone. In terms of video delivery, today one effectively has all users connected to some of the same video sources. VDT will change that paradigm by connecting, on an individual basis, a specific individual to all video sources and, further, by enabling that individual to select that portion of the information universe that is of interest to that user. This makes the TV a universal but positionable window (through interaction and control) on the world. Video dialtone is defined by both technical considerations and regulatory considerations. Although the regulatory aspects of this topic are considered in more detail in Chap. 3, an initial definition is provided here.

The FCC described video dialtone in their November 1991 Further Notice of Proposed Rule Making as: "An enriched version of video

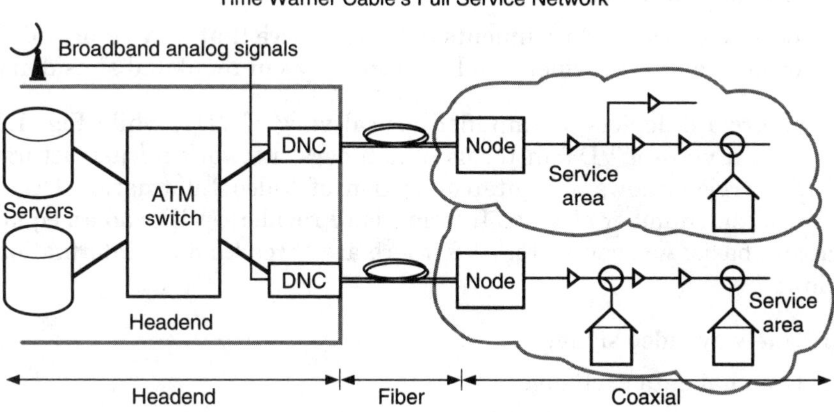

Figure 1.7a A high-end cable TV network supporting sophisticated features (part 1).

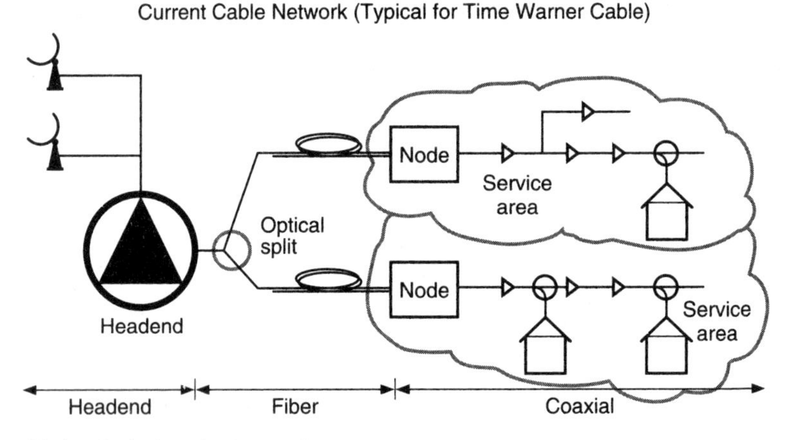

Figure 1.7b A high-end cable TV network supporting sophisticated features (part 2).

common* carriage under which the telephone company will offer various non-programming services in addition to the underlying video transport." In July, 1992 the FCC made a decision on video dialtone that encompasses the aspects described in the paragraphs that follow. The FCC described a two-level model for video dialtone:

1. Level 1 Basic Transmission, Level 1 Gateway, and a number of enhanced functions
2. Level 2 services: Enrichments to basic service that may be provided by any service provider or a local company's unregulated subsidiary

Figure 1.8 depicts a simplified logical view of VDT, while Fig. 1.9 depicts a view of VDT in terms of a general network infrastructure. VDT service allows the interconnection of Video Information Users (VIUs) to a number of Video Information Providers (VIPs) on an equal access basis. As seen in Fig. 1.8, there are three kinds of information flow:

1. One-way video signals
2. Optional data exchange
3. Signaling among VIUs, VIPs, and the network.

---

* A cable TV company is not a common carrier and does not provide a common carrier service, namely, it does not have to support multiple information providers. VDT service is a common carrier service.

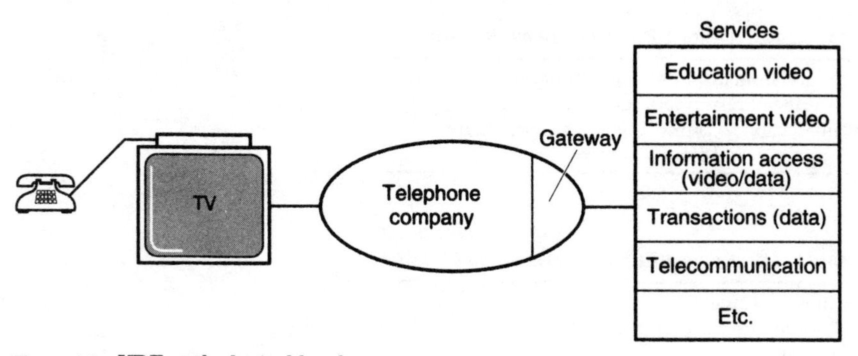

Figure 1.8 VDT at the logical level.

The allowed VDT Level 1 services are discussed next. A local telephone company can provide the following, on a common-carrier basis, with equal access to all service providers:

- Basic service: service connection, directory, help, and other basic and enhanced functions
- Nondiscriminatory basic transport and routing
- Collection of network information for billing
- Basic (see Table 1.3) and enhanced (see Table 1.4) gateway functions

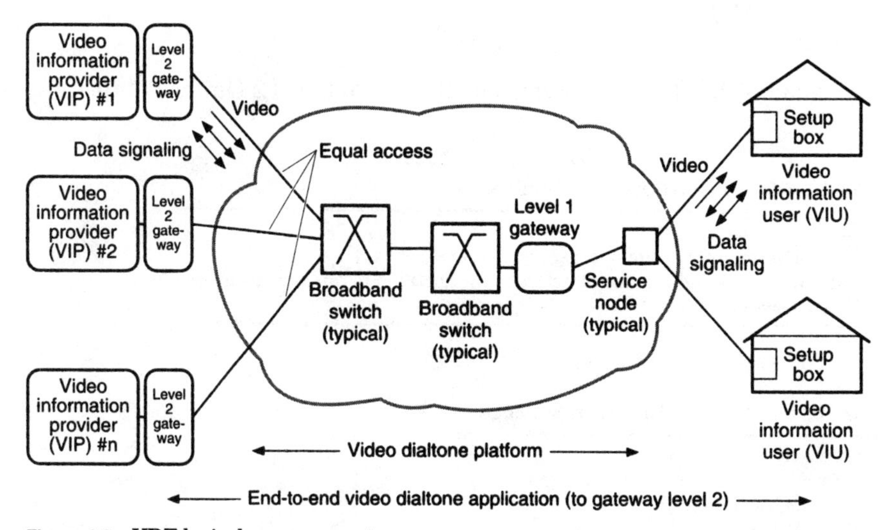

Figure 1.9 VDT logical arrangement.

TABLE 1.3 Level 1 Basic Gateway Functions

Establishment of network connections
Interface for billing and provisioning
Navigation to connect to service providers
Transport and switch control

TABLE 1.4 Level 1 Enhanced Gateway Functions

Channel guide
Channel identification
Lock out
Picture-in-picture control
Sequence programming
Sleep timer
VCR programming

The allowed VDT Level 2 services are enrichment to basic service, including enhanced and other non-common-carrier services may encompass:

- Enhanced video gateway to advanced services
- Advanced services
- Advanced navigation and research
- Video processing services
- Non-common-carrier services such as billing, collection, and order processing
- Video user equipment and inside wire
- Basic (Table 1.5) and advanced (Table 1.6) Level 2 Gateway functions.

TABLE 1.5 Level 2 Gateway Functions

Advanced navigation
Downloading applications to user equipment
Establishment and release of network connections to advanced services
Video server control

TABLE 1.6 Level 2 Advanced Services

Interactive education
Interactive entertainment and games
Interactive shopping
Video Yellow Pages

At that time, the FCC did not allow telephone companies to provide programming directly to subscribers.

A VDT system designed to carry full-motion video at, say, 1.544 Mbps, will also carry other types of information, such as image (still pictures), multimedia (e.g., CD-I), and games. As long as there is some return capacity for signaling, even an asymmetric system such as ADSL, can support multimedia and games since the upstream commands will consist of relatively short data bursts. It should be noted that there are several technical currents at play within the context of VDT:

1. The cable TV companies wish to upgrade their systems to carry more channels, support interactivity, and carry telephony services (at least outgoing calls to a collaborative interexchange carrier, or full-service via PCS).
2. The RBOCs wish to get into the residential TV distribution business, at least in terms of providing the underlying structure, but perhaps also the content portion.

Based on the preceding discussion, video dialtone is not simply an upgraded cable TV system based on the latest ATM, SONET, and fiber technologies but a mechanism for open access to a variety of video information providers. Some claim that video dialtone must go beyond just video and provide the equivalent of a point-of-sale (POS) function in every home.

Typically, a VDT platform consists of a newly deployed broadband network, along with elements from the public switched telephone network (e.g., ISDN and packet switching). In traditional carrier terminology, the VDT network enables two services:

1. Information exchange service to VIUs
2. Information exchange access to VIPs

See Fig. 1.10. The term *exchange access system* (EAS) is also used by carriers to refer to the VDT network. Within the video dialtone context, "downstream" refers to transmission of information toward the VIU, while "upstream" refers to information from the VIU. To subscribe to VDT services, the user needs video delivery equipment in conjunction with the setup box; the local telephone company (or other provider for that matter) makes available the Level 1 gateway. The VIP would provide the Level 2 gateway as well as the video server (to store video programming). Video servers, which are an integral component of a video dialtone system, are expected to experience major market growth in the next few years, from $133 million in 1994 to $5.2 billion in 1997.[13]

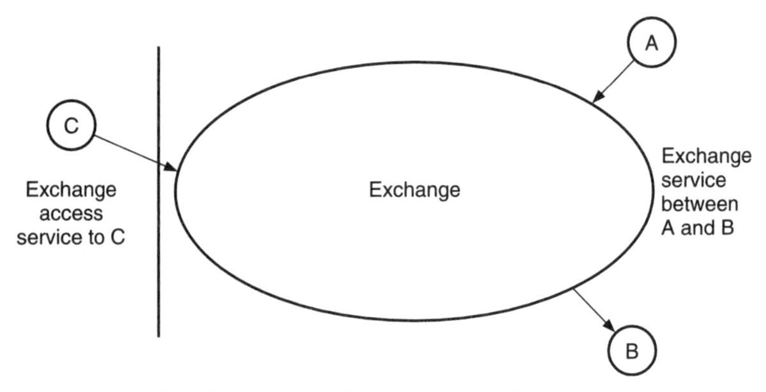

Figure 1.10 VDT exchange and exchange access service.

Figure 1.11 depicts the type of physical architecture that the RBOCs may view as a candidate VDT platform. This topic of VDT platforms is examined in greater detail in Chap. 5; video servers and related technology are discussed in Chap. 7.

1.3.4 Signaling and control in video dialtone

Next to actually delivering the video signal itself over an appropriate transport platform, the most important aspect of a VDT platform relates to signaling and control. This enables the user to communicate with the Level 1 and Level 2 gateways previously described. In addition to allowing VIUs to obtain video, image, voice, and information, the VDT service must support network and application signaling.

Table 1.7 depicts the types of VDT signaling required to provide the desired interactivity. In contrast to the isochronous nature of buffered-compressed video, signaling can be bursty in nature. However, based on the experience of ISDN and ATM, relatively low-capacity channels can be employed. Typical video dialtone systems under development use 16 to 56 kbps (unmultiplexed VIU interfaces tend to need the lower speed, while multiplexed signals from several users might need more bandwidth).

Figure 1.12 depicts two important interfaces that come into play in designing a VDT system, including the signaling apparatus. These are the VIU Network Interface (VIU-NI) and the VIP Network Interface (VIP-NI). For example, a VIU-NI based on ADSL-1 will support a single user-to-application session based on MPEG-1. This implies that a single video channel is transmitted to a particular user at a particular time. Although multiple televisions in a home may view a single program, the televisions cannot independently view multiple applications. ADSL-3 supports the simultaneous delivery of four MPEG-1 signals.

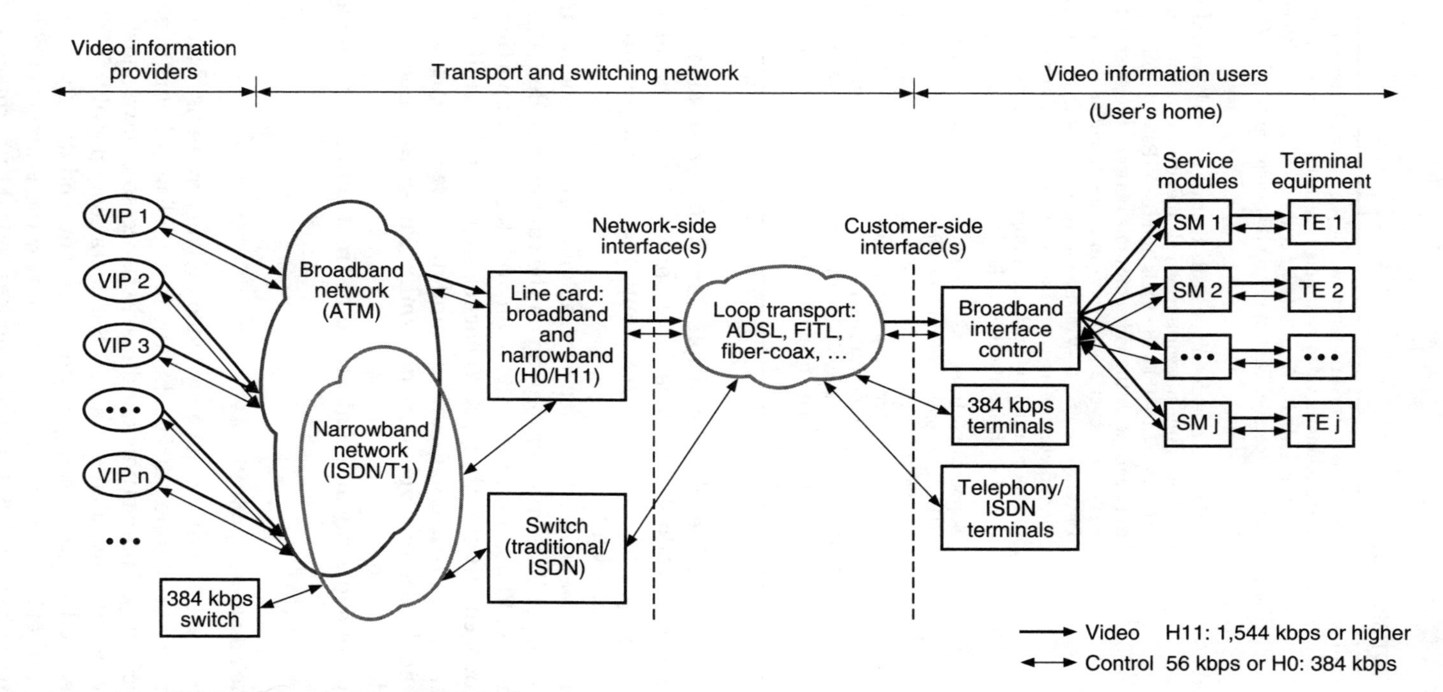

Figure 1.11 Physical network view of VDT platform.

TABLE 1.7 VDT Signaling

Network signaling	Supports setup and teardown of the network element-to-network element video channels (the network element is typically a switch). Typically, this type of signaling is switch-vendor-specific. This signaling is transparent to the VIPs and to the VIUs.
Network access signal	This type of signaling entails routine network control information and network-provided menus. This provides an infrastructure that the VIUs and the VIPs will utilize to communicate with the video dialtone network (more specifically, with the Level 1 gateway) in a multivendor environment. Developers of VIP-VIU applications, as well as equipment manufacturers, must make their products compatible with the specification of such signaling. Several carriers have published such specifications. See Ref. 14.
Application signaling	Supports postconnection communication between VIPs and VIUs (e.g., pause, rewind, fast forward). It entails application control information that is transacted by applications executing on the user equipment (typically in the setup box) and in the video server. It is desirable to define a basic set of signaling capabilities available to VIP-VIU application developers and equipment manufacturers, to support a multivendor environment.

As the VDT network evolves, multiple simultaneous sessions over a VIU-NI will be available. Studies have shown that a typical home needs the capacity for at least three simultaneous channels and, perhaps, as many as five.[15]

As mentioned earlier, VDT services an platforms are being introduced in phases. These phases depend on the carrier. For Bell Atlantic, as an example, Phase 1 corresponds to an architecture supporting DS1 (1.544 Mbps) circuit switching, X.25 packet switching for bidirectional signaling, a 1.544 ADSL loop, and a stand-alone Level 1 gateway; the primary application is video on demand.[14] Phase 2 supports other advanced functions.

The topic of signaling and control will be treated in greater detail in Chap. 5.

1.3.5 Video dialtone applications

Applications are the driver for new video delivery systems. More than ever, consumers are time-starved, having to work longer hours in a competitive economy. Many studies and market successes have shown that consumers want convenience. Table 1.8 depicts some possible VDT services listed in decreasing order of video, content, and density, and then current market reality. In the United States, there are over 225,000 medical practices, 105,000 dental practices, 60,000 pharmacies, 6000 hospitals, 12,500 university-related research labs, 11,300 commercial

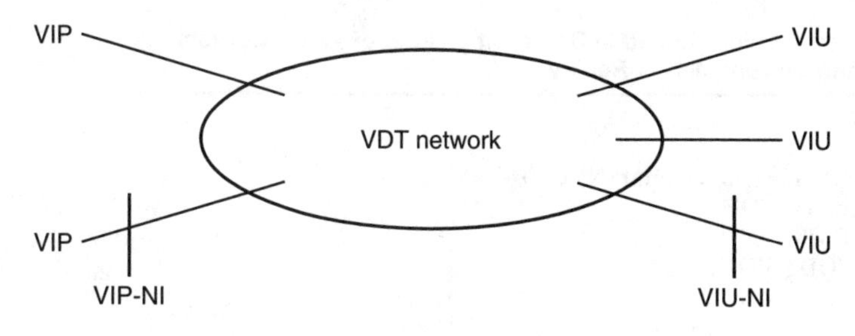

Figure 1.12 Key interfaces in a VDT network.

labs, 726 federal labs, 20,000,000 retail establishments, 600,000 corporate offices, and 371,000 factories. There are over 58 million privately owned homes and 33 million leased apartments. There are 249,000 daycare centers, 110,000 K–12 schools, 28,600 libraries, and 2807 universities and colleges.

In spite of the technology push, supported by the dozens of vendors who stand to greatly increase revenues as processor-intensive, memory-intensive, and communications-intensive hardware is deployed to support end-to-end digital video, the expected transition, when considering the cost of the service (see Chap. 8), is from multiple channel PPV today, to NVOD in the immediate future, to VOD at some future point. There are now a number of VOD technology trials underway. The purpose of these trials is strictly to validate the technology—even if these technology trials are completely successful, it will be a challenge to bring the cost in line with what the consumer pays at a rental store.

Issues related to VDT markets are reexamined in Chap. 2.

1.3.6 Video-VDT industry constituents

Table 1.9 depicts a panoply of parties interested in video distribution; this list identifies the industry constituents. Table 1.10 depicts equipment and components of a video distribution plant. Table 1.11 lists the key vendors of this convergence industry.

1.4 Emerging Switching Technologies: ATM

Asynchronous Transfer Mode refers to a high-bandwidth, low-delay switching and multiplexing technology now becoming available for both public and private networks. ATM principles and ATM-based platforms form the foundation for the delivery of a variety of high-speed digital communication services aimed at corporate users for high-speed data, local area network (LAN) interconnection, imaging, and multimedia applications. Residential applications, such as video dis-

TABLE 1.8 Video Services Listed in Decreasing Order of Video, Content, and Density, and Current Market Reality

Entertainment
Broadcast
Near real time video on demand (NVOD)*
Video on demand (VOD)†
Impulse VOD (IMVOD)‡
Interactive VOD (IVOD)§
Games¶
Education
Access to libraries and other repositories
Access to video archival (e.g., stored historical, network broadcast clips)
Courses
Tutoring
Home degrees
Shopping
Retail goods
Services (theater, travel tickets, travel reservations, doctor appointments)
Real estate
Information
Bartering
Personals
Products search
Information search
Job recruiting
News on demand (tailored news delivery)

* NVOD implies the consumer's ability to select any movie for near-immediate (say within 5 or 15 minutes) viewing (also called near-instantaneous VOD).

† This also called Movies On Demand (MOD)—VOD implies the consumer's ability to select any movie at any time for immediate viewing. Pay Per View (PPV) implies paying to see a well-publicized event, e.g., a boxing match occurring at a specific time and day. Cable operators also use PPV to show a selected movie that plays at a specific time and day.

‡ This service entails providing a few (typically 6) minutes of a movie free, but immediately after the free viewing period (say after 6.01 minutes), the movie is billed. The notion of impulse arises from the consumer's tendency to make an impulse decision after having enjoyed the first few minutes free.

§ There are contradictory definitions of what IVOD means. We mean a VOD program where the user can interact during the course of the movie to obtain VCR-like functions such as pause, rewind, and fast-forward. Another definition is as follows[16]: IVOD-i (interactive VOD with instantaneous access) is a service where users can retrieve and individually control program information from a library instantly, with instant control response (e.g., fast forward, pause, rewind); IVOD-d (interactive VOD with delayed access) is a service where users retrieve and individually control program information from a library, each access being characterized by a nonzero waiting time approaching instant viewing (e.g., one minute), but limited to an engineered fixed waiting time value depending on the available bandwidth resources in the network, as well as on the popularity index of the requested program.

¶ Over 33 percent of U.S. households have purchased a video game console, many of which are presumably still working—the video game industry exceeds $4.5 billion annually.

TABLE 1.9 Key Constituents of the Video Distribution Industry

Advertisement agency
Cable manufacturers (all types of physical cables)
Cable TV component manufacturer
Cable TV contractor
Cable TV investor
Cable TV program network
Commercial TV broadcaster
Educational TV station, school, or library
Independent cable TV operator
Interexchange carrier (telco)
Lawyers
Local exchange carrier-RBOC (telco)
MDS, STV, or LPTV operator
MSO (multiple system operator)
Program producer, packager, or distributor
Regulators
SMATV or DBS operator
Software developers for video dialtone systems
Technical consultants
Telephone company component/equipment manufacturer
Video servers manufacturers

TABLE 1.10 Elements of a Video Distribution System

Traditional	Satellite equipment
Two-way radios	Security equipment
Amplifiers	Splitters
Antennas	Subscriber-addressable security equipment
Cable tools	Telephone equipment
Cable TV passive elements including cable	Vehicles
Cable TV RF distribution-distribution electronics	VideoCiphers
Compression-digital equipment	Emerging*
Computer equipment	ADSL equipment
Connectors	ADSL-configured setup boxes
Controllers	ATM switches
Converters	ATM-configured setup boxes
Descramblers	Circuit switches (DS1 and DS3)
Fiber-optic cable	Digital Crossconnect systems
Fiber-optic electronics	Digital Loop Carrier (DLC) equipment
Headend equipment	Digital video servers
Interactive software	Fiber To The Node (FTTN) equipment
Lighting protection	Fiber To The Curb (FTTC) equipment
Microwave equipment	Host Digital Terminals (HDT)
MMDS transmission equipment	Level 1 and Level 2 hardware
PCS equipment	Level 1 and Level 2 software
Power supplies	MPEG-1 decoders
Receivers and modulators	MPEG-2 decoders
Remotes	Optical Network Units (ONU)
Safety equipment	Real-time MPEG-2 encoders

* In addition to those listed in the traditional section above.

TABLE 1.11 Key Players in the Convergence Industry (Partial List)

Service Providers	*Network Components*
Leading North American Cable MSOs	Data Network Access Equipment
TCI	Digital Equipment Corp.
Time Warner	General Instrument Corp.
Continental	Scientific-Atlanta, Inc.
Comcast	Zenith Electronics Corp.
Cablevision	Transmission Media/Systems
Cox	AT&T GBCS
U.S. RBOCs/LECs	AT&T Multimedia Services
Ameritech	AT&T Network Systems
Bell Atlantic	Antec
BellSouth	Broadband Technologies Inc. (BBT)
Pacific Telesis	C-COR Electronics
NYNEX	Digital Equipment Corp.
Southwestern Bell (now SBC Communications)	General Instrument
	Northern Telecom
U S WEST	Philips
GTE	Scientific-Atlanta
Global Telecom Operators	Texscan
BT	Network Switches
DBP Telekom	Alcatel
France Telecom	AT&T Network Systems
NTT	Ericson
Telstra	Fujitsu
U.S. Alternate Access Providers (AAPs)*	General DataComm
Metropolitan Fiber Systems (MFS)	IBM
The Teleport	NEC
U.S. Interexchange Carriers (IXCs)	Northern Telecom
AT&T	Newbridge Networks
MCI	Olivetti
Sprint	Stratacom
UK/European MSOs and Cable Operators	Siemens
Companie General des Eaux	Video Servers
DBP Telekom	Digital Equipment Corp.
France Telecom	Hewlett-Packard
Nynex	IBM
TeleWest	Microsoft
Telia	Oracle/n-Cube
Videotron	Silicon Graphics
Value-Added Networks (VANs)	Setup Boxes
Advantis	Apple
GEIS	Divicom
	General Instrument
	Hewlett-Packard
	Olivetti/Acorn
	PictureTel
	Scientific-Atlanta
	Satellite Delivery/Headend Systems
	AT&T GBCS
	Hughes
	Scientific Atlanta

* Sometimes also called Competitive Access Providers (CAPs).

tribution, videotelephony, and other information-based services, are also planned. ATM is the technology of choice for evolving B-ISDN public networks, for next-generation LANs, and for high-speed seamless interconnection of LANs and WANs (wide area networks). ATM supports transmission speeds of 155 and 622 Mbps and will be able to support speeds as high as 10 Gbps in the future. As an option, ATM will operate at the DS3 (45 Mbps) rates; some proponents are also looking at operating at the DS1 (1.544 Mbps) rates. ATM and the many supporting standards, specifications, and agreements constitute a platform supporting the integrated delivery of a variety of switched high-speed digital services.

The telephone carriers are seriously looking into delivering video dialtone services over an ATM platform to take advantage of the multiplexing and switching capabilities of this technology. This topic is treated in greater detail in Chap. 8.

1.5 The Evolving TV Set and Setup Box

The evolving TV set, perhaps better renamed point of video delivery (PVD), is the product of the confluence of high-power hardware, software, and networking technologies. The marriage of the TV set, the personal computer, and the communication equipment into one device, called by some *teleputer*, is being prognosticated for the late 1990s.

Closer to the immediate future, the TV set and/or the setup box will have varying degrees of intelligence. At the very least, the setup box needs to decode the video signal and the signaling information; it must also be able to transmit signaling messages in response to the viewer's actions. The activities can be modeled as falling in what have been called the *user plane* and the *control plane*. The user plane deals with movement of user information such as video and possibly other information (e.g., screens for video games); the control plane deals with signaling aspects, namely, interactions dealing with ordering a movie, pausing, browsing, etc. The complexity of the device depends on the "amount" of video decompression that is needed and on the complexity (or lack thereof) of the signaling protocols (this is covered in Chap. 5).

One critical factor dealing with setup boxes, relating to its cost and therefore the eventual penetration of video dialtone services, is the openness (or lack thereof) of the interface. At this time, the interface is closed and, therefore, each system has a vendor-dependent piece of equipment. Efforts are underway to publish some aspects of the interface to foster open production of such equipment. The IEEE 802.6 Committee has indicated an intention to develop standards for this interface.

Another consideration is that the use of the setup box should not eliminate or limit functions that may already be available on a TV set, such as recordability of programs, picture-in-picture, or channel surfing.

This topic is revisited in Chap. 13.

1.6 Video Compression

Compression algorithms are critical to the viability of digital video services, both in terms of storage as well as communication. Video and other multimedia objects are large: digitized speech, music, still images, and, even more so, full-motion, full-quality video all require large amounts of storage and communication bandwidth. For example, uncompressed digital TV video requires 140 to 270 Mbps. It follows that compression schemes are the only hope for widespread deployment of digital video in general and multimedia in particular. Fortunately, video signals contain a substantial amount of intrinsic redundancy, so that compression can be undertaken prior to storage or transmission. Compression schemes that are able to reduce the raw bits by a factor of 100 or 200 are being sought. Some have even claimed compression ratios of 1:2500 using *fractal* methods. Besides vendor-proprietary methods for both TV and industry applications and for videoconferencing applications, the most important standards are MPEG-1 and MPEG-2, Motion JPEG, and ITU-T H.261.

Table 1.12 provides an approximate view of the various data rates for the algorithms discussed in this book. The view is approximate because each algorithm may support multiple coders or encode at different resolution levels, including view area (a-by-b pixels), color level, and frames per second.[17] There is a technology push in some quarters for VDT to use MPEG-1 at 1.5 Mbps; although this may be a technical necessity because of the existing PSTN, MPEG-1 is not sufficiently flexible to support all types of video signals currently on TV (e.g., sporting events). It is advocated here that MPEG-2 be the basis for digital NTSC television (conceivably, in advanced systems that can decode both MPEG-1 and MPEG-2 systems, it may be possible to mix the two systems using MPEG-1 for talk show-like programs and MPEG-2 for more action-oriented programs). What has held back the videoconferencing market until now, besides cost, is the extremely poor quality of the video at the lower information rates (e.g., at 112 kbps).[18] VDT will run the same risk if the quality to which the consumers are accustomed is not there. This topic is revisited in Chap. 6.

1.7 Options for Digital Video Distribution

This section summarizes the technology options that have been discussed or implied in this chapter that apply to digital video distribu-

TABLE 1.12 Approximate Encoding Rates for Various de jure-de facto Standards

Standard	Approximate range of data rate	Compression*
CCIR (now ITU-R) 601/D-1	140–270 Mbps	reference
CCIR 723	32–45 Mbps	3–5×
CD-I (for locally based TV-delivered interactive programming)	1.2–1.5 Mbps	100×†
DVI/Indeo (for multimedia and CD-ROM-based PC applications)	1.2–1.5 Mbps	160×
H.261 (videoconferencing and video telephony, e.g., ISDN)	64 kbps–2 Mbps	24×†
Motion JPEG (e.g., medical applications)	10‡–20§ Mbps	7–27×
MPEG-1	1.2–2.0 Mbps¶	100×
MPEG-2	3–10 Mbps for regular TV; 15–20 Mbps for high definition TV	30–100×
Software compression (small windows)	2 Mbps (approximate)	6×
U.S. commercial systems using mild compression	45 Mbps	3–5×
Vendor-specific methods	0.1–1.5 Mbps	100×

* Compared to broadcast quality.
† Not same quality in terms of pixels and colors as reference.
‡ 640 × 480, 24 bit color, 15 fps, 1:10 compression; or, 640 × 480, 24 bit color, 30 fps, 1:20 compression.
§ 640 × 480, 24 bits color, 30 fps, 1:10 compression.
¶ Baseline standard; other rates are also possible.

tion in general and VDT in particular. See Table 1.13, adapted from Ref. 19. These topics are examined in more detail as the discussion proceeds.

1.8 Additional Reading

This text examines the many issues involved in the technological enhancement of video delivery discussed in this chapter, including technical considerations, regulatory considerations, market considerations, and service development considerations. However, no book can cover all topics at the ultimate detail level. There are many books on the market covering related topics that the reader may wish to consult. However, given the speed at which technology is being generated, it is fairly hard to get pertinent and up-to-date texts that try to apply some order and pedagogy to the avalanche of material that one reads in the trade press on a daily basis.

Among other texts, the reader may want to consult the following references: general video principles, Ref. 20; video from a multimedia perspective, Ref. 21; video from a distance learning perspective, Ref. 22; still video, Ref. 23.

TABLE 1.13 Technology Options for IVSs

Server	Video system
Robotic mass storage	ADSL
Multiprocessor supercomputer with disk storage	HFC
Burst transfer, distributed cache	FTTN
Download, user-based storage	FTTC
Storage	FTTH
Tape	Other (e.g., wireless, satellite)
Magnetic (Redundant Array of Inexpensive Disks)	Feeder
Optical	Digital fiber (SONET)
Gateways	Analog fiber
Software defined (carrier controlled)	Fiber and coax
Proprietary	Coax
Switch	Twisted-pair cable
Digital crossconnect system	Wireless
ATM, true	Distribution
ATM, circuit emulation	Digital fiber (SONET)
Backbone	Analog fiber
DS1/DS3	Fiber and coax
SONET	Coax
ATM/SONET	Twisted-pair cable
Encoding	Wireless
MPEG-1	Setup box
MPEG-2 main profile (B frames)	Simple
MPEG-2 simple profile	With multiple MPEG-1-MPEG-2 decoders
Motion JPEG	With ATM processing
	With or without sophisticated user interface

References

1. *Multimedia Week,* March 14, 1994, p. 1.
2. The IEEE Institute, "Fear, Greed Breed New Technology Use, Keynoter Tells U.S. Broadcaster," vol. 18, no. 3, May/June 1994, pp. 1 ff.
3. T. Miki, "Toward the Service-Rich Era," *IEEE Communications Magazine,* February 1994, pp. 34 ff.
4. *USA Today,* April 26, 1994, B1.
5. A. Reinhardt, "Building the Data Highway," *Byte,* March 1994, pp. 46 ff.
6. G. Lawton, "Interoperating Cable Into The Great Opportunity," *Communications Technology,* November 1993, pp. 26 ff.
7. R. Hay, "PCS Via Cable: An Enabling Technology for Cable's Future Opportunities," *Communications Technology,* April 1993, pp. 48 ff.
8. P. Schaller, M. Shafer, "Preparing Your Plant for PCS," *Communications Technology,* April 1993, pp. 52 ff.
9. L. K. Snow, "Video Services and Trials," *Proc. Eastern Communications Forum,* 1994.
10. *Wireless PCN Telecommunications,* Information Gatekeepers, Boston, January 1994.
11. *Communications Technology,* April 1993, pp. 15 ff.
12. D. Ingelbrecht, "Opening Up Japan," *Cable World,* February 28, 1994, p. 2.
13. *Multimedia Week,* February 21, 1994, p. 5.
14. Bell Atlantic, *Signaling Specification for Video Dial Tone,* TR-72540, Issue 1, Release 1, August 1993.
15. A. Lakhani, *Winning With Video Dialtone,* The Network Knowledge Company, Rancho Palos Verdes, Calif.

16. D. Deloddere, et al., "Interactive Video on Demand," *IEEE Communications Magazine,* May 1994, pp. 82 ff.
17. D. Minoli, "Digital Video Compression: Getting Images Across a Net," *Network Computing,* July 1993, pp. 146 ff.
18. D. Minoli, "1995: The Year of Video in Enterprise Nets?" *Network World,* Dec. 5, 1994, p. 21.
19. J. R. Jones, "Baseband and Passband Transport Systems for Interactive Video Services," *IEEE Communications Magazine,* May 1994, pp. 90 ff.
20. O. Eldib, D. Minoli, *Handbook of Digital Video,* Artech House, Norwood, Mass., 1995.
21. D. Minoli, R. Keinath, *Distributed Multimedia Through Broadband Communications Services,* Artech House, Norwood, Mass., 1994.
22. O. Eldib, D. Minoli, *Distance Learning Technologies and Applications,* Artech House, Norwood, Mass., 1995.
23. D. Minoli, *Imaging in Corporate Environments,* McGraw-Hill, New York, 1994.

Chapter

2

The Service Drivers and Markets

VDT is impacted by competitive, regulatory, technological, and market factors. This chapter examines some market data and some market drivers in support of VDT services and the entry into this market by the RBOCs and other players, such as VIPs. This chapter is not meant to be exhaustive nor to supply enough of a view to furnish a complete business case in favor of one VDT service or another. It simply aims at providing a macroview of the market possibilities.

Proponents claim that the average RBOC could generate several billion dollars of revenue per year from pay-per-view (PPV) alone,* as well as other revenues from additional VDT-based services. Such numbers need to be firmly established. Clearly, the decision of whether or not to deploy a new broadband residential infrastructure at this time, or at some point in the near future, is not a trivial one, given the investments required. Without demonstrable markets, the carriers are likely to limit the deployment of broadband technology to large commercial users of fastpacket services. Market information, when reliable, can prove useful. Unfortunately, a lot of such information advanced by market researchers is not statistically valid, and therefore is suspect as a decision support resource.

2.1 Video Market Statistics

This section provides some information related to video services that sets the stage for the discussion to follow in this chapter and in the rest of the book.

---

* We account later for a smaller figure of about $20 million; the entire PPV business in the United States in 1993 was $377 million.[1]

Table 2.1 depicts some penetration data for video-related hardware.[2] According to proponents, there is an abundance of opportunities for video dialtone services; there are also many potential players. Major areas advanced as VDT opportunities, as already alluded to in Chap. 1, include entertainment, education, information access, and transaction (e.g., teleshopping) support. Entertainment is a $225 billion per year business in the United States, of which $83.4 billion are in the video-cinema arena. The cable TV revenues were around $25.5 billion per year in 1994. These revenues are expected to grow to $35 billion per year by 1998; for comparison, the local telecommunications revenues (for RBOCs and wireless services) were around $106 billion per year in 1994* and are expected to grow to $126 billion by 1998.[3] Table 2.2 depicts some of the video revenues that VDT may impact or affect. Table 2.3 depicts a number of additional statistics relating to the industry that can be used to get a feel for various market aspects of video dialtone. Table 2.4 depicts revenues for a number of areas beyond video distribution that may be impacted by VDT.

The fundamental cable TV services to date have been the basic service (basic set of local and network channels), premium service (basic service plus channels such as HBO), and PPV service. VDT aims at providing these and many other services. One key issue in VDT is the perennial question about what consumers are willing to purchase in terms of video services and at what price. For example, according to some studies by the Network Television Association, people do not watch more TV channels when more are available.[5] The average household in the United States receives 36 TV channels but watches fewer than half. The data shown in Tables 2.5 and 2.6 were collected. As a data point, an average cable TV subscriber spends $30 for the service and $15 for videotape rentals per month. On the average, customers that have the PPV capability and use it, use it to order about three movies a year. Some trials have indicated that as the system is made more flexible, consumers tend to use PPV more frequently (for example, 6 to 12 times a year).

For comparison, Table 2.6, based on a number of sources,[3] provides an assessment of the local telephone, wireless, and cable TV market.

VDT may find applications in institutional delivery of video, for example, to schools. Cable TV companies already have a high market share: 78 percent of all public elementary and secondary school districts and 62 percent of all public school locations use cable for distance-learning applications (in addition, almost half the schools have

---

* For comparison, the total U.S. telecommunications business, including interexchange carriers, was $154 billion in 1994.

TABLE 2.1 U.S. Household Penetration of Video Products in 1994

TV (monochrome and color)	98%
Color TV	97%
VCRs	91%
Monochrome TV	49%
Color TV with stereo	44%
Video game software	42%
Home computers	37%
Camcorders	21%
Projection TVs	9%
LCD-based TV	9%
TV-VCR combination	6%
Home satellite earth stations	4%
Laser disk players	1%

TABLE 2.2 Estimates of the 1994 Video Entertainment Industry Revenues*

	Movie theater ($ billion)	Broadcast TV ($ billion)	Cable TV ($ billion)	Videocassette ($ billion)	Total ($ billion)
Consumer expenditures	5.90	0.00	19.37	13.60	38.87
Advertiser expenditures	0.30	29.00	2.34	0.00	31.64
Programming bundled by packagers	0.00	9.20	3.79	0.00	12.99
Total	6.20	38.10	25.50	13.60	83.40

* Based on a variety of secondary sources[4] and some primary research.

TABLE 2.3 Statistics Relating to the Video Industry

Film income (typical)	
U.S. theater rights	4%
Pay cable rights	10%
U.S. and Canada home video rights	11%
TV rights	8%
Foreign theater rights	67%
Entertainment video market revenues (1992)	
Cable	28%
Theater	7%
Tape rental-purchase	17%
Advertising (based on broadcast TV)	58%

satellite dishes for increased service capabilities). However, the cable TV companies have a goal to add the capability to handle two-way switched communications; VDT technology provides such capability. Table 2.7 depicts statistics on the number of schools in the United States. Studies show an expected classroom penetration for interactive video of 1.5 percent in 1995, 10 percent in 1998, and 25 percent in 2003.

TABLE 2.4 Areas That May Be Impacted by VDT

Area	Revenue, $ billion
Catalog shopping	71.4
Broadcast advertisement	31.6
Cable TV	25.5
Home video	13.6
Information services	10.0
Tapes and CDs	9.0
Theater	6.2
Video games	4.7
Messaging	3.3

TABLE 2.5 Channel-Watching Habits

Channels received	Channels watched
36	13
58	14
80	14

TABLE 2.6 Local Telephone, Wireless, and Cable TV Market

	1994, $ billion	1998, $ billion
Local calls	41	46
Toll calls (interLATA* and intraLATA apportioned revenue)	16	18
Network access charges	28.7	29.6
Private lines and private virtual networks	6	8
AAP	2.3	4.4
Wireless (personal communication services and cellular telephony)	15	25
Total	109	131
Cable TV	25.5	35

* LATA = Local Access and Transport Area.

2.2 Video Dialtone Markets

In the next few years, there may be increased consumer demand for interactive entertainment, education, information services, and transactional services (reservations, shopping, etc.), as consumers become more comfortable with computer-based technology. The TV set has the advantage of being a universally deployed device that can become an access terminal supporting these and other services. Market opportunities are afforded to VDT as cable TV companies are newly reregulated and as traditional video service providers (e.g., traditional television networks) lose market share. Technology Futures Inc. prognosticates that there will be 10 million video access lines in the United States by 2000. In late 1994, Bell Atlantic, GTE, Pacific Bell, and NYNEX filed with the FCC papers indicating a VDT market of $790

TABLE 2.7 Public Schools Distribution by RBOC

	Elementary	Secondary	Postsecondary
Total public schools	62,350	20,375	10,875
Ameritech	18.9%	18.7%	20.8%
Bell Atlantic	12.3%	11.4%	23.3%
BellSouth	17.4%	15.6%	16.5%
NYNEX	11.5%	5.9%	14.7%
Pacific Bell	8.9%	10.0%	16.9%
SBC	14.2%	17.9%	3.6%
U S WEST	16.8%	20.5%	4.2%
Total private schools	16,475	2,100	1,775
Ameritech	22.9%	19.9%	17.3%
Bell Atlantic	16.6%	17.8%	14.7%
BellSouth	15.8%	12.1%	18.3%
NYNEX	12.5%	19.9%	20.2%
Pacific Bell	11.9%	12.2%	7.8%
SBC	8.1%	7.6%	10.2%
U S WEST	12.2%	10.5%	11.5%

million by 2000. However, the National Cable TV Association considers this number very optimistic. Bell Atlantic supplied the number of $583 million, GTE $102 million, Pacific Bell $80 million, and NYNEX $25 million. NYNEX and Bell Atlantic postulated penetrating 50 percent of the cable TV market; Pacific Bell assumed a 10 percent penetration initially, growing to 50 percent by 2000; GTE postulated 32 percent penetration. Table 2.8 depicts the industries that will be affected by activities and deployment plans related to video dialtone.

In addition to customers' willingness to pay (covered in a later subsection), two other factors will impact the VDT market, as follows:

1. Additional regulatory activity can be expected in 1995 or beyond (perhaps further legal clarification by the Supreme Court).
2. A plethora of different technologies is possible for distribution of video information. The technical choice will determine the types of

TABLE 2.8 Industries Affected by Video Dialtone

Broadcast TV networks (ABC, CBS, etc.)
Cable television
Computer industry
Education
Electronic equipment manufacturers (e.g., ATM equipment)
Entertainment
Information services
Manufacturing
Print information providers
Retail and wholesale trade
Satellite service providers
Telecommunication carriers and providers

services that are possible (e.g., interactivity, bandwidth, and channel gamut choice), determining, in turn, the ultimate success or failure of a video dialtone undertaking by a carrier.

The current regulatory climate appears propitious for the introduction of VDT services. Legislation in the U.S. Congress (see Chap. 3), if passed, would allow total competition for local telephone service and cable television in the future. The FCC rulings on the books permit RBOCs' delivery of VDT services (but not content), as regulators look to facilitate the availability of at least two providers in all (or at least major) metropolitan markets. (This concept, already introduced in Chap. 1, is also referred to as *two-wires.*) At press time, there was a requirement for the unbundling of the transport of the service (content) from the service itself, but it is possible, as noted, that more deregulation (even total deregulation) may be possible over time.

By 1996 or 1997, consumers in some areas could be able to choose among two or three local telephone providers and six or more television and video information providers. These video information providers would be in competition to the established cable TV company in the area and would use (in some instances) the infrastructure of the local telephone company to deliver video to the consumers. VDT brings up the issue of franchises: usually a cable TV operator pays a certain fee to the local municipality or township for the privilege of serving the area in question. VDT service does not entail paying such franchise fees. Therefore, it is expected that some new entrants (typically VIPs) will prefer the VDT route to the establishment of market presence.

2.2.1 Nature of the market

Proponents see VDT as being capable of "delivering and transacting everything except physical goods." The changes that have been seen in the past 20 years (e.g., cash machines, self-serve portfolio management in walk-in branch offices, cellular phones, home fax machines, pagers used while waiting to be seated at a restaurant, and use of home PCs) point to the fact that the *public will readily accept services that enhance the convenience of tasks they already perform or must perform, and ignore those that do not.** For VDT to be successful beyond the technical elegance of an integrated broadband network that delivers voice, data, and interactive video, it must enhance quality of life while compensating for the reduced disposable time available to the individual as the world becomes increasingly complex.

* In the opinion of this author, this point has all but been missed by the communications researchers who try to develop new PSTN-based services. Generally these researchers are totally technology-driven.

Since video dialtone service is aimed primarily at the residence market, it could provide the opportunity for the RBOCs to deploy a ubiquitous broadband network supporting new information-age services to their subscribers. The following video dialtone service applications could see near-term market acceptance:

- Movies on demand, including instantaneous VOD, as well as NVOD where the movie may be broadcast every (say) 15 minutes
- Video games, particularly multiplayer games (MPGs)
- Interactive television
- Distance learning
- Access to databases and libraries (particularly multimedia archives)
- Point-and-click shopping (e.g., see Fig. 2.1)

Longer term markets include:

- Broadcast television
- Telecommuting services
- Transaction services (e.g., reservations)
- Video telephony

Today, no single network supports all of these services. The VDT opportunity stems from the fact that one advanced broadband network with bidirectional information flow and sophisticated signaling capabilities can indeed support the aforementioned services and others that may emerge in the next few years. In some recent trials conducted by VideoWay, I-TV and other interactive services proved to be of interest to the consumers. Subscribers spent 56 hours a month on offerings such as video games (22 hours), interactive programs (21 hours), and information services (13 hours). Revenues can be generated from service subscription (fixed monthly fees), usage fees, advertisers (e.g., targeted advertisement), direct marketing avenues, and VIP and ISP contribution for the use of the platform and for other ancillary services (e.g., billing creation and delivery).

Another characterization of the market is by stratification into business areas, residential areas, institutional areas (public agencies, schools, hospitals, etc.), and rural areas.* Each of these market segments can be both consumers of information, with concomitant carri-

* As covered in Chap. 3 (Sec. 3.3), there are certain regulatory waivers for serving rural areas.

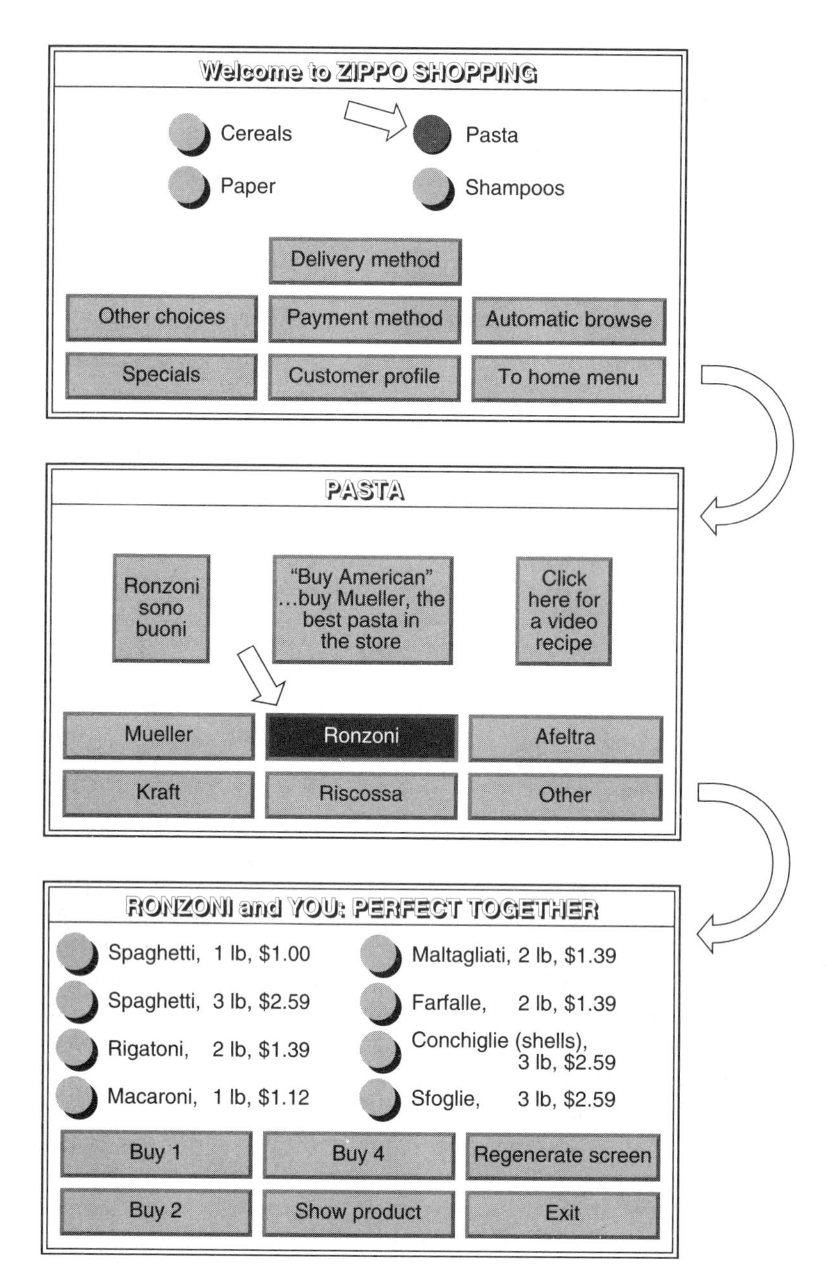

Figure 2.1 Home shopping of tomorrow.

ers' opportunities, and generators of information (à la ISP/Minitel) with other carriers' opportunities.

Table 2.9 depicts some market opportunities for carriers in the area of video dialtone.

In 1980 about 55 percent of the film industry income came from theaters, while 5 percent came from pay cable and 5 percent from home video. In 1990, only about 25 percent of the revenue came from theaters, about 35 percent came from home video, and 5 percent came from pay cable. This indicates that there have been major changes in the industry, although it should be noted that cable TV service did not experience a significant change. However, one can interpret the data to

TABLE 2.9 Market Opportunities for Video Dialtone

Video dialtone opportunities in the user market	
	Converter boxes
	Credit card readers
	Household integrator systems
	Inside wire
	Integrated computer-TV sets
	Signaling equipment
	Software associated with VDT access
	Viewer signaling devices
Video dialtone opportunities in the carrier market	
	Carrier operations systems software and hardware, to manage service (installation process, monitoring process, billing process, etc.)
	Distribution of video dialtone information, covering the local loop, interLATA service, and interregional service
	Switching devices
	Transmission equipment (multiplexing equipment, cable, transmitters, receivers, hardware for copper loops, hardware for fiber coax loops, etc.)
Video information provider (VIP) opportunities	
	Broadcast of time-scheduled events and programs
	Computer-based training (remotely archived)
	Delivery of time-shifted programs
	Entertainment on demand
	Games and contests
	Information archiving-warehousing
	Information gathering (market statistics)
	Information services
	Interactive entertainment
	Multimedia archive access
	Near-VOD movies
	PPV for special events
	Program guides
	Programming packaging
	Shopping
	Targeted advertising
	Targeted entertainment
	Targeted news
	VOD movies

say that consumers prefer to stay home rather than going to a theater; then one can infer that although cable did not fare well, any technology that improves one's ability to receive home-based video may find fertile ground in the market.

VDT can provide more efficacy for advertisers through targeted advertising and video mailboxes. Advertisement is estimated at $150 billion annually, of which about 43 percent is spent on local media (broadcast, newspapers, yellow pages, etc.), and 57 percent is spent at the national level. About 12 percent of total advertisement revenues (that is, $18 billion) are spent on local broadcasters* and 18 percent (that is, $27 billion) on national broadcasters (for a total of 30 percent, that is, $45 billion). For comparison, national print is about 8 percent ($12 billion), and local newspapers is about 20 percent ($30 billion). About 33 percent of the TV advertisement dollars are spent on network television, 27 percent on local spot ads, 27 percent on national spot ads, 6 percent on syndication outlets, 5 percent on national cable networks, and about 2 percent on local cable ads (the local cable expenditures then would be $0.02*33 = $660 million). Some claim that *new* targeted advertising (corresponding with local-level ads) could generate more than $1 billion per year nationwide. Some are already experimenting with interactive advertising (e.g., Reuters); here a consumer may be asked to play along, perhaps answer a few questions, and obtain a discount coupon for the product in question.

One area being given consideration for VDT is nonstore shopping (also referred to as video retailing or electronic retailing). This is based on the fact that catalog sales continue to show a 10 to 15 percent increase per year (over $70 billion in 1994). Additionally, the shopping channels on both broadcast TV and on cable are very profitable, generating over $7 billion a year. Observers see the "home point of sale market as virtually untapped." Video dialtone service could make inroads in this market, particularly in conjunction with razor-sharp targeted marketing and advertisement.

Transactional services in support of video retailing are viewed by some as benefiting from interactive VDT at an early point in time, since many VIPs have expressed interest in using teleshopping to sell promotional items related to the content shown in a program segment (e.g., sports items, music recordings, or exercise equipment). Transaction-based services include catalog-merchandise shopping, food shopping (see Fig. 2.1 as an example of what a future shop-at-home

---

* This includes both radio and television; TV is about 22 percent of the total (that is, $33 billion); the other 8 percent is for radio.

TABLE 2.10 Shopping Channels Being Launched at Press Time

BET
EMI
Home Shopping Spree
MOR Music
MTV-VH-1 Music
Nordstrom
Philips Electronics
QVC 2
Sony
Spiegel Catalog
Television Shopping Mall
TVMacy's
ValueVision
Warner Music

experience may look like*), banking, managing an investment portfolio, real estate,† and travel planning. Studies show more than 30 percent of the people would buy merchandise (such as books, recordings, and videos) or arrange travel electronically; more than 20 percent of the people would buy clothing; about 10 percent of the people would do home banking and portfolio management. With interactive video retailing, the consumer can bring up a file and obtain more information on products of interest; this can compensate for the low level of in-store service. Table 2.10 depicts companies, concerns, and networks that were planning at press time to become operational as home shopping TV channels. Bell Atlantic and U S WEST are both working with retailers such as JCPenney, Land's End, and Nordstrom to develop video catalogs. Competition, the availability of two-wire video platforms, interactivity, and consumer convenience are of key interest to these providers.[6]

As an example, during 1995, Interactive Systems Inc. was planning to install a trial version of a device called *InTouch TV* in 1500 homes in Portland, Oregon, which enables subscribers to place orders at the touch of a button for takeout foods and other goods which are advertised on TV. The system also lets consumers reserve tapes at the video store, order movie tickets, and print out local news or weather on a TV-attached printer. A special setup box, which connects to the TV, com-

* The consumer could just drive by the store later and pick up the purchase already packed and ready to go, or the goods could be delivered to the buyer's home.

† Some high-end real estate has already been sold through video "infomercials" that run on local cable channels.

municates over a modem with the processing site. Local TV stations KOIN, KATU, KGW, KPDX, and KOPD signed on as participants.

Education and distance learning are advanced by proponents as other market opportunities for VDT, as the quality of public schools continues to decline. Much of the *institutional-based* education will probably be carried on by existing (e.g., broadcast television, cable systems, and satellite systems) services or by evolving ATM-based backbone services,[7,8] although the proponents see a market for VDT. There are approximately 120,000 schools in the United States, as noted earlier, with an annual budget approaching a half trillion dollars. However, *in-home* distance learning (which is either point-to-multipoint or point-to-point) could be based on VDT. Applications could include advanced placement courses, professional preparation courses, continuing education and enrichment, specialized industry training, and off-campus completion of courses or degree programs, particularly in exurban and rural communities. Universities such as New York University, Purdue University, University of Phoenix, University of Maryland, National Technological University, and several others offer on-line educational services. Corporations are reportedly investing $300 billion a year to train and retrain workers (about 85 percent of that figure is spent internally, and 15 percent is spent in external courses and seminars). Library services and tutoring are other conceivable opportunities for VDT that go beyond entertainment.

Other market opportunities may materialize. For example, film libraries are reportedly interested in VDT as a business opportunity for using VOD to make their products available to a wider market. Interactive video database providers foresee an enlarged market becoming available under VDT. These providers may bundle movies, advertising, and other services. Publishers of intellectual information may need to move to electronic publishing, including VDT outlets, to retain market share. Access through TV may expand the customer base for services such as Prodigy and other information providers. There also may be opportunities for VDT in the telecommuting arena for videoconferencing, information service access, transaction services, and multimedia mailboxes.[9]

Customers are hopeful that radically new services will improve their lives; this, in the view of proponents, gives telecommunication carriers a window of opportunity, since only sophisticated terrestrial networks can support the widest possible range of services.[10] For example, a system prototyped by Interactive Health Network allows subscribers to interactively obtain health information over the TV. The provider has compiled a computerized database of text and videos on topics such as diseases, injuries, fitness, nutrition, preventive medicine, and mental health. The system shows icons at the bottom of the screen. Viewers with a remote-control device can select icons of interest and appropriate submenus. The information is stored in a video server. The Interactive Health Network is trying to sell this VIP content function to one or more of the I-TV systems to go on-line in 1995 (Bell Atlantic, TCI, Time Warner, U S WEST, and Viacom). It must be kept in mind, however, in this discussion of the market that, as of early 1995, the RBOCs were precluded from actually owning the content that supports these services; they are only allowed to deploy the platform to support VIPs and other providers (i.e., ISPs). Finally, estimates are that less than 30 percent of U.S. households will have access to interactive broadband services by 1998, putting an upper bound on the potential market opportunities.[1] (Only about 25 to 37 percent of U.S. households had a computer in 1995, depending on whose estimate one uses.)

2.2.2 Societal changes and market forces

There are a number of societal changes underway that will ultimately impact daily life and the economic outlook of people in the United States (and elsewhere). In addition to the underlying technological changes, these changes include:

- Demographic changes
- Reengineering of business, industry, government, and other institutions
- Cultural values shifts
- Migration to information-based culture and economy
- Social dynamics

It is debatable how these will alter the market landscape in the next five to eight years, vis-à-vis the current assessment by the carriers on how quickly to deploy VDT services and the type of supportive platform needed. However, a quick assay of these changes and an extrapolation of the potential impact is in order.

Demographics. There has been a slowdown in the population growth in the United States. The population is aging,* as measured by the median age. Consequently, the workforce is currently shrinking. There also has been a shift toward emphasizing lifestyle and ethnic diversity. Over one-quarter of all households in 1995 are single-person house-

---

* The 18-to-34 age group will decline by several million by 2000, while the over-50 group is growing three times faster than other groups. Also, people tend to live longer, increasing the ranks of the older age brackets.

holds, and close to 60 percent of the households now have two wage earners. These factors may determine the type of programming, the type of programming delivery, and the modality of programming delivery that will be successful in the future.

Workforce. Over one-third of the workers joining the workforce in 2000 will be from minority groups (as a group, women will account for 48 percent of the workforce). The number of people who work at home is expected to double from 20 million today to 40 million by 2000.[9] These factors may determine the amount of time and disposable income that can be funneled into video and information services. Also, they could impact the type of programming, the type of programming delivery, and the modality of programming delivery that will be successful in the future.

Job factors. The 1990s are characterized by business restructuring and downsizing. The ranks of middle management are being reduced as supervision is reduced to an optimal minimum. Many workers are finding jobs in smaller companies. People have to work longer hours and/or hold multiple jobs. This implies a climate of time scarcity. Also, smaller companies may not have the resources to install expensive video-based, company-developed distance learning. However, this may open the door for opportunities in the home-delivered training and education market. These factors may determine the type of programming, the type of programming delivery, and the modality of programming delivery that will be successful in the future.

Consumerism. The culture is based on consumerism. With multiple workers per household, there is increased disposable income. Without accessing all the implications of consumerism, one can note the fact that consumers are fickle: an item that is considered "hot" today may not move off the shelf tomorrow. This attitude can carry over to network-provided services: A service under VDT may have initial market appeal but later may not fare well. Another aspect of consumerism and related marketing is that the marketplace is becoming increasingly segmented (negating the concept of "mass market"); the new trends are micromarketing and micromerchandising. Besides applying to video retailing, this concept also applies to entertainment: One has narrow niche markets with a shift from broadcasting to narrowcasting and from multicasting to unicasting. These trends may determine the type of programming, the type of programming delivery, and the modality of programming delivery that will be successful in the future.

Values. As a proxy for a measure of values, one can point to the fact that U.S. citizens now spend about $370 billion annually* in gambling (e.g., casinos, lottery, horse racing). Although no games of chance are allowed over the TV, some proponents say that VDT should try, in whatever manner possible, to capitalize on such largess.

2.2.3 Market dynamics from cable operators' point of view

Looking at the industry from the cable company's point of view, the actual cable for cable TV hookup now passes by 97 percent of the residences in the United States and has an overall subscriber rate of 60 percent.† Observers expect future growth for the cable industry to come from PPV and I-TV. Given the competition likely to affect the cable industry and price regulation, these services will not be able to sustain the high growth rates that the industry enjoyed in the 1980s. Therefore, cable operators are being advised by financiers to seek strong growth in the telephony arena, possibly via Personal Communication Services (PCS).‡

About 50 percent of the local exchange carriers' and cellular companies' business§ (estimated at $95 billion annually¶) is open to competition (with the "right" regulatory changes).[11] One of the challenges, however, is the fragmented nature of the cable TV network; for example, an area such as greater Chicago has over 20 operators. This implies that it is difficult to get a cohesive LATA-wide service (e.g., telephone service).

Cable operators see themselves as being up for grabs as the telephone carriers offer up billions of dollars to finance acquisitions and mergers. For example, in 1993, $52.8 billion in acquisition money was

---

* That is a per capita figure of $1500 per person per year.

† There are 92.1 million households with TV sets (also see Table 2.1). Cable TV systems pass about 89.5 million homes, giving a pass penetration close to 97 percent. There are about 56.8 million subscribers served by over 11,250 systems; the subscription penetration is close to 60.7 percent. About 15 percent of the subscribers have setup boxes.

‡ Conglomerates of cable TV companies (banned under the auspices of WirelessCo) made strong showings during the 1994 FCC auctions of PCS bandwidth at the 1- and 2-MHz range.

§ As noted in Chap. 1, there is interest in PCS on the part of the cable companies. For example, in 1993, Cable Television Laboratories, in conjunction with Rogers Cablesystems Ltd., issued a Request for Proposal (RFP) to telecommunications vendor companies that may want to build equipment for cable operators' use in PCS applications.[12]

¶ This includes $40 billion in local calls, $15 billion in toll calls, $28 billion in access charges, and $12 billion in wireless services in 1993.[3]

floated.* For years, federal and state regulators limited competition in cable and shielded cable TV franchises from market forces; however, regulation is being reexamined (see Chap. 3) to promote competition among the potential providers.[11]

As RBOCs look at taking market share from the cable companies through VDT, the cable companies look at taking market share from the videotape rental business through simple VOD. It is estimated that if video rental stores could continuously stock the 10 most popular videos, consumers would rent them 80 percent of the times; however, since they run out, they only actually account for 30 to 40 percent of the rental volume.[12] As noted in Table 2.2, the video rental business is almost as large as the cable industry. Some prognosticate that an investment to upgrade the cable network for VOD (particularly for the top 10 hits), could skim 50 to 80 percent of the video rental business.[12]

2.3 Roadblocks or Competition to VDT

2.3.1 Willingness to pay

Table 2.11 depicts typical willingness-to-pay results that have been used for planning purposes, compiled from primary and secondary sources. Willingness to pay has been a thorny issue over the years. In addition to time constraints, consumers have only a limited disposable income budget that they like to spend on video and information services. And, although some knowledge workers (e.g., writers) can put information to work (profitably) for them, many consumers are not able to use information in a revenue-generating manner. As an example of this issue, a cable system may pass many homes, but the penetration could be small (say, 15 to 30 percent). Also, 63 percent of the households with basic cable service have access to PPV, but less than 30 percent have *ever* ordered a PPV program.† Over 75 percent of cable TV's revenue of $22.9 billion (total) is generated by basic and premium subscriber fees, making the PPV portion small (less than $0.4 billion,[13] or less than $60 million per RBOC region—even assuming that the RBOC could capture an ambitious 50 percent of the market, that would represent about $30 million‡ per year, less than 1 percent of their total

* The figure is arrived at as follows: Bell Atlantic/TCI (defunct): $35 billion; AT&T/McCaw: $12.6 billion; U S WEST/Time Warner: $2.5 billion; NYNEX/Viacom: $1.8 billion; SBC Communications/Hauser: $0.65 billion; BellSouth/Prime Management: $0.25 billion.

† Because of these "abysmal" results, producers have been reluctant to use a PPV distribution channel, vis-à-vis a more lucrative channel such as theaters, home video, premium television, network television, basic Cable TV, syndication, and foreign markets.

‡ This is the figure alluded to at the beginning of this chapter.

TABLE 2.11 Typical Willingness-to-Pay Numbers

Service	Payment method	Range
Broadcast television	Per month	$4–$7
Distance learning	Per hour session or class	$1–$2
Distance learning—high-end seminar	Per seminar (4–8 hours)	$50–$200
Interactive TV session	Per 2-hour session	$1–$2
Movie on demand	Per movie	$3 (older movie) to $9 (newer movie)
Special events	Per event	$15–$35
Telecommuting services	Per day	$3–$10
Teleshopping	Per shopping event	$1 or 1 percent of bill, whichever is smaller
Video games	Per 2-hour session	$2
Videoconferencing	Per hour	$10 local, $50 long distance
Videomail	Per 2-minute video message	$2–$6

revenue). The other revenue possibility is from penetrating the videocassette rental market, which was estimated previously at $13.6 billion total (of which approximately 30 percent is from adult entertainment—these market realities need to be given consideration by VIPs and RBOCs).

There is also an issue of scale. For example, some carriers now offer to automatically dial a number requested from directory assistance for a fee of 35 cents. Occasionally, this may be an issue of convenience, for example, from a public telephone booth, where one may not have a pen. However, at least from a cost-scale point of view or perhaps from a cultural point of view, one may not be willing to pay the equivalent of $315 per hour* for a service that is not life-sustaining (one may have no choice but to pay those fees to a brain surgeon or a cardiologist), but not for a service that the consumer can perform himself or herself.

As of today, the PPV data are not too encouraging in terms of motivating VDT-based PPV.† Therefore, the issue of video program distribution is not totally driven by the distribution technology (e.g., HFC, ADSL, and ATM), as some carriers and communications researchers tend to believe. Producers generate a greater revenue stream, needed to cover their investment into the film or program, by releasing programs to different markets at different times. The video industry aims

* This is calculated as follows: one must listen to the end of the message anyway, so the incremental time to transcribe the number is postulated at 1.5 s, while the dialing time is postulated at 2.5 s. This means that the equivalent per-hour cost of the service is $\$(3600/4)*(0.35) = \$315$. An individual making $5 an hour may not want to pay somebody else $315 an hour to do something this person could also do.

† As of press time, the highest-grossing nonsport event was Howard Stern's 1993 New Year's Eve party.

at maximizing profits by charging viewers what they are willing to pay, through a process called *windowing.** This implies approaching four segments of the market at the appropriate "window of opportunity," as follows:

1. Movie theater (0 to 12 months from inception)
2. Home video, plus premium television, plus PPV (13 to 30 months from inception)
3. Network television (31 to 60 months from inception)
4. Basic Cable TV, domestic syndication, and foreign syndication markets (61 months from inception and onward)

This implies that the RBOCs need to consider appropriate contractual or equity agreements to encourage a VIP to use a VDT distribution channel over another distribution channel.

Observers reach the following conclusion:

> Interactive systems can be implemented in various ways, and the next few years will be a turbulent period as various types and levels of interactivity are tested. This turbulence may spawn technical innovation, but until hardware and software standards become more settled, it will be difficult for interactive content producers to accommodate numerous systems.[14]

Much of the recent focus of VDT, in general, and of interactive television, in particular, has been on the transmission technology and the TV setup boxes, but to understand where the opportunities might be for developers of multimedia-rich content and services, a closer examination of the home environment is needed.[15] This includes information on the number of distinct simultaneous sets active in a home, the type of video programming, and the services of interest. Several players in the I-TV arena have taken such an approach, including AT&T and Videoway Communications (Montreal). They ran a two-year employee trial (of 50 people). They found that instead of a single "silver bullet," what sells I-TV services is packaging of several services under the I-TV umbrella. Such services must include entertainment, transaction processing (e.g., teleshopping), communication, and information. It was found that when such popular services are integrated (bundled), the application found consumer appeal. Videoway Communications was planning I-TV service to 30,000 homes by 1995. The service contemplated was pay-per-view with direct sale orders (at the click of a

---

* Video programming has a high production cost, ranging from $2 million per show hour for medium- to high-quality network television material, to $50 million or more for medium- to high-quality films.

mouse) for items advertised by the sponsors (for example, snack items); it would also include distance learning, electronic catalog, banking transactions, and electronic mail. Again, diversity is what appears to have the maximum consumer appeal. However, the realization arises that the "TV set is where people go to relax. . . . [T]rying to bring in . . . other applications and choices is going to make it a much more complicated activity."[14] This factor may also impact the ultimate market acceptance of I-TV.

Many, if not most, trial operators indicate that the real point of their trials is not to test the technology itself, which will change substantially anyway between the end of the trial and the commercial service rollout, as much as to find out what subscribers will watch and how often. Rochester Tel and USA Video are even collecting information about the correlation between the local weather and the viewers' service demands. For example, trials have already demonstrated that subscriber demand falls unless programming is refreshed with new features and options, such as new films, comedy shows, TV reruns, time-shifted national news, etc. An AT&T trial conducted in Chicago in 1993 indicated that interest diminishes after about three months, unless new material that goes beyond basic news and sports events (obviously refreshed on a daily basis) is routinely added. Another trend that has been noted is that consumers may give up premium channels (such as HBO) and simply rely on VOD for the nonstandard programming; this service "cannibalization" has revenue implications for service and content providers.

2.3.2 Competitors

There are some formidable competitive forces to RBOCs' VDT deployment. The conventional wisdom dispensed by VDT seminar leaders is that there will be *two* guided transmission (wire line) networks into every home: the RBOCs will retain the *narrowband* wire and the cable TV companies will retain the *broadband* wire (this is the two-wire scenario). These observers see RBOCs as being at a strategic disadvantage, which will ultimately prevent them from becoming the purveyors of video entertainment services. Some analysts also share such a view, seeing the "cable TV industry as able to offer a more attractive solution."[16] Figure 2.2 depicts a view of the growth of the communication-information industry in the next few years, as seen by some analysts.[16] The two obvious points in the diagram are that the local exchange carriers (LEC) and RBOCs are at one extreme with negative growth, while the cable TV companies are at the other extreme with high growth. Although this view has now been impacted by the reregulation of the cable industry, a relatively similar outlook is probable. By examining the diagram, one could see why, in theory, an RBOC-cable TV

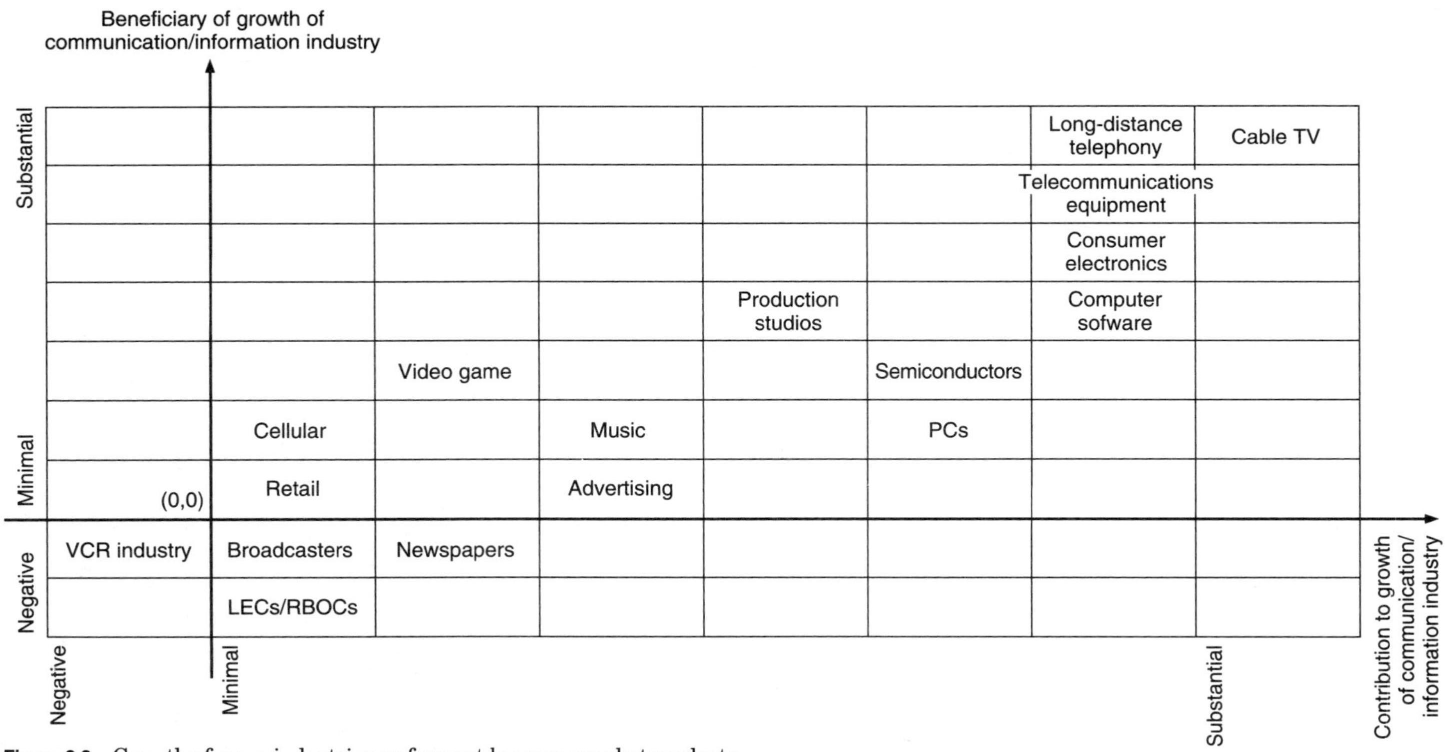

Beneficiary of growth of communication/information industry

Substantial
Minimal
Negative

Long-distance telephony
Cable TV
Telecommunications equipment
Consumer electronics
Production studios
Computer sofware
Video game
Semiconductors
Cellular
Music
PCs
(0,0)
Retail
Advertising
VCR industry
Broadcasters
Newspapers
LECs/RBOCs

Negative
Minimal
Substantial

Contribution to growth of communication/ information industry

Figure 2.2 Growth of some industries as forecast by some market analysts.

partnership makes some sense, if it could be worked out. However, the view of these seminar leaders, already making an early profit on VDT, does not need to be given immediate and final credence, because, under the right circumstances, it appears that the RBOCs could make an effective entry into the video distribution market. In particular, RBOCs are hiring talent from the cable and entertainment industries to secure the video and programming expertise that they lack.

The competitor roadblocks to the RBOCs' success in video delivery stack up as follows (also see Fig. 2.3):

1. There are the entrenched cable TV providers. Although these operators do not necessarily enjoy the highest customer satisfaction, they are already in the home, and they are positioning themselves, through public relations, as the carriers that already have a broadband "pipe" into the home (although their service is currently one-way). They are looking at both increasing the bandwidth they deliver and adding upstream bandwidth capabilities. A few cable TV

	RBOCs	Cable TV
+	• Five times as much revenue • Profitable • Low/no debt burden • High level of experience with high-availability networks • Technical knowledge • Standards-based • Familiar with switching	• Have a deployed broadband network • Have relationships with video content providers
−	• Currently do not have broadband network in place • Regulated vis-à-vis content • New entrant: new relationships needed with VIPs • May be technology- rather than service-driven • Banking on totally new technologies (ATM, MPEG-2, FITL/WDM, etc.) • Rarely hire end-users that can educate RBOCs on real end-user needs and perspectives	• Lower cash flow • Not generally profitable • High debt • Legacy of unswitched/one-way networks • Legacy of older technology • Networks not open/not standards-based • National network fragmented • Poor market perception • Reregulation • Subject to competition from new entrants

Figure 2.3 A comparison between telephone carriers and cable TV companies.

operators have indicated interest in joint cable TV-RBOC VDT trials. Prior to the TCI/Bell Atlantic and Cox/SBC Communications partnership raptures, one could hope for cooperation in this arena, at least within the scope of the partnerships.

2. As noted earlier, the revenues from PPV are fairly small. Theaters clearly are going to retain their first-showing window of 12 months or so. After that, the producers are not going to jeopardize the tried-and-true videocassette market. Also note that in 1983, reputable analysts predicted that the interactive video services market would be worth $8 billion to $10 billion in six years. Within those six years, however, nine out of ten ventures had failed and had accumulated losses of $1.25 billion, according to a study, *Video-On-Demand/ Interactive Television,* published in late 1994 by Decision Resources Inc. (Waltham, Mass.).[14]
3. There will be new entrants in the video distribution environment. For example, with new compression and modulation schemes, even traditional broadcasters will be able to deliver as many as eight channels over their allocated bandwidth, some of which may be reserved for PPV services via addressable setup boxes (addressed over the "air").* Direct broadcast satellite (DBS) is also assumed to increase availability and thereby put a downward pressure on the price of delivered video. Consumer equipment in the $500 to $700 range was available as of the end of 1994; the typical entry-level monthly fee was targeted in the $15 to $25 range. Although more than half of the satellite programming is now scrambled, there are about 4 million C-band television receive-only (TVRO) receivers in the United States. These consumers have been shown to be willing to pay the $20 to $30 monthly fees to unscramble the service. Other new entrants include alternate access providers (AAPs)† and even long distance telephone companies.
4. There are other wireless technologies (DBS, STV, cellular, etc.) that may saturate the video entertainment market.
5. There are instances in the inter-RBOC competition when one RBOC gets involved (through acquisition, partnership, or construction) in video delivery in another RBOC's territory (e.g., SBC Communications in Bell Atlantic's area). More recently, there has been inter-RBOC *telephony* competition over a cable system (e.g., SBC Communications delivering telephony services over the Hauser

---

* See Sec. 4.3.3.

† These are companies providing local loop competition to the RBOCs, usually through a terrestrial fiber-based network.

Communication cable system in Bell Atlantic's area).[17] Another example of an RBOC providing out-of-region telephony services via a cable TV company acquisition is U S WEST. With its planned purchase of Wometco and Georgia Cable Television, the company was planning to offer voice services in the greater Atlanta area currently served by BellSouth.

There are other industry-intrinsic roadblocks to VDT. For example, movie studios are concerned about the market dominance of cable TV networks and PPV packagers; VDT is viewed as yet another technology in the hands of these industry elements (some argue the opposite, namely, that studios could use RBOC-based VDT as a way to bypass the traditional cable TV networks and PPV packagers).

Anecdotally, broadcasters have thus far been lukewarm to VDT opportunities. Local television stations see themselves already reaching the market that the sponsoring advertisers had in mind—since neither broadcasters nor local TV stations do much in the area of targeted advertisement,* they do not perceive VDT's capabilities in this regard as being important at this time. One possibility for them is to use VDT to deliver archival material and/or new content to specialized (niche) audiences.

At another level, the LECs are facing competition in those service arenas, specifically, those of exchange and exchange access services, where they have had a monopoly for years. AAPs, wireless services providers (providing PCS), and perhaps cable TV providers will make inroads in these services in the next few years. This will impact not only the business customers, as has been the case up to now, but also the residential customers, who will be able to use AAPs, PCS, and cable TV companies to make and receive telephone calls. The impact will be both an erosion of revenues† as well as a general distraction and preoccupation. Such preoccupation has historically taken a disproportionate dimension vis-à-vis the level of revenue diversion up to the present time. Such preoccupation could cause carriers to lose the focus needed to deploy a new, complex, integrated broadband network supporting the plethora of services identified in this chapter.

An interesting synthesis of what might happen in the next few years is provided by the city of Rochester, New York. In 1994, Rochester Tele-

---

* Targeted advertisement has a lower granularity than the total area of coverage (which for a local station would be a circle 40 miles in radius): it is town-specific, neighborhood-specific, or even consumer-specific.

† Exchange access represents about 25 percent of the RBOCs' revenues and about 60 percent of their profits; therefore, it is a very lucrative segment of the market. For every $1 lost in access charges, there is a 45-cent loss in overall profit.

phone Corporation, the local exchange provider, opened up the local loop to competition, by "welcoming" Time Warner Cable to provide telephone service on a fully interconnected basis, where residents can freely call each other regardless of who is their telephone service provider. It is interesting to note that as Time Warner Cable prepared to offer telephone service over its cable system, MFS Communications Inc., an AAP, announced it was planning to start offering switched telephony services in the same region.[18] Also, suppliers such as Motorola are building devices that adapt cable networks for multimedia use, supporting voice, video, data, and wireless service at the TV set over cable.[19] Furthermore, there is the inter-RBOC telephony competition mentioned earlier.[17] These market dynamics, if replicated on a national scale in the next three to five years would have major repercussions on all interested parties. (These topics are discussed further in Chaps. 14 and 15.)

Finally, some argue that in the future there may be a considerable level of market inertia, both in terms of the service provider as well as in terms of the services (namely, that there will be a continued acceptance of noninteractive TV and a preference for tape rentals at local outlets). This would imply that, at least for the next five years, interactive entertainment may gain only a fraction of current TV revenues, as a result of slow market penetration. Some proponents are beginning to realize that, as we have argued for a number of years, "expectations that consumers will be able and willing to spend more than they currently do on entertainment programming as a result of new services appear unrealistic, given current and anticipated levels and growth of income."[14]

2.4 Synopsis of the On-line Information Services Market

This section provides an encapsulated market and strategic assessment of the on-line information services market. Some position this market as being cross-elastic with a number of interactive VDT services, as described subsequently. The market can be partitioned into Value-Added Networks (VANs) and Internet access providers. These are examined in turn.

2.4.1 Value-Added Network providers

The major consumer on-line services are, in order of market share: Prodigy (a joint IBM-Sears venture), CompuServe (a division of H&R Block), America On-line, GEnie (a division of General Electric Information Services), and Delphi. A gamut of other providers is available. These services count about 5 million subscribers in 1994. Major growth has been experienced in this industry in the recent past: from a revenue of about $100 million in the United States in 1988 to a revenue of $570 million in 1992, and about 120 percent growth per year over a number of years. The number of subscribers is forecast to grow at a compound rate of 35 percent per year, from 5 million in 1993 to 35 million in the year 2000. In spite of the growth, only 9 percent of the households with PCs access on-line services. Continued sales of PCs and modems for the foreseeable future will provide ample opportunities for information providers. In addition, Microsoft is planning to include an on-line service with the 1995 release of Windows; the software will eventually support on-line banking, financial information, travel services, chat bulletin boards, and electronic magazines.[20] Ventures expected to make an impact in the future include Ziff-Davis with Interchange Network (later acquired by AT&T), News Corp. with Delphi Internet, Apple with eWorld, Bell Atlantic with Stargazer, AT&T with ImagiNation and PersonaLink, as well as Microsoft with Microsoft Network, as mentioned. Table 2.12 compares the growth of on-line services with other related information delivery mechanisms and media.

With the market opportunities afforded by the commercialization of the Internet, a considerable number of new entrants appear ready to approach the on-line services market. The largest players in tradi-

TABLE 2.12 Growth of On-Line Services Compared with Other Consumer Information Systems

Consumer item	Nature	1992 expenditures	Projected yearly change
Analog musical record	Analog	$ 3,650,000,000	–3.05
Audio-stereo	Analog	9,360,000,000	–0.03
Books (consumer)	Analog	13,430,000,000	9.18
Box Office	Analog	4,870,000,000	2.33
Cable television	Analog	17,500,000,000	17.45
CD players	Digital	3,000,000,000	19.25
Consumer PCs	Digital	5,880,000,000	11.00
Consumer software	Digital	784,000,000	20.50
Digital-CD music	Digital	5,375,000,000	39.00
Home video	Analog	11,970,000,000	12.88
Laser disk hardware	Digital	90,000,000	39.75
Magazines (consumer)	Analog	6,400,000,000	5.05
Newspapers	Analog	10,275,000,000	3.35
On-line services	Digital	570,000,000	123.25
Television sets	Analog	9,600,000,000	1.15
VCR	Analog	6,300,000,000	–0.03
Video game hardware	Digital	2,125,000,000	31.25
Video game software	Digital	3,780,000,000	44.75
Total		$114,959,000,000	

NOTE: Information compiled from a variety of sources; yearly changes apply only to the immediate future.

tional media are going on-line, from magazines, newspapers, cable programmers, and game developers to the film and recording industry, including Time Inc., *The Washington Post, The New York Times,* ESPN, Times Mirror, Tribune Company, Hachette Fillipache, Ziff-Davis, Scholastic, CMP Publications, The Home Shopping Network, QVC, and Viacom.[21] Content-based companies are going on-line for three basic reasons[21]:

- The relatively low cost of entry
- The assurance of finding a ready-made market from one of the major operators such as Prodigy and America On-line
- The evolution of the personal computer as a household multimedia device offers, in the view of proponents, more assurances, in the near term, than the various interactive TV solutions that have been advanced thus far

Interactive, multimedia services, such as access to video databases, video mail, and videoconferencing, are emerging at the technology level. However, "killer applications" in multimedia have yet to be discovered, and the on-line service industry is debating what approach will be required to deliver higher bandwidth and more costly consumer-oriented multimedia services. Observers note that consumer research still needs to focus on the consumer's willingness to purchase and use a range of devices needed to access broadband services.[21] Even when focusing on narrowband, building an on-line system from the ground up to compete with networks such as Prodigy, CompuServe, and America On-line can cost millions of dollars. Furthermore, there is no assurance of success, because of the increased competition that is expected in the next two to five years. Financial advisors make the statement that "on-line services investments are not for the faint of heart."[22]

The on-line services that are thriving are those that have facilitated the formation of communities of interest and have enabled those users to communicate in an unimpeded manner. "Chat rooms" and bulletin boards are commercially successful applications found on narrowband networks. Table 2.13 depicts what are considered to be viable on-line service opportunities in the next couple of years.[23]

Digital information services have been used in the business sector for a number of years, mostly for information retrieval or distribution. The trend now is for consumers to begin using such services. The availability of communication services at relatively low rates, particularly when measured in constant dollar terms, and the embedded base of home computers have sustained the growth experienced by services such as Prodigy, CompuServe, GEnie, and America On-line. In addition, the emergence of GUI software that incorporates "intelligent agents" is improving ease of use, stimulating increased penetration.

TABLE 2.13 On-Line Service Opportunities

Information publishing
Repositioning-repurposing existing content; developing new content; packaging information for narrowcasting (e.g., MusicNet, Infonautics, Hearst Publishing's HomeNet)
Interactive entertainment
Developing games that can be played on the network by groups of people (multi-player games); creating on-line communities of interest (e.g., ImagiNation Network, Crossover Technologies, 3DO, plus games on America On-line, Prodigy, and CompuServe)
On-line shopping
Creating an on-line "place" where customers shop in an interactive, efficient, and entertaining manner (e.g., Eshop, Redgate Communications, QVC, HSN)
Electronic delivery
"Shipping" of digital information (electronic mail and bulk digital products), including EDI
Enabling technologies
Authoring tools, navigation tools, transaction processing systems, networking hardware, video compression hardware, multimedia production equipment

NOTE: See Ref. 23 for a description of the products listed in this table.

On-line service providers appear to realize that four key factors come into play: content richness and specificity, ease of navigation, consumer engagement, and affordability. Existing on-line services, such as CompuServe, Prodigy, and America On-line have been optimized to enable information providers that already have publishing interest in print or other media, to generate a new ancillary revenue stream. To be successful in the market, however, a new paradigm for delivery of information is likely to be required. On-line services should begin to migrate to platforms that support an entirely new market segment requiring narrowcast, special interest information; in the future, an increasing portion of the consumer market will move in this direction. Prodigy, CompuServe, and America On-line have already undertaken extensive testing of ISDN and cable-modem delivery of information to support the throughput needed for video, quality audio, image, and graphics.

There is a trend toward narrowcasting; consumers and knowledge workers now expect to be able to receive highly focused information. The goal should be to approach vertical rather than horizontal markets. Merely listing access to thousands of databases is no longer an end goal but only a necessary point of departure. Even the print media experience supports this trend: targeted publications are doing better than general publications. In the video delivery context, network television is losing market share to special interest cable programming. Therefore, to experience long-term success, on-line services must tap into the desires of specific audiences and provide deeper and richer information that is found on other services.[22]

Navigability is equally important: consumers want to be able to find the information they need in an effective, directed way. At the same time, the ability to perform unconstrained browsing must be retained. Sophisticated indexing mechanisms must be supported to expedite access to specific information, allowing the user to assimilate a plethora of information coming from a gamut of heterogeneous sources. The need to be engaged is related to visually rich interfaces, including at least quality graphics and imaging, and, in the future, full-motion video. Traditional information services, including mainframe-based LEXIS-NEXIS and Dialog, are seeing their market eroded by simpler, easy-to-use, cost-effective services that can provide a larger number of users with the communication capabilities and information access they want and need.[23] Prognosticators see a movement away from such traditional single-host architecture toward an open, distributed architecture of multiple hosts and servers connecting publishers and service providers directly to consumers. Affordability entails rethinking the charge based on connect time; today, most services are billed on a connect-time basis. As connectivity becomes more affordable, there is a need to find more service-driven billing mechanisms that encourage rather than discourage use. On-line services are viewed by some as the ideal mechanism to empower publishers to deliver these features to the consumer.

Numerous on-line services are expected to be launched over the next few years, which will drive increased consumer awareness and stimulate growth of the ISP industry. Services expected in 1994 include Apple's eWorld and AT&T's PersonaLink. eWorld is targeted at Apple Macintosh and Newton users and is built on technology that Apple is licensing from America On-line. AT&T's PersonaLink is supposed to be an advanced messaging service as well as an electronic shopping metaphor. Other companies known to be investing in on-line services include News Corp., Bell Atlantic, IBM, Sears, General Electric, and Knight-Ridder.[21] For example, the Interchange On-line Network, developed by the Ziff-Davis Interactive Company (now owned by AT&T) integrates the work of multiple publications and sources with customer editorial specifically designed for the on-line environment; a new software and services infrastructure was developed in support of the network.[22] The electronic publishing subsidiary of the *Washington Post* is going on-line on the Interchange On-line Network.

The on-line information service provider companies are being offered the following advice[20, 23]:

- Build competitive barriers by either accessing and packaging exclusive content, or by building and controlling a distribution channel, or by focusing on technology opportunities.

- Select the best possible distribution network. The distribution network determines how much staying power and cash are needed for a profitable business.
- Balance the opportunity of being first to market on a particular network or medium with the lack of penetration of the distribution network or of the service (particularly for new, ground-breaking services).
- Consider the application envisioned in the light of truly available network connectivity, particularly for the access portion of the network. While narrowband services based on analog lines are ubiquitous, services such as PCS, broadband, and interactive TV on cable systems may be years away, at least in terms of a critical mass of deployment.
- Target focused markets (consumer, business, or technical) and specialty areas (chat, technical, or shopping), in addition to horizontal markets such as electronic mail. As a related point, note that the top four applications for home PCs are, in order, word processing, spreadsheets, entertainment-games, and personal finances.
- Focus on improving the customer-to-network interface through the use of sophisticated GUI-based software.

Some observers focus on the future opportunities of teleshopping. As the communication bandwidth and the processing power on the PCs increases, these proponents see the entire shopping experience changing to an on-line event. Consumers can view the article, ask information about the product, and place an order. With broadband networks, consumers get high-resolution graphics of the products in question; as a short-term measure, some service providers use a combination of consumer-based CD-ROMs with network-provided information, e.g., pricing. An example includes CompuServe/Metatech CD-ROM/network integration.

Some within the on-line services industry see the introduction of broadband networks, broadband service provider content and equipment, and consumer equipment (e.g., setup boxes) as having a slow penetration cycle: only 10 percent of the U.S. telephone network is to be rewired for residential video services by 1998.[24] These observers claim that even with total cooperation and under the best of prevailing market conditions, it would take about five years to expand bandwidth and communications for television and build the links that are missing from the cable, phone, and data communications networks.[21] On the other hand, PC-based on-line services could be immediately successful; as a proxy, one can note that about 6 PCs in 10 being purchased now have modems, and 4 PCs in 10 being purchased have a CD-ROM drive.

By press time, 17 percent of households had modems, and 7 percent had CD-ROM drives.

2.4.2 The Internet and Internet access providers

The Internet is a networking and application solution that is becoming increasingly popular among researchers, developers, telecommuters, and even the general public. Several vertical industries are currently strong supporters of the Internet, including higher education, government agencies, and major corporations. The Internet is also becoming increasingly popular among healthcare providers, k-12 schools, and community colleges.

The Internet is an international network that supports thousands of computer networks and millions of computers worldwide. The Internet supports research, development, and engineering functions and is sponsored by a variety of federal agencies. Estimates of the number of Internet users vary from 7 million to 12 million; the number of subscribers has been growing at 100 percent per year for the last three years. In the United States, the Internet is made up of three tiers of networks: a T1/T3 backbone, regional networks, and local networks. In addition to these three tiers, a number of commercial networks (e.g., CompuServe) have established gateways or bridges to the Internet.

Traditionally, Internet services were available to institutional users (e.g., academics, researchers, government agencies). Now there is an effort underway to commercialize its services. Commercial Internet services are offered by a number of providers. These providers can be categorized into two major groups: national Internet providers and regional Internet providers. Each of these groups of providers differs in terms of their geographic presence, the number of services they support, and the network solutions that they offer.

National Internet service providers operate in multiple cities across the United States. These companies provide participating members with access (gateways) to other domestic and international networks. They include the Advanced Network and Services Inc. (ANS), California Education and Research Federation Network (Cerfnet), Performance Systems International, UUNet Technologies Inc., and Sprint Corporation.

Regional Internet providers include the San Francisco Bay Area Research Network (BARRNet), Committee on Institutional Cooperation Network (CICNet), and Southwestern States Network (WestNet), all of which operate in the western region of the U.S. Global Enterprise Services Inc./Northeast Regional Network (JvNCnet) and New England Academic and Research Network (NEARNET) operate in the Northeast. The Committee on Institutional Cooperation Network (CICNett) operates in the Midwest, while the Southeastern Universities Research Association Network (SURANet) operates in the Southeast.

Traditionally, the user had to employ UNIX-based commands to access and navigate the Internet. With the recent introduction of GUIs for the Internet (the best example being Mosaic), access to the network and its resources is becoming much more intuitive, thereby opening up access to the general public. One of the challenges faced by all the GUI-based approaches, however, is the throughput required by such interface. This requirement is driving the access away from traditional dial-up lines and toward new wideband and broadband systems, such as cable access and video dialtone access.

By accessing the Internet, users (particularly telecommuters who are being targeted by cable TV companies and video dialtone providers) can receive some or all of the following services: logon services (e.g., TELNET and rlogin), e-mail, file transfer, host-to-host communications, bulletin boards, directory services, browsing, and access to on-line catalogs. In effect, one can view the services provided by the Internet as being supported by network-resident servers; the user employs client software to access these servers. Four key servers are available, as follows:

1. *Anonymous FTP (File Transfer Protocol) server.* This server allows users on other systems (e.g., PCs) to log on to the Internet computer in question and to retrieve files that are identified as being public. This enables a user to create and share public archives with others on the Internet. The guest user does not need a preallocated password, hence the term *anonymous.* An extensive amount of information is available to Internet participants through anonymous FTP, such as public domain or freely available software, abstracts and full papers, public domain books, and Internet standards. File transfer services are offered by most Internet access providers.

   Archie can be used in conjunction with this service. Archie is a program that can be used to locate files that are available by anonymous FTP. The service catalogs the contents of hundreds of on-line file archives; it gathers together the location information, names, and other details of files and indexes them in a dedicated database. Users can then contact an archie server and search this database for needed files. The archie service is accessible through a range of Internet services, such as TELNET and e-mail. There are over a dozen archie servers around the world. Archie services are currently supported by the major Internet service providers.

2. *Gopher servers.* Gopher allows the user to browse information across the network without having to login or know in advance

where to look for information. The Gopher system offers information as a simple hierarchical system of menus and files: using a Gopher client, the user can select a particular item on a menu and receive either a submenu or a file of information. Gopher services orginated at the University of Minnesota and are offered by CERFNet, CICnet, and NEARNet, among others. There is commercial-grade Gopher client software available for Windows and Macintosh System 7 operating systems.

3. *Wide Area Information Servers (WAIS).* WAIS provides an easy-to-use keyword searching system. A WAIS server contains a full-text index of the documents located in the server. The user of the WAIS server submits a keyword or phrase query and the server returns a list of the documents that contain the keyword. The user can then select one or more documents from the list for downloading and display.

4. *World Wide Web (WWW) server.* WWW is similar to the servers already described, but it offers hypertext technology that links together a web of documents so that these can be navigated in any number of ways, as well as a more sophisticated GUI, namely, Mosaic. Hypertext is any document that contains links to other documents; selecting a link automatically displays the second document. Some of the information on WWW consists of hypermedia (hypermedia is hypertext where the content includes some or all of the following: text, graphics, video, voice, and music). More specifically, the WWW is a set of public specifications and a library of code for building information servers and clients. WWW* is ideal to support cooperative work in complex research fields. WWW uses Internet-based architectures employing public and open specifications, along with free sample implementations on the client and server end, so that anyone can build a client or a server. The three key components of WWW are URL (Uniform Resource Locator), HTTP (HyperText Transfer Protocol), and HTML (HyperText Markup Language). A URL is the address of the document that is to be retrieved from a network server; it contains the identification of the protocol, the server, and the filename of the document. When the user clicks on a link in a document, the link icon in the document contains the URL which the client employs to initiate the session with the intended server. HTTP is the protocol used in support of the information transfer: It is a fixed set of messages and replies

*1995 saw the introduction of Thomas, a WWW server providing on-line access to U.S. House of Representatives and Senate bills, accessible by H.R./S. **siglum** or topic keyword.

that both the server and the client understand. The document itself, which is returned using HTTP upon the issuance of a URL, is coded in HTML; the browser interprets the HTML to identify the various elements of the document and render it on the screen of the client.

Mosaic was first released in April 1993 (Mosaic 1.0) and updated in September 1993 (Mosaic 2.0). It is a graphical browser for the World Wide Web server that supports hypermedia. It was developed at the National Center for Supercomputing at the University of Illinois. Mosaic is available free for X Window System, Windows, and Macintosh. In addition to this public domain software, there are now commercial versions of GUIs for WWW servers.

Security continues to be one of the major concerns related to the Internet.[24] Studies published in early 1995 indicate that neither software nor hardware "fire-walls" are totally adequate in guaranteeing total integrity.

The reader may refer to Ref. 9 for a more detailed description of Internet services and access technologies. The pertinent point, in this context, is that there is growing interest on the part of the user community to obtain access to the Internet, and on the part of the providers to facilitate such access. Both cable TV companies and RBOCs appear ready to pursue such market opportunities under the auspices of cable and video dialtone systems.

References

1. R. Brown, "Pay Per View: Driving the Superhighway," *Broadcasting & Cable,* November 29, 1993, pp. 54 ff.
2. *Video Technology News,* January 17, 1994, p. 5.
3. The Insight Research Corp. (Livingston, N.J.), Telephone Revenue Sources, *Cable World,* February 7, 1994, p. 25.
4. Veronis & Associates, *Communications Industry Forecast,* 1991.
5. *USA SNAPSHOTS,* March 23, 1994.
6. Comments by B. Diller, CEO of QVC, *Wall Street Week,* April 22, 1994.
7. D. Minoli, "Distance Learning Applications for Broadband Networks," Datapro Report, November 15, 1993.
8. O. Eldib and D. Minoli, *Distance Learning Technologies and Applications,* Artech House, Norwood, Mass., 1995.
9. O. Eldib and D. Minoli, *Telecommuting,* Artech House, Norwood, Mass., 1995.
10. T. Miki, "Toward the Service-Rich Era," *IEEE Communications Magazine,* February 1994, pp. 34 ff.
11. *CNYCA Newsletter,* February 1994, pp. 4 ff.
12. *Communications Technology,* April 1993, pp. 18 ff.
13. J. Cooper, "Waiting for Technological Godot," *Broadcasting & Cable,* November 29, 1993, pp. 57 ff.
14. A. Lindstrom, "The Business of Interactive Video," *Telephony,* November 1994, pp. 6 ff.
15. *Multimedia Week,* March 14, 1994, p. 3.

16. *Communacopia Investment Research Report,* Goldman Sachs, New York, N.Y., July 1992.
17. A. Bryant, "Southwestern Bell to Invade Bell Atlantic Phone Market," *New York Times,* May 23, 1994, p. D1.
18. G. Naik, "MFS to Offer Local Calling in Rochester," *Wall Street Journal,* May 19, 1994.
19. C. Hill, "Motorola Launching System to Convert Cable Networks For Multimedia Use," *Wall Street Journal,* May 20, 1994.
20. B. J. Lyther, "Online Services for Personal Communicators," *The Red Herring,* April 1994, pp. 50 ff.
21. G. DeRose, "How Far Can Consumer Online Services Go?" *The Red Herring,* April 1994, pp. 52 ff.
22. M. E. Kolowich, "How New Online Architectures Create New Business Opportunities," *The Red Herring,* April 1994, pp. 42 ff.
22. *The 1994 Telco Video Services Report,* Jupiter Communications, New York, N.Y., 1994.
23. A. A. Piol, "Digital Information Services: Here Today and More Tomorrow," *The Red Herring,* April 1994, pp. 46 ff.
24. T. Clark, "New Systems to Make Internet Safer," *Inter@ctive Week,* Jan. 16, 1995, pp. 10 ff.

Chapter

3

Regulatory Aspects

3.1 Introduction

This chapter examines key regulatory aspects of video dialtone and national information infrastructure (NII). Policy makers are now revisiting regulatory and policy issues concerning communications in general, and the superhighway and video delivery in particular. There are efforts underway by Congress and the Clinton administration to eliminate regulatory barriers that prevent local telephone companies, long distance carriers, and cable TV companies from entering one another's business.* The intent of reducing these barriers is to foster competition that could expedite development of the superhighway, and to enhance the diversity of available video services.[1] Specifically in the context of VDT, the FCC wanted to encourage the development of a network or a network of networks that provides the service on a globally interconnected basis and enables entrepreneurs to provide telecommuting, education, and public services in "a dramatic new manner."

Some of the issues related to deregulation revolve around the concept of universal service, which was institutionalized in the 1934 Communications Act. Through cross-subsidization from business users, the telephone companies agreed to serve the rural, disadvantaged, or less lucrative users. The question under legislative consideration is what constitutes universal access in the context of the superhighway and in

* The Clinton administration is quoted as saying that "the administration's policy with respect to . . . communications industries . . . will be . . . to promote, where possible, healthy competition that can foster quality and efficient services . . . and to rely on government regulation only as a last resort. . . ." The stated cornerstones of these initiatives are to: (1) encourage private investment; (2) encourage and protect competition; (3) provide open access to the PSTN; (4) ensure universal service; and (5) facilitate flexible and responsive government action.

the context of (entertainment) video. At the technological level, there is a clear need to reexamine the 1934 Communications Act since neither video nor computers were around when that act was codified into law. It is likely that major regulatory changes will occur before the end of the decade; the information in this chapter provides a backdrop against which such changes can be examined a prior or a posteriori.

3.2 Regulatory Initiatives

3.2.1 Background activities

Two restrictions of relevance to video services that were imposed as part of the divestiture of the Bell System are*:

1. The RBOCs† cannot own information services.
2. The RBOCs cannot deliver video content within their service area.‡

The three branches of the U.S. government were, at press time, in the process of lifting or modifying these restrictions on video delivery as well as other restrictions (see Sec. 3.3.3). Table 3.1 depicts some of the recent judiciary, legislative, and executive activity in this area. The most fundamental action to date (press time) was The Cable Act of 1984 which precludes telephone companies from owning cable systems or offering video services in their service areas (Section 613 (b)). To balance the impact, the regulation bars cable companies from offering switched local telephone service. The 1984 Act also permitted cities and municipalities to charge franchise fees for granting rights to cable companies and to require public access channels. The liberalization process began in August 1993 when a federal court lifted the restriction on video content (in the context of Bell Atlantic§).

The FCC plays an important role in the video distribution arena. Sections 63.54 to 63.58 of the Commission's rules make it unlawful for a common carrier to provide video programming directly to subscribers in its telephone service area. However, the FCC has often revisited the issue of telephone company's distribution of video programming. For example, in 1987 the FCC issued a Notice of Inquiry (NOI) to review the telephone company/cable cross-ownership restrictions and solicit comments on regulatory changes that might permit telephone compa-

---

* Other restrictions included long distance telephone service and manufacturing.

† Some of the restrictions about video programming also apply to other large LECs, e.g., GTE; this discussion focuses on RBOCs.

‡ An RBOC is precluded from the selection, pricing, and packaging of programming.

§ This decision applied only to Bell Atlantic; other RBOCs must obtain relief on their own merit.

TABLE 3.1 Recent Judiciary, Legislative, and Executive Activity Applicable to Video Services

Year	Legislative act	Explanation
1984	Modified Final Judgment (MFJ)	Judiciary decision by Judge H. Greene to divest AT&T and create the seven RBOCs. RBOCs are forbidden from manufacturing equipment, providing long distance telecommunication services, delivering information (including video), and owning the content.
1984	Cable Act	Forbids telephone companies from owning cable systems or offering video services within their operating regions. The restriction applies whether they do so directly or through an affiliate. Deregulates the cable TV industry (except for local switched telephone service), but permits cities to charge franchise fees (up to 5 percent of their gross revenue). Cable TV companies must provide public access and public service capacity.
1991	Appeals to MFJ	On appeals to the court, Judge H. Greene lifts RBOCs' restrictions on providing information services, allowing them to own sports, weather, news, and other data services for distribution over phone lines.
1991	High Performance Computing Act	(Then) Sen. Gore's bill authorizes the establishment of the National Research and Education Network (NREN) and funds research on high-speed networking technology.
1992	Cable Act	Reregulates cable TV companies, including "must carry" mandates, service standards, and rate regulation (see Sec. 3.3.5).
1993	Antitrust Reform Act (proposed)	Aimed at phasing out RBOCs' MFJ limitations. U.S. Attorney General and FCC would grant the RBOCs the right to offer interstate and interexchange services and burglar alarm services, own a partial interest in electronic publishing undertakings (from 50 to 80 percent), and manufacture equipment. This allows competition between local and long distance companies. It also proposes the creation of a new, optional class of regulation (called Title VII) for broadband interactive services (Brooks and Dingell Bill).
1993	Bell Atlantic versus U.S.	August 1993 judiciary action where U.S. District Court Judge T. S. Ellis rules as unconstitutional the 1984 Cable Act, preventing local phone companies from providing TV programming in their service territories. Rule only applies to Bell Atlantic. Under appeal at press time.
1993	National Communications Competition and Information Infrastructure (proposed)	Major restructuring of the 1934 Communications Act, permitting phone companies to deliver video and other services, based on opening up the local loop to competition, while providing open platform and universal service (for more than just plain old

TABLE 3.1 Recent Judiciary, Legislative, and Executive Activity Applicable to Video Services (*Continued*)

Year	Legislative act	Explanation
		telephone service) (Markey, Fields, and Boucher Bill).
1993	NII	H.R. 1757 passes House in September of 1993; no Senate support, but portions are included in bill S.4, eventually not passed: Aims at expanding the High Performance Computing Act, establishing federal programs to develop and disseminate applications of high-speed networking in the fields of education, health care, and libraries (Boucher Bill).
1993	NII: agenda for action	Executive action of the Clinton administration that proposes the formation of the Information Infrastructure Task Force to promote private sector investment; reform communications regulation; ensure universal, secure, and reliable service; promote interworking standards; promote applications in education, manufacturing, health care, and government information. Other goals are to protect intellectual information and improve the management of the usable electromagnetic spectrum. The Information Infrastructure Task Force is to be composed of federal officials and of a U.S. Advisory Council on the NII that includes 25 private sector as well as public sector appointees.
1994	Vice Presidential Address on the Antitrust Reform Act (also, similar comments are made by Pres. Clinton in 1994 State of the Union speech)	The Clinton administration, through Vice President Gore, gives voice to its support of the Antitrust Reform Act on January 11, 1994. Highlights: ". . . connect all . . . classrooms . . . libraries . . . hospitals and clinics by the year 2000 . . ."; private investment, fair competition, and open access are given prominence; openness and universal service are promoted by the proposed legislation; the administration will support research on networking, development of applications, and electronic delivery of government services.
1995	Various bills including T. Bliley, J. Fields, and J. Dingell on the House side and L. Pressler on the Senate side	Action pending

nies to deliver video programming. In 1988, the FCC issued a Further Notice of Inquiry (FNOI) and a Notice of Proposed Rule-Making (NPRM). The FCC decided, through such process, that it should recommend to Congress that the cross-ownership ban be repealed or modified to allow local telephone companies to provide video programming in their service areas, subject to Computer Inquiry III-based safeguards. (Computer Inquiry III safeguards include a mandate for open network architectures* as well as cost-accounting protection against cross-subsidization of nonregulated activities.†)

In July of 1991, Judge Greene reluctantly modified the MFJ to allow the RBOCs to provide information services *except video programming.* RBOC entry of this market has been limited, but the ruling opens the doors to video dialtone (among other opportunities).

In November 1991 the FCC issued a Further Notice of Proposed Rule Making (FNPRM) to *amend its rules to permit telephone company entry into video dialtone services.* It also issued the First Report and Order (FR&O), where it interpreted the cable-telephone company cross-ownership ban and concluded that the ban does not apply to interexchange carriers. The FR&O concludes that neither a local telephone company providing video dialtone services nor its customer is required to obtain a local cable franchise. At the same time, the FCC issued a Second Further Notice of Inquiry (SFNOI), where it invited comments on its earlier conclusion that local telephone companies be allowed to provide video programming.

As noted in Chap. 1, the FCC described video dialtone in its November 1991 FNPRM as an enriched version of video common carriage under which the telephone company will offer various nonprogramming services in addition to the underlying video transport. The FCC goals for VDT were as follows (among others):

1. To make available nationwide, publicly accessible, advanced telecommunications networks, and an advanced infrastructure capable of providing integrated voice, data, and video services to serve the public interest
2. To foster competition in the video and communications markets so that free market forces rather than governmental regulation will determine the outcome of new services
3. To foster a diversity of information services for the American public

Other aspects of the FR&O included the following:

1. There would be no franchise requirements for the local telephone companies.

---

* In the definition of the FCC, the open network architecture (ONA) is the overall design of a carrier's basic network facilities and services to permit all users, including the enhanced service operations of the carrier and its competitors, to interconnect to basic network functions on an unbundled and equal-access basis.

† For completeness of discussion, Computer Inquiry I established that telephone companies cannot provide information services; Computer Inquiry II declared that telephone companies could deliver information services but only through a separate subsidiary.

2. There would be no franchise requirements for programmers.
3. The interexchange carriers are allowed to be distributors of video programming.

In July 1992, the FCC made a decision on video dialtone through the Second Report and Order, as follows (among other provisions*):

1. Local telephone companies are permitted to provide a basic platform for the delivery of video programing but may not provide video programming directly to subscribers (see Table 3.2).
2. It increased to 5 percent (from 1 percent) the allowable ownership of video program companies by telephone companies.
3. It allowed telephone companies to purchase and leaseback cable facilities to "sole providers."

The FCC† advanced the two-level model for VDT services and platforms already introduced in Chap. 1. The model includes Level 1 Gateways and Level 2 Gateways. However, the FCC did not allow telephone companies to be packagers or provide programming directly to subscribers. Hence, they could provide the network facilities but could not operate the system themselves.

The basic platform, which is allowed, is a common-carriage transmission service coupled with the means by which consumers can access any or all video program providers making use of the platform.[3] The basic platform falls under the regulation of Title II of the Communications Act. Video programmers can provide either single or multichannel services. Enhanced services can be provided by any service provider or by a local telephone company's unregulated subsidiary. By definition, *enhanced services* are services offered over common-carrier transmission facilities used in interstate communications, which employ computer processing applications that act on the format, content, code, protocol, or similar aspects of the subscriber's transmitted information; provide the subscriber additional, different, or restructured information; or involve subscriber interaction with stored information.[3] Enhanced services are not regulated by Title II of the Communications Act.

---

* There also was a recommendation to Congress to repeal the statutory restriction on telephone company-cable television cross-ownership.

† Interestingly, the FCC has been reported as stating that it does not perceive VDT as the principal competitive force against cable companies, but just as a potential competitive force; DBS, Multipoint Multichannel Distribution Systems (MMDS), or wireless are cited instead as more imminent competitive forces.[2]

TABLE 3.2 RBOC-Permitted VDT Services*†

Electronic video-based Yellow Pages
Interactive advertisements
Interactive education
Interactive entertainment
Interactive games
Interactive information
Interactive multimedia session
Interactive news, weather, stock tracking
Interactive transactions
Interactive video-based shopping (e.g., as shown in Fig. 2.1)
Other broadband services

* To be provided by an unregulated subsidiary. Must be provided on an equal-access basis with other VIPs and ISPs.
† Video programming not permitted.
NOTE: See Table 3.6 for more details.

At least in terms of public pronouncements, these rules received mixed reactions from both the RBOCs and the cable TV companies, where one industry may have hoped for more scope, and the other industry may have hoped for more restrictions. For example, the National Cable TV Association (NCTA) is pushing for a seven-year moratorium on RBOC entry into in-regions video services, whether through construction of a new network or by a buyout, until competition is introduced in the local loop; the United States Telephone Association (USTA) has recommended that the RBOCs get the choice between having their rates regulated like cable companies, with a price cap, or getting rate of return regulation, while at the same time being able to enter the voice and video market as a whole, without separate subsidiaries.[4]

On August 24, 1993, the U.S. District Court in Alexandria, Va., declared the cross-ownership provision of the 1984 Cable Act to be unconstitutional.* Before this ruling, cross-ownership was permitted only outside the Bell Atlantic region; this ruling allows Bell Atlantic to provide cable programming within its telephony service areas. However, as it stood, the ruling is limited to Bell Atlantic and not to other RBOCs. These RBOCs will likely file suits to extend the decision to their regions, unless more general legislative "relief" emerges from the legislative actions listed in Table 3.1. As might be expected, the reaction of the cable TV industry was, again, one of concern.

* Judge T. S. Ellis III ruled that the 1984 Cable Act violates the First Amendment's rights of a carrier's ability to "express ideas by . . . a significant mode of communication: video programming."

Until new regulatory apparatuses are in place, the conventional wisdom was that the RBOCs may consider purchasing cable companies that operate outside their regions to enter the video field; the inconclusiveness of the Bell Atlantic/TCI and SBC Communications/Cox Enterprises deals dampens such merger or acquisition expectations. It now appears that the RBOCs' road to video, VDT in particular, will be through upgrade of the existing network to support digital video over ATM, transported through a variety of new loop infrastructure such as ADSL, HFC, FTTN, FTTC, or even FTTH.

As a press-time summary of the regulatory environment, we note that the 103rd Congress failed to pass Sen. Ernest Hollings (D-S.C.) comprehensive telecommunications bill. Hollings's bill, the Communications Act of 1994 (S. 1822), would have brought with it sweeping reforms, essentially rewriting the Communications Act of 1934 for the first time in 60 years. In a formal statement, Hollings noted that his bill had broad public support, but that "only one sector of the industry continues to oppose the bill—the telephone companies, especially the regional companies." However, state and local regulators indicated that they would have been "preempted," that is, limited, in their role of local administrators of the NII. Among other provisions, the bill would let telephone companies get into other businesses on an equal schedule as the companies competing with them. For example, it allowed cable and telephone company competition; however, it would have allowed the RBOCs' entry into long distance telephony. In the final days of the 103rd Congress, Senate GOP Leader Robert Dole, of Kansas, had sent substantial revisions to the bill to Hollings with conditions that portions of Dole's amendments were "nonnegotiable." These changes came from Dole's office after the bill had been passed in committee and had included provisions to which the RBOCs had agreed. House Telecommunications Subcommittee Chairman Edward Markey (D-Mass.) also expressed dismay at the demise of the Hollings bill, calling this "bad for the country . . . we had an historic opportunity to pass legislation updating our telecommunications law . . . that opportunity is gone." Markey indicated at that time that he was going to "fight" for legislation in the 104th Congress. If telecommunications reform does not happen in 1995, then likely it will not happen until 1997.

3.2.2 Definition of video programming

In the previous discussion, it was noted that currently RBOC cannot own video programming. The FCC provides a definition of *video programming* as follows[3]:

> Video Images which are severable from the interactivity in a program service: . . . to the extent a service contains severable video images capable of

> being provided as independent video programs comparable to those provided by broadcast stations in 1984, that portion of the programming service will be deemed to constitute video programming . . . the mere inclusion of some interactive capability would not be sufficient to transform other video programming into non-video programming . . . Many of the video services that could be provided over a video dialtone network involve a high degree of interactivity . . . that would enable the subscriber to tailor the video images to his or her specific requests . . . these services fall outside the scope of video programming . . .

Applying these definitions and concepts, one can easily label all television programs and most cable TV programs (e.g., PPV, VOD, conventional home shopping, PPV-VOD with embedded VCR controls*) as being video programming. Programming that would fall outside the definition of video programming include (see Table 3.2) video search and retrieval, video games, user-tailored news, weather, and stock market quotation. It is likely that some confusion could arise in the future as to what constitutes a "high degree of interactivity" in a specific situation; such confusion would have to be cleared up through a petition to the FCC.

3.3 Related Regulatory Aspects

This section covers some related regulatory facets.

3.3.1 Waiver in rural areas and for good cause

Currently, as noted, a telephone common carrier is precluded by the FCC from owning a cable company in its own territory. However, a carrier is exempt from this telco-cable cross-ownership in locations where there are fewer than 2500 inhabitants in a nonurbanized area (there are approximately 4500 such places in the United States). There are proposals (first surfaced as part of the Second Report and Order) to raise that threshold to 10,000 inhabitants, thereby encouraging more service providers to make necessary investments.

There also is a good cause waiver to the FCC rule "in those areas where . . . video programming . . . could not exist except through a cable system owned by . . . the local telephone . . . carrier. . . ." This waiver, however, has been difficult to secure.

---

* These services include some interactive capabilities, but not enough to escape the FCC ban. Even new "gimmicks" such as letting the user select a camera angle for a sporting event, or a low level of interactivity such as picking from one of, say, 10 different endings to a movie, fit the definition of severable video images.

3.3.2 Local loop competition

In 1992, the FCC adopted an interrelated policy that collectively opened up the local exchange market "to the benefits of competition." The September 1992 ruling stated that the LECs must allow central office access to alternate access providers for collocated equipment supporting nonswitched services. The FCC also indicated that it was contemplating extending the rule to switched services. This comes against a backdrop of proposed legislation and other recommendations on competition discussed in the next section.

3.3.3 Recent regulatory activities on RBOCs' deregulation

This section provides in encapsulated form some of the recent proceedings to eliminate and/or reduce regulation on the RBOCs. Table 3.3 depicts some of these recent legislative efforts.[5]

In March 1994, two bills were making their way through Congressional committees (see footnote to Table 3.3). The House Energy and Commerce Committee and the House Judiciary Committee passed different versions of a bill that would allow the seven RBOCs to offer long distance service and to manufacture telephone equipment. The key difference between the two bills is that one calls for the RBOCs to seek Justice Department approval before providing long distance service.[6, 7]

Table 3.4, resulting from a National Telecommunications and Information Administration (NTIA) infrastructure study, might be considered an RBOC's wish list (except possibly for the local loop competition which so far has been advocated by only a handful of LECs and RBOCs).

3.3.4 Other legal drivers for broadband services

In addition to the activities listed thus far, there are a number of other legal initiatives and drivers for broadband. A number of regulators and legislators realize that telephone companies need to be able to deliver video to make the investments in broadband economically viable. Thousands of mission-critical networks of *Fortune* 1500 companies still use modems operating at 9.6 or 19.2 kbps—corporate data communications can well exist without 155-Mbps access or backbone links* for years to come. Table 3.5 encapsulates some of these drivers.

---

* At most, in the opinion of this author, one could make an exception to this statement for the small (3- to 6-node) backbone network of *Fortune* 50 companies.

TABLE 3.3 Recent Regulatory and Related Efforts on Loop Competition (Not Enacted)

H.R. 5096 (1992)	Reverse the restrictions on information services, manufacturing, and long distance.
H.R. 2546-S. 1200 (1992)	Establish objective for national broadband network by year 2015.
H.R. 5559 (1992)	Reverse the restrictions on information services.
S. 173 (1992)	Reverse the restrictions on manufacturing by separate subsidiary.
H.R. 3515 (1992)	Reverse the restrictions on information services by separate subsidiary.
H.R. 3515-S. 2112 (1992)	Revision of ONA to guarantee equal access.
S. 1086 (1993) [Hollings-Danforth-Inouye], in conjunction with a new 1994 version replacing S. 1086*	Allow RBOCs to offer video distribution services within their serving areas (through separate subsidiary), as well as manufacturing and long distance service (in 6 years); carriers required to subsidize universal service.[4]
H.R. 1504 (1994)	Rescind ban against RBOCs' delivery of video in its own region but requires separate subsidiary.
H.R. 3626 (1993) [Brooks-Dingell]*	Eliminate MFJ restrictions on RBOCs' long distance telephone service (in-region right-of-way, out of region in 5 years); allows electronic publishing (through separate subsidiary, in 6 years).[4]
H.R. 3636 (1993) [Markey-Fields]*	Rescind telephone company-cable cross-ownership, allowing RBOCs to offer cable in local service areas. Also allows RBOCs to offer video services (under the VDT model), through a separate subsidiary—must support VIPs in open manner and reserve 75 percent of capacity for them.[4]
Clinton administration plan (1994)	Allow RBOCs to provide video services in their regions, with provisions such as providing equal access and channel capacity, implementing ONA, precluding RBOCs from buying existing cable systems in own area, eliminating state barriers.

* These bills were not passed by the 103rd Congress.

Observation: Table 3.3 provides only highlights of major provisions—many ancillary provisions are contained in these bills (interested reader should refer to bills for a complete picture).

TABLE 3.4 Possible Deregulation Directions

Allow competition in local switched services by Alternate Access Providers (AAPs) and support ONA fully.
Allow telephone companies to be video common carriers.
Define video programming as an unregulated enhanced service.
Modify depreciation rules to reflect economic value.
Remove restrictions on information services.
Remove restrictions on manufacturing.
Remove restrictions on telephone company-cable company cross-ownership.
Use incentive regulation instead of rate-of-return regulation.

TABLE 3.5 Related Drivers for Deployment of Broadband

California Public Service Commission	Commission to allow competition in all areas with the goal of achieving statewide access to full-motion switched video services by 2000.
NII	Administration's vision for the superhighway. Aspects of the vision include competition, private investments (RBOCs, cable TV, AAPs, utilities, broadcasters, satellite carriers), access to government information, universal service. Suggests funding (about $2 billion annually) to fund research and construction.
North Carolina's Superhighway	ATM/SONET network for distance leaning to start service in 1994 (100 sites) and expand to major network in future (3500 sites by 2004). Interconnects secondary schools, universities, community colleges, medical centers, government or public safety sites, and industrial parks.
Partnership for Progress	National Governors' Association is seeking to "foster the development of cutting-edge telecommunication technologies" at the national and state level.
Pennsylvania's Fiber-Optics Law	Mandates that Pennsylvania's LECs and IXCs must have a broadband network deployed by 2015—regulatory incentives offered.
State-level incentives	Various state reliefs (regulatory, taxation, etc.) to foster deployment of open in-state fiber-optic facilities (e.g., in addition to the ones previously listed, Illinois, Nebraska, New Jersey, Tennessee).

3.3.5 Cable TV regulation

The first commercial cable TV system appeared in 1950. The industry was highly regulated until 1984, when it was deregulated. The Cable Act of 1984 precluded cable TV companies from offering switched local telephone service (only a handful of states currently allow such competition). The industry was reregulated in 1992 to 1993 through the 1992 Cable Act. The FCC enumerated requirements in the following areas (in alphabetical order):

- Antibuythrough
- Customer service standards
- Electronic equipment compatibility
- In-house wiring
- Indecency
- "Must carry"-retransmission consent*

*H.R. 525 of 104th Congress, "To Repeal the Must-Carry Provisions of Title VI of the Communications Act of 1934," was introduced in early 1995.

- Ownership limits and carriage agreements (no single operator can pass more that 30 percent of homes in area; the figure is relaxed to 35 percent if it entails a minority area)
- Program access
- Public interest-home shopping study
- Rates regulation (10 percent decrease mandated based on September 1992 levels—added 7 percent decrease mandated the following year[8])
- Sports programming migration

In spite of the new restrictions, cable operators are not governed as common carriers; therefore, they do not have responsibilities with regard to equal access (to other VIPs and providers) and with regard to availability of programming to customers. They may own, have financial interest, and exercise editorial control in video programming. They can, therefore, select, package, bundle, create tiers, and establish prices for the video programming they deliver.

The FCC was motivated, through both the Cable Act of 1992 and the VDT regulation, to foster competition in the video distribution arena. As of press time, there was cable TV competition in less than 75 communities, covering about 2 percent of the total cable TV subscriber population (the availability of DBS service will clearly alter the situation).*

3.3.6 Administration plan for amending the 1934 Communications Act

As noted at the beginning of this chapter and in the intervening discussion, the Clinton administration is seeking to eliminate regulatory barriers that prevent providers of all kinds (local telephone companies, long distance carriers, and cable TV companies) from entering one another's business. In terse terms, the stated goals are:

1. Amend Title VII of the Communications Act of 1934.
2. Create a voluntary, streamlined, regulatory framework to encourage the development of full-function broadband networks.
3. Empower the FCC to ensure an open, interoperable network supporting universal service and authorize the FCC to reduce regulation on (some) carriers.
4. Regulate rates consistent with the 1992 Cable Act.

---

*Access to Thomas (WWW) in January 1995 showed 12 H.R./S. bills related to the cable TV industry.

5. Allow state and local governments to regulate, subject to FCC guidelines, noncompetitive services.

As part of this process, the FCC mandated two cable rate cuts and approved 700 pages of rules. These rules, still under "digestion" at press time, were widely blamed for slowing down the recent trend for mergers and acquisitions, including the Bell Atlantic/TCI and SBC Communications/Cox deals.

3.4 Video Dialtone

As noted previously, cable TV service is a non-common-carrier service. It supports one-way transmission of video programming and other services. As discussed in Sec. 3.2.2, video programming is a service that is equivalent to TV broadcast service, namely, that involves a one-way transmission and is available to a universe of subscribers; subscriber interaction is limited to service selection.

In contrast, VDT is a common-carrier service that, as implied in Chap. 1, provides a regulated, nondiscriminatory (i.e., guaranteed access) basic platform to service providers for *video programming* (as previously defined) and other services. It also provides an unregulated

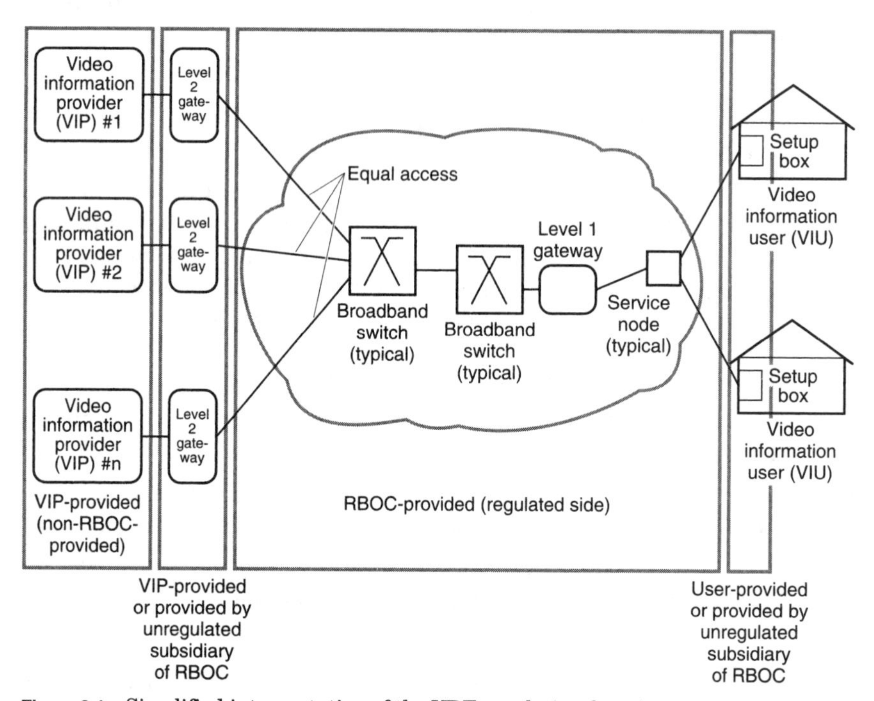

Figure 3.1 Simplified interpretation of the VDT regulation function.

TABLE 3.6 Regulatory View of VDT Services and Gateways

	Level 1	Level 2
Services	RBOC provided Common-carrier basis Equal access Basic transport and routing Broadcast or switched Service connection Directory Help "Basic enhanced" functions Collection of billing measurements (for VIPs)	Provided by nonregulated RBOC's subsidiary or other provider Enrichment to basic services, such as enhanced and noncarrier services, to be provided by unregulated subsidiary of the RBOC or other service providers on an equal-access basis Advanced services, including interactive shopping, interactive games, interactive entertainment, interactive education, broadband information services, and video Yellow Pages Enhanced video gateway for advanced services Advanced navigation and search capabilities Video processing services Billing and collection Video consumer equipment Inside wiring
Gateway	(L1F)* Menu for ISPs and VIPs (L1F) Establishing selected connections (L1F) Basic control for transport and switching (L1F) Interface to billing information (L1EF)† Channel lockout (L1EF) Display sequence establishment (L1EF) VCR programming (L1EF) Picture-in-picture	Establishment/disestablishment of connections to advanced service nodes Server control Advanced navigation Interface to databases Downloading of application software to setup box Authoring tools

* L1F: Level 1 Function.
† L1EF: Level 1 Enhanced Function.
NOTE: RBOC bills VIP for use of VDT platform; VIP bills customer for the service it receives from the VIP.
NOTE: "Enhanced *Services*," for the established meaning of the word in regulatory parlance, are not allowed for the regulated side of an RBOC—enhanced *functions* are allowed.

mechanism for telephone companies and other ISPs and Enhanced Service Providers (ESPs) to deliver *nonvideo programming* services (e.g., interactive shopping, interactive games, interactive entertainment, interactive education, broadband information databases). The VDT regulation is consistent with the Cable Act of 1984 and with the Open Network Architecture. Table 3.6 is a summary of Level 1 and 2 services and gateways discussed in Chap. 1. (Again note that at writing time RBOCs cannot provide video programming to customers.) Figure 3.1 depicts a graphical interpretation of the VDT regulation.*

* Interpretations of the various rulings have on occasion been more than 600 pages long; Fig. 3.1 is only a high-level, single-view, interpretation.

References

1. "Administration to Lift Speed Limits on Info Superhighway," *Video Technology News,* January 17, 1994, p. 1.
2. Comments by the FCC Chairman, *Broadcasting,* December 21, 1992.
3. *FCC Second Report and Order,* FCC, Washington, D.C., July 1992.
4. V. Pasdeloup, "Telecoms Reform Advanced by Hollings' New Bill," *Cable World,* February 7, 1994, p. 2.
5. "What the Feds Have in Store for 1994," *Cable TV and New Media,* February 1994, pp. 4 ff.
6. *New York Times,* 3/17/94, D3.
7. A. Reinhardt, "Building the Data Highway," *Byte,* March 1994, pp. 46 ff.
8. Cable World Staff, "Cable Execs See Telco Fingerprints All Over New FCC Rate Roll-backs," *Cable World,* February 28, 1994, pp. 1 ff.

Chapter

4

TV and Cable Technology—a Primer

4.1 Introduction

This chapter covers, in abbreviated form, some of the fundamental background technical issues that come into play in video, digital video, and video dialtone. Some of these concepts are used in the rest of the book, in describing the significant innovations in video transmission that can be expected in the next few years.

4.2 Television Basics

The electronic image-scanning techniques used today were invented by Zworykin in the United States in 1929. Broadcasting systems were introduced in Germany and in England in 1935. Since then, television broadcasting has gone through many phases of innovation, improving picture quality, available services, and distances covered. Commercial TV broadcasting started in the years following WWII; after a war-induced hiatus, by 1948, the United States, France, Great Britain, and the USSR had TV service. Services had only local reach, and various countries chose different TV schemes: the United States selected 525 lines per picture frame, France selected 819 lines, and the United Kingdom selected 405 lines. Particularly in Europe, as the radio links started to be broadcast with enough power to cross national borders, the issue of differing standards became a problem. Thus, the European system converged to the current system of 625 frame lines, recommended by the Consultative Committee on International Radio (CCIR; now ITU-R) in 1970, as a compromise between vertical definition and channel bandwidth.[1]

TABLE 4.1 Key Families of TV Standards Worldwide

NTSC	Analog U.S. and Japanese format
PAL	Analog European format
SECAM	Analog French and Eastern European format

Given the genesis just described, one should not be surprised to find out that at this juncture, when considering the existing analog broadcasting systems, there are more than a dozen systems used worldwide, falling into three major families: NTSC, PAL, and SECAM (see Table 4.1). All three of these schemes evolved from black-and-white formats and considered it a mandatory requirement to support backward compatibility. In North America, the television scanning and transmission standard is known as the *National Television System Committee* (NTSC). The standard takes its name from the organization that developed the standards currently used in the United States, Canada, and Japan. In terms of resolution, there are two approaches in use worldwide: 625/50 and 525/60 (actually 59.94), where the first number refers to the number of lines per frame, and the second number refers to the (interlaced) frames per second.

The 1980s saw progress in two directions: HDTV and digital video processing along with compression. HDTV provides a major improvement in the quality of the delivered video: It has a 16:9 aspect ratio compared with 4:3 for regular TV,* and it supports 1125 (or 1250) frame lines. Digital compression is enabling three new categories of services: PC-based video (e.g., downloading video from a server or from a broadcast program); videoconferencing and videotelephony; and new video distribution-VDT systems.

4.2.1 Scanning process

TV signals are generated and transmitted in a fundamentally different way from the techniques used in film making[1]: in cinema, a complete, two-dimensional image is recorded on film and is projected onto a screen as a unit; in TV transmission, the information must be carried as a one-dimensional signal, because of the characteristics of the transmission channel. Hence, the pictures must be converted into a serial stream of information. This is accomplished through the process of picture scanning (see Fig. 4.1*a*): The electronic TV camera scans the picture, the signal is transmitted, and the receiving station reproduces the scanned picture. The picture is scanned in a series of horizontal

* This supports a viewing angle of 30° rather than 10° for regular TV, at normal viewing distances.

sweeps that encode the luminance (amount of reflected light) at each horizontal spot of the picture. Hence, the transmitted signal represents the "instantaneous" amount of reflected light. Regeneration of the picture at the far end takes place by varying the intensity of a light beam projected onto a screen, in proportion to the received signal, as seen in Fig. 4.1*b*.

In addition to the luminance signal, one needs *sync pulses,* which are used to mark the beginning of the picture, and *blanking pulses* to turn off the beam at the reproduction end, while the scanning beam returns to the beginning of the next scanning line. These three signals are combined for transmission, as seen in Fig. 4.1*c*. The resulting signal is

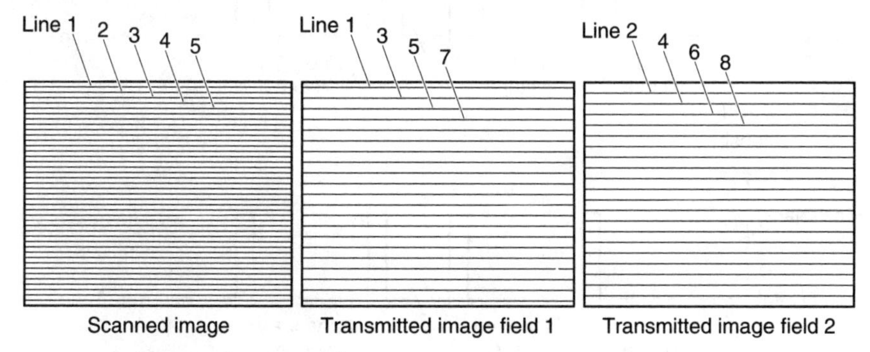

Figure 4.1*a* TV scanning approach.

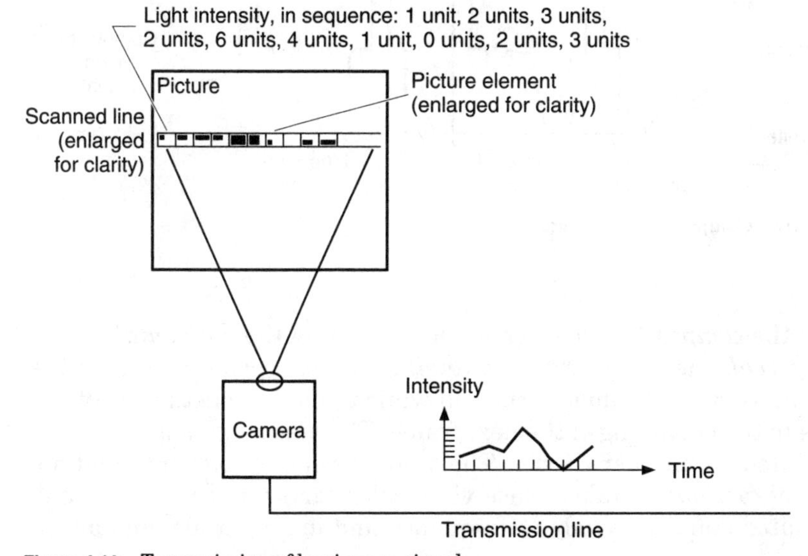

Figure 4.1*b* Transmission of luminance signal.

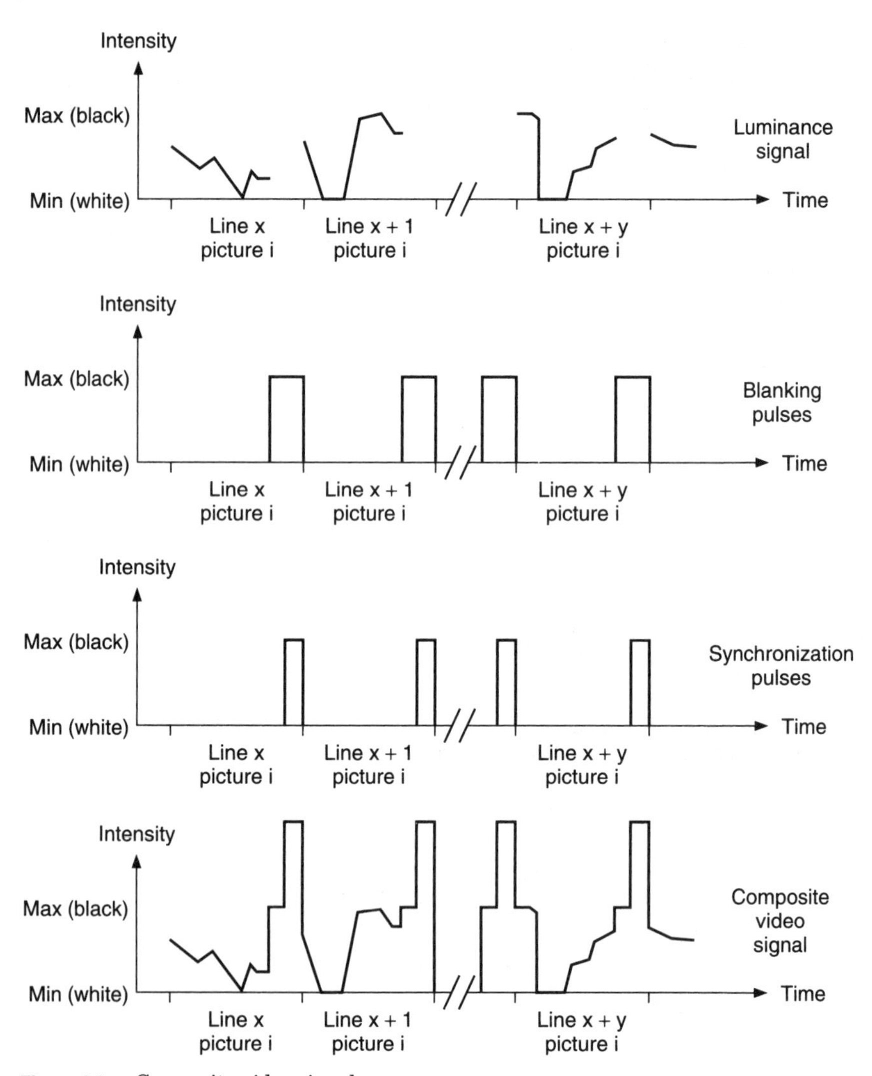

Figure 4.1*c* Composite video signal.

called the *composite video signal.* In addition to the *horizontal blanking interval,* there is a *vertical blanking interval* corresponding to the interval when the beam is turned off while it moves from the end of one frame to the beginning of the next frame. The audio signal is frequency modulated on another carrier. The transmission of color information, called *chrominance,* takes place via a color carrier in the video band, modulated simultaneously in frequency and in amplitude, and added to the luminance signal.[1] The signal is broadcast by amplitude modulating the composite signal on an appropriate carrier wave.

As noted, in the NTSC system used in the United States, there are 525 scanning lines, constituting an image that is scanned at 30 *frames* per second. However, in order to reduce flicker, alternate lines (i.e., lines 1, 3, 5, . . . and then lines 2, 4, 6, . . .) are delivered in a sequence of *fields;* each field consists of 262.5 lines. Since alternate lines are scanned in sequence and are interlaced, the screen is refreshed 60 times a second. In such interlaced scanning, all odd-numbered lines are scanned first, and then all even-numbered lines are scanned. This completes two fields of information or one complete picture frame. The process repeats ad infinitum. It should be noted that not all lines are visible. For example, in a 525 system only, 483 lines actually contain direct video information intended for the (general) user (the other 42 lines, 21 lines per frame, contain other information, e.g., captions).

For reconstruction of the picture, an electron beam of a traditional *cathode ray tube* (CRT)* "shoots" electrons at the coated surface of the tube. The surface is covered with phosphor, which glows when radiated by electrons. The image of the electron beam is usually called the "dot." Once the dot reaches the right side of the CRT screen (it takes approximately 52 μs for the dot to reach the end of a line), the horizontal sync pulse is received by the TV set instructing the dot to travel back to the left side of the CRT screen, through what is called *horizontal retrace,* and paint the next line. The interval of time the dot is retracing to the left is also called the *horizontal blanking interval.* The horizontal retrace lasts for approximately 10.2 to 11.5 μs.† During this time, the electron gun in the CRT display is turned off. The process is repeated until the spot reaches the bottom of the CRT screen, with a travel distance of 243.5 lines. A vertical sync pulse is transmitted, instructing the dot to return to the top of the CRT screen. It takes the equivalent of 21 lines travel time for the dot to return to the beginning of line one or two (this is about 1300 μs). The electron gun is turned off during the vertical blanking interval or *vertical retrace.*

Based on this discussion, it should be clear that absence of accurate system timing causes video picture degradation. For example, an attenuation of the horizontal sync pulse can keep people from making copies of VCR tapes (a time base corrector can be used to rectify problems with system timing, by regenerating the sync pulses).

---

* LCD technology works differently.

† It is common to refer to the horizontal interval as being 63.5 μs; this is the sum of the line-scanning time plus the horizontal retrace time.

4.2.2 Aspects of color TV

The range* of colors available on a CRT depends on the phosphors used as well as on the method of beam superimposition. Phosphors are usually specified in terms of the 1931 Commission Internationale de L'Eclairage (CIE) chromaticity coordinates (the 1976 Uniformity Chromaticity Scale is also used because of the uniformity of the chromaticity space). Figure 4.2 depicts the color space of a typical monitor that uses phosphor P43 for green and P22 for red and blue on the 1931 CIE xy space. The corners of the triangle correspond to the chromaticity coordinates of the primaries realized by specific phosphors used in the construction of the CRT. The colors inside the triangle correspond to physically achievable colors.

Voltage and luminance of a CRT are normally plotted on a logarithmic scale. The *slope* of the resulting function is called the *gamma of the display.* For monochrome CRT, a relatively high gamma level (2.8 to 3.0) is desirable; this allows the dark levels of the display to appear a deep black. For a color monitor, a high gamma produces strong color distortions as luminance is increased or decreased. Hence, it is necessary to modify the signals to each of the display's three guns; this procedure is known as *gamma correction.*

Additive color mixture methods form the underlying principle by which CRT monitors "mixed" colors. The inside surface of the color monitor is composed of hundreds of dots of phosphor (see Fig. 4.3). Phosphors are compounds that emit light when radiated with electrons. The amount of light emitted depends on the strength of the beam. The phosphors on the screen are in groups of three: red phosphors, blue phosphors, and green phosphors. Because the phosphor dots are small, the output of the three elements of the triad appears amalgamated when viewed from a distance, so that the result is a field of color that appears homogeneous.

To display a green object on the screen, for example, all the green phosphors forming the outline and the interior of the object are made to emit light. Other hues are produced by making two or more of the three phosphors in a triad emit light simultaneously. For example, the simultaneous activation of a green and a red phosphor field in the immediate proximity results in the perception of a yellow or orange color, depending on the luminance of the individual phosphors. When the energy distribution of a blue phosphor and the energy distribution of a green phosphor are additively superimposed, the mixture consists of a broader set of wavelengths; this implies that the mixture will be less saturated than the blue or green alone. When the energy distribu-

* See Ref. 2 for a more extensive treatment of this topic.

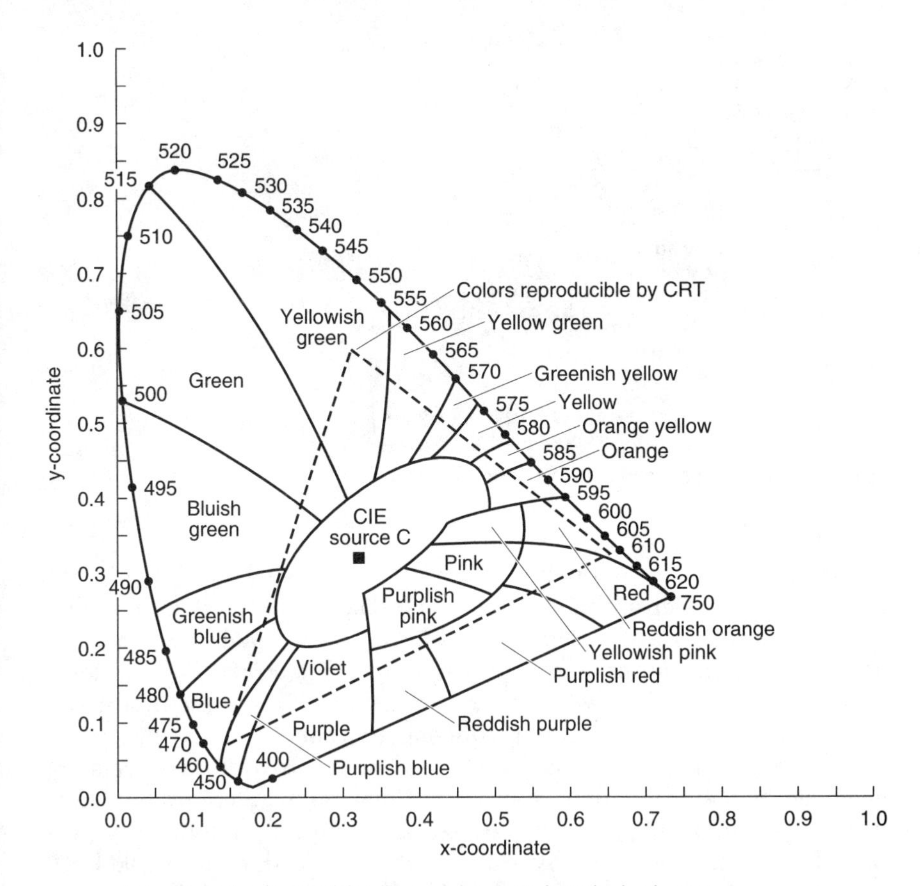

Figure 4.2 Color space of a typical monitor.

tion of a red phosphor is additively superimposed with the energy distribution of a green phosphor, the resulting hue is yellow. All three phosphor signals superimposed together produce a distribution containing all of the visible wavelengths that are perceived as white. Varying the intensity of the three phosphors produces differing levels of lightness.

The total number of colors that can be produced by the monitor depends on the number of steps of gray level obtainable for each phosphor. If the electron gun can be stepped over four levels, the resulting palette has 64 colors. Many (PC or workstation) monitors are capable of 256 steps of gray from each gun, resulting in a palette of over 16 million unique combinations. However, the eye is not capable of discrimi-

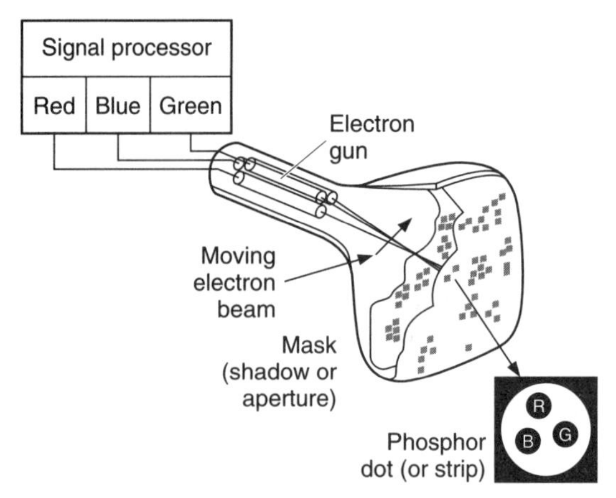

Figure 4.3 A typical CRT.

nating many of the small changes in color. The viewable palette has fewer colors: Under optimal conditions, about 3 million discriminable colors (colors that are recognizably different when placed adjacent to one another) can be produced in a monitor. The scope of the palette decreases to about 7000 when colors located at different monitor areas must be recognized as different from one another.

A common CRT technology is shadow mask. In this case, color is achieved by a spatial additive process: Color mixture occurs by juxtaposition of small primary color dots that cannot be individually perceived (that is, resolved) by the observer. The shadow mask CRT consists of (1) three electron guns in close proximity, (2) a shadow mask, and (3) a screen painted with regular zones of three types of phosphors. The guns can be arranged in a straight line (called in-line), in a delta arrangement (called delta gun), or as a single three-beam gun (see Fig. 4.4).

4.2.3 Video bandwidth

The video bandwidth of an NTSC signal is 4.2 MHz; the FCC has allocated a transmission spectrum of 6 MHz for a broadcast TV signal. The bandwidth directly impacts the quality of the video signal in terms of resolution. A bandwidth of 4.2 MHz supports a maximum of 336 lines of horizontal resolution; hence, even if a camera produced higher resolution, NTSC limits the quality of the received signal. Cable-based systems are not obligated to use the same bandwidth; however, because of practical considerations (source material, satellite feeds, live shots from existing cameras, even signal spillage, etc.), the same 6 MHz of nominal bandwidth is also (almost universally) used. In theory, the increased

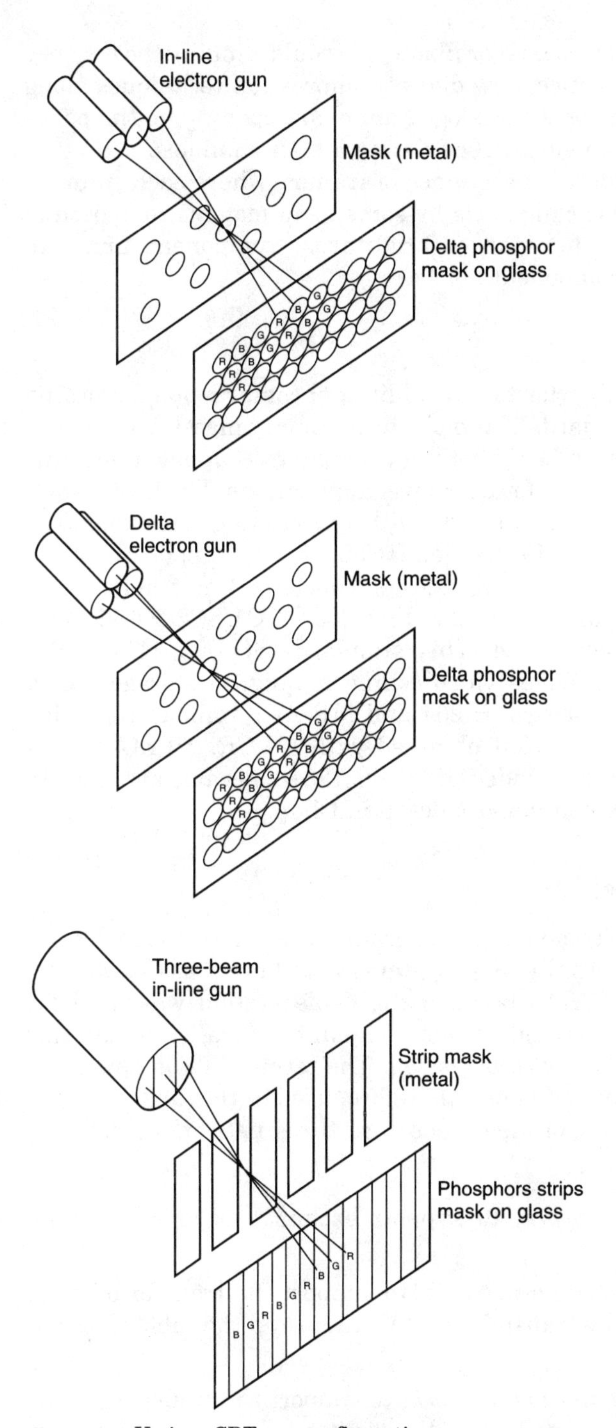

Figure 4.4 Various CRT-gun configurations.

bandwidth available on coaxial or fiber cable could simplify the delivery of advanced TV; in practice, new digital compression techniques along with multilevel coding of the analog channel are opening up the possibility of delivering such video over existing 6-MHz channels.

Video can be considered a sequence of frames, where each frame is an array of *pixels* (also called *pels* by some). An image is composed of three components: a luminance (brightness) component and two chrominance (color) components.

4.2.4 Digitization

Video can be digitized, prior to any additional compression, according to the CCIR 601 standard. The bit rate is approximately 165 Mbps. Although useful in a number of high-end commercial applications, this data rate is simply too high for user-level applications. The first accomplishment of standardization efforts with respect to compression was ITU-T (formerly CCITT) Recommendation H.261, supporting video coding at $p \times 64$ kbps ($p = 1, 2, \ldots, 30$). The second accomplishment was a standard originated by the ISO/IEC JTC1/SC2/WG11 (also known as MPEG) Committee. This standard ISO/IEC 11172 (also known as MPEG-1) provides video coding for digital storage media with a rate of 2 Mbps or less. H.261 and MPEG-1 standards provide picture quality similar to that obtained with a VCR. MPEG-2 standards for "network" quality video (ISO/IEC 13818) became available in 1994. This topic is treated in more detail in Chap. 6.

4.3 Cable TV Basics

This section provides some basic background information on cable TV networks. Note that initially these systems were known as *Community Antenna Television* (CATV) because the cable system was used for those communities that could not get a signal or a quality signal; this service is now known as *retransmission.* The term CATV is now effectively an anachronism and is not, therefore, used in this text.

From a general point of view, there are three networks related to video transmission[1]:

1. The *contribution network,* to support exchange of video between television studios
2. The *primary distribution network,* to support the transfer of video information from the transmitting TV studio and the cable distribution center
3. The *secondary distribution network,* to support the transport of the video signal from the headend to the viewer

The emphasis of this text is on the secondary distribution network. However, some quick observations about the contribution network are in order. The contribution network provides connectivity between sources of video and between studios in order to support production and postproduction requirements. This network needs to support the highest transmission quality possible. The 4:2:2 digital *component* video standard (see Chap. 6) can be used, although widespread acceptance of the standard has been relatively slow because of the cost of replacing embedded equipment; consequently, many contribution networks carry *composite* NTSC, PAL, SECAM, while aiming at maintaining the highest quality possible. For more discussion see, e.g., Ref. 3.

4.3.1 Cable headend and trunk network

A cable headend is the central point at which TV signals, downlinked from a satellite or received from a terrestrial microwave system using the equipment of the local cable TV company, are multiplexed with signals provided by local stations* and then are multiplexed onto the cable for local distribution. Hence, the headend receives, processes, organizes, and distributes video signals. The headend may also have automatic commercial insertion equipment that triggers on tones embedded in the satellite-delivered signal and starts a locally stored tape or optical drive for injection of local advertisements.

The headend converts the received video signals†, which might be in various formats (e.g., AM, FM, digital, prestored analog, prestored digital, etc.), processes them as needed, modulates them over a *radio frequency* (RF) carrier at the appropriate frequency, and finally combines them in a *frequency division multiplexing* (FDM) manner, through a *combiner.* As noted, a headend can inject local programming such as local commercials, public service programming, and other programming of particular relevance. See Fig. 4.5.

The headend typically contains addressable control processors that communicate with the customer's addressable setup box to enable channels, premium services, or PPV events. A headend (in a large metropolitan area) typically serves from 10,000 to 100,000 customers.

---

* Over-the-air TV signals are collected using a Yagi or a log periodic antenna and are fed to a heterodyne processor that tunes the signals by frequency, reamplifies them, and converts the higher frequency to the appropriate frequency for transmission over the cable. For example, WNYC Channel 31 in New York City, received in the UHF band at 572–578 MHz, may be converted to frequency band 198–204 MHz for cable delivery at Channel 22.

† For example, signals received from a satellite have to be down-converted to a useful intermediate frequency.

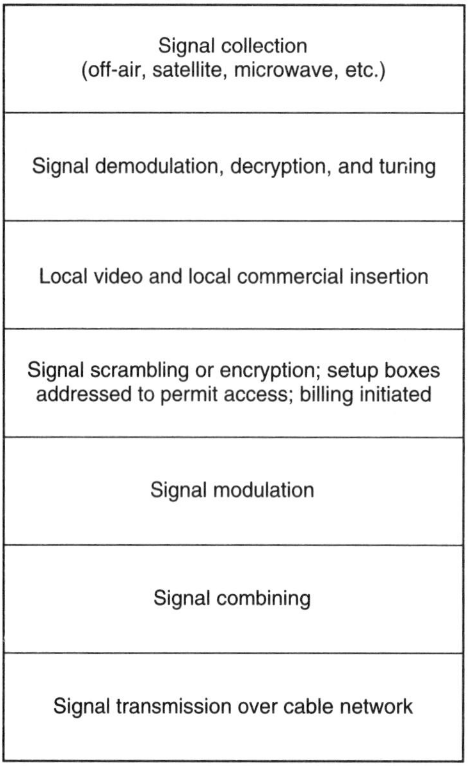

Signal collection (off-air, satellite, microwave, etc.)
Signal demodulation, decryption, and tuning
Local video and local commercial insertion
Signal scrambling or encryption; setup boxes addressed to permit access; billing initiated
Signal modulation
Signal combining
Signal transmission over cable network

Figure 4.5 Functions performed by a traditional cable TV network headend.

There could also be more than one headend belonging to a cable company in a large metropolitan area. Many cable TV networks also employ subsidiary facilities called hubs; a *hub* is a signal reception and retransmission facility supporting a range of functionality, including optical to electrical conversion, or processing and control functions similar to those of the headend but on a smaller scale.

The distribution of video is accomplished in traditional cable networks by modulating the signal to be compatible with existing TV sets using *amplitude modulation-vestigial sideband* (AM-VSB) techniques (if the distribution is not directly compatible with the TV set, a conversion is needed in the setup box; an example of this is DBS transmission).

A traditional cable TV network is optimized for one-way transmission of a common set of video signals.[4] In such a cable system, the distribution network consists of a trunk line, connected to a *trunk-bridger amplifier* (TA-BA), connected to a feeder line, (optionally) connected to a *line extender amplifier* (LEA), connected over a tap to a drop line, and finally connected to the residence's TV and/or setup box. Larger trunk cables carry signals from the headend to the various neighborhoods

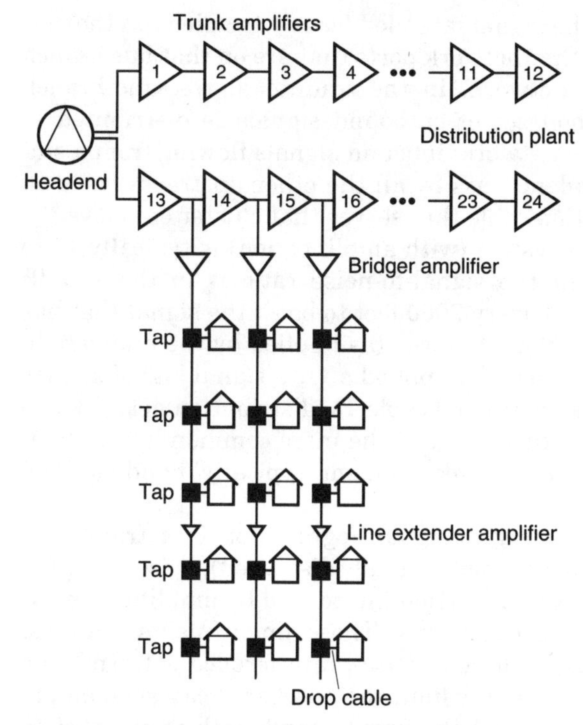

Figure 4.6 Traditional coaxial distribution system.

around town; this main section of the network is called the *trunk network* (in telephony, this would be called the *feeder network*). The transportation network in the feeder often consists of a 0.750-in gas-injected dielectric cable.[5] RG-59U and RG-6U are also used in portions of the network (see Table 4.4). Figure 4.6 depicts a traditional coax-based *tree-and-branch* (also referred to as *trunk-and-branch*) system. Every couple of thousand feet, the main cable is tapped to carry off the TV signal for local distribution within a certain area of a given neighborhood.

This bus-based cable TV distribution network contains trunk amplifiers, bridger amplifiers, and taps; line extender amplifiers may also be used for longer runs.* The demarcation point between the trunk network and the local distribution network (which in the cable TV industry is called the *feeder network* or, for short, the *feeder*) is realized in a bridger amplifier.†

---

* The quality of the received signal, therefore, depends on how many amplifiers and taps that are traversed; clearly, a user at the end of a long and crowded run may experience substantial degradation.

† A bridger amplifier is often located inside of the housing of a trunk amplifier.

In any bus network, the signal intended for any recipient on the network passes by all the other network ports that are on that bus branch of the tree-and-branch network. In the traditional tree-and-branch cable network, a common set of outbound signals is distributed to every home (port) on the network; inbound signals flowing from a single port toward the headend pass by all the other upstream ports on the segment of distribution cable that serves that cluster of homes.[6]

The reach of the cable system (with amplification) is typically 15 to 20 miles, at which point the signal-to-noise ratio is in the –40-dB range. Amplifiers are used every 2000 feet to boost the signal that has experienced attenuation along the way (as implied by the numbers in Table 4.4, amplification is usually applied after a signal loss of 28 dB). Figure 4.7 depicts how a traditional cable TV distribution network can be upgraded to use fiber; this is one of the more common hybrid fiber coaxial architectures. Each trunk segment typically handles 2000 homes.

Some of the factors leading to signal degradation in a traditional coaxial system include attenuation (already discussed), noise along the distribution path, nonlinear distortion introduced by amplifiers, noise ingress at the home, and reflections. These impairments manifest themselves as picture degradation (noise), interference patterns, and poor audio. These factors not only limit the number of cascaded amplifiers but also negatively impact the usable bandwidth in the system and the use of the upstream channel.*

4.3.2 Frequency division multiplexing of channels in coaxial systems

Traditional coaxial-based systems use FDM to carry multiple video channels over cable. Each channel occupies a certain frequency band allocated exclusively to that channel. As indicated, an NTSC channel requires a nominal 6 MHz (since the actual signal is only 4.2 MHz, this arrangement provides the needed guardband separation between channels). Table 4.2 depicts the traditional TV channel allocation for broadcast transmission. Coaxial-based systems follow similar allocation at the low end but diverge at the higher end. Table 4.3 depicts the FDM allocation used on cables, for systems with up to 80 channels (80 channels employ downstream system bandwidth of 500 MHz, with upper frequency to 550 MHz; many systems in place at this time employ a downstream system bandwidth of 400 MHz, with upper fre-

---

* In a system, the upstream channel typically would have to be shared by 20,000 users. Because of the accumulated ingress noise as well as the noise due to amplification, this channel is hard to manage. Some of the new proposals now being advanced limit the number of users that share an upstream channel.

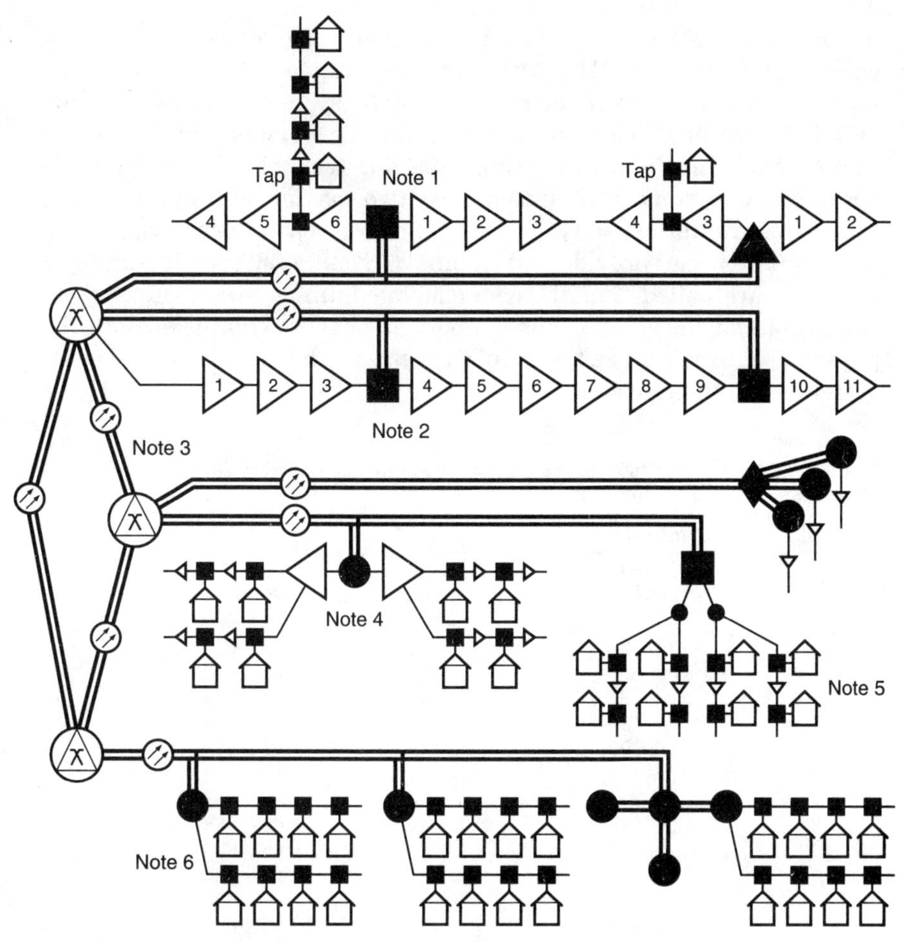

Notes:

(1) Fiber backbone used to reduce amplifier cascades
(2) Fiber backbone complemented with coaxial cable: protection switching used to increase availability
(3) Digital self-healing fiber ring interconnecting various video sources/headends
(4) Fiber to the feeder; eliminates trunk amplifiers and coaxial trunks
(5) Star feeder; utilizes optical trunk repeaters for economic reach
(6) Fiber and passive networks: eliminate trunk coaxial cables and RF amplifiers

Observation:
All equipment shown in this figure is commercially available

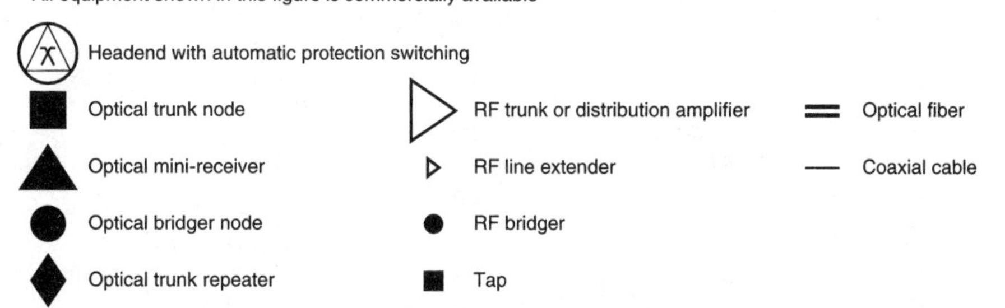

Figure 4.7 A partially upgraded cable TV distribution network (traditional).

quency to 450 MHz, supporting 60 channels). There are also systems operating up to 750 MHz. Although current-generation cable equipment has a maximum capacity of 750 MHz, systems that can operate at 1 GHz are now becoming possible, and developers hope to extend that to 2 GHz in the future using emerging technologies.[7] Some cable operators (e.g., some TCI systems) use two coaxial cables into a home to achieve greater capacity. For example, two coaxial cables operating up to 450 MHz can provide 120 channels. Systems supporting up to 22 channels are called "small" systems; "medium" systems support 40 channels; and "large" systems support 52 or more channels. Both the physical cable as well as the amplifiers need to be upgraded for a cable system to carry more channels.

TABLE 4.2 Television Channel Frequency Allocation for Over-the-Air Transmission in North America

Channel no.	Band (MHz)	Channel no.	Band (MHz)
2	54–60	36	602–608
3	60–66	37	608–614
4	66–72	38	614–620
5	76–82	39	620–626
6	82–88	40	626–632
7	174–180	41	632–638
8	180–186	42	638–644
9	186–192	43	644–650
10	192–198	44	650–656
11	198–204	45	656–662
12	204–210	46	662–668
13	210–216	47	668–674
14	470–476	48	674–680
15	476–482	49	680–686
16	482–488	50	686–692
17	488–494	51	692–698
18	494–500	52	698–704
19	500–506	53	704–710
20	506–512	54	710–716
21	512–518	55	716–722
22	518–524	56	722–728
23	524–530	57	728–734
24	530–536	58	734–740
25	536–542	59	740–746
26	542–548	60	746–752
27	548–554	61	752–758
28	554–560	62	758–764
29	560–566	63	764–770
30	566–572	64	770–776
31	572–578	65	776–782
32	578–584	66	782–788
33	584–590	67	788–794
34	590–596	68	794–800
35	596–602	69	800–806

TABLE 4.3 Cable TV Channel Frequency Allocation

	Band (MHz)	System channels*	Downstream system bandwidth
Upstream bandwidth	5–30		
Channel no.			
2	54–60		
3	60–66		
4	66–72		
5	76–82		
6	82–88		
7	108–114		
8	114–120		
9	120–126		
10	126–132		
11	132–138		
12	138–144		
13	144–150		
14	150–156		
15	156–162		
16	162–168		
17	168–174		
18	174–180		
19	180–186		
20	186–192		
21	192–198		
22	198–204		
23	204–210		
24	210–216		
25	216–222	From 12 to a maximum of 22	170 MHz
26	222–228		
27	228–234		
28	234–240		
29	240–246		
30	246–252		
31	252–258		
32	258–264		
33	264–270	30	220 MHz
34	270–276		
35	276–282		
36	282–288		
37	288–294		
38	294–300		
39	300–306		
40	306–312		
41	312–318		
42	318–324		
43	324–330	40	280 MHz
44	330–336		
45	336–342		
46	342–348		
47	348–354		
48	354–360		
49	360–366		

TABLE 4.3 Cable TV Channel Frequency Allocation (*Continued*)

	Band (MHz)	System channels*	Downstream system bandwidth
Upstream bandwidth	5–30		
Channel no.			
50	366–372		
51	372–378		
52	378–384		
53	384–390		
54	390–396		
55	396–402	52	350 MHz
56	402–408		
57	408–414		
58	414–420		
59	420–426		
60	426–432		
61	432–438		
62	438–444		
63	444–450	60	400 MHz
64	450–456		
65	456–462		
66	462–468		
67	468–474		
68	474–480		
69	480–486		
70	486–492		
71	492–498		
72	498–504		
73	504–510		
74	510–516		
75	516–522		
76	522–528		
77	528–534		
78	534–540		
79	540–546		
80	546–552	80	500 MHz

* For traditional systems, the last two frequency bands are used for Channels 98 and 99; hence, the number of "channels" is as shown.

In a traditional cable system, the frequency band from 5 to 30 MHz is reserved for upstream traffic (if this could be achieved in spite of the return path accumulated noise problem); the downstream bandwidth actually starts at 54 MHz. In *hybrid fiber coaxial* (HFC) systems, the upstream bandwidth typically covers the 5- to 42-MHz range, or even 5 to 45 MHz.[4]

4.3.3 Characteristics of coaxial cables

Table 4.4 depicts some characteristics of coaxial cables. Notice that as the frequency increases so does the attenuation. This implies that it is somewhat challenging to carry more bandwidth into the home.* Also note that some cables either do not support the higher bandwidths, or the attenuation is simply too high, which implies that a cable plant based on these cables would have to be replaced if the operator wanted to deliver more channels. Note that the bigger the cable (in terms of diameter), the better the performance as measured by across-the-board attenuation. For example, an RG-19A cable, which has a diameter exceeding 1 in, provides an attenuation of 1.1 dB/100 ft, or 22 dB in a 2000-ft run. In any communication system, one needs reamplification after a loss of 25 to 40 dB (28 to 30 dB is typical). Figure 4.8 depicts the effect of signal attenuation through a coaxial cable. Figure 4.9 depicts the physical construction of two types of coaxial cables used in video distribution: distribution cable and drop cable.[8] As seen in Fig. 4.9, the coaxial cable has a center conductor made of aluminum or steel (these are copper-coated). This conductor has a low DC resistance so that power (60 Hz AC) can be carried to power the amplifiers along the way. The center conductor also carries the RF signal. The center conductor is held in place by a low-loss polyethylene material; the foam material is designed to provide mechanical support (air dielectric cables using disk insulators are also common). The outer conductor provides a return signal path and serves to terminate the internal electromagnetic field. The outer conductor is covered with a polyethylene jacket to protect the cable from its surroundings.

Proponents call coaxial cable the ideal broadband medium for the last mile, being good, cheap, and easily installed. In fact, coaxial does have good characteristics, further discussed in the paragraphs that follow; however, it is very unlikely that the RBOCs will use it to wire millions of new homes, for the following reasons:

1. An historical trend since the early 1960s has been toward digital technology because it is more reliable, more compact, easier to maintain, and costs less to operate.
2. The RBOCs do not have expertise, although easily obtainable, on coaxial technology, TAs, TBs, LEAs, cable plants, radiation management, etc.
3. The reliability of analog-based amplifiers is not up to the standards of telephone service.

* RF attenuation is frequency-dependent, decreasing with the square root of the frequency; e.g., attenuation at 216 MHz (cable TV Channel 24) is twice that experienced at 54 MHz (TV Channel 2), since the frequency went up by a factor of 4.

TABLE 4.4 Characteristics of Coaxial Cables

RG- type (/U)	Nominal impedance, Ω	Nominal diameter, in	Attenuation 200 MHz, dB/100 ft	Attenuation 400 MHz, dB/100 ft	Attenuation 900 MHz, dB/100 ft
6	75	0.290	3.1	—	6.9
6A	75	0.336	4.3	6.5	10.1
8	52	0.405	3.0	4.7	7.8
8A	52	0.405	3.0	4.7	7.8
9	51	0.420	2.8	4.1	6.5
9B	50	0.430	2.8	4.1	6.5
11	75	0.405	2.9	4.2	6.5
11A	75	0.405	2.9	4.2	6.5
12A	75	0.475	3.2	4.7	7.8
14A	52	0.558	2.3	3.5	—
17A	52	0.885	1.5	2.4	—
19A	52	1.135	1.1	1.8	—
22B	95	0.420	4.5	6.8	11.0
55	53.5	0.206	7.0	10.5	16.0
55B	53.5	0.206	7.0	10.5	16.0
58	53.5	0.195	6.2	9.5	14.5
58 (double shielded)	53.5	0.206	6.6	9.2	13.4
58A	50	0.195	8.2	12.6	20.0
58A (double shielded)	50	0.195	6.9	10.1	15.5
58C	50	0.195	8.2	12.6	20.0
59	73	0.242	4.9	7.1	—
59B	75	0.242	4.9	7.1	11.1
62	93	0.242	4.4	6.3	11.0
62A (mil spec)	93	0.242	4.4	6.3	11.0
62A (U/L)	93	0.260	4.4	6.3	11.0
62B	93	0.242	4.4	6.3	11.0
63B	125	0.415	2.9	4.1	—
71B	93	0.250	3.9	5.8	—
122	50	0.160	11.0	16.5*	28.0
141A	50	0.190	—	9.0*	—
142B	50	0.195	—	9.0	—
174	50	0.100	13.0	20.0*	—
178B	50	0.070	—	29.0*	—
179B	75	0.100	—	21.0*	—
180B	95	0.140	—	17.0	—
187A	75	0.110	—	21.0*	—
188A	50	0.110	—	20.0*	—
195A	95	0.155	—	17.0*	—
196A	50	0.080	—	29.0	—
212	50	0.336	3.6	5.2	—
213	50	0.405	3.0	4.7	7.8
214	50	0.425	3.0	4.7	7.8
215	50	0.412	3.1	5.0	—
217	50	0.555	2.3	3.5	—
218	50	0.880	1.5	2.4	—
219	50	0.880	1.5	2.4	—
223	50	0.216	7.0	10.5*	—
316	59	0.098	—	20.0*	—

* Maximum attenuation.

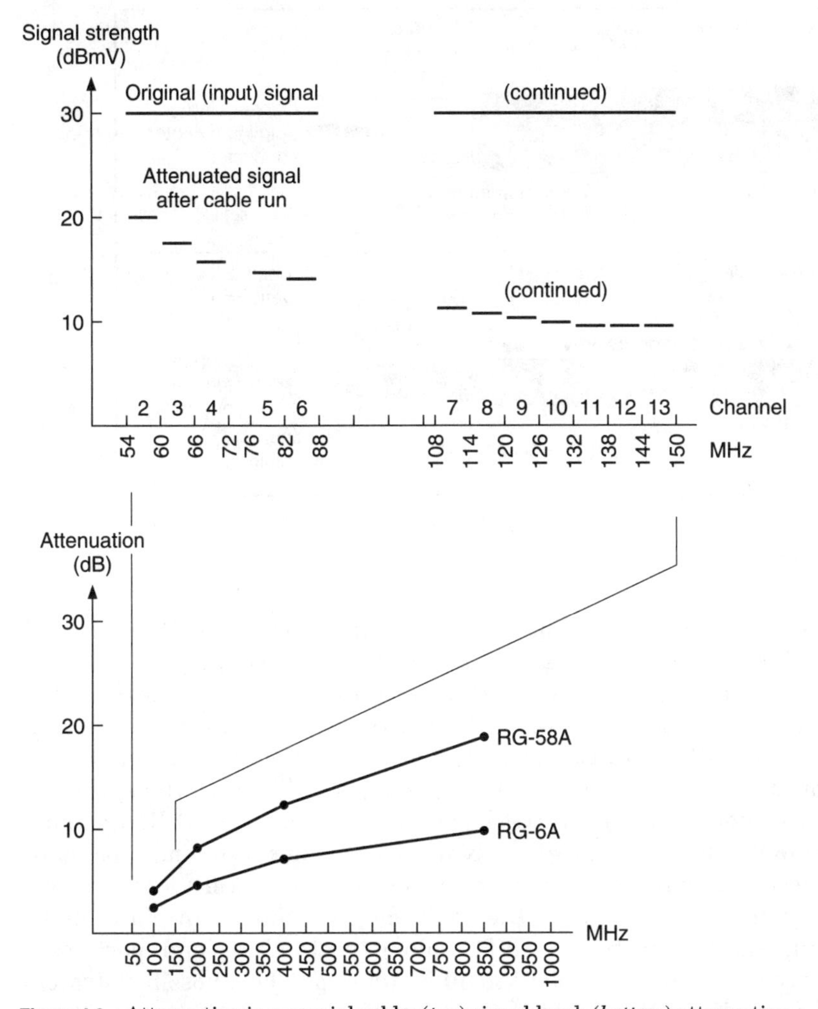

Figure 4.8 Attenuation in a coaxial cable: (*top*) signal level; (*bottom*) attenuation.

4. Two-way systems, whether mostly asymmetric or symmetric in terms of bandwidth, are more difficult to manage, and there is little practical experience, even within the cable TV industry.
5. Fiber technology is here, it is getting cheaper, and it is highly flexible.
6. A massive overhaul requires considerable investment. Interestingly, a major portion of the cost of such an overhaul (estimated to be as much as 80 percent) comes from the labor involved, not the cable. Therefore, if such an overhaul were to be undertaken, one might as well deploy the more flexible and "future-proof" fiber.

Consequently, coaxial cable may well serve a transitional function, being used by the RBOCs to deliver video where it is already in place

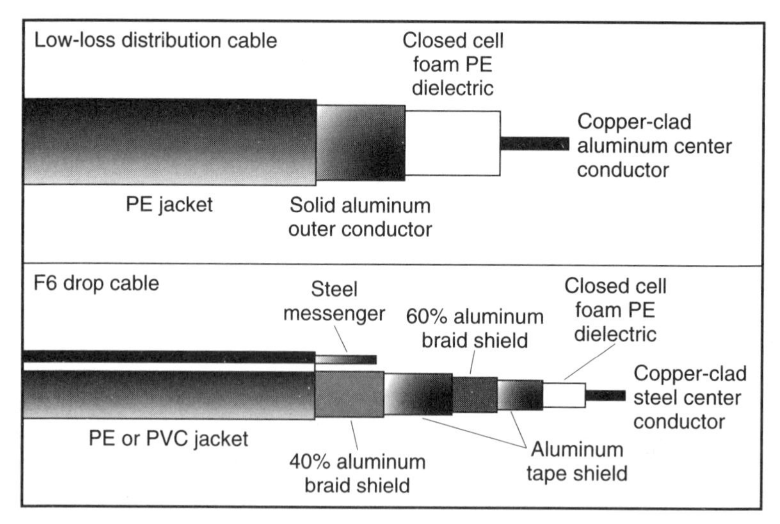

Figure 4.9 Physical construction of typical cables used in cable TV networks.

(naturally, the cable TV industry has a large embedded base and will continue to use it to the extent possible). The preceding discussion is not meant to minimize the "good" transmission quality of coaxial cable. The issue investigated here is whether the RBOCs will replace twisted pair cable with coaxial cable as the last link in the existing network, as suggested by some (e.g., see Ref. 8). This is unlikely. In the opinion of this author, from an RBOC's point of departure, HFC systems, although able to provide an early incarnation of end-to-end broadband connectivity, will see deployment as an overlay network for those out-of-region situations where the cable may already be in place, or for small, short-term trials; FTTC, along with ATM, and perhaps a twisted-pair drop cable of 10 to 50 meters, may be a possible strategy before implementing FTTH (note that there is work underway in The ATM Forum to deliver 52 or even 155 Mbps to the desktop using twisted-pair cable). From a cable TV company point of departure, a migration to HFC is a very useful and reasonable strategy, providing immediate and measurable benefits.

At the analog transmission level, coaxial cable offers high bandwidth, low leakage levels (although leakage can occur at the connectors), relatively low signal loss, and low return loss. A well-designed coaxial cable has good shielding characteristics.

4.3.3.1 Distribution cables. In cable TV applications, coaxial cables are used from the headend or from the fiber termination node into the various neighborhoods. The required distribution cables are available in a number of sizes, depending on the amount of signal loss that is tolerable; as noted in Table 4.4 (also see Fig. 4.10), a larger diameter cable (as measured by the inner diameter of the outer conductor) has a lower attenuation. The distribution cables range in diameter from 0.5 to 1 in. These cables are designed to have a life approaching 30 years.[8]

4.3.3.2 Drop cables. The drop cable connects a customer to the tap. It has different transmission requirements than the distribution cable. There are four standard sizes of drop cable (from smaller diameter-larger attenuation to larger diameter-lower attenuation): F59, F6, F7, and F11 (F6 is the most common).

4.3.4 Digital modulation over coaxial cables

New techniques for the use of coaxial cables in the delivery of video are now emerging. These techniques use digital modulation of the underlying analog carrier.

As noted earlier, in NTSC, 30 frames per second of 525 lines are transmitted as two consecutive fields of 256.5 lines, where each horizontal line consists of discrete *picture elements* (pixels). Each pixel is defined by an analog signal representing its brightness and color using amplitude modulation. When transmitted in a coaxial cable, this analog signal degrades as it is reamplified from 20 to 30 times between the headend and the domicile (fiber technology, as covered in the next section, reduces that regeneration process to four cascaded

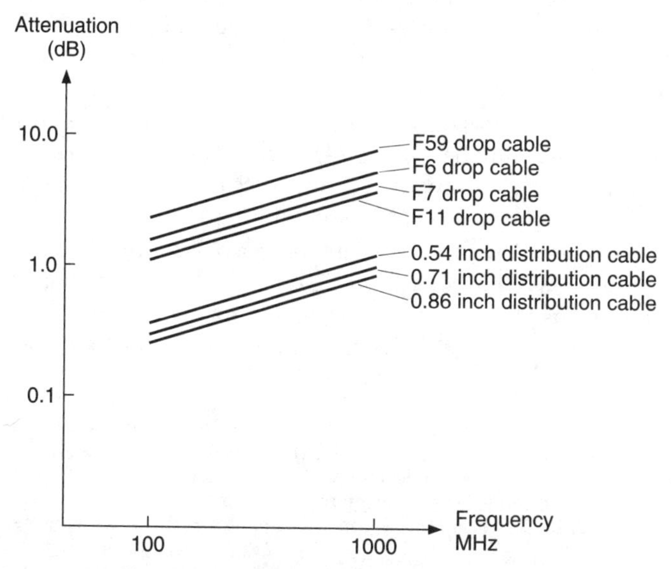

Figure 4.10 Comparison of characteristics of distribution and drop cables.

TABLE 4.5 Some Characteristics of VSB and QAM

	4-VSB	16-QAM	8-VSB	64-QAM	16-VSB
Data rate for a 6-MHz channel	21.5	21.5	32.25	32.25	43
Encoding levels	4	4	8	8	16
Number of 1.5-Mbps MPEG-1 channels in a 6-MHz channel	13	13	20	20	27
Number of HDTV channels	1	1	1	1	2
Number of 6.0-Mbps MPEG-2 channels in a 6-MHz channel	3	3	5	5	8
Phase-noise and continuous wave interference rejection	Excellent	Good	Good	Poor	Good
Required carrier-noise on channel					
Without forward error correction	22 dB	22 dB	28 dB	28 dB	34 dB
With forward error correction	17 dB	17 dB	23 dB	23 dB	29 dB
Equipment cost (operator and consumer equipment)	Lowest	Low	Low	High	Low

elements—even less in the future). Therefore, there is interest in sending digital TV signals over coaxial cable (or over fiber). Not only does this produce a cleaner signal, but it also facilitates compression of the signal (Chap. 6).

Two methods have emerged for digital modulation out of the HDTV efforts which can be used over cable: *quadrature amplitude modulation* (QAM) and *vestigial sideband* (VSB). Table 4.5 compares some characteristics of the two schemes at 4, 8, and 16 encoding levels.[9] (Other schemes have been proposed, e.g., *Quaternary Phase Shift Keying,* or QPSK, by TimeWarner Cable.[6])

16-QAM (General Instrument) and 4-VSB (Zenith/AT&T) techniques have been proposed for over-the-air transmission (see Fig. 4.11 for the QAM constellation). Schemes that are adequate for HDTV include 64- and 128-QAM, and 4-, 8-, and 16-VSB. It appears that the 8-VSB approach, first introduced by Zenith/AT&T, will become the HDTV standard.[10, 11] The consortium of companies developing the U.S. HDTV standard is known as the *Grand Alliance* (GA). In February 1994, the FCC's Advisory Committee Technical Subgroup agreed to approve ιhe proposals made by the GA based on 8-VSB.[10]

The same principles and technology can be used for digital cable transmission. In a cable system, despite its signal and noise problems (which can be less serious than over-the-air problems), one can use 64-QAM or 16-VSB.* Some favor 16-VSB for cable systems, which has

* The peak carrier-to-noise ratio must be in the 40 dB range; in a cable system, the interfering signals are less severe than in a terrestrial transmission environment. The greater noise margin on the cable can be used to increase the digital capacity of the signal, thereby obviating the need to perform even more compression of the video signal than is already achived with MPEG-1 or MPEG-2 methods to obtain increased channel capacity.

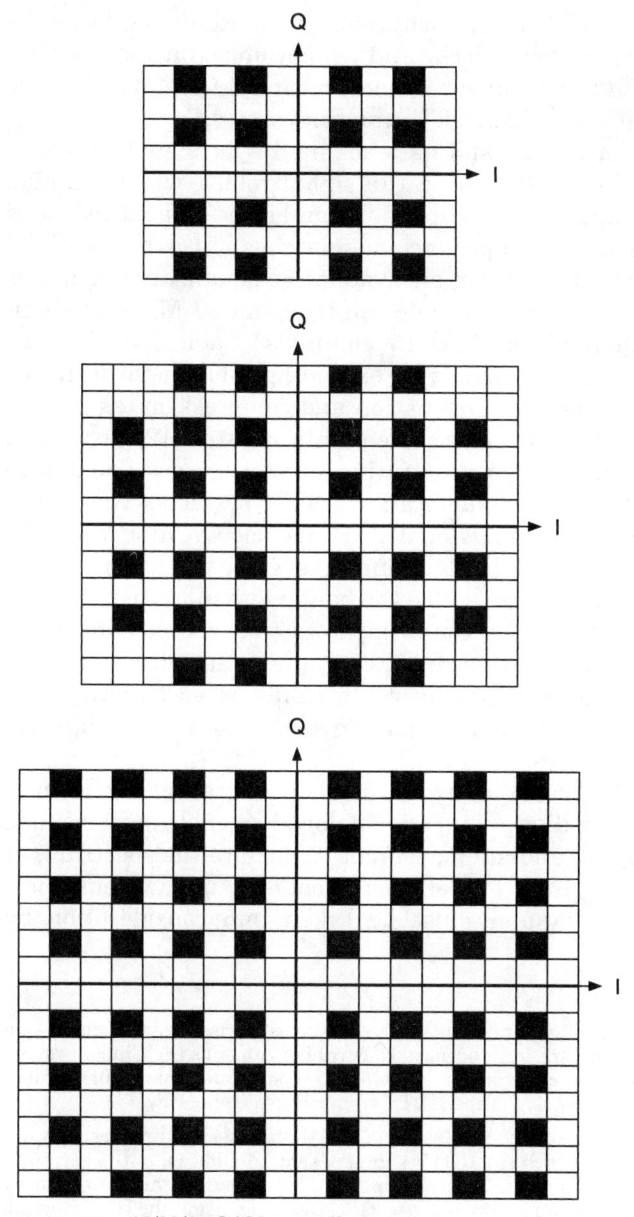

Figure 4.11 16/32/64 QAM constellation.

been proposed by Zenith/AT&T, because one must acquire and lock the modulated RF carrier, data clock, and synchronization information, and this is more difficult (and/or expensive) in 64-QAM since phase shifts and phase jitter act to mask the distinct states.*

In 16-VSB, a pilot carrier and data segment sync signal are used, independently of the data, making the signal robust while enabling signal acquisition under typical cable noise and interference conditions (e.g., microreflections and impedance mismatches). Also, the vestigial sideband electronics are simpler, since adaptive equalization is not as complex. As seen from Table 4.5, one can transmit 27 MPEG-1 channels, 8 MPEG-2 channels, or 2 HDTV channels.† Cable operators see "the expansion of channel capacity by extending cable bandwidth to 1 GHz and the use of digital transmission and compression" as greatly enhancing the usefulness of their cable plant.[9] Figure 4.12, based on a variety of sources, depicts, for illustrative purposes, a possible allocation of bandwidth on next-generation cable systems. Note that although this proposed arrangement does not use QAM or VSB, the idea of mixing analog and digital channels, as well as including return bandwidth, is the same.[6] Also, notice the bandwidth allocation for PCS over the cable (in theory, the coax repeaters that are present every 2000 ft could be replaced with a repeater that also acts as an antenna to aggregate local mobile traffic over a wireline-based trunking system). A high-end 2.4-GHz system under development at NTT supports 1000 channels.[12]

Many traditional cable TV networks are now migrating to HFC networks. As compressed digital video is deployed on cable systems, one can expect that digital and analog signals will share the spectrum, at least for a period of time. The effects of channel impairments in a mixed digital-analog system must be taken into consideration by

---

* In 1994, General Instrument and Zenith agreed to cross license digital transmission and compression technologies; Zenith licensed General Instrument's DigiCipher system, including the compression, access control, and QAM transmission; General Instrument licensed 16-VSB, the cable version of the HDTV transmission technology.[10]

† As noted, 8-VSB provides 32.25 Mbps over a 6-MHz bandwidth. Broadcasters are assigned a new channel in which the HDTV signal is sent, in addition to the currently assigned channel to be (continued to be) used for NTSC broadcasting (this is to occur by 2002); in the new channel, broadcasters will use 21.5 Mbps to transmit the HDTV signal. Broadcasters were lobbying through the National Association of Broadcasters for the opportunity to use the other portion of the 32.25 Mbps (namely, 10.75 Mbps) as desired; they seek "regulatory flexibility" to do many kinds of digital broadcasting besides HDTV in the second channel they will be granted for HDTV broadcasting (e.g., six MPEG-1 channels or two MPEG-2 channels).[10] Also, there is some interest in the NTSC channel, when NTSC ceases to be broadcast in 2008. (The HDTV modulation standard was adopted in 1994; other aspects of the system-level standardization process were due for completion in 1995.)

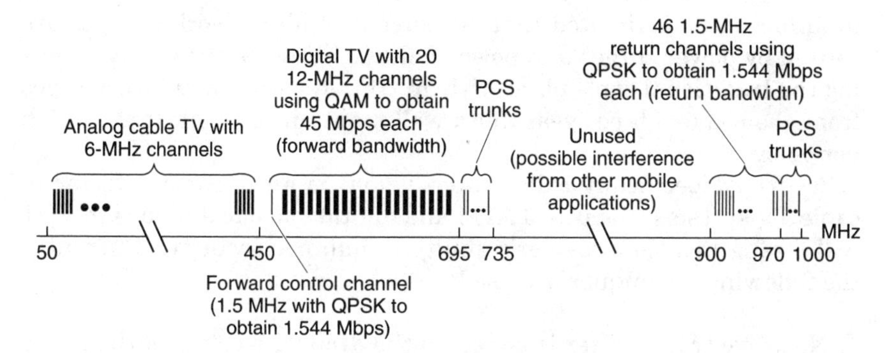

Figure 4.12 Proposed Time Warner use of cable bandwidth. *Note:* There are many proposed ways of allocating higher frequency bandwidth; this figure depicts one example for illustrative purposes.

designers trying to fit the largest possible bandwidth over these systems. The most important effects are the distortion introduced by the various amplifier systems (TA, BA, and LEA). For analog video signals, distortions due to amplifier nonlinearities result in the *composite second order* (CSO) and *composite triple beat* (CTB) effects produced as a result of video carrier sum and difference frequencies.[13] The distortion effects produced by mixed digital transmission, however, are different. Here the distortions are a result of three kinds of interactions:

1. Analog carriers beating with analog carriers
2. Analog carriers beating with digital channels
3. Digital channels beating with digital channels

The bottom line is that when operating a mixed cable system, one needs to balance the goal of using as high a digital signal power as possible to provide good signal quality while maximizing the amount of data overhead required for (forward) error correction, with the goal of keeping such power within reason to preclude it from affecting the noise floor of the adjacent analog channels. Note that a digital signal appears as random noise and, because of its uniform power distribution, produces a spectral energy distribution that, in the case of the third-order distortion, spreads into the adjacent analog channels.[13]

4.3.5 Other facets of cable systems

Reliability is an important factor in relation to cable networks. Traditional cable TV networks, being bus- and amplifier-based, are subject to failure. It is estimated that 50 percent of all network outages are caused by power surges and power failures.[6] Studies show that replacing the traditional cable plant with an HFC system can reduce outages from about 2 to 3 h per year (for a well-maintained plant) to 0.5 to 1 h per year.

Security (specifically called *interdiction*) is an important factor in cable TV systems (the use of ATM and digital video techniques in VDT will strengthen the provider's ability to improve security). Currently, the following techniques are used:

1. Negative trap: A filter is added on the drop wire or tap of the house in question. (This is possible since the programming is distributed on an FDM basis.)
2. Positive trap: The signal contains a jamming frequency; a notch filter must be inserted at the drop wire, tap, or setup box to remove the interfering signal.
3. Addressable setup boxes are used to enable the signal to be transferred or descrambled. (A drawback of the addressable setup box is that it typically outputs its signal onto Channel 3 or 4 of the TV's input, making TV-remote control features difficult to retain.)

I-TV in general and communication with a setup box in particular need a usable upstream channel. Some also plan to use that channel in the 5- to 30-MHz range to carry telephony services (500 simultaneous calls could be handled in that bandwidth). However, noise injected by the multitude of potential users of such channel and signal reflections have, in the past, limited its usefulness. New techniques are now being advanced. For example, Unisys, Time Warner Cable, and Cable Television Laboratories have conducted joint tests on *code division multiple access* (CDMA) technology, also known as *spread spectrum,* for the return path. CDMA has been developed for military communications where man-made noise may be added on purpose on a channel to create jamming. Spread spectrum technology scatters digital information in a pseudorandom way over the allocated channel using low-power signals. Therefore, this technology could be used in cable systems to[14]:

- Make signal impervious to noise
- Provide increased security
- Enable multiple access
- Dynamically assign available bandwidth

4.4 Synchronous Optical Network (SONET) and Fiber Transmission

This section provides a short synopsis of fiber-optic transmission methods. For a more complete treatment, the reader is referred to an introductory text on this field (including Ref. 15).

A fiber provides a low-loss channel (better than 0.5 dB/km) for the transmission of information that has been converted to optical energy. This loss implies that repeaters are needed only every 35 to 50 km. For local loop applications, which typically imply spans of 3 to 5 miles, repeaters are not required.

4.4.1 SONET

A fiber carries 622 Mbps, 1.244 Gbps, 2.488 Gbps, or even 9.95 Gbps of information (a pair of fibers is needed for the send and receive side). Today, all newly introduced fiber follows the SONET standards. SONET enables the interconnection of equipment from any vendor that follows the standard. It specifies the clock rates and bandwidth levels, the optical signals, the fiber characteristics, and the operations procedures, including mechanisms for self-healing, which is achievable when fiber rings are used.

The basic building block on a SONET frame is an 810-byte frame transmitted every 125 μs* to form a 51.840-Mbps signal known as a *synchronous transport signal level 1* (STS-1). See Fig. 4.13. Because of the international agreements, known as *synchronous digital hierarchy* (SDH), that build on a more fundamental block of bytes called *synchronous transport module* (STM), which correspond to the concatenated payload of an STS-3 envelope, the terms STS-3c (STS-3 *concatenated*), STS-12c, etc., are now used in describing SONET at these rates. (STS-nc are now also called EC-nc for Electrical Carrier-level-n.)

The functionality of SONET is achieved by defining the basic STS-1 signal, and an associated byte interleaved multiplex structure that creates a group of standard rates at N times the STS-1 rate. In other words, one must define frame structures that describe how bits are assembled into standard units fit for transport. N takes selected integer values from 1 to 255; currently, the following values are defined: $N = 1, 3, 9, 12, 18, 24, 36$, and 48; trials for $N = 192$ equipment were being planned at press time. For transmission over fiber facilities, an optical counterpart of the STS-1 signal, called the *optical carrier*

* It repeats 8000 times per second.

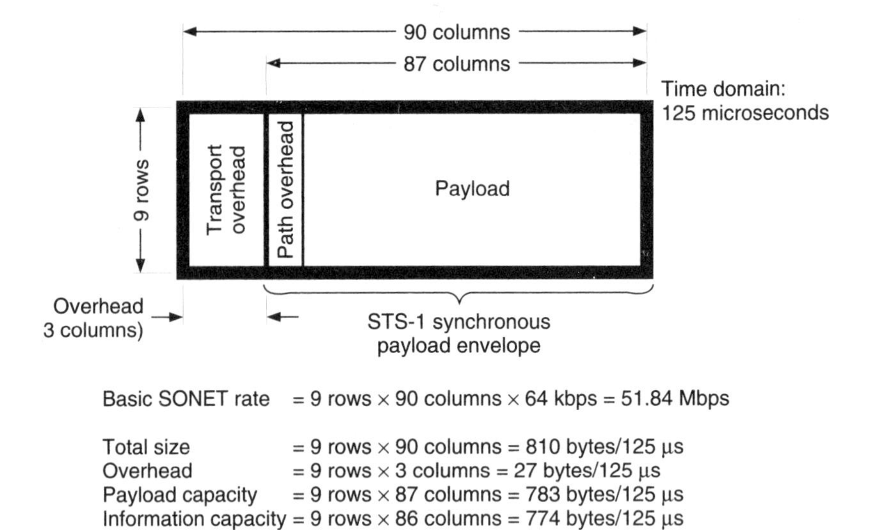

Basic SONET rate = 9 rows × 90 columns × 64 kbps = 51.84 Mbps

Total size = 9 rows × 90 columns = 810 bytes/125 μs
Overhead = 9 rows × 3 columns = 27 bytes/125 μs
Payload capacity = 9 rows × 87 columns = 783 bytes/125 μs
Information capacity = 9 rows × 86 columns = 774 bytes/125 μs

Figure 4.13 SONET STS-1 frame.

Level-1 signal (OC-1), is defined. It is an electrical-to-optical mapping of the STS-1 signal. The OC-1 signal forms the basic SONET transmission building block from which higher level signals, such as OC-3 and OC-48, are derived. For example: the OC-3 signal operates at 3 × 51.84 Mbps, or 155.52 Mbps. The line rates are defined in Table 4.6. Effectively, the suffix represents the number of equivalent DS3s that the fiber-optic system can carry. For example, OC-48 systems carry 48 DS3s (1344 T1s).

The STS-1 signal is divided into a portion assigned for transport overhead and a portion that contains the synchronous payload. *Synchronous payloads* are payloads derivable from a network transmission signal by grouping integral numbers of bits in every frame. The synchronous payload carries the user's data stream. The *synchronous*

TABLE 4.6 SONET-SDH Rates

STS-1	OC-1	51.840 Mbps
STS-3c	OC-3 (STM-1)	155.520 Mbps
STS-9c	OC-9	466.560 Mbps
STS-12c	OC-12 (STM-4)	622.080 Mbps
STS-18c	OC-18	933.120 Mbps
STS-24c	OC-24	1.244160 Gbps
STS-36c	OC-36	1.866240 Gbps
STS-48c	OC-48 (STM-16)	2.488320 Gbps
STS-96c	OC-96 (STM-32)	4.976640 Gbps
STS-192c	OC-192 (STM-64)	9.953280 Gbps

payload envelope (SPE) is a 125-μs frame structure composed of STS path overhead and bandwidth for a payload. A *payload pointer* in the transport overhead indicates the location of the beginning of the STS SPE. In a synchronous transfer mode, the payload can be used to transport asynchronous DS3 signals or a variety of sub-DS3 signals.

Transport overhead is overhead added to the SPE for transport purposes. Transport overhead is, in turn, composed of line overhead and section overhead. The transport overhead has been designed to accommodate several different functions, including maintenance, user channels, frequency justification, orderwire, channel identification, and growth channels.

SONET provides a synchronous transfer mode that is similar to traditional *time division multiplexing* (TDM) methods. This implies that either the bandwidth is allocated to a pair of users for an interval of time that is very long (months or years) or a circuit-switching capability is overlaid on SONET to assign the channel as needed.* Therefore, if the bandwidth is "permanently" allocated, one wants at least to assign just about the right bandwidth and no more to keep the "waste" to a minimum. This gave rise to a complex mechanism to carry DS1s, DS2s, etc., over SONET, called *virtual tributaries*. This approach to bandwidth conservation did not lead to viable commercial products. Hence, the entry of ATM techniques riding on top of SONET: ATM allows dynamic bandwidth allocation to the users of the SONET channel, so that bandwidth is used as effectively as possible. Figure 4.14 depicts the carriage of 53-byte ATM cells over SONET to achieve the desired sharing of the bandwidth among various users (e.g., various video channels). The reader may wish to consult Ref. 16 for a more extensive treatment of SONET.

4.4.2 Digital modulation

Fiber is neither an analog medium nor a digital medium: Fiber can *carry* an *analog-modulated* signal or a *digitally modulated* signal. See Fig. 4.15.

A digital signal of a given data rate, say 155 Mbps, may need to be carried over a fiber (this signal may be the result of multiplexing several lower speed signals together, e.g., 22 6-Mbps MPEG-2 video streams, plus SONET overhead). This signal is used to directly modulate the (light) carrier (generated by the laser diode) using amplitude

* Some efforts in support of this approach to switched SONET were undertaken but did not result in commercially available products.

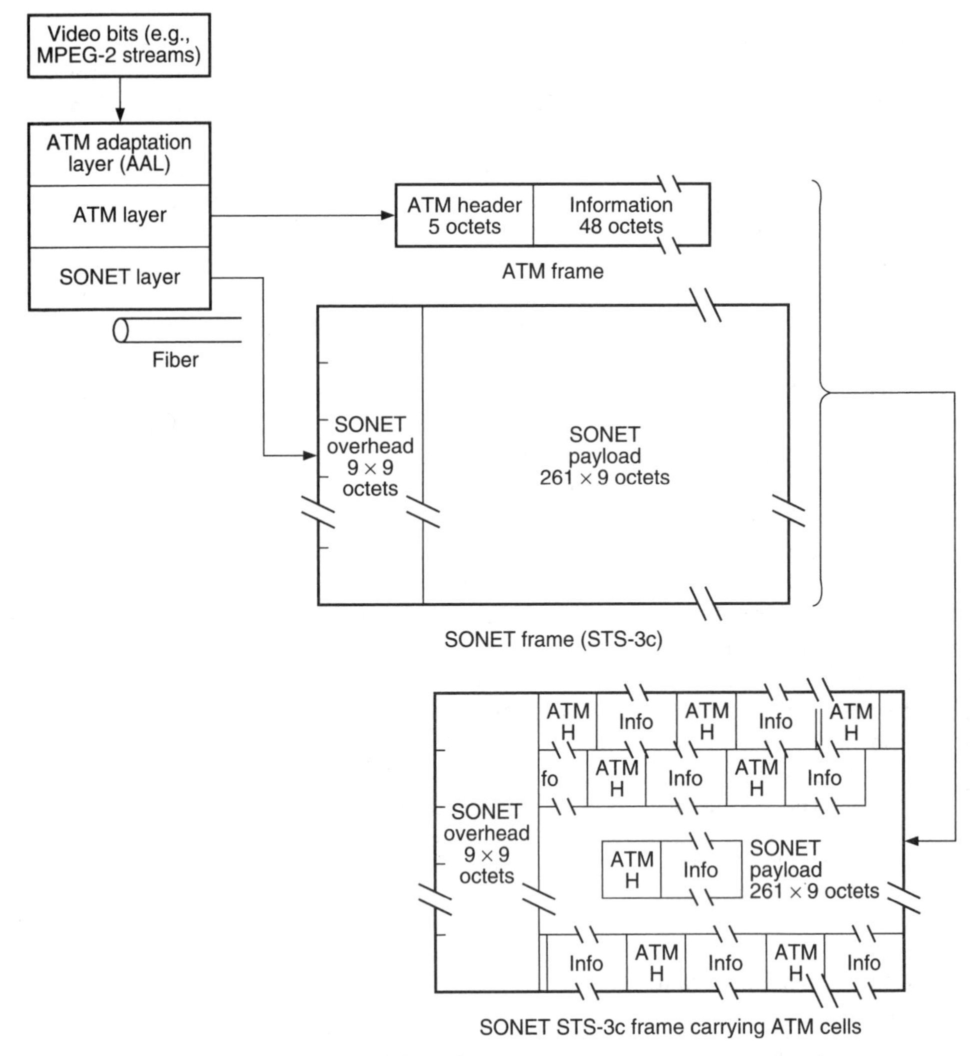

Figure 4.14 Carriage of ATM cells over SONET.

shift keying* techniques. Typically, a single-level coding is used, where a high value of light emission represents a 1, and no light represents a 0. The laser light is pulsed on and off at the desired clock rate (e.g., 155 Mbps, 622 Mbps). In the context of fiber optics, this modulation process is called *intensity modulation* (IM). In theory, multilevel coding could be used, where no light represented 00, some light represented 01,

* Amplitude modulation.

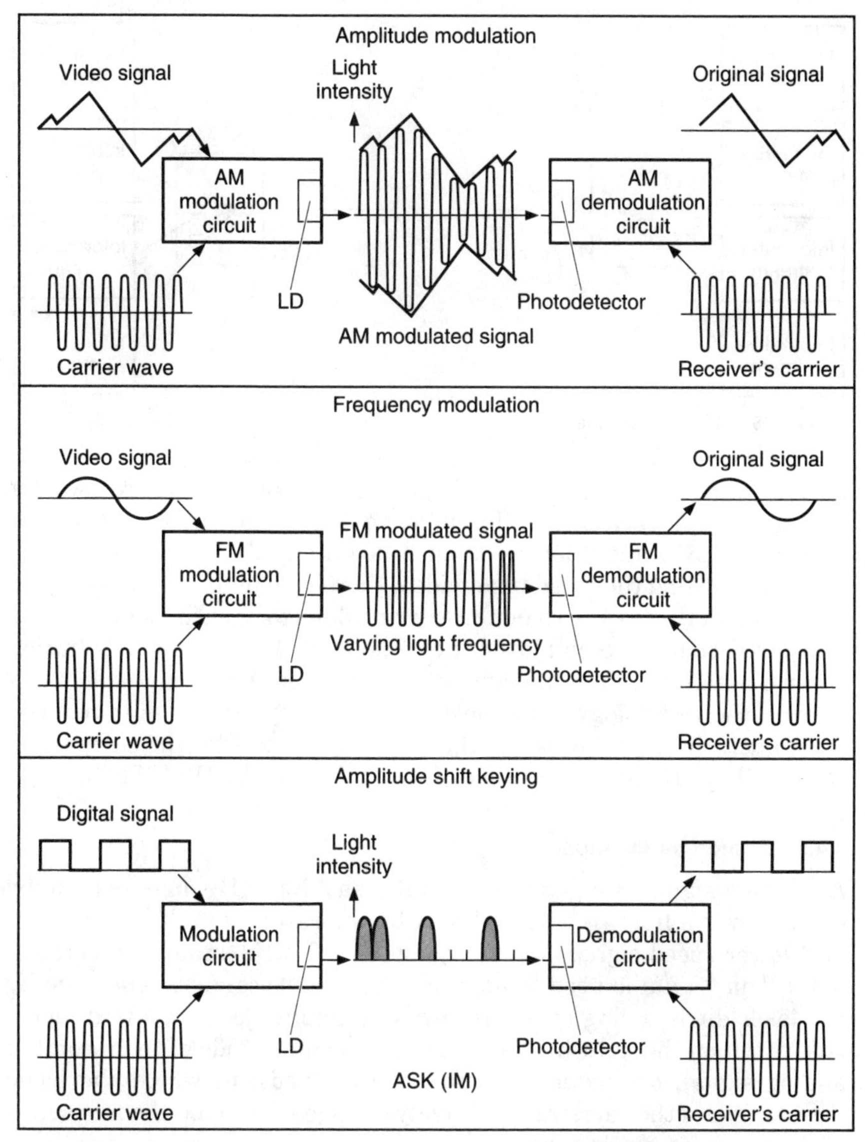

Figure 4.15 Modulation processes.

more light represented 10, and the maximum light represented 11. This technique is not yet in widespread use because of the implications of sensitivity with regard to the photodetector. At this time, the preferred ways to achieve increased throughput are:

1. Increase the clock rate so that the duration of the interval to represent a bit, during which the light is on or off, is decreased in half, quarters, etc.

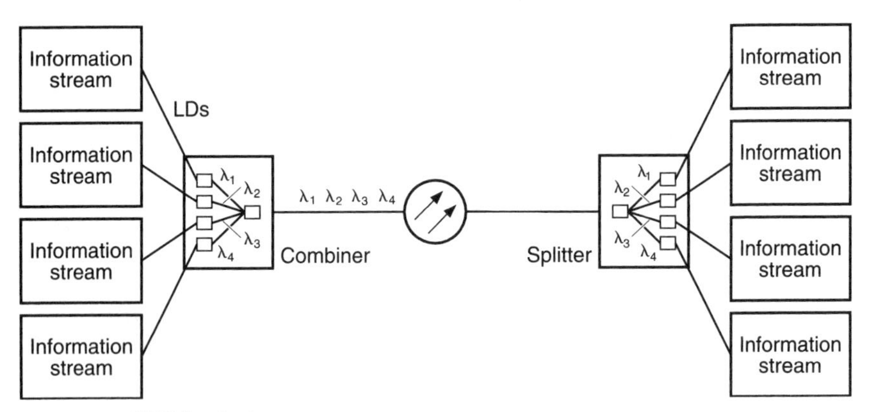

Figure 4.16 WDM technique.

2. Use wavelength division multiplexing techniques, where two (or more) distinct beams at different wavelengths are carried in the fiber (e.g., at 1310 and 1550 nm). Each beam entails the on/off process, each affording the rated bandwidth, thereby doubling the bandwidth carried in the fiber; two noninterfering detectors, each "tuned" to the desired beam, are employed. See Fig. 4.16. WDM loop distribution with a high number of beams is one of the proposed methods; however, the technology is too new and relatively expensive to be used extensively so that other methods are being used for the early trials (see Chap. 14 for one example of an application by US WEST).

4.4.3 Analog modulation

An analog signal of a given bandwidth, say 160 MHz, may need to be carried over a fiber (this signal may be the result of multiplexing several lower speed signals together, e.g., 16 10-MHz analog video channels). This signal is used to modulate the (light) carrier (generated by the laser diode) using either traditional amplitude modulation methods, in which the light intensity follows the amplitude of the incoming analog signal, or frequency amplitude methods, in which the basic wavelength of the carrier is additively changed, based on the frequency of the incoming analog signal.

4.4.4 Commercial fiber-based video systems

At the present time, the use of fiber in the cable TV context tends to rely on analog modulation techniques* to avoid employing codecs at

* As noted in Chap. 1, carriers have transported digital video at the 45 Mbps rate for some time.

the termination points of the fiber. Several analog video products are on the market; some are single-channel systems, and others are expandable to 16 channels. Both AM and FM systems are available. AM-based systems are more susceptible to noise, and AM lasers are more expensive. FM systems experience less noise, but the signal must be converted to AM to interface to an existing (coax-based) distribution system (in many systems, this conversion is done internally to the system). A typical system might have 16 10-MHz-wide video channels along with 16 100-kHz-wide audio channels.

One application of WDM is to carry two signals in a fiber: a digital to transport traditional telephony and digital television (for those users who might support it), and the other to carry analog TV for existing cable TV applications (some of the proposed architectures discussed in Chap. 5 use two or more distinct fibers).

4.4.5 Optical network unit and host digital terminal

Many VDT proposals use the terminology *optical network unit* (ONU) and the companion *host digital terminal* (HDT) in discussing the configuration for the distribution of the video signals to the residences. This terminology is based on the RBOCs' Technical Advisory TA-NWT-000909,* which specifies generic requirements for Fiber In The Loop (FITL) systems (including next-generation digital loop carrier systems). An FITL is composed of an HDT at the CO and a subtending ONU in the field. The HDT is used on the feeder routes and provides an integrated interface to the CO switch, using the principles of the RBOCs' Technical Requirements TR-TSY-000303. The ONUs are located close to (or even within) the user's location; they directly terminate service lines (e.g., a telephony line). A passive optical network provides the connection between the ONU and the HDT.

FITL† is an enabling technology in support of video services, including networked multimedia. Although some low-end services are possible using traditional loop technology operating at T1/DS1 (1.544 Mbps) or T2/DS2 (6.132 Mbps), many applications under consideration require the type of bandwidth afforded by fiber-based facilities and ATM. These high-capacity facilities have been available for years in the interoffice and feeder portion of the network; however, end-to-end, premises-to-premises fiber is required to support data-intensive multimedia applications. Although some large businesses may already have fiber loops, the majority of the users currently do not have such end-to-

---

* *Generic Requirements and Objectives for Fiber In The Loop Systems.*

† This section is based on Ref. 17.

end facilities. FITL efforts aim at extending the fiber connectivity all the way to the user or to a pedestal close to the user. FITL-based systems are being introduced at this time and should experience increased penetration in the future. One possible application is delivery of video services. Figure 4.17 shows a generic FITL system.

4.4.5.1 Host digital terminal. The HDT supports the subtending ONUs and facilitates in interfacing the FITL system to the remainder of the RBOC's transmission and operations network. The HDT may concentrate telecommunications traffic from all of its subtending ONUs for efficient feeder transport and to present a highly utilized interface to the local switching system (i.e., there is more bandwidth at the distribution side than at the feeder side). The HDT may typically have the ability to associate portions of feeder bandwidth with customer service interfaces dynamically (on a call-by-call basis). In terms of traditional telephony functions, the HDT may separate or groom locally switched traffic from nonswitched and nonlocally switched traffic so that the latter can be routed away from the local switching central office. New functions will be required for VDT service.

4.4.5.2 ONU. The ONU is the network element that provides the tariffed network interfaces, as well as, potentially, future service interfaces for several residential and small business customers. Traditional

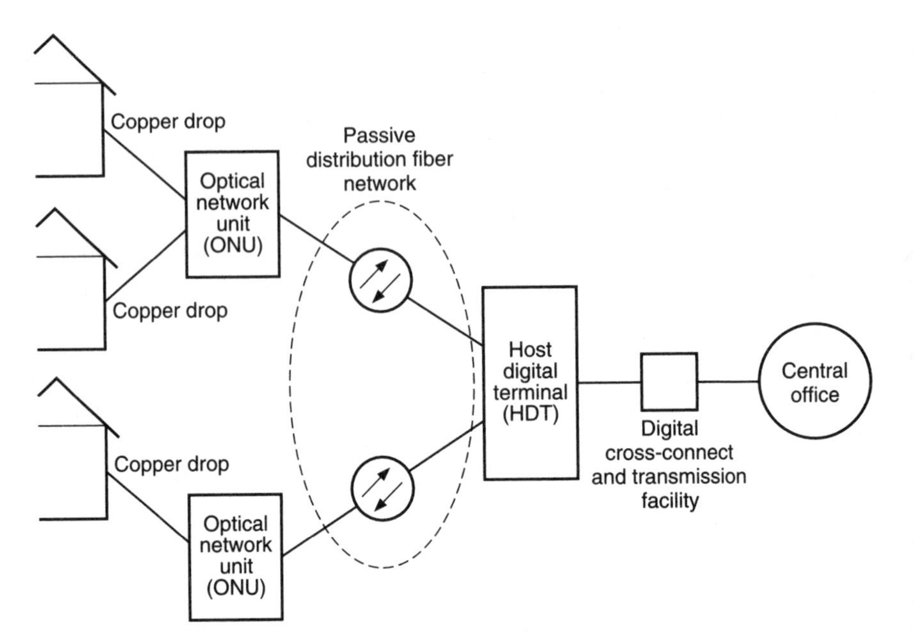

Figure 4.17 FITL architecture.

telecommunication services are provided over metallic twisted pairs (drops) to a network interface where they are handed off to the customer's equipment. With VDT coaxial or fiber cable, drops can be used. Some of the traditional telephony functions performed by the ONU are:

- Electrical-to-optical conversion
- Analog-to-digital conversion of voice frequency signals
- Multiplexing of individual services onto the high-speed optical facility
- Maintenance for both individual services and the optical transmission facility

New functions will be required for VDT service.

There are three different sizes of ONUs, with each product targeted at a different residential application. ONU Type A supports 12 users, ONU Type B supports 24 users, and ONU Type C supports 48 users. Each ONU accepts a specified number of DS0s from the *passive distribution network* (PDN*) (12, 24, or 48), and is capable of using each DS0 (64-kbps channel) to provide an individual telecommunication service. Upgradability to provide DS1-based services (e.g., switched fractional DS1, ISDN primary rate access, video distribution) is an objective for the near future. Typically, the ONU must support a service drop length of 500 ft.

4.4.5.3 PDN topologies. The PDN can be implemented in a number of ways:

1. Point-to-point; one fiber to the ONU (as seen in Fig. 4.17)
2. Point-to-point; two fibers to the ONU
3. Multipoint; one fiber to the ONU, using passive splitters
4. Multipoint; two fibers to the ONU, using passive splitters

Point-to-point information transport involves information transfer between the HDT and ONUs over a dedicated fiber path for each ONU. The implementation of point-to-multipoint information transport allows the sharing of an HDT optical line unit and a portion of the fiber facility by multiple ONUs. In this case, passive optical splitter technology is used, and signals are broadcast from the HDT to the subtending ONUs. The fiber link between the HDT or splitter and subtending ONUs could consist of either one or two fibers.

---

* This also known as *passive optical network* (PON).

4.4.5.4 Drop wire options. The last few feet from the ONU to the residence could employ shielded or unshielded twisted-pair cable, fiber-optic cable, or coaxial cable, based on economic considerations. It is already possible to deliver 100 Mbps (or more) up to 300 ft on twisted-pair cable. Coaxial cable supports DS3 rates up to 450 ft. Of course, fiber provides the highest bandwidth.

4.4.5.5 Examples of FITL equipment. Several vendors manufacture FITL equipment, including ADC, AT&T, DSC, Northern Telecom, and BroadBand Technologies (BBT), particularly with an eye to video services (see Chap. 11 for some examples).

4.4.6 All-optical networks and other optical technologies

As both telephone carriers and cable companies enter the FTTC, FFTN, and the FTTH arena, one needs some leading-edge optical technologies to obtain the necessary economic justification. Some lasers, although powerful, still cost around $10,000.[7]

One technology that is being watched is all-optical networks using optical amplifiers. Such a system will allow carriers and cable companies to improve reliability, quality, and service in a cost-effective manner. All-optical networks have been under development for long-haul, point-to-point networks; the next step is to deliver the technology for multipoint local applications. Although the cost of optical amplifiers is high, it is expected to fall in the future. When the price makes them more cost-effective, these amplifiers will find immediate use: until now carriers have increased the network capacity by increasing the clock rate of the components, as previously discussed; now one can use WDM in a more effective manner. In a traditional fiber system with electronic regenerators, each beam of light needs its own regenerator. Amplifiers can amplify an entire band so that all WDM beams (up to 10 at this juncture) are automatically amplified without added electronics or photonics. Hence, given optical amplifier technology, the only cost to add bandwidth to a fiber is incurred in the transmit/receive optics; there are no new costs along the path. In addition, to increase the bandwidth by an order of magnitude or more, this technology allows the operator to mix analog and digital signals in a single fiber.

Fiber-optic systems now under development use *solitons* as carriers. Solitons are a method of coding laser pulses that enables them to hold their shape over very long distances. Experiments have already achieved a throughput of 16 Gbps over 20,000 km of fiber.[7] The technology is being developed for long-haul submarine cables to be installed in the 1997 to 1998 time frame. Once the research for the long haul is completed, engineers see the deployment of this technology in commercial networks around 1998 or 1999.

4.5 Pre-ATM Switched Digital Services for Video

ATM-cell relay service is the switching technology being proposed by proponents as the ideal method to carry video for both business and residential VDT and generic video distribution applications. However, until ATM is widely deployed in the residential market, other switched technologies such as Switched T1 and Switched T3 may be employed (e.g., Bell Atlantic uses Switched T1 in the ADSL trials).

RBOC requirements have been published to support Switched T1 in a non-ISDN environment (document TR-NWT-001068), as well as in an ISDN environment (document TR-NWT-001203). A number of vendors have made plans to support ISDN-based Switched T1 (e.g., NTI's Fiber World and AT&T's Service Net 2000).

References

1. L. Bellato and M. Dinaro, "The Fundamentals of Video Transmission," Electrical Communication, 3rd Quarter 1993, pp. 214 ff.
2. D. Minoli, *Imaging in Corporate Environments,* McGraw-Hill, New York, N.Y., 1994.
3. J. Anderegg, et al., "Transparent Video Transmission In Contribution Networks," *Electrical Communication,* 3rd Quarter 1993, pp. 227 ff.
4. R. Pinkham, "Combining Apples and Oranges," *Telephony,* January 24, 1994, pp. 32 ff.
5. J. Monteforte, Storer Cable, Bricktown, N.J., personal communication, April 1994.
6. R. Pinkham, "Combining Apples and Oranges: The Modern Fiber/Coax Network," *Telephony,* February 7, 1994, pp. 28 ff.
7. *Communications Technology,* April 1993, pp. 18 ff.
8. B. Carlson and J. Chamberlain, "Coax: The Ideal Broadband Medium for the Last Mile," *Telephony,* March 14, 1994, pp. 20 ff.
9. V. Bugliera, "Digital Modulation and Transmission Technologies for Cable Applications," *Communications Technology,* April 1993, pp. 44 ff.
10. *Video Technology News,* Phillips Business Publishing, February 28, 1994, pp. 1 ff.
11. L. Lockwood, "Washington Demo of an Improved DSC-HDTV," *Communications Technology,* April 1993, pp. 26 ff.
12. T. Miki, "Toward the Service-Rich Era," *IEEE Communications Magazine,* February 1994, pp. 34 ff.
13. J. B. Waltrich, "Distortion Produced by Digital Transmission in a Mixed Analog and Digital System," *Communications Technology,* April 1993, pp. 40 ff.
14. R. Karpinski, "Can CDMA Work Its Magic for Cable?," *Telephony,* January 24, 1994, pp. 16 ff.
15. D. Minoli, *Telecommunications Technology Handbook,* Artech House, Norwood, Mass., 1991.
16. D. Minoli, *Enterprise Networking, Fractional T1 to SONET, Frame Relay to BISDN,* Artech House, Norwood, Mass., 1993.
17. D. Minoli and R. Keinath, *Distributed Multimedia Through Broadband Communications Services,* Artech House, Norwood, Mass., 1994.

Chapter

5

Video Dialtone Transport and Signaling Architectures

5.1 Overview

Any telecommunication network is composed of four logical elements as follows:*

1. *Termination modules.* User equipment connected to the network. In the VDT context, these consist of VIP's video servers, and VIU's setup boxes.
2. *Transmission modules.* These include both the access segment (typically called local loop) and the interoffice backbone segment (transmission between the first serving element and the far end). This includes ADSL systems, HFC, SONET-based fiber, etc.
3. *Switching modules.* These enable time-limited connections to other users of the networks. This includes DS1 circuit switching, DS3 circuit switching, and ATM switching.
4. *Signaling modules.* These include signaling over the access segment as well as over the interoffice segment. In Chap. 1, the concepts of VDT network signaling, network access signaling, and application signaling were introduced.

This chapter examines a number of video dialtone architectures with repect to these four elements of the supporting access and backbone network, focusing on signaling. A more detailed treatment of the access network apparatus is found in Chaps. 9 to 12.

* In specific networks, not all four of these elements may be explicit or fully manifested.

5.1.1 Required specifications

Carriers are now publishing specifications to make VDT possible in a multivendor environment. Openness is a much-needed VDT goal (even a mandate), fostering or retarding penetration of the service. Such specifications typically include:

- Definition of the video service function supported (e.g., video on demand).
- Definition of the physical infrastructure for transport of the video signal, looking toward the users (including definition of the VIU-NI) and looking toward the information providers (including the VIP-NI). For example, the infrastructure could be ASDL to the VIUs and traditional T1 to the VIPs.
- Definition of the signaling capabilities to be supported.
- Definition of the physical infrastructure for transport of the signaling information, including definition of the signaling interface (e.g., X.25 systems, ATM).

A number of carriers and providers have already published some specifications, and more are expected in the near future, although some have taken the approach of developing closed systems. These specifications issued by the carriers are directed to vendors of Level 1 gateways, loop equipment vendors (e.g., ADSL, hybrid fiber coaxial), user device vendors (e.g., setup box, home PC software, TV manufacturers), developers of Level 2 gateways, vendors of video servers, and to VIPs.

5.1.2 Logical transmission channels

In VDT, the transmission channel can support:

- *Single-cast communication:* unidirectional transmission of (video) information from a VIP to a single VIU with bidirectional signaling
- *Multicast communication:* unidirectional transmission of (video) information from a VIP to a set of VIUs with bidirectional signaling
- *Broadcast communication:* unidirectional transmission of (video) information from a VIP to all users in the subscriber base with bidirectional signaling
- *Two-way video communication:* for applications such as videoconferencing and continuous presence distance learning

These channel characteristics are used in the service definition of a VDT platform. Figure 5.1*a* depicts a logical view of a video dialtone

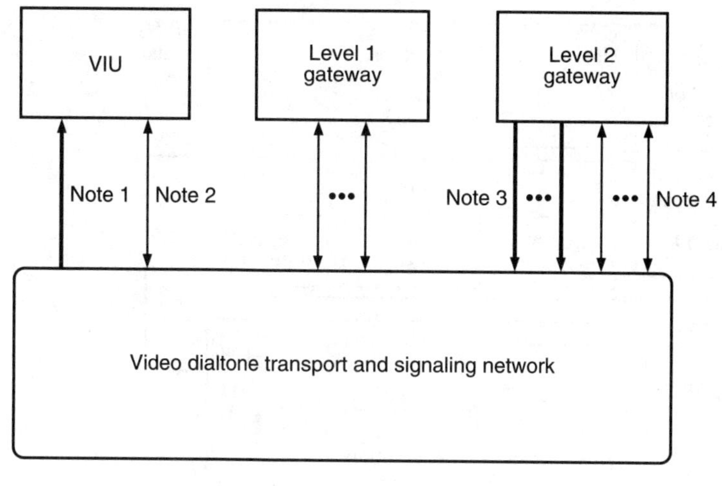

Video signal (in indicated direction)

Signaling signal (in indicated direction)

Note 1: Bandwidth of channel depends on underlying platform technology, type of video compression, and number of simultaneously delivered programming channels

Note 2: Number of links depends on load; bandwidth of channel depends on underlying platform technology (multiple virtual channels over one facility may be supported)

Note 3: Number of links depends on load and whether multiplexing is used or not

Note 4: Number of links depends on load and whether multiplexing is used or not

Figure 5.1*a* Logical view of a (phase 1) VDT architecture.

platform, while Fig. 5.1*b* shows a functional view. As an example, traditional video transmission is (typically) characterized as downstream single-cast. Signaling is typically characterized as bidirectional and symmetric (with more bandwidth on the VIP-NI link than in the VIU-NI link); for some applications, signaling over the VIU-NI can be characterized as multicast downstream.

As an example, the Bell Atlantic Signaling Specification for Video Dial Tone, calls for the following VIU-NI logical configuration*[1]:

* Refer to the actual specification for any development details; the information contained herewith is for pedagogical purposes only.

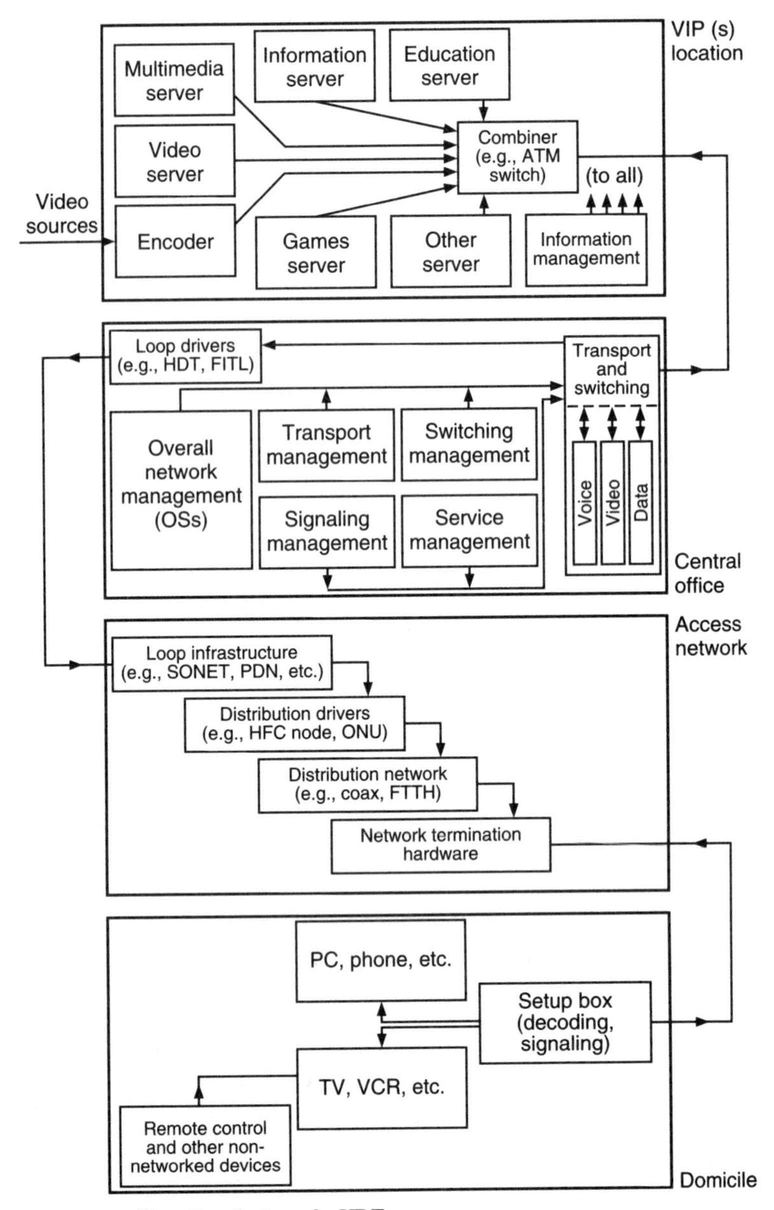

Figure 5.1*b* Functional view of a VDT.

1. One downstream single-cast video channel
2. One two-way signaling channel configured as
   a. Two two-way signaling virtual channels
   b. One downstream multicast signaling virtual channel

5.2 Network Infrastructure

5.2.1 VDT platform functional level

This section provides a short overview of the physical infrastructure that can be used to support VDT services. Chapters 9 through 12 provide more in-depth coverage.

Figure 5.2 depicts a VDT platform at the topological level. At this junction, the CO includes only the equipment necessary to support VDT service (e.g., DCSs or ATM switches); in principle, it could also support other equipment to use the bandwidth and signaling facilities of a VDT network for services, such as video mail and videoconferencing bridging.

From the consumer's point of view, it does not matter where the (video) information is actually stored. Typically, a VIP's video archive would not reside in a carrier's CO, although it would use equipment at the CO, such as digital cross-connect systems and core* ATM switches. At press time, storing an entire movie at the CO was subject to regulatory restrictions. A solution advanced by some is to store only a few seconds of a movie in the CO, while actually storing the movie itself with an information provider (not necessarily a VIP). In this proposed framework, the information provider serves as a broker for those institutions that wish to place their products in an information warehouse (IW). The IW could act as a centralized location to support a variety of video-information services. As long as an open interface to the IW is available, it is possible to easily deliver the information to all consumers of such information. (Some, in the carriers' sphere of interest, hope that in the future the IW functionality could reside in the CO.)

Figure 5.3 depicts a simplified diagram of a VDT system at the physical level. Two major *subnetworks* are evident in Fig. 5.3: the backbone subnetwork and the local distribution subnetwork.

5.2.2 Backbone subnetwork

The technologies used in each subnetwork shown in Fig. 5.3 can be different, each optimized to the most critical functions. Although one could use AM-FDM, FM-FDM, and TDM (e.g., DS3, SONET) in the backbone, a likely approach to be used by the RBOCs† is ATM over SONET; see Fig. 5.4. ATM technology, which is designed to carry voice, data, and video, is covered in Chap. 8. The backbone portion could employ OC-24

---

* ATM planners now are in favor of placing smaller ATM switches closer to the user location. Hence, a carrier may wish to place an "edge" ATM switch in the proximity of (possibly in the building of) the VIP.

† Several cable TV companies are also looking at ATM; see Chaps. 10 and 15.

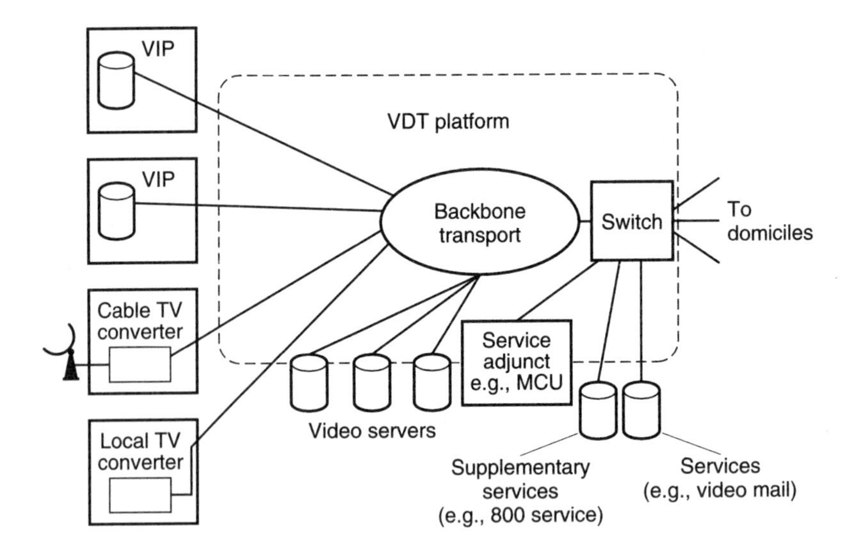

Figure 5.2 Topological view of VDT platform.

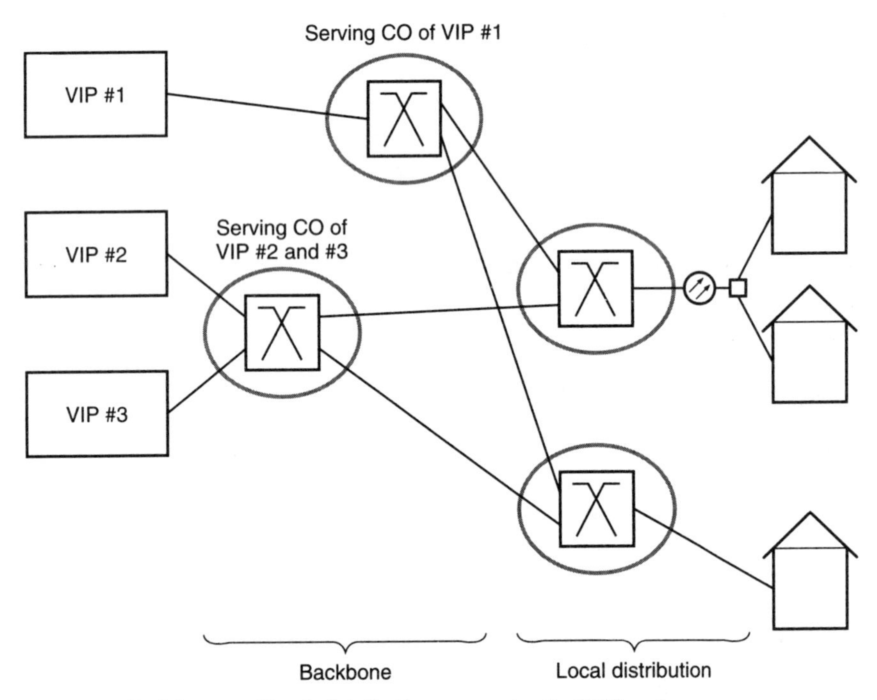

Figure 5.3 Backbone and local distribution segments of a VDT system.

(1.2 Gbps) or OC-48 (2.4 Gbps) transmission, while the distribution network could operate at 155 Mbps (45 Mbps and 622 Mbps could also be used in the immediate future and in the longer term, respectively). Perhaps as an early deployment, a *Synchronous Transfer Mode* (STM) approach could be used in conjunction with SONET, with an eye to later migration to ATM. (See Fig. 5.5 for a topological view and Fig. 5.6 for an operational view of this approach.) The local distribution subnetwork

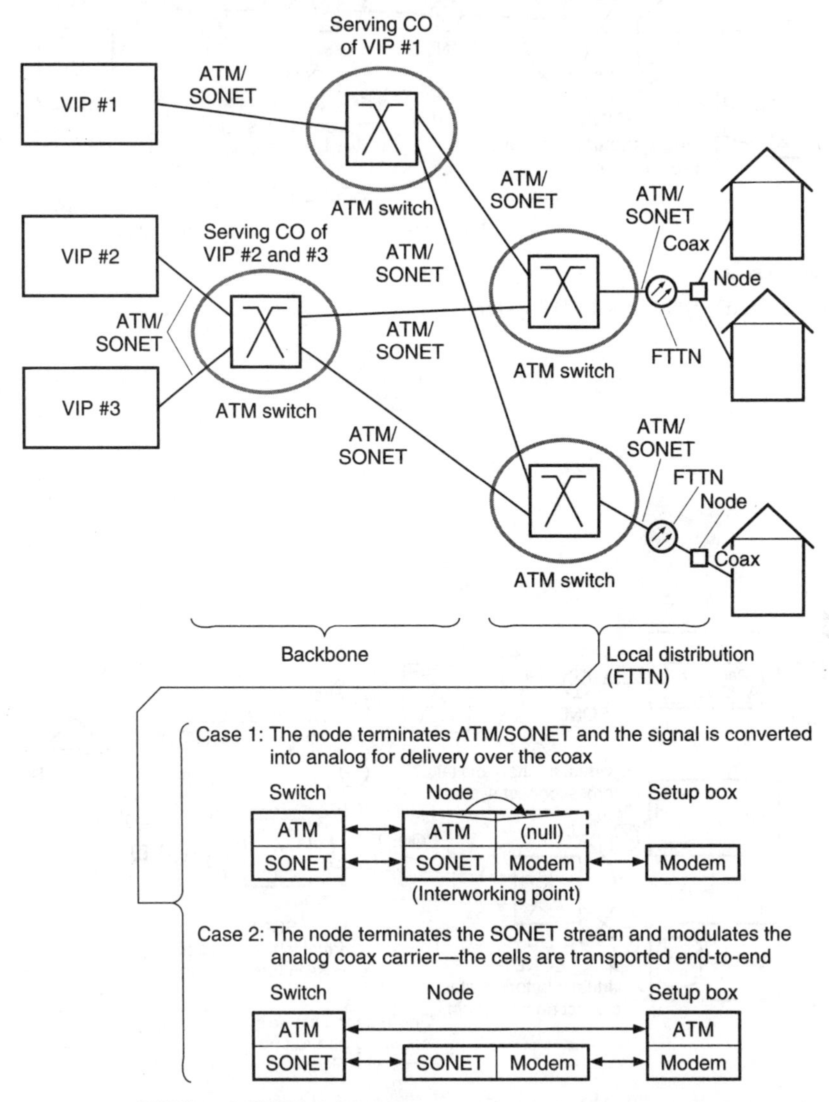

Figure 5.4 ATM-based FTTN architecture.

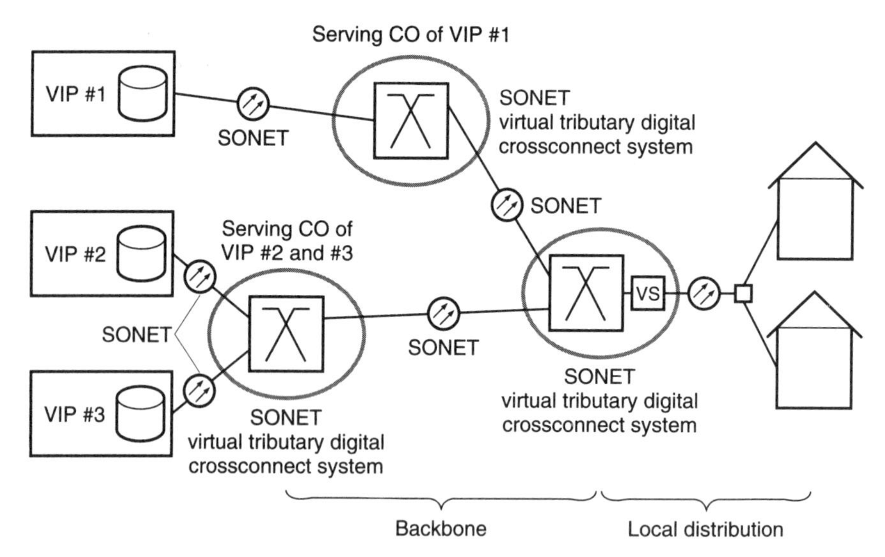

Figure 5.5 Early STM-based VDT system.

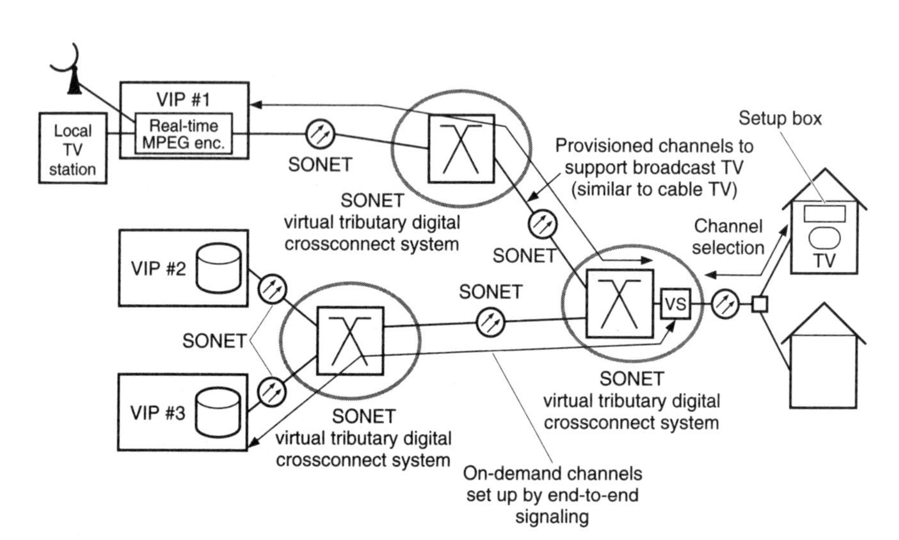

Figure 5.6 Signaling in STM-based VDT.

can also consist of a variety of technologies already identified in Chap. 1 (e.g., FTTN, FTTC, ADSL, and wireless).

5.2.3 Switching technology

The major difference between existing cable TV networks and VDT video delivery is the fact that in VDT only the channels of interest are *switched* into the user's access facilities (in some arrangements, the broadcast network TV channels are all made available, but the premium and/or pay TV channels are switched as needed). Three switching technologies are available in PSTNs:

1. Voice switches (DS0/64 kbps digital and/or ISDN). These switches do not have the port or backplane capacity to support video, even for MPEG-1 signals.
2. DCSs operating at the DS1 or DS3 rates. These switches are planned to be used (by some VDT implementations) but have limitations.
3. ATM switches (Chap. 8). These are positioned by proponents as being ideal for VDT; however, a number of technical issues (i.e., the ultimate ability to carry video, voice, and data efficiently), standardization, network integrity, cost issues, throughput considerations, consumer equipment inventory, and geographic availability remain to be solved in toto and/or validated.

DCSs are slow in terms of switching: they used to operate at the "minutes" level; they are now faster (operating at the "seconds" level) but could not support an environment where the DCS is used to switch channels in real time, to support customer channel surfing. DCSs in use today are relatively small in terms of ports, compared to the large number of subscribers who may subscribe to VDT services. DCSs have the advantage that they support multicasting. A packet-switching network can be employed to carry low-speed (two-way) signaling and other data between users and the VIP (this is the approach followed by Bell Atlantic).

There have been early discussions in the VDT context about the development of a device called a *video selector*. Selectors allow consumers to select among the available channels as they are delivered across the backbone network and are available at the customer's serving CO. A selector would be designed to terminate a relatively large number of broadcast-multicast video sources (on the VIP side of the selector), as well as specific-niche VIPs. The selector would reduce the amount of user signaling that has to be carried deep into the VDT network, while still permitting on-demand services. When users request video programs "present" at the selector, user signaling does not go any farther than the selector itself. Requested video channels not present at the selector need to be put on line through the appropriate VIP; hence the signaling is carried farther into the network (see Fig. 5.6; signaling issues are covered in Sec. 5.3). Whether video selectors will be developed at the commercial level remains to be seen.

In an ATM-SONET platform, an ATM-configured, cross-connect system could be used initially to perform the functions of a selector; ATM switches could be deployed at a later date. Work is underway to develop effective support of point-to-multipoint ATM connections. Much remains to be done to develop fully flexible multipoint-to-multipoint connectivity required for full-function I-TV, in support, for example, of group games, cooperative work, and continuous presence distance learning.

5.2.4 Distribution subnetwork

As discussed earlier in this text and as illustrated in Fig. 5.7, delivering video to consumers in a VDT environment can be done in a number of ways. The guided transmission methods to be discussed here, and in more detail in Chaps. 9 to 11, include copper (twisted-pair), coaxial cable overlay, and fiber. Many variants of these configurations have been advanced by vendors and carriers. In nearly all of the proposals and trials underway, the signal is being delivered to the consumer over a copper, coaxial, or part-fiber facility, whether the video signal is analog or digital.

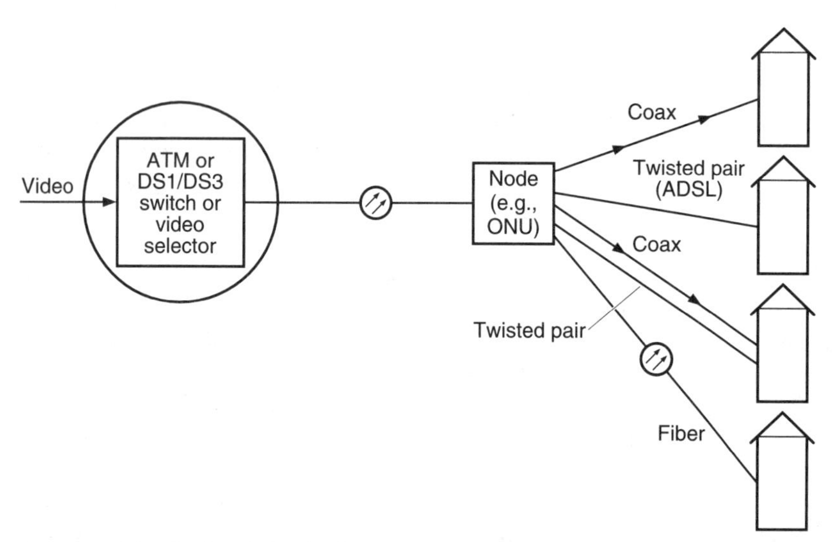

Figure 5.7 Distribution technologies—macroview.

As seen in the protocol model of Fig. 5.4, one must distinguish between the physical delivery mechanism and the TV signal itself. One can use an analog fiber and/or an analog coaxial cable and, through a modem function, deliver digital-compressed video to the television (e.g., one can obtain 1.2 Gbps over a coaxial cable). Or, one can use an analog fiber and/or an analog coaxial cable and deliver an analog signal to the TV set. Or, finally, one can use a digital fiber-ADSL link and deliver digital video to the setup box or TV. See Table 5.1.

The techniques advanced by the suppliers would allow the coaxial cable to either deliver an analog TV signal or a digitized-compressed signal using a modem function. The coaxial or coax-copper architectures (listed later), based on 1.2-Gbps digital coaxial capacity, can support several digital channels to the home. In addition to broadcast channels always present on the coaxial, these architectures can also deliver a number of switched channels to the customer. For example, to keep some consistency with SONET, four 155-Mbps channels could be delivered to a customer in an HFC architecture; in turn, each of these channels could support approximately 20 to 25 MPEG-2 multiplexed channels or 80 to 100 MPEG-1 channels, or combinations thereof. Two or three of these 155-Mbps channels could contain the digitized broadcast channels; alternatively, one could use a portion of the coaxial spectrum to deliver analog channels and a portion of the bandwidth (e.g., above 450 MHz) to deliver digitally modulated-digitally encoded channels. Figure 5.8 depicts graphically the use of digitally modulated coaxial cable with payloads that fit into SONET, to retain a high degree of SONET compatibility in the feeder and backbone portion of the VDT network. (Note that the actual payload is less than the SONET line rate.)

5.2.4.1 Two-wire, twisted-pair-coaxial cable. A number of proponents advance a side-by-side, twisted-pair/coaxial combination, with two wires into the home on the part of the RBOC or cable TV company. In this arrangement, the coaxial cable carries the video (the number of channels depending on the coaxial cable or system), and the twisted pair carries the telephone traffic. When this approach is used by the RBOC, the optical network unit can physically terminate the two "wires" to the home and can combine the traffic toward the CO switch.

TABLE 5.1 TV Signal-Transport Combinations

	Medium	
TV signal	Analog	Digital
Analog	e.g., traditional coax	e.g., cannot be carried unless it is digitized
Digital	e.g., use of modem	Straight MPEG-2-ATM streams

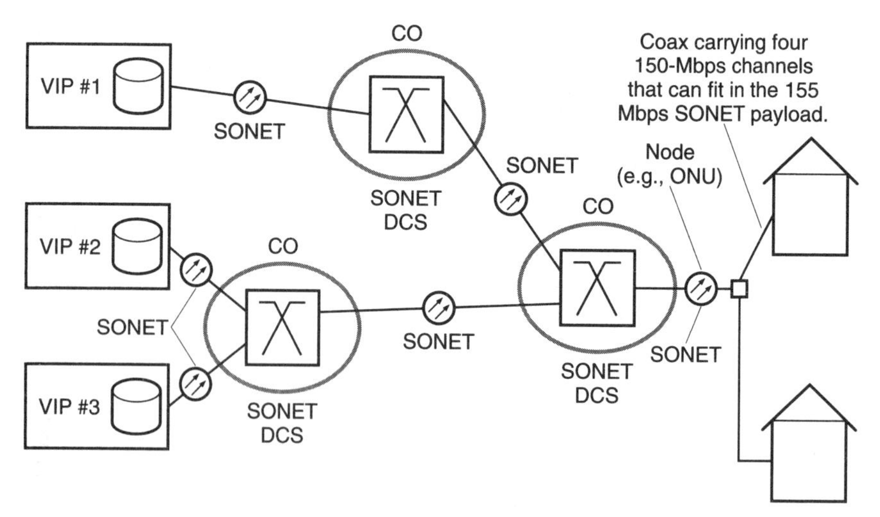

Figure 5.8 Use of digitally modulated coaxial cable with payloads that fit into SONET to retain a high degree of SONET compatibility in the feeder and backbone portion of the VDT network.

The issue with this approach, besides the cost, present mode of operation technology, and dual plant, is that, unless the ONU takes on the function of converting the digital video stream from the CO to an analog format (and also does the decompression) for delivery to the home, the signals to the CO must really be carried on two pairs of fiber: a digitally modulated SONET fiber and an analog-modulated TV signal fiber. Therefore, this approach entails maintaining four plants in the local loop: digital fiber, analog fiber, twisted pair, and coaxial cable.

Some practitioners believe that "new builds" should be installed using copper pair(s) and a coaxial cable side-by-side. Other practitioners propose a coaxial cable overlay architecture for existing plant and fiber for new builds.

5.2.4.2 One-wire coaxial cable. This arrangement uses a single coaxial cable that supports both video and telephony. Video signals are delivered to the Optical Network Unit via an HDT (these network elements were described in Chap. 4). The (video) signals could, in theory, pass through passive optical splitters so that the same signal could be sent to several ONUs, in support of delivering broadcast TV programming. In this arrangement, the ONU performs optical-to-electrical, fiber-to-coaxial cable conversion and, possibly, the digital decompression as well as the digital-to-analog conversion. The video (TV) signals are then sent to multiple domiciles (4 to 12) over a coaxial cable. The ONU also provides the interdiction function to ensure that homes receive only the programming they are entitled to (interdiction could be controlled using a low-speed downstream channel.) The PPV-VOD function can be implemented by "broadcasting" the program to all homes on the segment (4 to 12) but using interdiction with addressable setup boxes to identify only the party of interest.

At the customer site, telephony services are separated from the video by the setup box (conversion to analog is required for both the video and telephone signal if the incoming signals were digital). The setup box is also used to signal the customer's service requests to the network. Telephony and low-speed signaling can be carried at lower frequencies of the coax (50 MHz and below), although it can also be carried above 50 MHz if desired.

5.2.4.3 Coaxial-twisted-pair arrangement for video and telephony over coaxial cable. This low-end, two-wire proposal is a variation of the coaxial approaches previously discussed. It employs a coaxial cable to carry downstream video signals and a dedicated twisted-pair cable to handle signaling; telephony would be carried by a completely different twisted-pair cable. (Alternatively, telephone service could be provided on the same twisted-pair facility, but service could not be supported while the facility was in use for signaling or ordering, without the additional complexity of a device to demultiplex telephony from signaling.) Some existing cable TV systems use this approach to provide PPV, but from the posture of an RBOC, customers may not perceive this approach as value-added over "plain old cable TV service." This approach could support a number of other interactive applications.

5.2.4.4 Twisted-pair approach (no coaxial cable). ADSL-1 or ADSL-3 technology can be used to support digital video delivery to the customer over a twisted-wire pair. ADSL-1 provides for the delivery of 1.2 Mbps for video, telephony services, and a two-way 16-kbps data channel. This permits delivery of one MPEG-1 video channel. ADSL-1 is limited to 18 kft from the node, perhaps an ONU.

ADSL-3 provides for the delivery of 6.0 Mbps for video, telephony services, and a two-way data channel of 16 kbps. This permits the delivery of one MPEG-2-encoded video channel at the 6.0-Mbps rate or four MPEG-1 channels. ADSL-3 is limited to 9 kft from the node-ONU.

ADSL-1 and ADSL-3 require equipment at the CO and ONU with line cards implementing the ADSL encoding, as well as equipment at the customer site to provide decoding of the physical ADSL signal and to provide the multiplexing or demultiplexing of the video, telephony, and data channels.

5.2.4.5 Fiber To The Home. FTTH is viewed as the ultimate VDT-superhighway path into the home. Currently, the issue is one of cost. If cost were no issue, there are clear technical merits to this solution, particularly bandwidth availability. Existing hardware can support 1.2 Gbps of low-noise digital capacity, and systems expected to reach the market in the next couple of years can provide 10 Gbps.

To minimize the amount of fiber that needs to be deployed (in particular, the active components of the fiber systems, e.g., expensive LDs and photodetectors), FTTH systems have been proposed using *passive optical splitter* (POS) technology. A POS takes a signal coming from the CO and can replicate it on a number of outgoing fibers. POS can be designed to have split ratios as high as 1:128 (one signal is distributed to 128 different ONUs). This method works well for the replication of broadcast TV programming. See Fig. 5.9.

In place of a POS-based architecture, WDM could also be used for FTTH. As covered in Chap. 4, a WDM combines optical signals of differing wavelengths onto the fiber to increase the capacity of the fiber system. There is interest in deploying such a system that supports two-way communication. To achieve this, each customer is assigned two specific wavelengths (one to receive and one to transmit). Appropriate equipment at the CO multiplexes and modulates the wavelengths corresponding to the individual video and information channels. At the remote end, somewhere close to the curb of the cluster of homes in question, the signals are demultiplexed and delivered to the appropriate customer over a dedicated fiber.

5.2.4.6 Fiber To The Curb. In FTTC architectures, more likely to be deployed in the next couple of years than FTTH, the fiber is brought to

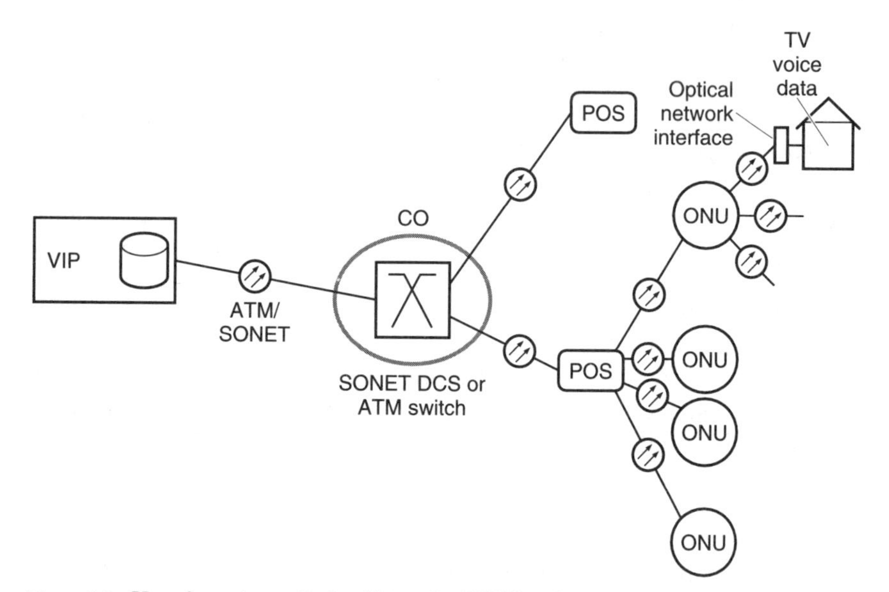

Figure 5.9 Use of passive optical splitters for FTTH architectures.

a point close to the home but is shared with a number of homes in the cluster (say 2, 4, 8, or 16). Coaxial cable can be used into the home, using the HFC approach. This approach can also use POS technology, but in this case, the POS is at the "curb" rather than at the feeder level, as shown in Fig. 5.9. The POS would have a 1:32 split ratio where 32 curbside ONUs are fed by a POS (each curbside ONU may serve from 4 to 12 customers). Note that the WDM technique previously discussed could also be used for the FTTC architecture. The cost per subscriber of an HFC system varies from a low of $1400 to $2000 or more (1994 costs).

5.3 Signaling and Control

5.3.1 Overview

The sections that follow focus on the important issue of signaling. Signaling enables the customer to communicate with the local switch, the Level 1 gateway and the Level 2 gateway, to secure the desired video service. For a comprehensive treatment of signaling from a general telecommunications point of view, see Ref. 2.

Signaling is not as visible to a casual observer as the physical architecture of the fiber-coax infrastructure, or even as visible as the new video distribution elements such as the digital video server and the (ATM) switch. Yet, signaling is one of the most critical underlying technologies, without which some of the interactive services of VDT, including VOD, IVOD, I-TV, MPG, and even simple video transmission and reception are not possible. In fact, as described in Ref. 2, such technology is fundamental to all telecommunication systems.

Signaling is pervasive in a VDT network. It applies to the following areas: VIU-to-VDT network; VIP-to-VDT network; VIU-to-VIP; intra-VDT and interVDT network transmission and switching elements (network element-to-network element); VIP-to-VIP; and VIU-to-VIU interactions. The first four kinds of interactions are the most basic (except that internetwork signaling would be more applicable to future interconnected VDT networks—for example, to support access to programming not on the VDT network on which the specific consumer is located). VIU-to-VIU signaling may come into play for MPG, videotelephony, fully interactive distance learning, and other interactive applications. VIP-to-VIP signaling may take place when one VIP chooses to make available a program to a customer, where the VIP needs to obtain the program from another VIP (this is an alternative to the customer going directly to a remote VIP over the broadband infrastructure). Early efforts in VDT signaling are being directed to the first four areas just listed.

As noted in Chap. 1, signaling falls into three categories, as follows (see Fig. 5.10):

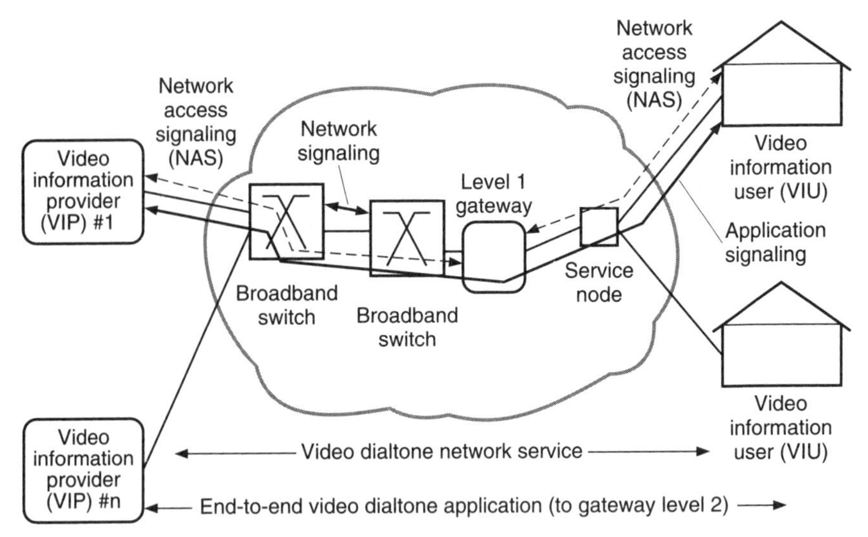

Figure 5.10 Types of signaling and control in VDT.

1. *Video application signaling* (VAS): Mechanism to enable end-to-end signaling between applications executing in the VIP and VIU nodes over an established communication path.
2. *Network access signaling* (NAS): Mechanism allowing a user of VDT services (VIP or VIU) to interact with the Level 1 gateway to request the establishment or disestablishment of network paths and other functions. It can be partitioned into an NAS-VIU (VIU-to-VDT network signaling) and NAS-VIP (VIP-to-VDT network signaling). This type of signaling is analogous, in functionality, to the Q.931, Q.2931, or DTMF of an ISDN, ATM, and public switched telephone network, respectively.
3. *Network signaling* (NS): Mechanism to set up and tear down network element-to-network element transmission paths through the network in response to a user request. The path could be a circuit-switched DS1 (such as in the ADSL approach used by Bell Atlantic), a circuit-switched DS3 (such as in some existing digital-based, distance-learning networks), or an ATM virtual connection. This type of signaling is analogous, in functionality, to the Signaling System No. 7 in the public switched telephone network.

VDT control, therefore, is the set of interactions and protocols that enables the VIUs and VIPs to:

- Access network services
- Manage network services

- Manage video applications (VDT application control includes menu navigation, movie VCR-like controls, etc.)

An important issue under consideration at this time is if an industrywide standard for signaling (e.g., ITU-T Q.931, Q.2931) is needed and/or can evolve, or if each RBOC and the chosen server-setup box partner will develop a separate signaling protocol. At least in the short term, the signaling may be specific to each carrier, but at least the signaling specifications are being published.

This signaling discussion proceeds on two fronts. In Sec. 5.3.2, we provide a general modeling overview of signaling at the functional level, in a platform-independent manner. In Sec. 5.3.3, we summarize, for pedagogical and illustrative purposes, one published VDT signaling specification. Additional information on signaling, from a setup box point of view, is provided in Chap. 13.

5.3.2 Signaling and control approaches: a general model

As implied by the previous discussion, two key areas of VDT control involve VDT application control (specifically called VAS in the previous section) and video session and connection management, including signaling* and/or configuration† management. This section presents a generalized view of signaling and control that captures the necessary functionality, without being tied to a specific implementation (e.g., as described in Sec. 5.3.3).

5.3.2.1 Application control. Application control is the hierarchically highest interaction between entities involved in video and related VDT services. This type of control is directly related to VIP selection, service selection, real-time service customization, and application supervision (e.g., VCR-like controls) on the part of the customer. VIP selection interactions on the part of the VIU take place between the VDT network (also called the exchange access system in Chap. 1) and the VIU, across the VIU-NI; application selection interactions take place between the Level 2 gateway and the VIU. Application control includes functions, such as VIP selection, access control interactions‡ (e.g., VIU

---

* Here "signaling" refers to the real-time engagement of resources needed to support the service, e.g., the use of switched virtual connections.

† Here "configuration management" means the pre-real-time allocation of resources to support the service, e.g., the use of provisioned or permanent virtual connections.

‡ Traditionally, screening has been done at the headend; this would imply screening at the VIP location. In VDT, it is possible to do screening in the network (as an analogy, the network currently can support screening callers to block 900 access).

screening, transfer passwords, or interdiction), request and download menu or list files, menu navigation, application selection, upload list and service files (e.g., list of selections), VOD video programming selection, change of VOD selection*, cancellation of movie and/or other application, VCR-like control, and reservation or allocation in support of PPV events. Application control may be viewed as covering both interactions between VIUs and the VDT network, and between VIPs (server or Level 2 gateway) and VIUs. Functions related to VIP selections are typically implemented in the VDT network; functions related to application selection (e.g., supported by a Level 2 gateway) may be implemented in the VDT network or at the VIP site. Some add the capabilities to support on-line help and subscription account management, such as automating changes and access to customer billing records. These interactions take place between the VIU and a service control bureau.

This type of control is relatively new for common carriers, since, traditionally, carriers support only physical connectivity. Therefore, the signaling protocols that have been designed by carriers focus completely on managing bandwidth and connections. In addition to having to enlarge the functional scope of signaling, it is most critical that this signaling, at least in terms of what is visible to the user, be simple and intuitive. The ease of use and the way a consumer is able to interact with the Level 2 gateway, through a consistent and perhaps already familiar look and feel (e.g., a cash machine interface, a TV remote control interface, a VCR or CD player-like interface, or a PC-like windowing interface such as Macintosh or Windows) will go a long way to foster the commercial success of VDT†.

In terms of signaling protocols, the approaches listed in Table 5.2 are possible at the practical level. In addition to the ergonomics issues just listed, some important selection criteria include cost of development of the protocols‡, cost of development of hardware and software supporting the protocols, protocol complexity, time to market, protocol performance, expandibility, and flexibility. It is also conceivable that a carrier would use some proprietary or closed elements in the control architecture, as well as some open or published elements.

---

* Some also include the change of a broadcast channel (i.e., channel surfing). Major performance considerations have to be taken into account if this type of signaling is also supported.

† At least, if the protocol is not "simple," it must allow the development of an application programming Interface (API)-like capability to insulate the customer from the underlying complexity that may be introduced by protocol designers.

‡ By one estimate, 1000 staff years have been spent in developing ATM signaling standards (i.e., an average of 100 people full-time, worldwide, over a period of 10 years). This equates to $250,000,000 when using a loaded salary of $250,000 per person.

TABLE 5.2 Possible Approaches to VDT Application Control

Proprietary-VDT-specific	Proprietary or closed protocols developed by a vendor or alliance of vendors. These vendors typically manufacture equipment and/or software in support of VDT.
Open but VDT-specific	New protocols and/or extension of existing protocols, published by carriers so that the suppliers manufacturing equipment and/or software for VIPs can connect to the VDT network. In addition to carriers, vendors of VIP equipment can specify protocols. Alternatively, third parties could develop and publish specifications.
Open, industrywide	New industrywide protocols specifically developed from the bottom up for VDT and cable TV VOD service. Alternatively, the capabilities, functionality, and features of existing open software platforms (e.g., client-server remote procedure calls, OSI remote operations service element, Open Software Foundation-distributed computing environment) can be used, since the specifications and supporting software modules are already in the public domain.

Finally, it is critical, in our opinion, that any protocol that is developed follow the layered principles and structure of the excellent protocol guidelines embodied in the ISO's open systems interconnection reference model. This model has been specifically developed to simplify protocol development by simplifying, streamlining, and modularizing the required protocol functionality to make things simpler, not harder, as some protocol "designers" totally unfamiliar with the open systems interconnection reference model try to make things appear.

5.3.2.2 Video session management. VDT networks will very likely use connection-oriented communication services. Connection-oriented services require signaling, unless they are permanently invoked. These signaling capabilities are related to the establishment, alteration, suspension and resumption, and disestablishment (release) of a video session. Alterations could include addition of parties (particularly in the context of videotelephony and MPG) and changes of some connection attributes (perhaps important for future services more so than for traditional video services). Some changes in capabilities, implementation, or perspective will be needed with regard to traditional switched services: usually, the sessions in the PSTN last a few minutes; an often quoted study purports that the average length of a voice call is just a few minutes. Video services and MPG may require long sessions. Sessions may last from 1½ hour per night to maybe 8 hours per night. (One is often reminded that by the time an average student in the United States leaves high school he or she will have watched 25,000 hours of television; that is a lot of long sessions.)

There are many well-developed signaling protocols for a variety of traditional communications services (e.g., Q.931, Q.933, Q.2931, CCS7).[2] However, additional capabilities are likely to be required* to support the video session (which equates to what is traditionally referred to as a *call* in telecommunication parlance) and the connection management needed in VDT. Furthermore, new capabilities may be needed to support some of the application control functions listed above†.

From the VIU perspective, at least for traditional video services, the connections are point-to-point, VIU-to-VIP; some future services such as MPG and videoconferencing will involve point-to-multipoint and multipoint-to-multipoint connections. From a VIP perspective, some connections may be point-to-multipoint, e.g., multicasting of a special event over PPV service. Some work has been done, to date, to support multipoint service (e.g., ATM), but much work remains to be done at this juncture to make the service flexible enough to support real services. Also, some protocols support, at least in theory, "call setup" negotiation of parameters (e.g., ITU-T Q.2931); however, vendor products and/or actual carrier networks may not implement these features. Although there has been some discussion about in-session change of connection parameters—e.g., bandwidth, *quality of service* (QOS)—not much concrete support has emerged. The ATM Forum version of ATM signaling supports in-session adding and dropping of "legs" to the unidirectional point-to-multipoint service that has been defined. However, as noted, from a VIU perspective, this is not a critical feature in terms of traditional video services. There is, however, interest in the capability to put a session on hold and resume it later. For example, in VOD, one may watch part of a movie, and then take a few minutes' break to answer an incoming phone call, answer the door bell, or have a snack. This type of capability requires not only a connectivity-level signaling indication and possibly a temporary relinquishing of network resources, but one also needs to interact with the video server to pause the movie at a determined point.

In addition to adding in-session signaling, two features of existing telecommunication signaling protocols (such as ITU-T Q.2931 and Q.931) that need to be expanded to support (some) VDT features are:

---

* Even if a protocol like ATM's Q.2931 had all the needed features, it may not be widely implemented within the time frame of initial interest for the deployment of VDT networks. For example, an extensive analysis of ATM products and switches undertaken in 1994 indicated that signaling in an ATM context will not be available in a critical mass of products until mid-1996.

† Although application control lies at a higher hierarchical level, it may rely on the signaling capabilities of the lower level, much the way a service at layer *n* of the Open Systems Interconnection Reference Model relies on the services of layer *n*-1.

1. Eliminate restrictions on allowing only real-time, call-associated signaling: The action desired can only be requested at the time it is needed and not at a prior time. A possible extension mechanism that is being investigated in some distance-learning networks that use ATM is a non-call-associated signaling approach; however, no standardized, feature-rich extensions exist yet.
2. Eliminate restrictions on allowing requester services only: Generally, but not always, the party asking for a service is the party receiving the service (this is also called *first-party*). For some VDT services and situations, there is the need for *third-party signaling,* where the entity requesting the service via a signaling interaction may not be the entity that actually needs the service or the connectivity. Some early work in this direction has been done under the auspices of ISDN signaling, particularly in primary rate access, where a channel (on a T1 access facility) can be used to set up connections for other T1 access facilities. Issues of privileges and security need to be taken into account in designing third-party signaling.

There has been interest expressed by the RBOCs in using ATM for VDT. As noted earlier, ATM can be carried end-to-end, even over a network that has a fiber coaxial cable component, by having the node in the FTTN architecture perform the appropriate conversion. The node also would need to support the signaling aspects of ATM (i.e., Q.2931) if that is the approach chosen, at least for network access signaling. (Internetwork element signaling could use the B-ISUP signaling, or it also could use Q.2931; video application signaling is in addition to network signaling.) The reader is referred to Ref. 2 for a complete treatment of the features of Q.2931, B-ISUP, Q.921, ISUP, and Q.933; a detailed analysis of the features will enable the developer to assess the applicability of these available standards. A short synopsis of Q.2931 and B-ISUP from Ref. 2 follows.

Q.2931. Q.2931 has been developed by ITU-T with input from all the world's major carriers and telecommunication equipment vendors. It is an application* layer protocol in the open systems interconnection reference model language; it relies on a data link layer protocol (AAL 5 over ATM; see Chap. 8) and on an appropriate physical layer protocol (e.g., SONET). In the United States, work has been funneled to the ITU-T through ANSI T1S1. A standard for the support of what is called *capability set 1* features appeared in 1994; extensions to support capability set 2 were underway at press time. Q.2931 is an extension of Q.931, which is the ISDN access signaling protocol. Capability set 1

* Do not confuse this use of the word *application* with its use in the rest of the chapter.

supports, among other things, first-party connection control for point-to-point ATM connections over a public ATM network. The signaling protocol is independent of the actual access channel, whether it is DS3, STS-3c, or STS-12c, except that the appropriate physical layer protocol must be used. Capability set 1 does not support multipoint connections of any kind. It does not support third-party signaling. Some negotiation capabilities are available but may not be fully developed or implemented; hence, for all intents and purposes, bandwidth and quality of service* negotiation is not supported†.

Another area where there has been interest is in the separation of call-from-connection control. In traditional telecommunications, each call has a connection, and each connection corresponds to a call. Advanced new services, particularly multipoint and/or multimedia, may involve a connection with multiple calls. Capability set 1 does not support this separation. Some of the features described here as not being available in capability set 1 will be available in capability set 2. Protocols for capability set 2 are expected for 1995, but implementation and deployment will take additional time.

---

* As discussed in Chap. 8, which covers ATM service at length, ATM supports a number of different quality of service classes (five in some specifications; two in others); an "unspecified or best-effort class as well as an available bit rate class are also emerging. Different classes are used for different applications. When taking the traditional approach of delivering video over ATM using circuit emulation with AAL 1, the highest quality of service class is needed. If one chose the approach of delivering video over ATM using cell-relay service and AAL 5, then a class other than the highest could be used. The "unspecified or best effort" class cannot be used for real-time video services but could perhaps be used for store-and-forward video.

† A "manual" way of accomplishing negotiation has been suggested as follows: the calling party sends a connection request (a SETUP message) at, say, a given bandwidth; if the called party is not able to accept the call, it could return a RELEASE COMPLETE message, thereby informing the calling party about the inability to support the requested bandwidth. The calling party could then send in another request asking for less bandwidth and continue the process iteratively until a mutually agreeable bandwidth is found. This approach is slow and inefficient. Bandwidth negotiation may not make much sense anyway in a VDT network, at least for the VIU-NI interaction. In this scenario, the user would request the VIP to see a movie using MPEG-2 at 6 Mbps. The VIP may find that it has run out of bandwidth in its connection to the VDT network, say, because the PPV program was popular, or it was Mother's Day, or because of bad weather a lot of people were using the system. Upon being informed of the lack of bandwidth, the user could then request to see the movie using MPEG-1 at 1.5 Mbps. This is somewhat impractical for two reasons: (1) The setup box may not be sophisticated enough to deal with dynamic bandwidth situations to keep its cost down, both in terms of the communications as well as in terms of the multiple decoding boards; (2) The bandwidth granularity is too small (i.e., two points only, 6 Mbps and 1.5 Mbps), to make the system work effectively in a VDT VIU-NI environment. Note, for example, that in an LAN interconnection situation, the LAN-to-LAN link would operate well at any (large) number of bandwidth choices, making negotiation worth the effort. It is possible that bandwidth negotiation could be of some value over the VIP-NI.

One consideration about potential applicability is the complexity of the protocol and the ensuing cost of developing the software to run the protocol, as well as the cost of the fast processor needed to run the procedures and manage the state machine of the protocol. There has been some skepticism about the use of Q.2931 for VDT applications. Some feel that it is overkill and does not meet the market imperatives of expedited time to market, inexpensive setup boxes and video server signaling peers, and will lack real network support for some time to come.

The ATM Forum Specifications. The ATM Forum, in an effort to accelerate deployment of ATM products, has published a protocol for NAS (User-Network Interface Specification) as well as a protocol for NS (Broadband InterCarrier Interface Specification), which are based on and are extensions of international and/or national standards. The NAS protocol supports a basic multipoint capability (a unidirectional point-to-multipoint service); however, this service has a number of drawbacks, for example, its inability to support leaf-initiated joins and drops to and from a preexisting connection. In simple terms, these drawbacks preclude a user from obtaining I-TV services in a direct and effective manner. The unidirectional point-to-multipoint service could be used, perhaps, to support PPV, if the customers could communicate, in an out-of-band fashion, their requests to the VIP. The specification also fails to support third-party signaling and QOS or bandwidth negotiation. Some of these features may be available in the 1995 reissue of the specifications. However, it is not clear if The ATM Forum specifications will be optimally adequate for VDT, since the emphasis has been, to a large extent, on the movement of computer data and the "best effort" service class.

B-ISUP. This ITU-T-developed ATM signaling standard is used to support network element-to-network element signaling (what has previously been called NS). It is an extension of the ISDN User Part (ISUP) and supports CCS7 signaling. Depending on the VDT implementation, CCS7 may or may not be used for video distribution. Common channel signaling has been around since the 1970s, and only in the recent past has it seen more widespread introduction. CCS7 is ideal when there is some sort of centralized database across an RBOC, which many users (calls) have to get access to; the best example of this database is the 800 service database. B-ISUP is a 1994 standard. It does not support multipoint, third-party control, bandwidth negotiation, or QOS negotiation (in fact, at press time, it did not even support the transport of the QOS parameter end to end). There has been general skepticism on the part of many members of The ATM Forum about the ease of incorporating B-ISUP capabilities in user devices such as $15,000 routers or $30,000 private ATM switches, much less in lower-end VDT equipment.

5.3.2.3 Video connection management. The meaning associated with this term in the present discussion refers to establishing connections in support of VDT services, without the use of automated signaling capabilities. Naturally, it is preferable to employ automatic signaling methods. However, because of time to market, technology maturity, product availability, and cost of deployment considerations, a carrier may opt to use this approach, at least initially. Historically, this has been the course of action: first *permanent virtual connections* (PVC) packet-switching services, then, later, *switched virtual connections* (SVC) packet-switching services; first PVC frame-relay services, then SVC frame-relay services; first PVC cell-relay services, then SVC cell-relay services.

Connection management, accomplished through existing carrier operations and/or management systems, supports functions such as non-real-time establishment, non-real-time alteration, and non-real-time disestablishment of connections. Therefore, at the macro level, the capabilities supported by connection management are similar to those supported by signaling, except that the speed of action is usually measured in hours or days (occasionally in minutes). This approach has been used for traditional telecommunication services where topologies change very infrequently (e.g., a private line environment). Note that it supports the third-party capability.

One could say that, in effect, video users now have a private line to the headend, and that, therefore, switching (real-time or otherwise) is not needed, since the customer does not want to reach two headends. However, these days, switching is used both to reach a geographically different location or entity (e.g., when an individual dials the number of a colleague in another part of town), as well as to reach a different virtual entity at the same geographic location (e.g., two processes running in a mainframe). First, in VDT, the customer does want to reach multiple VIPs so that rapid switching between these is needed*; second, even when connected to a single VIP, the customer wants to switch between different programs offered by the VIP†. Once the connection to the VIP is made in a permanent virtual connection mode, the customer can obtain real-time or scheduled video programming; since the communication channel is permanently assigned, NAS functions disappear, and the customer uses VAS to obtain all the video control capabilities listed previously in the appropriate subsection.

---

* Alternatively, the user can establish two (or n) virtual paths to the two (or n) VIPs and then "switch" the information path without further signaling, by putting the appropriate virtual label in the cell or frame header.

† However, some of this intra-VIP switching may be supported by the application control machinery previously discussed.

Permanent virtual connections are not the best way to support a multiple-VIP VDT. A single point-to-point connection would be simple to terminate in the setup box, but to support multiple VIPs, the setup box may have to terminate multiple connections, making it more complex. Also, some network resources are allocated to the user, whether the user is actually using the system or not. Typically, this requires tables in the nodes supporting the connectivity (e.g., switch or DCS). Because of the large number of potential users, these tables can become very large.

5.3.2.4 Generalized control model. Since there may be many VDT implementations emerging, a *generalized control model* (GCM) that defines control functions, aggregates these functions into groups that have logical affinity, and describes the relationships and interactions between these entities, would be useful for platform-independent comparisons. The GCM can assist protocol developers to identify important interfaces and the type of functional interactions that need to be supported.

Figure 1.12 depicted a generalized view of the VDT network; the discussion in Sec. 5.2 refined the view by introducing the access segment and the backbone segment of the network. Given this amplification, Fig. 5.11 depicts key signaling interfaces of interest in VDT. Figure 5.12, built on Fig. 5.11, shows the control model that we use. The model incorporates a number of key *logical control modules* (LCMs) needed to support VDT control. (This figure is not exhaustive and only depicts key functionality.) LCMs* are groups of control functions that have logical affinity (see Table 5.3). These functions include both VAS and NAS functions. Not all LCMs are present in every VDT implementation; for example, in an environment where NAS is supported, the connection control LCM would not be present because the required connectivity is supported by the bandwidth and connectivity control LCM, which is an abstraction for a switch that has signaling capabilities.

Figure 5.12 also shows key interactions of interest (note that with 8 blocks, 56 different interactions are possible; the model identifies only the subset of these interactions that makes sense); in a specific implementation, not all these interactions need to be supported. Some of this subsetting may be due to supported functionality (e.g., I-TV versus NVOD); other subsetting may be related to the technological implementation.

---

* No restrictions are placed here on the physical implementation of LCM functions: these could be implemented in a distributed, localized, or centralized fashion. The implementation is system-dependent, based on the complexity of the module, its performance in terms of response time and population support, the required availability, management of the equipment implementing the LCM, etc.

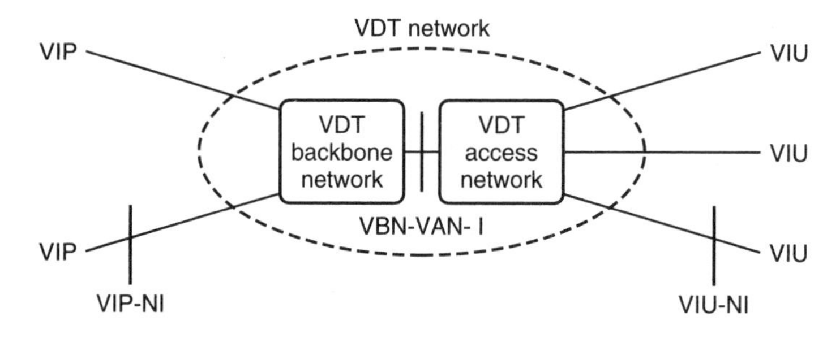

Figure 5.11 Amplified view of VDT network, with focus on important interfaces.

Given this reference platform, a general modeling effort proceeds from having identified LCMs (and real entities that embody or use these functions) to identifying which of the total pairwise interactions are pertinent, to determining what type of interface is required to support the interaction (interactions shown in Fig. 5.12 need to be supported across an appropriate interface). Finally, one needs to develop session models to adjoin the GCM to specific VDT architectures; these session models depict the temporal "syntax" and sequence of control events.

Starting with pairwise interactions, interaction i(1,8) can, for instance, support, among other activities, the request for a menu to select a VIP. If one thinks of these LCMs as being embodied in some kind of specific equipment, then the interface can be assumed to be a typical communication interface*, i.e., to have a physical-level characterization, a data link layer characterization, a network layer characterization, etc. For example, at the physical layer, one needs to determine if the signals will be carried over SONET, a T1 or DS1 line, a channel in the ADSL stream, etc. At the data link layer, besides other considerations, one needs to establish if the connection supporting the interaction is permanent or on-demand. If the connection is on-demand, then one needs to describe what events cause establishment (e.g., a human signal to the setup box for service), disestablishment (i.e., signal from the VIP that a movie is finished), suspension, resumption, etc., and which protocol will be used to accomplish this. For both types of connections, one needs to know the traffic profile and the expected performance. Table 5.4 depicts some of these characteristics for a generic VDT system.

* Although the functions carried across the interface need not be symmetric, i.e., i(1,8) $\neq$ i(8,1), the communication characteristics will usually be the same.

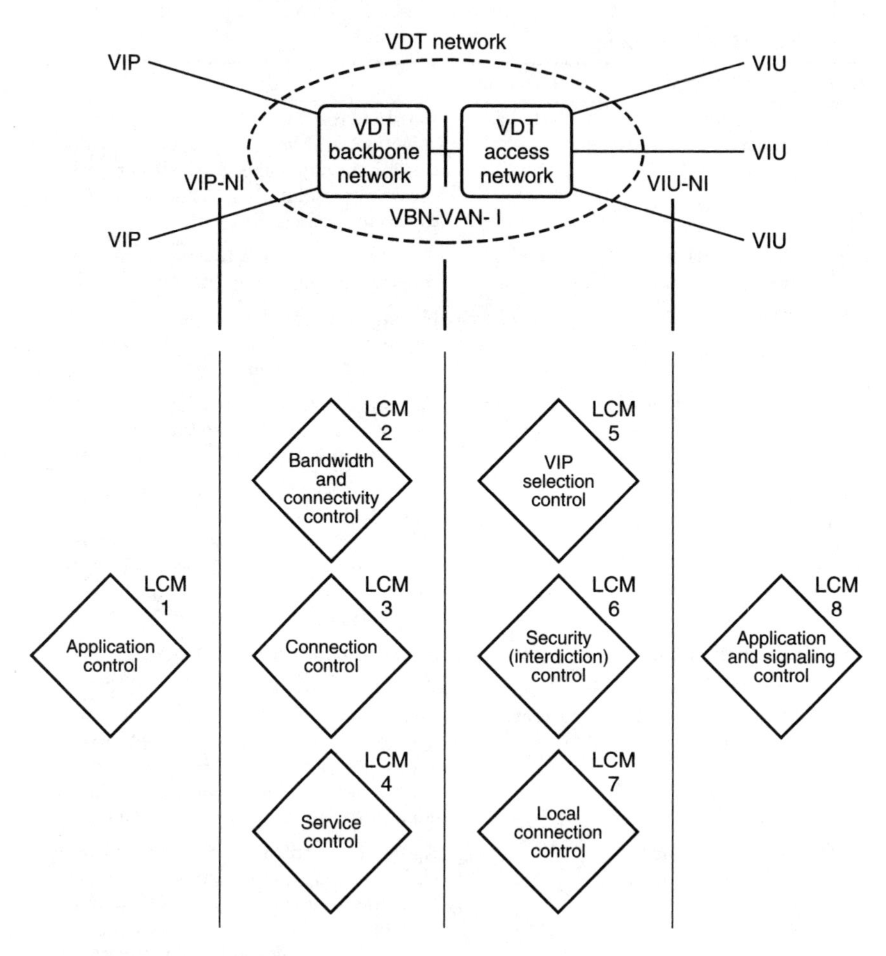

Key interactions of interest:

i(2,8), i(8,2) i(2,4), i(4,2)
i(1,8), i(8,1) i(1,2), i(2,1)
i(8,4), i(4,8) i(1,4), i(4,1)
i(8,5), i(5,8) i(7,3), i(3,7)
i(5,2), i(2,5) i(7,8), i(8,7)
i(6,2), i(2,6) i(8,3), i(3,8)
i(2,3), i(3,2)

LCM: Logical control module

Figure 5.12 Logical control modules of the GCM and key interactions.

Figures 5.13 and 5.14 depict two realizations of this GCM, as well as the session model, alluded to earlier. Figure 5.13 shows the interactions required to support broadcast video; note in particular that there is no video application control capability. Figure 5.14 shows a VOD example.

TABLE 5.3 LCM Functionality

LCM	Name	Location	Function
1	Application control	VIP	Level 2 gateway, including navigation, selection, and VCR-like functions; video session management; authentication, screening, and authorization; event notification and scheduling.
2	Bandwidth and connectivity control	Backbone	Equal access to VIPs through switching, routing, and signaling interpretation.
3	Connection control	Backbone	Nonsignaled control of sessions, including connection establishment, disestablishment, and reconfiguration.
4	Service control	Backbone	Variety of service-related functions such as service initiation, billing authorization, service reconfiguration, session management.
5	VIP selection control	Access	Level 1 gateway, including VIP listing and selection, establishment of signaled video connections, nonsignaled session management (first- and third-party), video channel change, basic screening, event notification and scheduling*.
6	Security (interdiction) control	Access	Network-based security, blocking, routing
7	Local connection control	Access	Support of nonsignaled connection management for cases where the supportive capability (e.g., LCM 3) is not implemented locally, but at some hub location. (This capability funnels the requests to LCM 3.)
8	Application and signaling control	VIU	Application control (VAS) at the initiator end, including navigation, selection, and VCR-like functions, event display, and scheduling. (This functionality is at a peer level with most of the LCM 1 functionality.) Signaling control (NAS); session management for nonsignaled connections. Also includes local channel switching.

NOTE: Many functions are technology-, implementation-, and service scope-dependent.
* This functionality is representative of the functionality of the video selector discussed earlier.

The broadcast video example in Figure 5.13 postulates that video selection functions are implemented in the VDT network*. As discussed in Sec. 5.2, the placement of the video selector at the "edge" of the VDT network is done to keep signaling from having to travel "deep"

* In this example, the *control* of the video selection function is done in the VIP selection control LCM, here materialized in the video selector. The actual *video path* is accomplished through the bandwidth and connectivity control LCM.

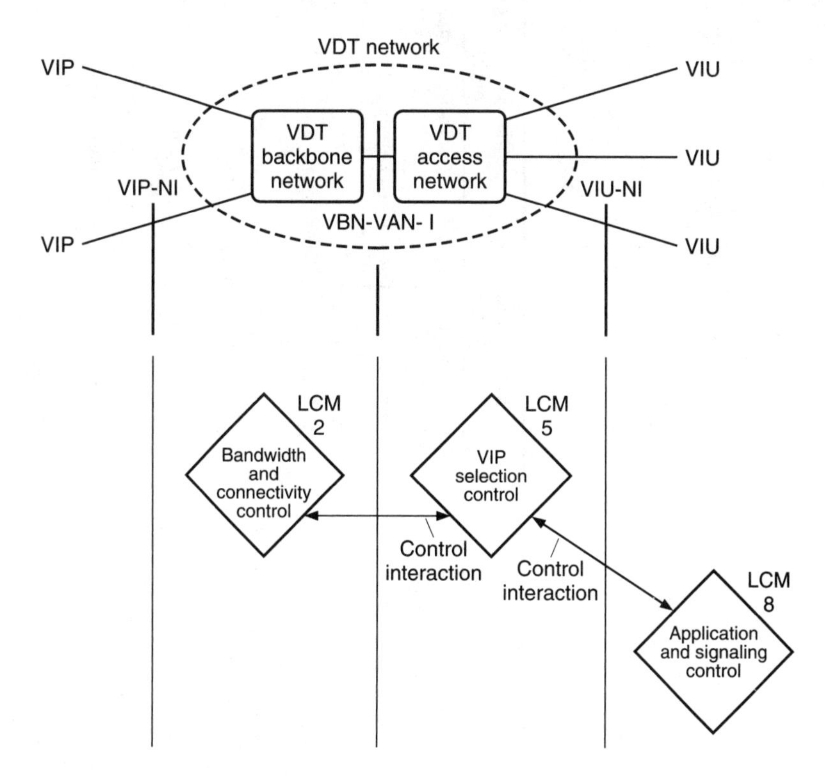

Sequence: #1 i(8,5)
#2 i(5,2)

VIU initiates the interaction i(8,5)/i(5,8) to request to be connected to a VDT system that supports transmission of digitized broadcast video. The VIP always has its feeds terminated on the RBOC's switch. The VIU just needs to be connected to the "right" signal present at the switch (which may be at some hub location, not necessarily at the serving CO). Once the LCM 5 determines the type of service requested, it communicates the need over interaction i(5,2)/i(2,5) to effect the connection. (Note: The video path is not shown in this figure.)

Figure 5.13 Example of particularization of the GCM to support signaled connection to a VIP-supporting broadcast video.

into the network, thereby impacting performance, particularly in terms of channel-switching time while channel surfing. The video selector terminates multiple *video* connections to appropriate VIPs. Typically, there is a permanent *control* connection between the VIU and the video selector, i.e., i(8,5), used to transmit requests for video channel switching. A separate permanent *control* connection, i(2,5), is maintained between the video selector and the VDT network (more specifically the bandwidth and connectivity control LCM; physically,

TABLE 5.4 Characteristics of GCM Interactions

Interaction	Source	Sink	Usage	Establishment events	Disestablishment events
i(1,4)	VIP (application control)	Service control bureau	—Service alterations —VIP access privileges alterations	VIP requests access to service control bureau	Interaction completion
i(2,1)	Bandwidth and connectivity control	VIP (application control)	—VIP-VIU session negotiation —Access to service control bureau	Usually permanent or VIP requests access to VIU or service control bureau	Permanent or interaction completion
i(2,3)	Bandwidth and connectivity control	Connection control	—Establishment-reconfiguration of VIU-VIP session and/or connections —Establishment-reconfiguration of VIU service control bureau session and/or connections	Permanent	Permanent
i(2,4)	Bandwidth and connectivity control	Service bureau	—VIU service control bureau session authorization —Notifications of service changes	Permanent	Permanent
i(2,5)	Bandwidth and connectivity control	VIP selection control	—Channel change-VDT network level	Permanent	Permanent
i(2,6)	Bandwidth and connectivity control	Security control	—Change of security status	Permanent	Permanent
i(7,3)	Local connection control	Connection control	—Establishment-reconfiguration of connections in access subnetwork	Permanent	Permanent
i(8,1)	VIU (application and	VIP (application	—Application control	VIU requests VIP to supply	Interaction completion

	signaling control)	connectivity control)	—Access to service control bureau	to VIP applications list	—VIU select service bureau
i(8,3)	VIU (application and signaling control)	Connection control	—Status	VIU requests connection	Interaction completion
i(8,4)	VIU (application and signaling control)	Service control bureau	—Service negotiation	VIU requests access to service bureau	Interaction completion
i(8,5)	VIU (application and signaling control)	VIP selection control	—Channel change for broadcast video	Permanent	Permanent
i(8,7)	VIU (application and signaling control)	Local connection control	—Status	VIU requests connection	Interaction completion

NOTE: Not all these flows are implemented in all VDT networks.

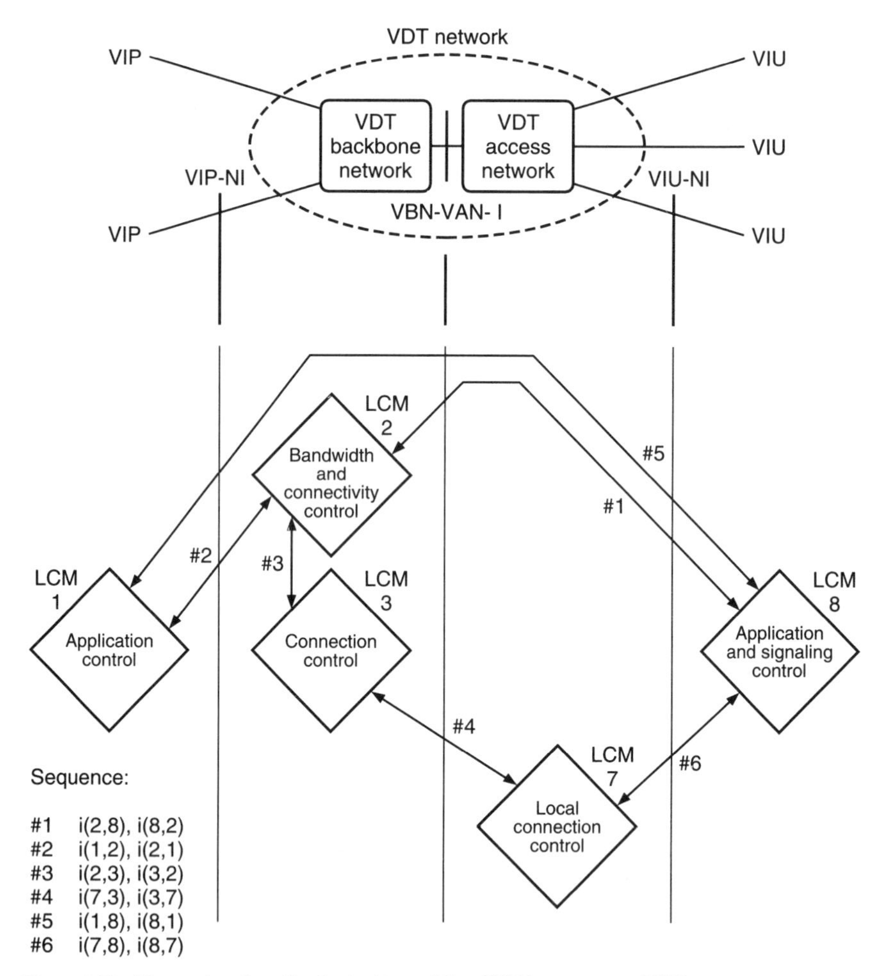

Figure 5.14 Example of particularization of the GCM to support VOD.

this is an appropriate switch). In this arrangement, it is likely that the video selector would keep tailored lists of video channels provided by each VIP. (This list is obtained by the selector over the VDT network.)

The sequence of control events in the VOD example of Fig. 5.14 is as follows. The first step entails establishing a *control* session between the VIU and the bandwidth and connectivity control LCM: i(2,8). Over the course of the interaction, the VIU may select a VIP in the menu provided by the bandwidth and connectivity control LCM (screening functions may also be performed at this juncture). A session between the bandwidth and connectivity control LCM and the VIP is established or may be already available: i(2,1). This session is employed to support aspects of the screening that need to be performed strictly by the VIP. At this juncture, a connection management session may be initiated between the bandwidth and connectivity control LCM and other connection management entities in the network, such as the connection control LCM and the local connection control LCM i(2,3) and i(7,3), to affect the VIU-VIP connection i(1,8). Initially, this VIU-VIP connection is a *control* connection, which is used by the VIU to select a movie or application. (This control connection can also be used to transmit VCR-like commands during the duration of the movie.) Once the VIU has made the selection of VIP-offered movies, the VIP can use the control connection that exists between it and the bandwidth and connectivity control LCM to request appropriate network resources, specifically a VIU-VIP *video* path. This triggers the bandwidth and connectivity control LCM to interact with the connection control entities [over the previously established connection with i(2,3)] to complete the required VIU-VIP video connection, over the interaction i(2,3).

5.3.3 Example of published VDT signaling specification

5.3.3.1 Overview. This section is a short synopsis of a published VDT signaling specification. The description is based on Bell Atlantic's, *Signaling Specification for Video Dial Tone,* TR-72540, Issue 1, Release 1, August 1993*. The summary is included here for pedagogical reasons to illustrate the concepts described in this chapter. Developers and other parties working on products of commercial value should refer directly to the original documentation, especially since only some key highlights are included. This discussion is not meant to imply that this approach is the best, the only one, or the recommended one. The discussion simply illustrates what such a protocol must encompass, how the various VDT elements (VIUs, Level 1 gateways, Level 2 gateways, VIPs, etc.) communicate, and so on.

Readers who do not require additional insight into signaling beyond the discussion provided in Sec. 5.3.2 may skip this self-contained section on first reading without a loss of continuity.

The discussion begins with a detailed technical description of a VDT session. Figure 5.15 illustrates the major network elements involved in providing such a session. Figures 5.16 to 5.18 provide high-level state diagrams illustrating an interactive VDT single-cast session, as well as more detailed state diagrams of the L1 gateway and L2 gateway functions, respectively. In the diagrams and the discussion that follow, the setup box is referred to as *customer premises equipment* (CPE). A VDT

* Material is not copyrighted.

session is defined in three subsessions, depending on the functions with which the VIU is interacting:

Subsession 1. L1 gateway: Provides signaling access to VIUs and VIPs and supervises video network control. VIU-controller functions include VIU authorization, menuing or navigation, and service selec-

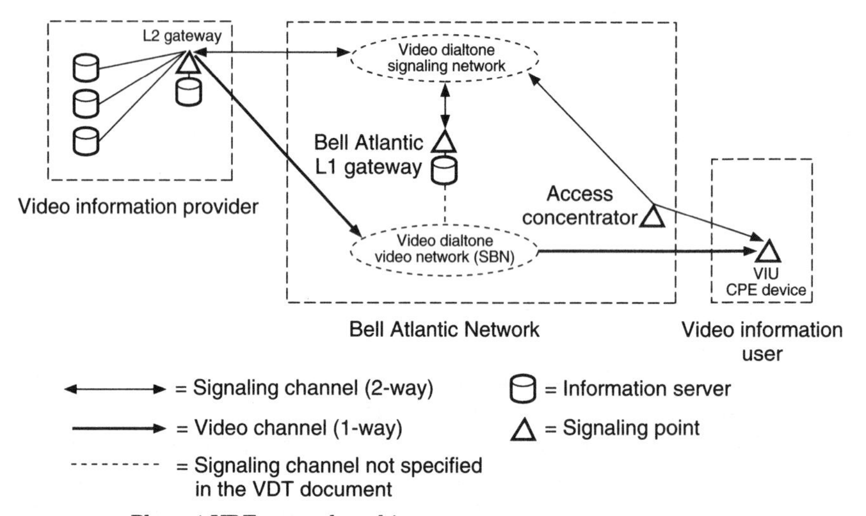

Figure 5.15 Phase 1 VDT network architecture.

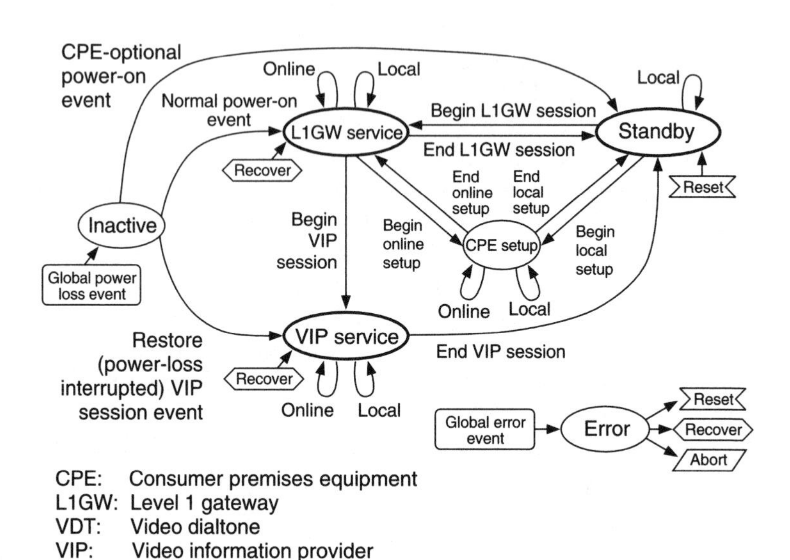

Figure 5.16 Setup box (CPE)—level state diagram—VDT session.

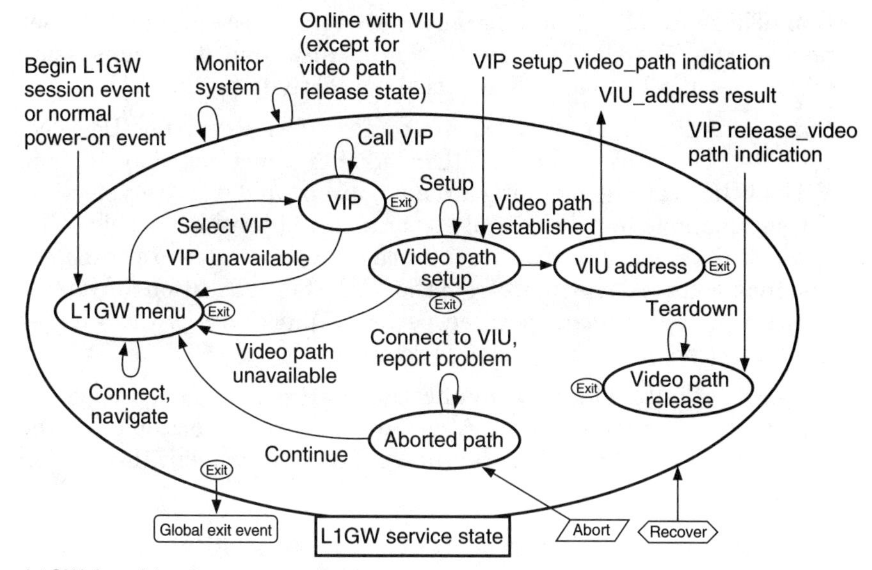

Figure 5.17 L1 gateway state diagram.

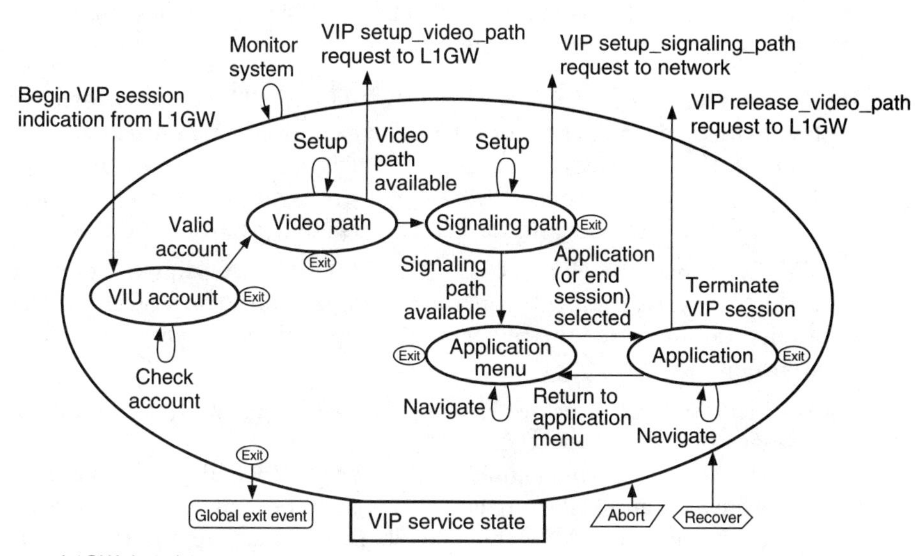

Figure 5.18 VIP—L2 gateway state diagram.

tion. These functions may be automated via a programmable user device. The VIU can exit the session at any time. This subsession will time-out after a specified period of inactivity by the VIU.

Subsession 2. L2 gateway (menus): Provides access to VIP selections (e.g., archived video). VIU-controller functions may include VIU authorization, menuing or navigation, and service selection. These functions may be manual or automated-preselected. The VIU can exit the session at any time. Since all menu states are transient and most expect some input from the VIU, this subsession may experience time-out after a specified period of inactivity by the VIU, at the discretion of the L2 gateway provider.

Subsession 3. L2 gateway (application): Provides the VIUs selection over the video channel. Application-specific functions may be available (e.g., video control features). The VIU can exit the session at any time.

The discussion then focuses on: (1) VDT network signaling (network signaling is switch-vendor-specific and, therefore, is not discussed in the specification); (2) VDT network access signaling; and (3) application signaling. The emphasis of the discussion is on VDT network access signaling.

This protocol specification for VDT network access signaling is broken down into the following three levels (see Fig. 5.19), following to the extent possible the open systems interconnection reference model:

1. *VDT message transfer part* (VMTP): The VMTP is the lowermost level and performs transport of VDT messages. It includes specification of the physical signaling network platform (network and access), as well as the network-level protocol(s). For Stage 1 Bell Atlantic VDT, this is standard X.25 plus some additional functions.
2. *VMTP translator (VMTP-T):* The VMTP-T translates between the VMTP (i.e., X.25 call-control messaging) and the VSCCP of the application level for the purpose of setup and teardown of signaling channels.
3. *Application level interface* (ALI): The ALI consists of three major subparts as follows:
   a. VDT signaling connection control part (VSCCP): The VSCCP performs signaling connection (i.e., switched virtual circuit) control. It uses the VMTP (via the VMTP-T) to set up and tear down *switched virtual circuits* (SVCs) as needed by the *VDT session control part* (VSCP).
   b. VDT session control part: The VSCP manages an interactive VDT session and includes the definition of a VDT message set.

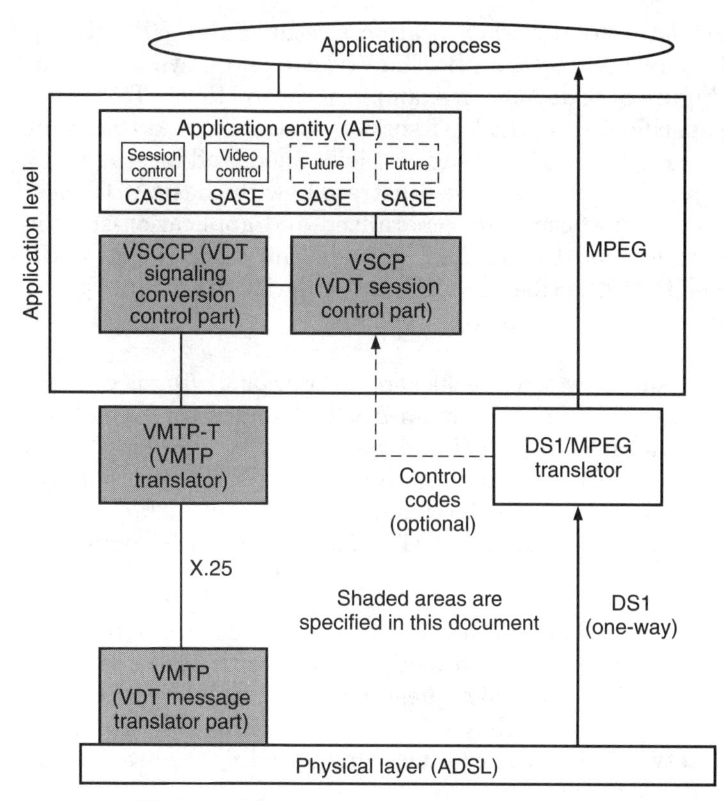

Figure 5.19 VDT signaling protocol—a structured approach.

The VDT message set consists of open systems interconnection reference model constructs such as *common application service elements* (CASE) and *specific application service elements* (SASE)*. For Phase 1 VDT, SASE will include a definition of video control features (SASE-VCF).

c. *Application entity* (AE): The AE provides the interface between the VSCP and the application process of the CPE device.

The VDT message transfer part is defined to be X.25; the service is realized over a packet-switched network. The next part of the specification deals with the VMTP-T function. This is followed by a description of the VDT signaling connection control part. As an overview, Fig. 5.20 illustrates VSCCP capabilities. The VIU-CPE initiates setup of a signaling circuit to the Level 1 gateway to begin a session. (The VIU

* New applications (SASE) will be added over time to the VDT platform.

cannot initiate setup to any other location.) Setup of the VIP-VIU signaling circuit is initiated by the VIP. The VIU can tear down the circuit at any time. Figure 5.21 depicts an example of control flow*. This is followed by the specification of the VDT session control part. At this point of the specification, the VDT message set is defined, and message flow diagrams are provided. These flows illustrate how the basic VDT message set is used to perform both generalized and application-specific VDT session functions. Figure 5.22 provides an example of VSCP flows†. One needs to describe the application entity (AE). Finally, the application-level VAS needs to be defined also.

5.3.3.2 VDT system for which specification is provided. To specify the VDT signaling capabilities, a common detailed understanding of the design and operation of the VDT network must be developed. The remainder of this subsection describes the Stage 1 VDT network architecture. Figure 5.15 depicts the VDT network. The network elements that require direct access to the VDT signaling network operate as follows:

L1 gateways. The Bell Atlantic L1 gateways interact with VIUs and VIPs (via L2 gateways) to provide the VIU with access to available VIPs and to allow VIPs to request network capabilities (i.e., network access signaling). They also interact with the switched broadband network (i.e., perform network signaling) to provide the VIU with video access to VIPs.

VIU-CPE device. The VIU is the end user of the VDT service. The VIU-CPE device typically is a device that is located on or near the VIU's television. The VIU enters commands via a handheld remote unit. The far end (L1 or L2 gateway) provides responses that are displayed to the user via the television (locally generated responses are also possible), or that are stored, or that trigger some other type of action. The VIU interacts with the Bell Atlantic L1 gateway (via the VIU-CPE) to select an L2 gateway. The VIU interacts with an L2 gateway to select an application, and then via the L2 gateway for application control features. The CPE is expected to be capable of generating a confirmation of command entry for all VIU-initiated commands, if such notification is to be given. Confirmation is not provided via signaling. Commands involving multiple key

* Refer to the actual specification for detailed VSCCP flows for the L1 gateway, VIU-CPE, and L2 gateway.

† Refer to the actual specification for detailed VSCP flows for the L1 gateway, VIU-CPE, and L2 gateway.

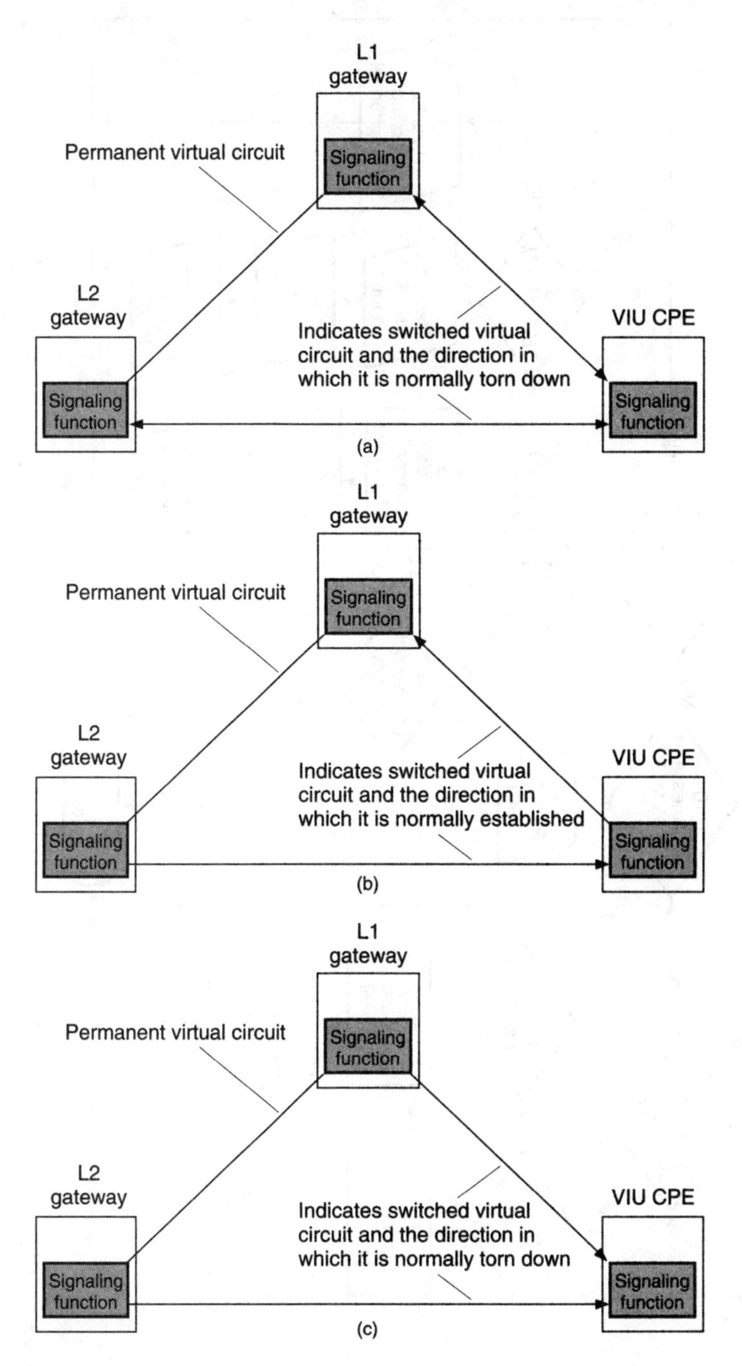

Figure 5.20 VSCCP capabilities: (*a*) VSCCP—network functional diagram; (*b*) Normal switched virtual circuit establishment; (*c*) Normal switched virtual circuit teardown.

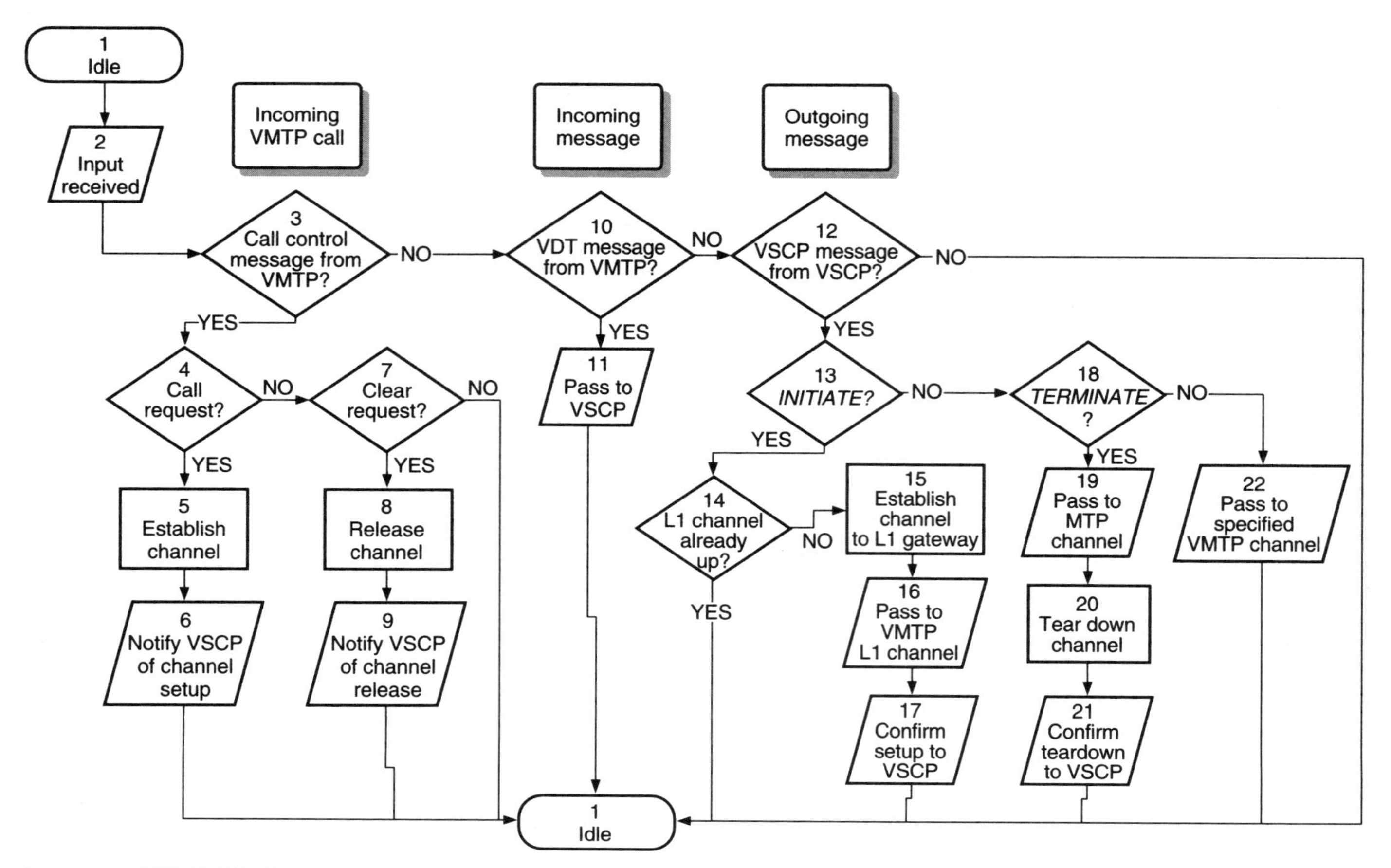

Figure 5.21 CPE VSCCP flow.

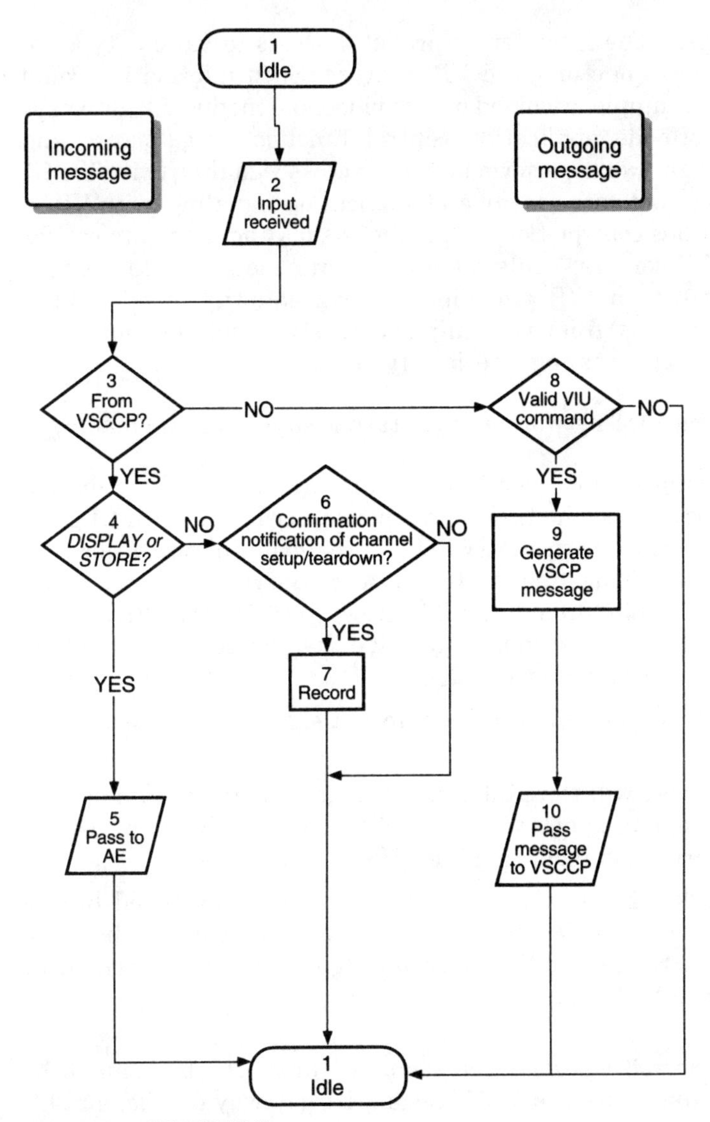

Figure 5.22 CPE VSCP flow.

presses (e.g., PIN) are not sent until all information has been entered by the VIU. Therefore, the CPE must buffer this information before handing it off to the signaling function. VIU messages may be generated based on a "hard" key (e.g., a pause button on the remote) or "soft" keys (e.g., multifunction keys assigned by the VIP). This may vary among CPE vendors.

L2 gateways. The L2 gateway provides access to various types of information via one or more VIP servers. Such information could include, for example, archived or live video and menus. L2 gateways may also provide application control functions (e.g., pause and restart). L2 gateways perform network access signaling (with the L1 gateways) for network control and application signaling (with VIUs) for applications control. Some L2 gateways may act as a gateway for multiple VIP servers, while others may provide access to a single VIP server (i.e., a VIP providing its own gateway functionality). These are identical from a signaling perspective, and, therefore, only one L2 gateway signaling interface type is specified.

The following VDT network architecture assumptions are made:

1. A VDT serving area is the area served by one L1 gateway. Initially, there will be at least one L1 gateway per server LATA. Each L1 gateway will interface with the switched broadband network (for network signaling) in its area. A VIU can receive service only from VIPs that are accessible through the L1 gateway serving that VIU (i.e., a VIP must interface to each L1 gateway, either directly or via another VIP, to serve all available VIUs).
2. There are two specified types of external signaling interfaces: VIP and VIU.
3. The L1 gateway will store information regarding the VIUs' network addressed (signaling and video) and a VIU profile. It will also store similar information regarding the VIPs.
4. The network signaling between the L1 and the switched broadband network is switch-vendor-specific and internal to the VDT network (for Stage 1 VDT). Network signaling is not specified in the Bell Atlantic VDT specification.

5.3.3.3 VDT interactive session operation. Figures 5.16, 5.17, and 5.18 provide state diagrams of a VDT session, L1 gateway details, and L2 gateway details, respectively. A high-level description of the VDT session depicted in these diagrams follows.

1. Typically, the VIU is automatically connected to the L1 gateway upon powering on the CPE device. Alternatively, some CPE devices may provide a local programming mode that may optionally be entered when power is turned on. The VIU is then prompted by one or more menus to ultimately select a VIP. Alternatively, a programmable CPE device may automatically specify VIU preferences and other information to the L1 gateway, resulting in a partial or total automation of this step.

2. L1 menuing is provided over the signaling channel.
3. All menu navigation or selection at the L1 gateway will involve the VIU making a selection from numbered menu items. The method for making these selections (e.g., number + ENTER, point and click) is independent of the signaling. However, the signaling message indicating the selection to the L1 gateway must contain the item number. Menu navigation or selection between the VIU and the L2 gateway is transparent to the network and may be provided using various methods.
4. When the VIU selects a VIP, the L1 gateway communicates this information to the VIP. The VIP will typically respond by requesting a video connection via the L1 gateway. Once established, the L1 gateway will acknowledge setup of the video channel, provide the VIU's signaling address to the VIP, and tear down the L1 gateway to VIU signaling channel. The VIP can then establish that the signaling channel and the session setup is complete. When the L1 gateway receives a request for video channel setup from a VIU, it first verifies that the VIU is requesting service from that VIP.
5. The application programming providing the VIP and VIU interfaces may vary among vendors; however, a standard command set is required of all VIP-VIU providers for network access signaling. Also, a standard command set is required for application signaling involving a Bell Atlantic video service L2 gateway. This application signaling is also proposed for use across all VIPs serving the Bell Atlantic region as a minimum standard command set.
6. Once a VIU-VIP signaling connection has been established, the VIU can return to the L1 gateway menus by terminating the current session, either by exiting via L2 menus or functions or by powering off the CPE device, and then beginning a new session. When terminating the session by powering off, the CPE device should first send an *end of session* indication. The VDT network provides no confirmation request before clearing the signaling channel. If a VIU confirmation is to be requested (e.g., *are you sure?*), this must be implemented within the CPE device.
7. If one or more VIU-VIP sessions are interrupted because of video network trouble, the L1 gateway will notify the affected VIU(s) by either setting up a signaling channel to each VIU or by using the multicast capability.
8. If an *end of session* indication is not received prior to a VIU-initiated teardown of the signaling channel, and the X.25 cause value is one of those that are recognized, the gateway should attempt to reestablish the connection. Reconnect attempts should be made no more often than every 15 seconds.

9. The L1 gateway and VIU network addresses are not known to the VIU-CPE device. All virtual channel setup requests are routed by the access transport system to the appropriate L1 gateway. Virtual channels to L2 gateways are initiated by the VIP. Therefore, no network initialization of the IV CPE device is required.
10. *Personal identification numbers* (PINs) may be used by the gateways to control access or to allow a VIU to control access.
11. There are three distinct subsessions within an interactive VDT session, already highlighted earlier. They are:

    Subsession 1. User interacts with L1 gateway to ultimately select a VIP and enter the Subsession 2 state.

    Subsession 2. User interacts with L2 gateway with features such as preview and search, and ultimately selects an application and enters the Subsession 3 state. Subsessions 1 and 2 may be automated or presubscribed using capabilities in the CPE device.

    Subsession 3. User receives application and may activate control features (e.g., pause), terminate Subsession 3, and return to Subsession 2 to choose another application or exit altogether (i.e., terminate the session).
12. The L1 gateway will be equipped with a time-out feature. After a specific period of inactivity (during subsession 1), the session is automatically terminated by the gateway. The gateway will first notify the VIU of a pending time-out via a signaling message. The VIU-CPE should monitor the signaling channel and notify the VIU of a loss of the signaling channel by displaying a loss of channel indication.

5.3.3.4 VDT signaling protocol architecture. The specification of VDT signaling requirements is partitioned into the three levels depicted in Fig. 5.19, already discussed.

5.3.3.5 Network signaling. Bell Atlantic does not define generic network control signaling protocols for Stage 1 VDT. This signaling is to be specified to the video switch vendor and is be provided as part of the switching product. Network control signaling for Stage 1 VDT involves the use of a separate switch controller device. This device takes commands from the L1 gateway and communicates these to the appropriate local or hub switches. For future stages of VDT, such signaling is expected to evolve toward signaling protocols now being developed for broadband ISDN (ATM) networks, discussed earlier in this chapter.

5.3.3.6 Network access signaling. Network access signaling, both VIU and VIP, is via the L1 gateway, as described in the subsection that follows.

VDT message transfer part (VMTP) specification. The VMTP is intended to provide a reliable means of information transfer among the VDT signaling nodes. Such information transfer must take place despite transmission distributed or network failures that may occur. Also, flexibility and scalability to a very large user base are important. Information transfer is required between a small number of information gateways and a large number of information users. Multiple, simultaneous sessions involving a single user will be required. Because these requirements have already been considered in the development of several "network link" protocols either in use or under development, it will not be necessary to specify a new VDT signaling network link protocol. Rather, an existing protocol will be used*. Hence, the VMTP is specified by Bell Atlantic to be the X.25 protocol as defined in ITU-T Recommendation X.25. X.25 has been selected because of its availability as a well-established, standard method of data communication, and its relatively low cost and high reliability in allowing multiple and easily reconfigurable sessions over a single access link. In conclusion, all signaling interfaces will operate using X.25, and all VMTP network addressing will be in conformity with ITU-T Recommendation X.121†.

VDT messages transfer part-translator (VMTP-T) specification. The VMTP-T provides a translation between the VMTP and the VSCCP of the application-level interface. This function is specified separately from the VSCCP to more easily accommodate future to VMTP (e.g., use of an non-X.25 network protocol). The details of the translator are not specified in the Bell Atlantic specification other than to require that the VMTP-T to VMTP interface conform to the VMTP requirements previously listed. VMTP-T, together with MTP, provide a transparent data transmission capability to the VSCCP of the application layer interface. The ALI to VMTP-T interface is not specified in this for Stage 1 VDT.

VDT signaling connection control part (VSCCP) specification. The VSCCP provides the setup and teardown of switched virtual circuits (SVCs) for the VSCP. VSCCP is a user of MTP (via the MTP-T). VSCCP will attempt to reestablish SVCs that are lost because of a network

---

* As noted in the discussion in the previous section, this has the advantage of requiring no protocol development time, meeting time-to-market considerations, and affording available implementations of the protocol.

† Refer to the original specification, Bell Atlantic TR-72540, for additional details on the usage of X.25 in the VDT context.

failure. Figure 15.20 illustrates VSCCP capabilities at a high level. The L1 gateway, VIU-CPE, and L2 gateway can initiate both setup and clearing of an SVC (only the VIU-CPE has some restrictions; see original specification). *Permanent virtual circuits* (PVCs) are used between the L1 gateway and the L2 gateways.

VDT session control part (VSCP) specification. This entails a description of VSCP messages at the VIU network interface, VSCP messages at VIP network interface, and parameters. VSCP messages at the VIU network interface follow:

INITIATE VDT. A message sent from VIU to L1 gateway to indicate that the VIU wishes to begin a VDT session. Information required to automate Subsession 1 described earlier (e.g., selected VIP, authorization information) can optionally be included.

SELECT. A message sent from VIU to L1 gateway to indicate a VIU selection from the current menu.

AUTHORIZE. A message sent from VIU to L1 gateway to provide authorization information (e.g., personal identification number).

DISPLAY. A message sent from L1 gateway to VIU to provide information for display by the CPE device. Included are parameters regarding how the information is formatted.

STORE. A message sent from L1 gateway to VIU to provide information for storage by the CPE device. Included are parameters regarding how the information is formatted.

VSCP messages at the VIP network interface follow:

SERVICE REQUEST. A message sent from an L1 gateway to VIP to indicate that a VIU wishes to begin a VDT session. Included in the message is the VIU identification.

VIDEO SETUP REQUEST. A message sent from VIP to an L1 gateway indicating a request to set up (or rearrange) a video channel from a designed VIP port to a designated VIU.

VIDEO SETUP ACKNOWLEDGE. A message sent from VIP to an L1 gateway acknowledging a successful setup of a video channel and providing the VIU's signaling address or the signaling unavailability of a video channel.

VIDEO RELEASE REQUEST. A message sent from VIP to an L1 gateway indicating a request to release a video channel, a designated VIP port to a designated VIU.

VIDEO RELEASE ACKNOWLEDGE. A message sent from VIP to an L1 gateway acknowledging a successful release of a video channel or indicating an error.

VSCP parameters follow.

INFORMATION. User information for use by the VIU-CPE device for storage or display to the VIU.

FORMAT PARAMETERS. Information for use by the VIU-CPE device in determining how to display or store the contents of an INFORMATION parameter to a VIU.

AUTHORIZATION INFORMATION. Information populated by a gateway in authorizing a VIU for receipt of a particular service (e.g., personal identification number).

SELECTION INFORMATION. Information populated by a VIU indicating a menu selection to a gateway.

VIU PREFERENCES. Information populated by a VIU for use by an L1 gateway in automating certain functions.

CPE TYPE. Information populated by a VIU-CPE device for use by an L1 and/or L2 gateway in identifying CPE capabilities.

CPE IDENTIFICATION. Information populated by a VIU-CPE device for use by an L1 and/or L2 gateway in identifying an individual CPE-VIU.

VIU GLOBAL ADDRESS. Information populated by an L1 gateway to identify a VIU to an L2 gateway.

VIU SIGNALING ADDRESS. Information populated by an L1 gateway to provide a VIU signaling address to an L2 gateway.

VIDEO PORT CHARACTERISTICS. Populated by an L2 gateway to communicate video setup information to an L1 gateway.

VIDEO CHANNEL IDENTIFICATION. Populated by an L2 gateway to communicate video setup information to an L1 gateway.

VIDEO PORT STATUS. Populated by an L1 gateway to communicate the result of a video channel setup or release request.

The message structure follows the message format of CCS7, for the ISDN User Part (see Ref. 2 for a discussion of this). The general message structure is as shown in Table 5.5. Table 5.6 depicts a mapping of the parameters previously described to the message set, that is, which messages carry which parameters. Each message and each parameter name is identified by a unique two hexadecimal digits code.

A number of parameters have been defined at the format level; others are still being defined. Table 5.7 depicts one example (refer to the specification for the other parameters). In the end, all parameters have to be clearly defined in terms of their format.

The next aspect of the definition is to define the message formats, in terms of the parameters that have been defined, amplifying the layout

TABLE 5.5 VSCP General Message Structure

Message type
Mandatory fixed part
Mandatory variable part
Optional part

of Table 5.5. Table 5.8 depicts one example (refer to the specification for the other messages).

Figure 5.23 is an illustration of a typical network access signaling message flow. A VIU initiates a VDT session by powering on the CPE device, which places an X.25 call to the L1 gateway. The L1 gateway identified the VIU from the X.25 address. Upon receiving an INITIATE from a VIU, the L1 gateway provides VDT information (e.g., welcome screen, menu of VIPs, request for PIN) to the VIU over the signaling channel (Subsession 1). Subsession 1 may consist of one or more *screens* of information. Each screen is sent over the signaling channel from the L1 gateway via a DISPLAY message(s), and the VIU will have the capability of moving between screens (via menu selections). One screen of information may require multiple DISPLAY messages. Optionally, based on VIU information in the INITIATE message, the L1 gateway may download multiple menu screens via one or more STORE messages, with navigation then occurring locally.

Based on VIU input (via the CPE device), a SELECT command is sent, indicating a request to be connected to a particular L2 gateway. The L1 gateway communicates this request to the indicated L2 gateway via a SERVICE REQUEST. The L2 gateway will typically respond with a VIDEO SETUP REQUEST for that VIU. Upon receiving a VIDEO SETUP REQUEST, the L1 gateway first verifies that the VIU is requesting service. If so, the video channel is set up, and the L1 gateway sends a VIDEO SETUP ACKNOWLEDGE to the L2 gateway. This includes the VIU's signaling address. The L2 gateway then places an X.25 call to the VIU, and Subsession 2 is established.

Upon receiving a VIDEO RELEASE REQUEST from an L2 gateway, the L1 gateway provides teardown of the video channel and responds to the L2 gateway with a VIDEO RELEASE ACKNOWLEDGE. During this operation, the L2 gateway has released the signaling channel to the VIU.

If a VIU-CPE device is unable to establish an X.25 connection with the L1 gateway, it should display a *network unavailable* indication to the VIU. The CPE device should wait 30 s, and, if no signaling message is received (i.e., DISPLAY or STORE), tear down the X.25 connection and give a *network unavailable* indication to the VIU. (Refer to specification for illustration of flow.)

TABLE 5.6 Mapping of Parameters to Messages

	INITIAL VDT	SELECT	AUTHORIZE	DISPLAY	STORE	SERVICE REQUEST	VIDEO SETUP REQUEST	VIDEO SETUP ACKNOWLEDGE	VIDEO RELEASE REQUEST	VIDEO RELEASE ACKNOWLEDGE
Message type	M	M	M	M	M	M	M	M	M	M
Format parameters				M	M					
Information				M	M					
Authorization information	O		M							
Selection information		M								
VIU preferences	O									
CPE type	O					O				
CPE identification	M					O				
VIU global address						M	M	M	M	M
VIU signaling address								M		
Video channel characteristics							M			
Video port information							M	M	M	M
Video port status								M		M

TABLE 5.7 Parameter Format for the Authorization Information

bit 8 7 6 5	4 3 2 1	
2nd address digit	1st address digit	Octet 1
4nd address digit	3rd address digit	Octet 2

NOTE: Address digits are 4-bit coded digits (0000 = 0, 0001 = 1, 0010 = 2, etc.).

If the L1 gateway receives no VIDEO SETUP REQUEST in response to a SERVICE REQUEST within a time-out window (setable in the L1 gateway), it will notify the VIU via a DISPLAY (e.g., *select VIP unavailable* and remain in Subsession 1). (Refer to specification for illustration of flow.)

If the video network is unavailable, the L1 gateway will notify the L2 gateway via the VIDEO SETUP ACKNOWLEDGE, notify the VIU via a DISPLAY (e.g., *network unavailable*), and tear down the VIU L1 gateway X.25 connection. (Refer to specification for illustration of flow.)

5.3.3.7 Application signaling. The VMTP, VMTP-T, and VSCCP for application signaling are identical to that defined for network access signaling previously described. The discussion that follows describes required VSCP messages for application signaling. Figure 5.22 depicts one example of VSCP flow. (Refer to specification for illustration of L1 and L2 flows.)

TABLE 5.8 Message Format of the INITIATE VDT Message

Field	*Octet*
Message type (M)	Octet 1
CPE identification (M)	Octet 2 Octet 3 Octet 4 Octet 5 Octet 6 Octet 7 Octet 8 Octet 9
CPE type	Octet 10 Octet 11
Authorization information	Octet 12 Octet 13
VIU preferences	Octet 14 onward (length TBD)

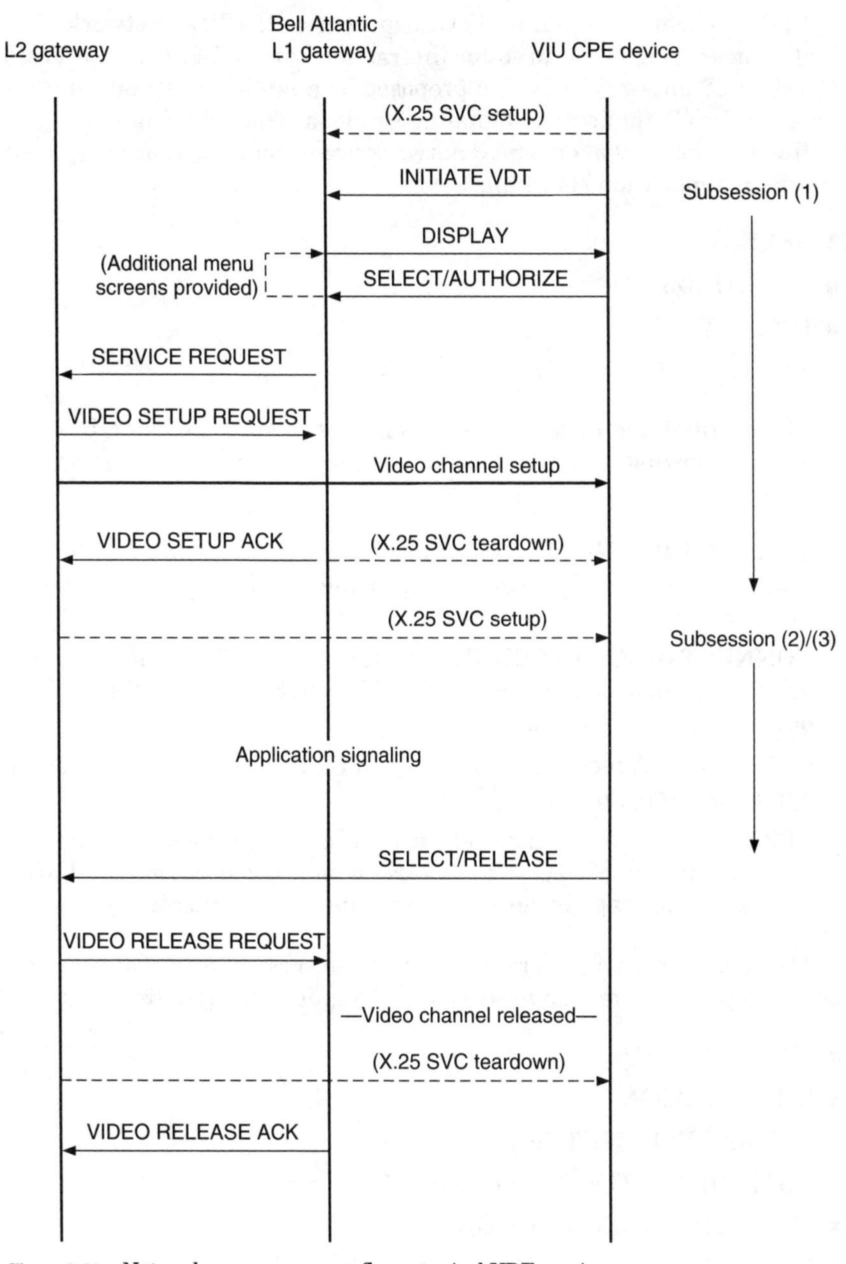

Figure 5.23 Network access message flow—typical VDT session.

VDT application signaling is transparent to the VDT network. The VSCP messaging is required for interaction with a Bell Atlantic Video Services L2 gateway; it is also proposed as a minimum standard message set for all application signaling sessions. The following messages, defined in the section on VSCP network access signaling, are also used in VSCP application signaling:

- SELECT
- AUTHORIZE
- DISPLAY
- STORE

Their format and purpose are the same as discussed earlier. In addition, the following new messages are defined for VSCP application signaling:

CONNECT REQUEST. A message sent from an L2 gateway to VIU that establishes Subsession 2 and requests CPE device attributes information.

CONNECT ACKNOWLEDGE. A message sent from VIU to an L2 gateway, in response to a CONNECT REQUEST, that includes CPE device attribute information.

CONTROL. A message sent from a VIU L2 gateway to control an application (e.g., pause a video).

TERMINATE. A message sent from VIU to an L2 gateway to indicate termination of a session. (A session may also be terminated with a SELECT message in response to an appropriate menu.)

The following parameters, defined in the section on VSCP network access signaling, are also used in VSCP application signaling:

- MESSAGE TYPE
- INFORMATION
- FORMAT PARAMETERS
- AUTHORIZATION INFORMATION
- SELECTION INFORMATION
- CPE TYPE
- CPE IDENTIFICATION

In addition, the following new parameter is defined for VSCP application signaling:

CONTROL INFORMATION. Information populated by the CPE device to control an application running on an L2 gateway. The CONTROL INFORMATION contains code points for the following actions:

Codepoint 1 = Pause
Codepoint 2 = Stop
Codepoint 3 = Rewind
Codepoint 4 = Fast Forward
Codepoint 5 = Slow Motion
Codepoint 6 = Play
Codepoint 7 = spare

Figure 5.24 provides an illustration of a typical application signaling message flow. Subsessions 2 and 3 are in almost all instances transparent to the VDT network. An L2 gateway may begin Subsession 2 by requesting (via CONNECT REQUEST) information on the VIU's CPE device (which can be provided via a CONNECT ACKNOWLEDGE).

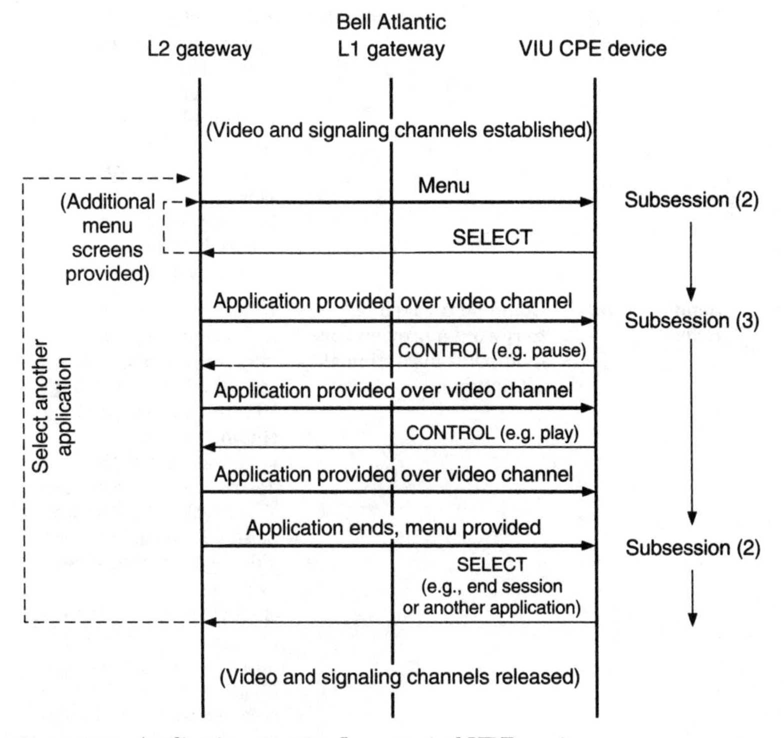

Figure 5.24 Application message flow—typical VDT session.

TABLE 5.9 VDT Video Control Features

	VIU CPE	Subsession 2	Subsession 3
PAUSE	Sends *Control-Pause* Displays confirmation to user	Could be invoked (e.g., to pause viewing of a preview) or ignored at discretion of L2 provider.	Suspends playing of the video selection, VIU remains in subsession 3. If VIU does not *un-pause* within a specified period of time (e.g., 15 min). *Stop* is automatically invoked at the L2. During a pause period, if not already on a dedicated channel, the L2 may initiate network commands to connect the VIU to a pause video channel (e.g., ads, logo). Ignored if received when not in normal play mode. The L2 may send the VIU audio or visual indications or alerts regarding time-out.
STOP	Sends *Control-Stop* Displays confirmation to user	Could be invoked (e.g., to stop viewing of a preview) or ignored at discretion of L2 provider	Suspends viewing of a video selection and terminates subsession 3. The VIU is returned to subsession 2 and receives an L2 menu. There will likely be a time period (e.g., 24 hours), administered by the L2 provider, during which the VIU can resume viewing at the point where *Stop* was activated. Ignored if received when not in normal play mode
REWIND	Sends *Control-Rewind*	Could be invoked (e.g., to rewind a preview) or ignored at discretion of L2 provider	If not already on a dedicated channel, the L2 may initiate network commands to connect the VIU to a dedicated video channel on which rewind is viewed. When the desired point is reached, VIU invokes *Play* at which point VIU receives normal viewing from that point over the existing video channel. If *Rewind* reaches beginning of video, *Play* is automatically invoked by L2 gateway. Ignored if received when not in normal play mode

TABLE 5.9 VDT Video Control Features (*Continued*)

	VIU CPE	Subsession 2	Subsession 3
FAST-FWD	Sends *Control-Fast-Fwd*	Same as *Rewind*	Same as *Rewind* except VIU is returned to L2 menu if *Fast-Fwd* reaches end of video
SLO-MO	Sends *Control-Slo-Mo*	Same as *Fast-Fwd*	Same as *Fast-Fwd*
PLAY	Sends *Control-Play*	Ignored if in normal mode. Return to normal mode if in another mode	Reestablishes normal viewing after a *Pause, Rewind, Fast-Fwd, Slo-Mo* or *FREEZE*. Otherwise ignored
FREEZE	If not already in *Freeze* mode, sends *Control-Pause* and local feature freezes current video display	Same as *Pause*	Same as *Pause*

A typical session will then involve, first, a number of SELECT or AUTHORIZE messages for the VIU to navigate menus and select an application. Then CONTROL messages can be used by the VIU to control the application.

A VIU may terminate a session in one of two ways. One is to select this option from an L2 gateway menu, in which case the L2 gateway tears down the X.25 channel. The other is to power off the CPE device, which will cause a TERMINATE to be sent from the L2 gateway by the CPE device, before the CPE device tears down the X.25 channel and enters the low-power mode. (If a TERMINATE is not sent, the L2 gateway may interpret the X.25 channel termination as an inadvertent VIU action and attempt to reestablish the signaling channel.)

If power is lost to the CPE device, the X.25 connection is torn down. The L2 gateway will recognize an abnormal termination and attempt to reestablish the connection. If successful (e.g., if power is restored to the CPE device), the session continues. If unsuccessful, normal session termination occurs.

At this juncture, one needs to define application entity functions. These functions are defined as appropriate for the common VDT session functions (CASE), as well as for each application running on the VDT CPE device-L2 gateway. The initial set of SASEs is for video control. The video control features for each VDT subsession are shown in Table 5.9.

References

1. Bell Atlantic, *Signaling Specification for Video Dial Tone,* TR-72540, Issue 1, Release 1, August 1993.
2. D. Minoli and G. Dobrowski, *Signaling Principles for Frame Relay and Cell Relay Service,* Artech House, Norwood, Mass., 1994.

Chapter

6

Digital Video Compression

6.1 Overview

Compression algorithms are critical to the viability of digital video, digital video distribution, video on demand, multimedia, and other VDT services. This chapter provides an overview of digital video compression as well as a synopsis of some MPEG-1 and MPEG-2 principles. The standards discussion is based directly on the ISO standards; such material is included here to promulgate the use of open standards; however, developers and other parties working on products of commercial value should refer directly to the original documentation, especially since only some key highlights are included. The overview material included in the early part of the chapter is based in part on Ref. 1.

At press time, digital video still tended to be used in specialized environments because of its relatively high cost. This results from the fact that chips implementing compression standard(s) are not yet being produced in quantities large enough to break a reasonable cost barrier of (perhaps) $100 to $300 for a complete PC or setup box plug-in board, and from the fact that software compression is not fast enough on most general purpose microprocessors. The $500 barrier was just about being broken in 1994.[2]

Over 12 video standards or formats are available worldwide for TV video* (without even counting variant broadcasting schemes beyond the basic NTSC, PAL, and SECAM methods). Table 6.1 summarizes the video standards discussed or alluded to in this chapter (it also

* Desktop video includes additional schemes, such as Ultimotion, Cinepak (Quicktime), TrueMotion, Laserdisc; videoconferencing products also use other schemes.

TABLE 6.1 Plethora of Video Formats and Coding

NTSC	Analog U.S. and Japanese format
PAL	Analog European format
SECAM	Analog French and Eastern Europe format
D-1/CCIR 601	Digital production standard
D-2	Digital production standard
DS-3 based	Digital U.S. commercial methods (nonstandard)
JPEG/Motion JPEG	Digital compression format (principally for still video, but also for some video)
MPEG-1	Digital compression full-motion video (low-end entertainment video and multimedia)
MPEG-2	Digital compression full-motion video (high-resolution)
DVI/Indeo	Digital de facto standard
CD-I	Digital consumer-electronic format
H.261 et al.	Digital videoconferencing format
HDTV	Digital scheme discussed in Chap. 4
Vendor-based	Digital vendor-specific videoconferencing formats

includes the HDTV proposed standard). In order for digital television to successfully enter the market, it is important that the audio and video compression technique agreed upon be used on an industrywide basis. It is also important that the transport approach also be standardized; standards such as MPEG-2 and ATM (Chap. 8) do exactly that.

6.1.1 Compression methods

Video can be considered a sequence of frames, where each frame is an array of pixels. The goal of a video coding algorithm is to remove redundant information and greatly reduce the data rate. Two types of redundancies exist in video: redundancy within a single frame and redundancy between adjacent frames. There are two classes of compression algorithms: *lossless* algorithms and *lossy* algorithms. Another way of classifying compression algorithms is as *entropy coding* and *source coding.*

Lossless compression is one where the entire information contained in the uncompressed message can be faithfully recovered by the decompressor. For example, instead of sending a 100-bit message 0111111111 . . . 111111, one could compress it as x0y1, where x and y are octets that take values 0 (base 10) to 255 (base 10). In this case, one would send (00000001)0(01100011)1, which is only 18 bits long, and yet the receiver is still be able to recover the message exactly. Lossless compression algorithms are symmetrical; namely, either the sender or receiver can perform the compression and decompression with the same level of computational complexity and without loss of data integrity. Compression of *data* material, either for transmission or for storage, clearly requires lossless methods. Many hardware and software products implement lossless compression. They typically double or quadruple the storage capacity on a disk (i.e., have a compression ratio of about 2:1 or 4:1), or double the apparent speed of a communication line. These algorithms can also be applied to files that represent voice or image information. Because the redundancy is higher, the compression ratios can be as much as 10:1. However, this is both less effective than the compression obtained with specialized lossy techniques and less than the information bandwidth reduction that is sought (typically 100:1 or even 200:1). Lossy compression algorithms do not aim at retaining the entire information, but just enough to be adequate for the task at hand. Lossy algorithms result in slightly degraded pictures. The advantage of these algorithms is that they can achieve 100:1 or 200:1 compression.

Going back to the other way of looking at compression, namely, source coding versus entropy coding, note that source coding deals with *features* of the source material and encompasses lossy algorithms. Source coding can be further classified as intraframe and interframe coding. Intraframe coding is used for the first picture of a sequence and for downstream pictures after some major change of scenery. Intraframe coding is used for sequences of similar pictures (even for those including moving objects). Intraframe coding removes only the spatial redundancy within a picture; interframe coding also removes the temporal redundancy between pictures. Entropy coding, on the other hand, achieves compression by using statistical properties of the coded signal and is, in theory, lossless.

Video can be compressed using lossless or lossy methods. For lossless compression, video can be digitized according to the ITU-R (formerly, CCIR: Consultative Committee on International Radio) 601 standard; the bit rate is approximately 165 Mbps. Although useful in a number of high-end commercial applications, this data rate is simply too high for user-level applications. Lossy methods, such as MPEG-1 and MPEG-2, are more appropriate for VDT and digital video distribution.

Two other methods are on the horizon: *fractal* and *wavelets**. The fractal transform uses Mandelbrot's approach of using simple equations to generate natural-looking images in a high-level of detail.[3] It is believed by experts to be a good compression scheme, particularly for still images of nature. Being based on equations, it can be expanded to

* At least three products were already available at press time, implementing these methods for PC-based delivery of video: Captain Crunch by Media Vision (Wavelet; 320 × 240 at 30 fps), VideoCube by ImMIX/Aware (Wavelet, 640 × 480 at 60 fps), and Pro-Frac by Total Multimedia Inc. (Fractal; 320 × 200 at 30 fps).

sizes even larger than the original, leading to claims of greater compression compared to other schemes. Images are segmented into domains that can be described as squeezed down, distorted versions of larger parts or "ranges" of the same image.[4] Artifacts include softness and substitution of details by other details typically undetectable in natural scenes. Packages ranging in cost from $1000 to $26,000 are available for desktop applications. The wavelet method is also based on mathematical techniques. A wavelet codec transforms a picture into a set of different spatial representations, some of which contain insignificant high frequencies, and one of which contains all the important low-frequency information. This scheme can also compress audio. Artifacts include softness, small random noise, and edge halos. Products based on this approach are also appearing.

6.1.2 Traditional digital video: broadcast quality

An image is composed of three elements: a luminance (brightness) element and two chrominance (color) elements. These elements come into play in the digitization process. There are two near-lossless methods of digitizing television signals: *digital component video* and *digital composite video.*

Digital component video, known as 4:2:2, is a time-multiplexed digital stream of three video signals: luminance (Y), C_r (R-Y), and C_b (B-Y). The 4:2:2 refers to the ratio of sampling rates for each component. This format is also often called D-1, referring to the tape format associated with the digital component recording. CCIR/ITU-R Recommendation 601 was adopted in 1982, after eight years of study and compromise among European, Japanese, and North American approaches. The standard accommodates equally well NTSC, PAL, and SECAM formats. Typical digital component video encoding systems have used the following parameters:

- NTSC: Luminance sampling frequency 13.5 MHz. Sampling frequency for color differences: 6.75 MHz. Pixels: 858 × 525 (about 720 × 484 for the active image area).
- PAL: Luminance sampling frequency of 17.734475 MHz. Sampling frequency for color differences: 8.867236. Pixels: 910 × 525 (about 768 × 484 for the active image area).

At the final stage, the word length for digital image delivery is usually between 8 and 10 bits, but to maintain precision, more may be used, particularly in the early stages of off-line processing (e.g., 16 bits). Since there is no one single sampling rate to obtain digital video, conversion between digital formats requires transcoding, not only of the formats, but also of the sampling frequencies.

The CCIR/ITU-R 601 standard supports both the 525-line, 60 fields per second format and the 625-line, 50 fields per second format. Table 6.2 depicts a summary of the key highlights of the CCIR/ITU-R 601 standard for the 525-line format.[5,6] (MPEG-2, covered later, aims at providing similar quality but at a much smaller data rate.)

The other encoding method, *digital composite video,* is known as $4f_{sc}$. This format, applicable to NTSC, PAL, and SECAM, also consists of three components: Y, I, and Q. However, I and Q are not multiplexed but quadrature-modulated and summed to the Y component. The result is a single information stream sampled at four times the color subcarrier rate. The term $4f_{sc}$ refers to "4x the frequency of the subcarrier." This format is often called D-2, referring to the associated tape format. Typical digital composite video encoding systems have utilized the following parameters:

- NTSC: Luminance sampling frequency of 14.31818 MHz (four times the frequency of the NTSC subcarrier). Sampling frequency for

TABLE 6.2 Summary of the CCIR/ITU-R 601 Digital Video Standard

Coded signals	Y, R-Y, B-Y (luminance and color differences)
No. of samples Luminance (Y) Each color difference (R-Y, B-Y)	 858 429
Sampling approach	Orthogonal, line, field, and picture repetitive. R-Y and B-Y samples are co-sited with odd Y samples in each line (1st, 3rd, 5th, etc. sample)
Sampling frequency Luminance (Y) Each color difference (R-Y, B-Y)	 13.5 MHz 6.75 MHz
Form of coding	Uniformly quantized *pulse code modulation* (PCM), 8 bits per sample for all three signals (luminance and color differences)
No. of samples per digital active line Luminance (Y) Each color difference (R-Y, B-Y)	 720 360
Correspondence between video signal levels and quantization levels Luminance (Y)	 220 quantization levels with the black level corresponding to level 16 and the peak white level corresponding to level 235
Each color difference (R-Y, B-Y)	224 quantization levels in the center part of the quantization scale with zero signal corresponding to level 128

color differences: 7.15909. Pixels: 910 × 525 (about 768 × 484 for the active image area).

The transmission in digital form of a composite television signal, whether NTSC, PAL, or SECAM, particularly for contribution networks (see Chap. 4), requires, according to the sampling rates used and the number of bits employed to represent the signal, a rate of 100 to 150 Mbps. In component video (e.g., 4:2:2), the bit rate can reach 216 to 270 Mbps; uncompressed HDTV requires about 1 Gbps.[7] Compression is unquestionably required.

In the United States, DS3 transmission facilities support about 45 Mbps; in Europe, E3 facilities support about 34 Mbps. Also note that DS1 supports 1.544 Mbps and E1 supports 2.048 Mbps. These rates have defined, at the pragmatic level, the boundaries for the commercially available video-encoding algorithms. For example, vendor proprietary methods have been used in the past in the United States to encode TV signals at 45 Mbps for remote delivery (e.g., between TV studios in two cities), utilizing commonly available telecommunication carrier services. Several suppliers have developed broadcast-quality DS3 video coders using relatively mild compression on 525/60-Hz video. Uncompressed pulse-code modulation of NTSC signals would require a minimum sampling rate of 8.4 MHz for a 4.2-MHz bandwidth, and 8 or 9 bits per sample, resulting in an uncompressed rate of 67.2 to 75.6 Mbps (in reality, higher data rates are needed, because the sampling rate can be higher). In Europe, the encoding of interest is 34 Mbps. These network rates clarify why the CCIR recently published Recommendation 723 that supports standardized video coding at 32 to 45 Mbps for full-resolution video or TV signals (720 pixels per line, 483/576 lines per frame, 59.94/50 2:1-interlaced frames per second). However, there is an interest in bringing the video data rate down further, as discussed in the next subsections.

6.1.3 Compression algorithms in common use

The video compression requirements vary between various applications, digital storage media, transport methods (e.g., terrestrial, DBS), and video programming types (e.g., talk shows versus sporting events). Nonetheless, it is important that easy interworking and movement between such media be accomplished. Many factors come into play, such as timing, program stream reconstruction, synchronization, demultiplexing/remultiplexing, packeting/repacketizing, and encryption.[8] It has been the objective of the standards study groups to limit the extent of the specifications to a minimum and to define only what it takes to accomplish meaningful interoperability.

The mid-1980s saw the emergence of ITU-T's Recommendation H.261*, supporting video compression and coding at $p \times 64$ kbps ($p = 1,2, \ldots, 30$). These standards are suited for videotelephony and videoconferencing but are not deemed appropriate for entertainment-quality VDT programming. (They could, however, be used in conjunction with VDT in support of these other services, perhaps in support of telecommuting.[10])

The late 1980s saw the emergence of a standard originated by ISO, which was formally adopted at the end of 1992. This standard, ISO/IEC 11172 (also known as MPEG-1), provides video coding for digital storage media with a rate of 2 Mbps or less. H.261 and MPEG-1 standards provide picture quality similar to that obtained with a VCR. Both standards are characterized by low bit rate coding and low spatial resolution. H.261 supports 352 pixels per line, 288 lines per frame, 29.97 noninterlaced frames per second (lower resolution is also supported). ISO/IEC 11172 typically supports 352 pixels per line, 240/288 lines per frame, 29.97/25 noninterlaced frames per second. Many useful video and multimedia applications require higher resolution than this to provide an acceptable level of quality to the user. Real-time encoding/decoding also introduces delays that increase with decreasing data rates. For example, current technology encoding at 128 kbps may result in unacceptable delays for a quality videoconference.

MPEG-1 was developed as a video compression standard to be used with CD-ROMs. The compression ratio is about 100:1; however, the quality of the picture is marginal for generic broadcast and cable applications of high-action movies and sporting events (the ADSL-1 industry will obviously object to this observation). MPEG-1 employs a source input format for motion video and associated audio with a data rate up to 1.5 Mbps. It provides a quality comparable to VHS.

The MPEG standard embodies the concepts of group of frames and interpolated frames (the presence of interpolated frames is optional). See Fig. 6.1. Each group of frames contains a frame that is intraframe coded only, to facilitate random access. There will also be predicted frames. Interpolated frames are formed from adjacent (past and future) *key frames* (both the stand-alone and predicted frames can be used for interpolation). As seen in Fig. 6.1, a group of frames consists of a single stand-alone frame, several predicted frames, and one or more interpolated frames positioned between key frames. The bandwidth allocated to each type of frame typically conforms to the ratio of 5:3:1 (intraframe coded, predicted, and interpolated). "Future" frames may

* Portions of the material that follow are based on Ref. 9.

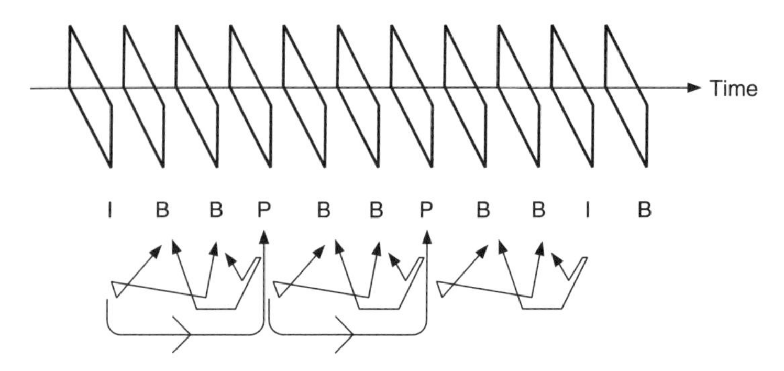

I: Intraframe coded frames (stand-alone)
P: Predicted frames (predicted from intraframes of previously predicted frames)
B: Bidirectionally interpolated frames (from nearest two I or P frames)

Figure 6.1 Group of Frames concept.

be employed to predict intermediary frames. (This can be done not only for stored material but also for real-time material by buffering a few frames for analysis.)

Both ISO and ITU-T are now working on and/or have completed new high-quality video-coding standards. These standards, however, generate more than the 1.5- to 2.0-Mbps output of the earlier standards: full-motion, reasonable resolution digital video is sought in the 4- to 20-Mbps range. MPEG started its study on the second-phase work (known as MPEG-2) in 1990, with completion in 1995. MPEG-2 supports both full CCIR/ITU-R 601 resolution as well as HDTV. The rates of interest range from 2 to 20 Mbps (as seen in Chap. 4, HDTV operates at 21.5 Mbps). Chips and systems are now available from several major vendors.

In a hybrid coder, a type of coder commonly used, an estimate of the next frame to be processed is created from the current frame. The difference between this estimate and the actual value of the variable(s) of interest contained in the next frame, when it arrives at the coder, is encoded by an appropriate mechanism. One of the more common examples of this type of coder is the motion-compensated *discrete cosine transform* (DCT) coder used in MPEG-1 and MPEG-2. Motion compensation capitalizes on the correlation between successive frames of a video sequence. A motion-compensation algorithm assigns a velocity (that is, a speed and direction to a moving object); constant velocity makes predicting the next frame of video fairly straightforward. Misprediction, however, can cause the loss of two or three frames of video. Even when true motion is not at play in a scene, motion compensation algorithms improve the data rate by seeking a block of identical (or nearly identical) values in the previous frame and spatially close to the block to be encoded; this block is then used to formulate a prediction. Only the information used to find the prediction block is sent to the decoder.[7]

The DCT concentrates the remaining signal into a small number (64 to be precise) of coefficients that can be quantized and efficiently represented. This coder is three to four times as efficient as one that uses no prediction, but it is sensitive to transmission errors and does not permit random access. This coder is used (with modifications to facilitate random access) in the MPEG-1 standard; the basic DCT method (see below) is also used in the *Joint Photographic Experts Group* (JPEG) still-image standard (as discussed later, some developers are using JPEG to also encode video).

As discussed earlier, encoding/decoding falls into two main types:[11]

Interframe. These encoders/decoders use combinations of key motion-predicted and interpolated frames to achieve high-compression ratios with low data rate. Examples of these types of algorithms include production level video (PLV)* and the various MPEG algorithms.

Interframe. These systems compress every frame (and sometimes every field) of video individually. These algorithms (e.g., Motion JPEG and TrueMotion) offer the advantage of frame-accurate editability; however, they produce from 2 to 10 times more data than the interframe algorithms.

Table 1.12 in Chap. 1 provided an "approximate" view of the various data rates for the algorithms discussed later. Some of the suppliers providing video compression equipment include, among others, AT&T, Compression Labs, Intel, LSI Logic, and Microsoft (which provides MPEG capabilities for the Windows system).

Until the recent development of chips able to support real-time encoding (see below), the issue of compression algorithm asymmetry was important. With symmetric algorithms, the compression process requires the same amount of clock time as decompression (playback); asymmetric compression requires considerably more clock time than decompression.[11] In theory, this makes the decompression possible on cheap, low-end equipment. Recent advances in chip power make this a less important issue, particularly since the suppliers are coming out with a single chip that can be programmed to be an MPEG-1 encoder, MPEG-1 decoder, MPEG-2 encoder, MPEG-2 decoder, a JPEG encoder, a JPEG decoder, or an H.261 encoder/decoder.

---

* PLV is a digital video algorithm that was developed in the late 1980s as part of Intel's Digital Video Interactive (DVI) technology. In the early 1990s, an improved version (PLV-2) was released. PLV provides VHS-quality video at 30 fps on a full screen of 256 × 240 pixels. It uses a modified vector quantization approach. It has a key (reference) frame every 120 frames, followed by predicted frames every third frame.

Powerful chips are being developed to support video processing by a number of vendors. As an example, in 1994, Texas Instruments unveiled a new chip called the *multimedia video processor* (MVP), which was called at the time the most powerful video-oriented *digital signal processor* (DSP) on the market. The MVP has the ability of performing over 2 billion operations per second (BOPS). The programmable DSP can be reconfigured to emulate chips such as C-Cube, AT&T, and IIT, and it can perform real-time compression and decompression of video and still images using MPEG, JPEG, or H.261 algorithms. The initial cost of the chip was $1500, with a target of $300 in 1995.[12]

6.1.4 A short discussion of DCT

This section provides a short description of some key DTC features, by discussing it in the context of still-picture encoding. Figure 6.2 depicts the key processing steps that are embodied in the DCT-based operation for the case of a single-component* (i.e., monochrome) JPEG image sequential codec (the functioning of the DCT in other codecs is similar). Notice a *forward DCT* (FDCT) function and an *inverse DCT* (IDCT) function.

One can think of DCT-based compression as compression of a stream of 8×8 blocks of gray-scale image samples. Each 8×8 block (represented by 64 point values known as $f(x,y)$, $0 \leq x \leq 7$, $0 \leq y \leq 7$) makes its way through each processing stage, yielding output in compressed form. For progressive mode codecs, an image buffer is placed between the quantizer and the entropy coding module; this enables the image to be stored and then sent out in multiple scans with follow-up information aimed at successively improving the quality of the received image.

Each 8×8 block of source image (frame) samples can be viewed as a 64-point discrete signal that is a function of the two spatial dimensions x and y. At the input to the encoder, these 64 source image samples are cranked through the following equation.[13]

For $0 \leq u \leq 7$, $0 \leq v \leq 7$, calculate the following 64 values:

$$F(u,v) = 0.25*C(u)*C(v)* \left\{ \sum_{x=0}^{7} \sum_{y=0}^{7} f(x,y)*\cos[(2x+1)*u\pi/16]*\cos[(2y+1)*v\pi/16] \right\}$$

* A source image-frame contains three image components (also called colors, spectral bands, or channels), e.g., RGB. Each component consists of an array of samples, to which the DCT can be applied in turn. A sample is expressed as an integer with precision P bits with values in the range $[0, 2^P - 1]$; all samples of all components within the same source image-frame must have the same precision (P can be 8 or 12), but image components may be sampled at different rates compared to each other.

where $C(u) = C(v) = 1/\sqrt{2}$ for $u = 0, v = 0$
$C(u) = C(v) = 1$ otherwise

Mathematically, the FDCT takes the input signal and decomposes it into 64 orthogonal basis vector signals. The output of the FDCT is a set of 64 basis signal amplitudes, that are known as *DCT coefficients.* The coefficient for the vector (0,0) is called the *DC coefficient;* all other coefficients are called *AC coefficients.* The DC coefficient generally contains a significant fraction of the total image energy. Because sample values typically vary slowly from point to point across an image, the FDCT processing achieves data compression by concentrating most of the signal in the lower values of the (u,v) space. For a typical 8×8 sample block from a typical source image, many, if not most, of the (u,v) pairs have zero or near-zero coefficients and therefore need not be encoded. At the decoder, the IDCT reverses this processing step. One can use 8-bit or 12-bit source image samples; 12-bit samples, however, require fairly large computational resources for FDCT or IDCT calculations.

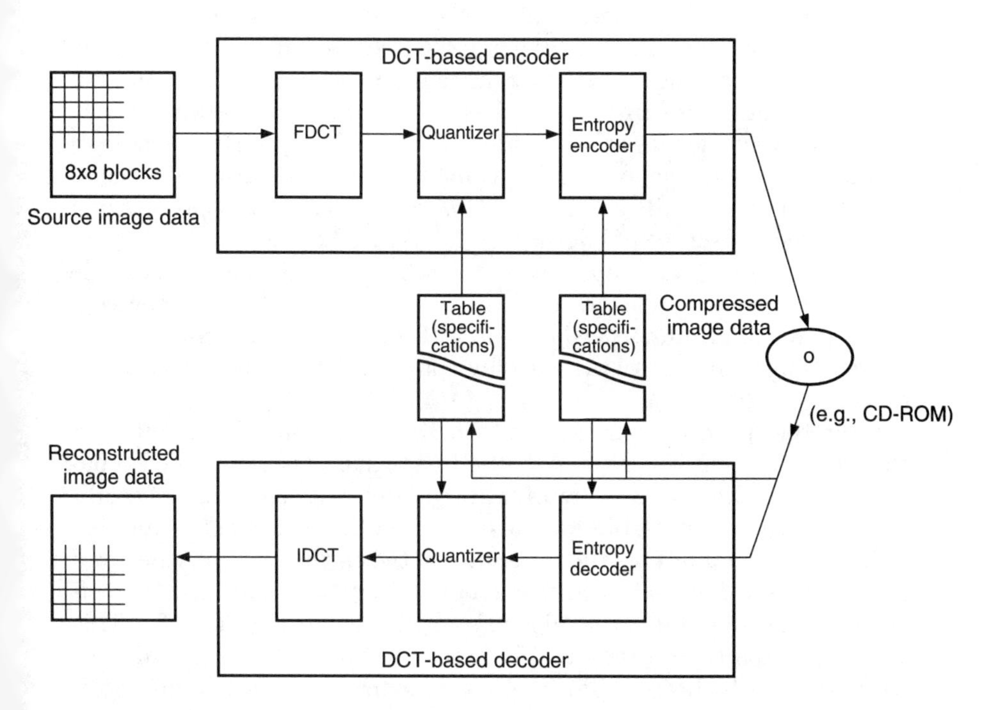

Figure 6.2 DCT usage in a JPEG codec (example).

In principle, the DCT introduces no loss to the source image samples; it just transforms them to a domain in which they can be more efficiently encoded. This means that if the FDCT and IDCT could be computed with perfect accuracy and if the DCT coefficients were not quantized, the original 8 × 8 block could be recovered exactly. But, as seen earlier, the FDCT (and the IDCT) equations contain transcendental functions (i.e., cosines). Consequently, no finite time implementation can compute them with perfect accuracy. In fact, a *number* of algorithms have been proposed to compute these values approximately. No single algorithm is found to be optimal for all implementations: an algorithm that runs optimally in software usually does not operate optimally in firmware (e.g., for a programmable DSP) or in hardware.

Given the finite precision of the DCT inputs and outputs, an interworking challenge arises: coefficients calculated by two different algorithms (e.g., one the sender and one the receiver), or even by independently designed implementations of the same FDCT or IDCT algorithm (which differ only minutely in the precision of the cosine terms or intermediate results) will result in slightly different outputs from identical inputs.

Each of the 64 DCT coefficients obtained at the output of the FDCT is then uniformly quantized by using a 64-element quantization table that must be specified by the application (or user). Each element can take an integer value from 1 to 255 (or 1023) that specifies the step size of the quantizer for its corresponding DCT coefficient. The purpose of quantization is to achieve further compression by discarding information that is not visually significant. Quantization is a lossy process and is the principal source of lossiness in DCT-based encoders.

When the aim is to compress the image or frame as much as possible but without visible artifacts, each step size is chosen to be the perceptual threshold of human vision. These thresholds are functions of the source image characteristics, display characteristics, and viewing distance.

After the quantization process, the DC coefficient, representing a sort of average of the value of the 64 image samples, is handled separately. Since there is usually high correlation between the DC coefficients of adjacent 8 × 8 blocks, the quantized DC coefficient is encoded differentially, namely, as the difference between the current value and the previous value. To facilitate entropy coding, the quantized AC coefficients are ordered into the "zigzag" sequence shown in Fig. 6.3.[13] This ordering helps the entropy-coding process by placing low-coordinates coefficients (which are more likely to be nonzero) before high-coordinates coefficients.

The last step for DCT-based encoding is entropy coding itself. This step achieves additional lossless compression by encoding the quan-

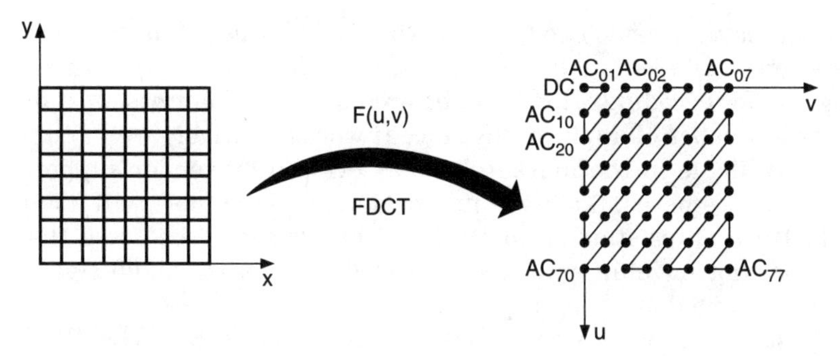

Figure 6.3 Ordering of quantized samples for entropy encoding.

tized DCT coefficients more compactly, based on their statistical characteristics. Entropy coding can be viewed as a two-step process. The first step converts the sequence of quantized coefficients (ordered as previously discussed) into an intermediate sequence. The second step converts the symbols to a stream in which the symbols no longer have externally identifiable boundaries. Two entropy-coding methods are in common use: Huffman coding and arithmetic coding. The sequential codec uses Huffman coding, but codecs with both methods are specified for all modes of operation. Arithmetic coding produces about 10 percent better compression than Huffman; however, it is more complex. Huffman coding requires that one or more sets of code tables be specified. The same tables used to compress an image are needed to decompress it. Huffman tables may be predefined and used within an application as defaults, or developed specifically for a given image in an initial statistics-gathering pass prior to actual compression.

6.2 JPEG and Motion JPEG

The JPEG standard has been developed jointly by both ISO and ITU-T (hence, the nomenclature "joint") for compression of still images. It can compress typical images from 1⁄10 to 1⁄50 of their uncompressed bit size without visibly affecting image quality. JPEG is the first international digital image compression standard for multilevel continuous-tone still images (both gray-scale and color). Some applications to which JPEG addresses itself include color facsimile, quality newspaper wirephoto transmission, desktop publishing, graphic arts, and medical imaging. Some have used JPEG to support video transmission (particularly in the medical industry where one wants to be able to adjust the resolution of the transmission in real-time, e.g., when viewing the display of a set of x-rays). JPEG plays a limited role in VDT at this time, and the discussion that follows is the only coverage contained in the book.

JPEG uses a methodology based on the DCT, discussed in the previous section. It is a symmetrical process, with the same complexity for coding and for decoding. JPEG will be an important compression standard for a number of applications since it works relatively well and is already available in the marketplace, as evinced by vendor support. Digital video cameras, fax machines, copiers, and scanners will likely include JPEG chips starting in 1994 or 1995. It should be noted that JPEG does not have an embedded encoded-compressed audio signal, since it was aimed at photographic material.

JPEG standards work was undertaken in the late 1980s. The JPEG algorithm aims to be at or near state of the art with regard to compression rate and image fidelity over a wide range of "originals" ratings. It also aims at having moderate computational complexity to permit software implementations with viable performance on a range of processors (PCs, workstations, etc.), as well as hardware implementations at reasonable cost. JPEG supports the following four modes of operation:

1. *Sequential encoding.* The image is encoded in a single left-to-right, top-to-bottom scan.
2. *Progressive encoding.* The image is encoded in multiple scans for applications in which transmission bandwidth is low and, hence, the transmission time may be long (the viewer can watch the image build up in multiple coarse-to-fine passes).
3. *Lossless encoding.* The image is encoded to guarantee exact recovery of every source image sample value (although this results in lower compression efficiency compared to lossy methods).
4. *Hierarchical encoding.* The image is encoded at multiple resolutions so that lower resolution displays may be accessed without having to decompress the image at full resolution.

For each JPEG mode, one or more codecs are specified in the standard. These modes of operation originated from JPEG's goal of wanting to be generic and to specify a flexible and comprehensive encoding family that can span a gamut of continuous-tone image applications. It is unlikely that low-end implementations will incorporate each and every coding mode listed in the standard; in fact, most of the implementations now on the market have implemented only the sequential codec. The sequential codec is a sophisticated compression method that will be sufficient for many applications. In practical terms, JPEG's mathematical processing makes it a challenge for real-time software-driven implementation on a PC. Fortunately, JPEG chips have become available. 10-MHz chips typically can compress a full-page 24-bit-color, 300-

TABLE 6.3 Quality of Compressed Image

Ratio of no. of bits in the compressed image to the no. of samples in the luminance component*	Quality
0.25 to 0.5	Moderate to good quality, sufficient for some applications
0.5 to 0.75	Good to very good quality, sufficient for many applications
0.75 to 1.5	Excellent quality, sufficient for most applications
1.5 to 2.0	Effectively indistinguishable from originals, sufficient for the most demanding applications

* Loosely known as "bits per pixel."

dpi image from 25 to 1 MB in about 1 s, or can compress a 640 × 480 pixel image, with 24-bit "true color," by a factor of 10 in 0.1 s. There are 25-MHz chips that can do the same job in 0.03 s. These chips cost around $150 to $250 in quantities.

For color images with moderately complex scenes, the various JPEG codecs normally produce reasonably good quality, as shown in Table 6.3.

Some vendors also use JPEG methods for encoding of full-motion video, NTSC TV signals in particular. This is known as *Motion JPEG.* While JPEG was not designed for full-motion video, it can accommodate it with some restrictions. For example, audio is not supported in an integrated fashion. One of the limiting factors of this use of the algorithm is that it works independently from frame to frame; hence, it cannot reduce the redundancies that exist between frames. Some view the fact that JPEG performs only intraframe compression as a benefit in the sense that it offers "fast" random access to any frame of the video material. Other full-motion video compression techniques performing interframe compression rely on periodic transmission of a reference frame, if the reference frame is sent every 20 frames, one may have to wait as much as 19 frames before the reference frame is received; this would equate to a wait of $^{20}/_{60}$ or 0.33 s. With JPEG, one needs to wait only for the time required to decompress one frame, that is, 0.04 s.

Network-based applications using JPEG for full-motion video are not likely to be widely implemented because JPEG is bandwidth-intensive. For video material displayable at a PC monitor at medium resolution, that is, 640 × 480 pixels, 24 bits for color representation, JPEG is required to compress about 1 MB per frame, or 30 MBps (240 Mbps) to a lower value. The downloading, displaying, and manipulation of full-screen video in digital form is a daunting task, even if one were to achieve a 50:1 compression; this is why standards such as MPEG-1 and MPEG-2 are needed for any digital video transmission except perhaps for desktop multimedia applications.

A number of computer systems can now decompress JPEG files in software. As a point of reference, a 24-bit 640 × 480 image takes less than 10 s on a Next computer with a Motorola 68040 main processor. As noted, the data rate for video compressed with JPEG varies. It is dependent on the source material and the quantization tables. "Reasonable" video is within the range of the SCSI disk write speed of approximately 600 kbps. ("Reasonable" means that although it is apparent the video is compressed, the artifacts are not objectionable.) "High-quality" video requires 1 Mbps, which calls for the use of fast SCSI disks or caching of short video sequences in large memory buffers. Another alternative is to lower the frame rate, allowing higher quality frames while keeping the data rate within the range of standard SCSI disks.

6.3 MPEG-1

As noted, the ISO/IEC/JTC1/SC29/WG11 MPEG working group has produced a specification for coding of combined video and audio information. It is directed at video display as contracted to still-image display to which JPEG addresses itself. MPEG specifies a decoder and data representation for retrieval of full-motion video information from digital storage media in the 1.5- to 2-Mbps range. Hence, it can be used in an ADSL-1 context (see Chap. 9). The specification is composed of three parts: Part 1, *Systems;* Part 2, *Video;* and Part 3, *Audio.* The system part specifies a system coding layer for combining coded video and audio and provides the capability of also combining private data streams and streams that may be defined at a later date. The specification describes the syntax and semantic rules of the coded data stream.

As mentioned earlier, MPEG-1 uses three types of frames: Intra (I) Picture frames; Predicted (P) frames; and Bidirectional (B) frames. I-type frames are compressed using only the information in that frame using the DTC algorithm. An incoming video signal of 1 s will contain at least two I frames. P frames are derived from the preceding I frames (or from other P frames) by predicting motion forward in time; P frames are compressed to approximately 60:1. Bidirectional B interpolated frames are derived from the I and P frames, based on previous and frame referencing; B frames are required to achieve the low average data rate.[11]

Since MPEG allows coding comparison across multiple frames, it can yield compression ratios of 50:1 to 200:1. The MPEG algorithm is asymmetrical. Namely, it requires more computational complexity (hardware) to compress full-motion video than to decompress it. This is useful for applications where the signal is produced at one source but is distributed to many. MPEG chips on the market at press time (a handful) already provide 200:1 compression to yield VHS quality at 1.2 to 1.5 Mbps; they also can provide 50:1 compression for broadcast quality at 6 Mbps.

One interesting application of MPEG-1 technology was recently reported in the press: a number of TV and cable TV stations are using circuit-switched analog cellular links to supplement their satellite, microwave, and manual-delivery methods of transporting full-motion video clips to the station's headend and/or transmission site. These stations use portable MPEG-1 real-time encoders which achieve a compression ratio of 167-to-1, along with a cellular system supporting inverse multiplexing. It takes about eight minutes to transmit a one-minute clip at Beta-quality levels (30 fps), using four simultaneous analog cellular dial-up lines, each achieving 9.6 kbps. The cost of a one-minute clip satellite transmission is $15; the cost over cellular links is $2. The additional advantage of this approach is the ability to supplement coverage to areas which would otherwise be hard to reach. The $40,000 MPEG-1 encoder, which has the size of a briefcase, compresses the video onto the hard drive of a PC. At the station, a $10,000 decompressor is used to regenerate an analog version of the video signal.

MPEG's system coding layer specifies a multiplex of elementary streams such as audio and video, with a syntax that includes data fields directly supporting synchronization of the elementary streams. The system data fields also assist in the following tasks[14,15]:

1. Parsing the multiplexed stream after a random access
2. Managing coded information buffers in the decoders
3. Identifying the absolute time of the coded information

The system semantic rules impose some requirements on the decoders; however, the encoding process is not specified in the ISO document and can be implemented in a variety of ways, as long as the resulting data stream meets the system requirements.

Figure 6.4 depicts an MPEG encoder at the functional level. The video encoder receives uncoded digitized pictures called *video presentation units* (VPUs) at discrete time intervals; similarly, at discrete time intervals, the audio digitizer receives uncoded digitized blocks of audio samples called *audio presentation units* (APUs). Note that the times of arrival of the VPUs are not necessarily aligned with the times of arrival of the APUs.

The video and audio encoders, respectively, encode digital video and audio as described in the MPEG Specification Parts 2 and 3, producing coded pictures called *video access units* (VAUs) and coded audio called *audio access units* (AAUs). These outputs are referred to generically as elementary streams. The system encoder and multiplexer produce a

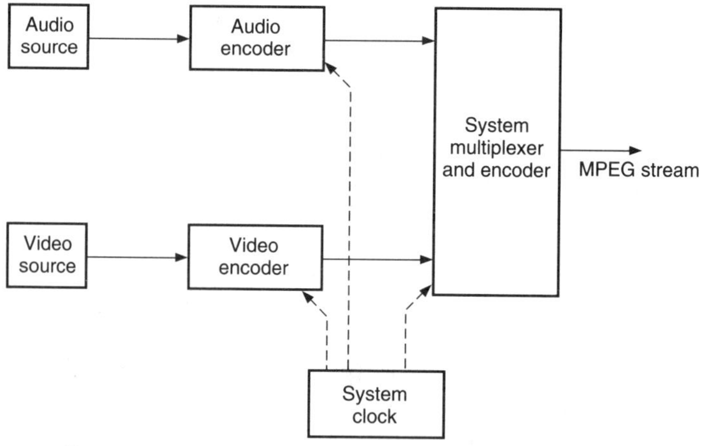

NOTE: Since the ISO 11172 specification specifies syntax and semantics of the coded data stream, the encoder model is not part of the specification itself; however, the reference decoder model is specified as part of the semantic definition of the information stream at the system layer.

Figure 6.4 Generation of an MPEG data stream.

multiplex stream, M(i), containing the elementary streams as well as system-layer coding, described later. Audio is supported from 32 to 384 kbps and can consist of a single channel or two stereo channels.

An important aspect of MPEG is synchronization. Synchronization is a fundamental aspect of communication. The principle may be known to the reader in the context of synchronizing various nodes on a network to enable character, block, or message recovery, as these data entities are transferred from the sender to the receiver. In a multimedia context, synchronization is even more "intimate" in the sense that the various signal objects comprising the combined signal must be stored, retrieved, and transmitted with precise timing relationships. The basic mechanism to achieve the desired synchronization is, to a large extent, the same in both contexts.

The *system time clock* (STC) is a reference time operating at 90 kHz. The STC is not necessarily phase-locked to the audio or the video sample clocks. The STC produces 33 bit time values (binary values from 0 to $2^{33} - 1$) incremented at 90 kHz. For some (but not necessarily all) VPUs and APUs arriving at the encoder, the value of the STC is determined and stored with the presentation units through the coding, transmission, and decoding processes. These values are called *presentation time stamps* (PTS). The time stamps, using a reference frequency of 90 kHz and unsigned binary values from 0 to $2^{33} - 1$, allow unique identification of operating time within the information stream over an interval exceeding 24 hours.[14,15]

In addition to the PTS, there are two other time stamps associated with the coded information stream itself, which are used to ensure synchronization between the various decoders and the information stream source in a decoding system:

1. *System clock references* (SCRs). These are generated as samples of the STC so that the value of the SCR equals the value of the STC at the time the last octet of the SCR exits the system encoder.
2. *Decoding time stamps* (DTS). These are similar to PSTs, except that the permutation ordering of pictures in the video coding process is reflected in the DTS values.

To achieve synchronization in multimedia systems that decode multiple video and audio signals originating from a storage or transmission medium, there must be a "time master" in the (decoding) system. MPEG does not specify which entity is the time master. The time master can be: (1) any of the decoders, (2) the information stream source, or (3) an external time base. All other entities in the system (decoders and information sources) must slave their timing to the master. If a decoder is taken as the time master, the time when it presents a presentation unit is considered to be the correct time for the use of the other entities. Decoders can implement phase-locked loops or other timing means to ensure proper slaving of their operation to the time master. If an information stream source is considered the time master, the SCR values indicate the correct time at the moment that these values are received. The decoders then use this information of what the correct time is to pace their decoding and presentation timing. If the time base is an external entity, all of the decoders and the information sources must slave the timing to the external timing source.

The MPEG system specification includes a syntax with three coding layers above the layer implicitly defined by the elementary streams. These are: (1) the ISO 11172 stream layer, (2) the pack layer, and (3) the packet layer. See Fig. 6.5.

- The ISO 11172 stream layer includes a sequence of packs followed by an end code.
- The pack layer includes the SCR field, the *mux rate* field, an optional system header packet, and the packet layer. (The mux rate field bounds the rate of bytes per second as measured by the current and succeeding SCR values and the number of coded bytes intervening.)
- The packet layer comprises packets containing information from individual elementary streams (there is information from exactly one elementary stream in each packet). The packet contents and all system-layer coding are octet-aligned, although the individual cod-

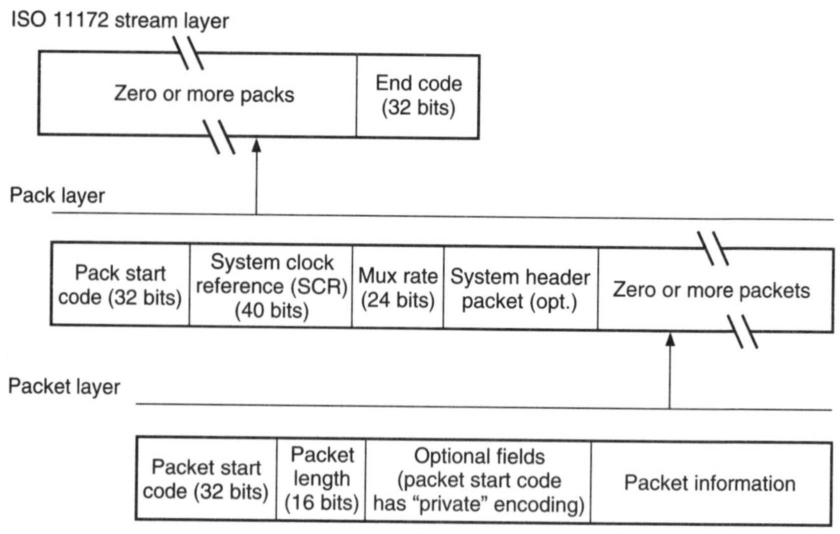

Figure 6.5 Simplified view of ISO 11172.

ing elements within elementary streams may not necessarily be octet-aligned. Each packet consists of a packet start code followed by the packet length (in octets, ranging up to $2^{16} - 1$). There are 69 different values of packet start code currently defined: 16 are for video, 32 are for audio, 2 are private, 1 is for padding; the remaining 18 codes are reserved for future use (of which the Multimedia and Hypermedia Experts Group may take 16). Private data are unrestricted, other than by the syntax and STD model that applies to the entire stream. Three optional fields may be included in the packet: STD buffer size, PTS, and DTS. Lastly, one finds the packet information from the elementary stream. The amount of packet information is limited only by the total available packet length (decremented by the data in the packet header itself) and by constraints imposed by the decoder.

The method of multiplexing the elementary streams (VAUs and AAUs) is not directly specified in MPEG. However, there are some constraints that must be followed by an encoder and multiplexer to produce a valid MPEG data stream. For example, it is required that the individual stream buffers must not overflow or underflow. The sizes of the individual stream buffers impose limits on the behavior of the multiplexer. Decoding of the MPEG data stream starting at the beginning of the stream is straightforward since there are no ambiguous bit patterns. Starting decoding operation at random points requires locating pack or packet start codes within the data stream.

6.4 MPEG-2

The compression schemes discussed so far do not produce adequate quality for full-motion video (although JPEG supports high quality, its data rate is too high). MPEG-2 was developed by ISO/IEC/JTC1/SC29/WG11 and is known as ISO/IEC 13118. This is probably going to be the most important video compression standard for VDT and VOD applications. The committee (WG11) consists of largely U.S. and European companies involved in or interested in video and audio compression.[5] MPEG-2 was formally accepted as an international ISO (from the French: International Organization for Standardization) standard in November 1994. As noted earlier, it aims at providing CCIR/ITU-R quality for NTSC, PAL, and SECAM, and also at supporting HDTV quality (this being a relatively newer requirement). MPEG-2 work is now driven by the desire to accomplish global unification of digital TV program generation, editing, storage, retrieval, transport, and display.[6,8] The standard provides a set of agreed-upon methodologies for audio and video compression and transport of complex multiplexes of associated data and related data services.

The objectives of newer video coding standards for full-motion video now under development, specifically MPEG-2, are as follows:

1. Picture quality should be higher than that of the current NTSC, PAL, and SECAM broadcast systems.
2. Compression to bit rates in the range of 4 to 6 Mbps for NTSC, PAL, and SECAM material and 6 to 10 Mbps for television signals conforming to CCIR 601. These target rates are already achievable in experimental systems.
3. The standard(s) need to be flexible enough to allow both high-performance/high-complexity and low-performance/low-complexity (e.g., intraframe mode only operation) codec systems.
4. The standard(s) should take into account existing standards. Compatibility consideration enables smooth migration of new standards while maintaining interoperability among equipment conforming to the old and new generation standards.
5. Compatibility should be maintained to the extent possible. There are two types of compatibility: upward/downward compatibility (addressing different picture format sizes), and forward/backward compatibility (addressing different generation standards). A system is upward compatible if a higher resolution receiver is able to decode material from the signal transmitted by a lower resolution encoder. A system is downward compatible if a lower resolution receiver is able to decode material from the signal or part of the signal trans-

mitted by a higher resolution encoder. A system is forward compatible if the new standard decoder is able to decode material from a signal or part of the signal of an existing standard encoder. A system is backward compatible if an existing standard decoder is able to decode material from the signal or part of the signal of a new standard encoder.

The MPEG-2 standardization agenda was set as follows:

- Phase 1 (1992): Set objectives of effort; evaluate subjective picture quality for various coding algorithms
- Phase 2 (1993): Evaluate functionality and implementation aspects of proposed algorithms
- Phase 3 (1994): Verify performance of selected algorithm in hardware and software systems
- Phase 4 (1994): Obtain official approval of standard

The following four layers come into play in the discussion of coding[16]:

Block: The smallest coding unit in the MPEG algorithm. It consists of 8 × 8 pixels and can be one of three types: luminance (Y), red chrominance (C_r), or blue chrominance (C_b). The block is the basic unit in intraframe coding.

Macroblock: The basic coding unit in the MPEG algorithm. It is a 16 × 16 pixel segment in a frame. Since each chrominance component has one-half the vertical and horizontal resolution of the luminance component, a macroblock consists of four Y, one C_r, and one C_b block.

Slice: A horizontal strip within a frame. It is the basic processing unit in the MPEG coding scheme. Coding operations on blocks and macroblocks can be performed only when all pixels for a slice are available. A slice is an autonomous unit, since coding a slice is done independently of its neighbors. Typically, each frame contains 30 slices of 512 × 16 pixels.

Picture: the basic unit of display. It corresponds to a single frame in a video sequence. The spatial dimensions of a frame are variable and are determined by the requirements of an application. Typically, these dimensions are 512 × 480 pixels, which is similar to NTSC broadcast quality.

MPEG-2 initially consisted of three *profiles* (simple, main, and next), each further divided into four *levels* (high-level Type 1, high-level Type 2, main level, and low level), the simple profile does not support bidi-

TABLE 6.4 Initial MPEG-2 Profiles and Levels

	Simple profile	Main profile	Next profile
High-level Type 1: supports 1152 lines per frame (lpf), 1920 pixels/line (ppl), and 60 fps. This equates with 62.7 Mpixels per second (pps) or 60 Mbps.		HDTV (U.S.)	HDTV (European)
High-level Type 2: supports 1152 lps, 1440 ppl, and 60 fps. This gives 47 Mpps or 60 Mbps.		HDTV (U.S.)	HDTV (European)
Main level: supports 576 lpf, 720 ppl, and 30 fps. This gives 10.4 Mpps or 15 Mbps.	Cable TV industry	Video broadcasting industry	
Low level supports 288 lpf, 352 ppl, and 30 fps. This gives 2.5 Mpps or 4 Mbps.			

rectional B* frames, shown in Fig. 6.1, eliminating the need to store them. See Table 6.4. The levels refer primarily to the resolution of the video produced; for example, the low level refers to a standard image format video with a resolution of 352 × 240 (also known as *source input format,* or SIF). The main level conforms to CCIR/ITU-R 601 quality. The high level is for HDTV.

The main level-main profile was standardized in 1993 (with chips available in 1994). Toward the end of 1993, the structure of profiles and levels was changed to reflect likely applications of the standard. The levels were modified to high level, high-1440 level, main level, and low level; the profiles were modified to simple profile, main profile, SNR† scalable profile, spatially scalable profile, and high profile.[17] A previous profile known as "main + profile" was split into two scalable profiles, anticipating situations of network congestion, whereby the scalable profiles can drop information if it cannot be transported. The "next profile" was renamed "high profile" to reflect the fact that it applies to high-resolution video applications. See Table 6.5.

* B frames are storage areas where the incoming frame is compared to the preceding frame and used to predict the next one. (This, however, requires that the decoder have more than two frames of video memory, thus more than doubling the memory requirement from 1 to 2 MB.)

† *Editorial note:* The acronym SNR is not defined even in the standard which says "SNR scalability: A type of scalability where the enhancement layer(s) contain only coded refinement data for the DTC coefficients of the base layer."

TABLE 6.5 MPEG-2 Conformance Points

	Simple profile	Main profile	SNR scalable profile	Spatially scalable profile	High profile
High Level		✓			✓
High-1140 Level		✓		✓	✓
Main Level	✓	✓	✓		✓
Low Level		✓	✓		

✓ = point of concordance (likely combinations of levels and profiles in actual applications).

The MPEG-2 main profile is the one of greatest current interest.[5] It can handle images from the lowest level with MPEG-1 quality, up through broadcast quality at the main level, to HDTV at the high level.

The expectation is that MPEG-2 will play a critical role in most industrial and consumer applications in the foreseeable future. While MPEG-1 was developed for computer applications and, hence, supports progressive scanning only, MPEG-2 is to be used in the TV world, and, hence, supports interlaced scanning. MPEG-2 will enable worldwide TV compatibility in transmission, products, and services. As noted earlier, currently, there are three major regional systems that are not directly compatible (NTSC, PAL, and SECAM).

The input to the MPEG-2 encoder is digital component video. The standard covers audio compression, video compression, and transport. In the transport area it defines[8]:

1. *Program streams:* A grouping of audio, video, and data elemental components having a common time relationship and being generally "associated" for delivery, storage, and playback
2. *Transport streams:* A collection of program streams or elementary streams (video, audio, data) that have been multiplexed in a nonspecific relationship for the purpose of transmission

These efforts at the system layer are aimed at providing a basic data structure; a data structure in this context is viewed as the *semantics* and *syntax* of a data stream that can serve as a common format for local usage (e.g., storage and edit) and for broadcast. A number of basic structural elements have been defined and are expected to become part of the system-layer syntax. Pivotal to this structure is the fact that the transport stream is based on packet principles. The packets (in the 130- to 192-octet range) contain digital information from a single elementary stream or data type*. Each packet has a header of up to four

* One advantage of this packetized approach, besides efficient transport (e.g., over ATM), is the fact that the packets can be encrypted in support of conditional access.

octets that provides information such as packet ID, clear/scramble indication, key (even/odd), and continuity counter.

Additional information on MPEG-2 is found in Sec. 6.6.

6.5 A Short History of Compression Products

This section provides a short summary of the recent history of video and image compression, particularly from a product point of view, and, more specifically, from a desktop computing perspective. Although this application is somewhat removed from the delivery of video, the same products, schemes, chips, etc., feed the same technology chain and ultimately the consumer demand side. Additionally, this technology is used to develop the interactive multimedia applications that so many providers hope to deliver over cable TV and/or VDT systems.

Desktop digital video is available in two major forms: *software-only playback* and *hardware-assisted playback*. Of more interest to broadcasters is hardware-assisted digital video: to decompress and display better quality digital video, computers (including setup boxes) use add-on hardware boards with fast, dedicated DSP chips. However, desktop video is also of interest. Most video on CD-ROM is now compressed/decompressed with software using Intel's Indeo or SuperMac's Cinepak, both supporting 30 fps for 160 × 120 pixels, 16-bit color, or 15 fps at 320 × 240 pixels (again, this is not adequate for entertainment video, but there is a convergence of technologies, also as discussed in Chap. 13).

MPEG-1 now offers an international standard capable of reasonable quality for slow-action movies, and one can achieve a display of 320 × 240 pixels at 30 fps on PC boards costing less than $400. MPEG-1-based encoding is superior in quality to software-based compression, even when this software compression is done on 100-MIPS workstations. Recently, MPEG-1 encoding took an evolutionary leap from a process- and time-intensive task undertaken on mainframes to a real-time task that can be accomplished via PC add-on boards ranging from $15,000 to $75,000. The key to MPEG-1's success may depend on whether consumers embrace the new MPEG-1-encoded VideoCD players that consumer electronic companies were introducing in 1994; if the VideoCD "revolution" manages to take off, with linear MPEG CDs and players replacing videotape cartridges and VCRs, the prices for MPEG-1 components will drop substantially.[4]

In 1993 there was only one vendor for playing back MPEG-encoded video on a PC (Sigma Design's ReelMagic); at the end of 1994 there were 40 vendors. Part of the success was due to the activities of the Open PC-MPEG Consortium, including the development of an appropriate Application Programming Interface. As noted at press time, these PC MPEG-1 boards could be purchased for $400.

In the meantime, MPEG-2 chips are already appearing (e.g., one by AT&T), although portions of the audio compression were the last to be finalized. MPEG encoding systems that were appearing at press time were able to compress video in less than 30 ms, making the algorithm's inherent encoding-time/decoding-time asymmetry a moot feature. C-Cube Microsystem's CLM-4000 chips can be assembled in parallel to compress MPEG-1 (using two CLM-4500 chips) or MPEG-2 (ten CLM-4600 chips). Toshiba has also announced a real-time MPEG-2 codec.

An industry group has been formed (late 1994) to develop implementers' agreements for the MPEG-2 standard and to extend its reach. More than 60 companies and organizations from industry segments such as programming, satellite, cable TV equipment manufacturers, consumer electronics manufacturers, component vendors, and cable TV operators, recently formed the North American Digital Group (NDAG). The group aims at focusing on the MPEG-2 standard and they are interested in interoperability and open standards. They seek to address modulation approaches, information datarate ranges, syntax for the transport of streams of data, service information tables, ancillary data services, and closed captioning. One of the initial goals was to gather information from the 28 existing U.S. standards groups* and assess the state of affairs. Europe already has a similar group, known as the European Digital Video Broadcast Group; over 150 companies from 17 countries support the activities of the organization.

In the final analysis, MPEG-1 will likely become the coding method of choice for CD-ROM program developers because of its reasonable quality video and audio, while MPEG-2 will become critically important to the television and entertainment video industry.

6.6 MPEG-2 Standard Details

This international standard† was prepared by SC29/WG11, also known as MPEG (*Moving Pictures Experts Group*). MPEG was formed in 1988 to establish an international standard for the coded representation of moving pictures and association audio stored on digital storage media. The MPEG-2 standard is published in four parts. *Part 1: Systems* spec-

* For example, the Advanced Television Systems Committee, the Electronic Industry Association, the Institute of Electrical and Electronics Engineers, the National Association of Broadcasters, the National Cable Television Association, and the Society of Motion Picture and Television Engineers.

† This section is based directly on ISO/IEC 13818 (Committee Draft, November 1993).

ifies the system coding layer of the MPEG-2. It defines a multiplexed structure for combining audio and video data and means of representation the timing information needed to replay synchronized sequences in real time. *Part 2: Video* specifies the coded representation of video data and the decoding process required to reconstruct pictures. *Part 3: Audio* specifies the coded representation of audio data. *Part 4: Conformance testing* (Part 4 was still in preparation at press time; it will specify the procedures for determining the characteristics of coded bit streams and for testing compliance with the requirements stated in Parts 1, 2, and 3). The specification is fairly complex; the standard is over 400 pages long.

Table 6.6 depicts some of the key terminology and concepts of MPEG-2 (items shown in brackets are specific to one part of the standard only).

TABLE 6.6 Key MPEG-2 Terms and Concepts

Access unit [System]	A coded representation of a presentation unit. In the case of compressed audio, an access unit is an audio access unit. In the case of compressed video, an access unit is the coded representation of a picture.
Bit rate	The rate at which the compressed bit stream is delivered from the storage medium or data link to the input of a decoder.
Channel	A digital medium that stores or transports an ISO/IEC 13818 stream.
Coded representation	A data element as represented in its encoded form.
Compression	Reduction in the number of bits used to represent an item of data.
Constant bit rate	Operation where the bit rate is constant form start to finish of the compressed bit stream.
Constrained system parameter stream (CSPS) [system]	An ISO/IEC 13818 program stream for which the constraints defined in Part 1 apply.
Data element	An item of data as represented before encoding and after decoding.
Decoded stream	The decoded reconstruction of a compressed bit stream.
Decoder	An embodiment of a decoding process.
Decoding (process)	The process defined in MPEG-2 that reads an input coded bit stream and outputs decoded pictures or audio samples.
Decoding time-stamp (DTS) [system]	A field that may be present in a PES packet header that indicates the time the an access unit is decoded in the system target decoder.
Digital storage media (DSM)	A digital storage or transmission device or system.
Editing	The process by which one or more compressed bit streams are manipulated to produce a new compressed bit stream. Conforming edited bit stream must meet the requirements defined in MPEG-2.

TABLE 6.6 Key MPEG-2 Terms and Concepts (*Continued*)

Term	Definition
Elementary stream [system]	A generic term for one of the coded video, coded audio, or other coded bit streams.
Encoder	An embodiment of an encoding process.
Encoding (process)	A process, not specified in MPEG-2, that reads a stream of input pictures or audio samples and produces a valid coded bit stream as defined in MPEG-2.
Entropy coding	Variable length lossless coding of the digital representation of a single to reduce redundancy.
Fast forward playback [video]	The process of displaying a sequence, or parts of a sequence, of pictures in display order faster than real time.
Layer [video, systems]	One of the levels in the data hierarchy of the video and system specifications defined in Parts 1 and 2 of the MPEG-2 standard.
MPEG-2 (multiplexed) stream [system]	A bit stream composed of zero or more elementary streams combined in the manner defined in Part 1 of the MPEG-2 standard.
Packet data [system]	Contiguous bytes of data from an elementary stream present in a packet.
Packet identifier (PID) [system]	A unique integer value used to associate elementary streams of a program in a single or multiprogram transport stream.
Packet [system]	A packet consists of a header followed by a number of contiguous bytes from an elementary data stream. It is a layer in the system coding syntax described in the MPEG-2 standard.
Packetized elementary stream (PES) [system]	A packetized elementary stream consists of PES packets, all of whose payloads consist of data from a single elementary stream, and all of which have the same stream ID.
Padding [audio]	A method to adjust the average length of an audio frame in time to the duration of the corresponding PCM samples, by conditionally adding a slot to the audio frame.
PES packet header [system]	The data structure used to convey information about the elementary stream data contained in the PES packet data.
Presentation time-stamp (PTS) [system]	A field that may be present in a PES packet header that indicates the time that a presentation unit is presented in the system target decoder.
Presentation unit (PU) [system]	A decoded audio access unit or a decoded picture.
Program-specific information (PSI) [system]	Normative data that are necessary for the demultiplexing of transport streams and the successful regeneration of program. In some cases, the nonmandatory information table is privately defined.
Program [system]	A collection of elementary streams with a common time base.
Random access	The process of beginning to read and decode the coded bit stream at an arbitrary point.
Side information	Information in the bit stream necessary for controlling the decoder.

TABLE 6.6 Key MPEG-2 Terms and Concepts (*Continued*)

Source stream	A single nonmultiplexing stream of samples before compression coding.
Start codes [system and video]	32-bit codes embedded in that coded bit stream that are unique. They are used for several purposes, including identifying some of the layers in the coding syntax.
STD input buffer [system]	A first-in, first-out buffer at the input of system target decoder for storage of compressed data from elementary streams before decoding.
Still picture	A coded still picture consists of a video sequence containing exactly one coded picture that is intra-coded; this picture has an associated PTS; and the presentation time of succeeding pictures, if any, is later than the still picture by at least two picture periods.
System target decoder (STD) [system]	A hypothetical reference model of a decoding process used to describe the semantics of an ISO/IEC 13818 multiplexed bit stream.
Time-stamp [system]	A term that indicates the time of an event.
Transport packet header [system]	A data structure used to convey information about the transport stream payload.
Variable bit rate	Operation where the bit rate varies with time during the decoding of a compressed bit stream.

6.6.1 Part 1: systems

The system part of MPEG-2 addresses the combining of one or more elementary streams of video and audio, as well as other data single or multiple streams that are suitable for storage or transmission. Systems coding follows the syntactical and semantic rules imposed by Part 1 of the MPEG-2 specification and provides information to enable synchronized decoding without either overflow or underflow of decoder buffers over a wide range of retrieval or receipt conditions.

Systems coding is specified in two forms: the *transport stream* and the *program stream*. Each stream is optimized for a different set of applications. Both the transport stream and program stream defined in MPEG-2 provide coding syntax that is necessary and sufficient to synchronize the decoding and presentation of the video and the audio information, while ensuring that coded data buffers in the decoders do not overflow or underflow. Such information is coded in the syntax using time stamps concerning the decoding and the presentation of coded audio and visual data and time stamps concerning the delivery of the data stream itself. Both stream definitions are packet-oriented multiplexes.

The basic multiplexing approach for simple video and audio elementary stream is illustrated in Fig. 6.6. The video and audio data are encoded as described in Parts 2 and 3 of the MPEG-2 standard. The resulting compressed elementary streams are packetized to produce the *packetized elementary streams* (PES) that are shown in Fig. 6.7*. Information needed to use PES packets independent of either transport stream or program stream may be added when PES packets are formed. (This information is not needed and need not be added when PES packets are further combined with system-level information to form transport streams or a program streams.) The systems standard (Part 1 of MPEG-2) covers those processes to the right of the vertical dashed line in Fig. 6.6.

The program stream is analogous and similar to MPEG-1 systems multiplex (ISO 11172-1 MPEG Systems). It results from combining one or more PESs that have a common time base into a single stream. The program stream definition can also be used to encode multiple audio and video elementary streams into multiple program streams, all of which have a common time base. Like the single program stream, all elementary streams can be decoded with synchronization. The program stream is designed for use in relatively error-free environments and is suitable for applications that may involve software processing of system information such as interactive multimedia applications. Program stream packets may be of variable and relatively long length.

The transport stream combines one or more programs with one or more independent time bases into a single stream. PES packets made up of elementary streams that form a program share a common timebase. The transport stream is designed for use in environments where errors are likely, such as storage or transmission in lossy or noisy media. Transport stream packets are 188 bytes in length.

Program and transport are designed for different applications, and their definitions do not strictly follow a layered model. It is possible and reasonable to convert from one to the other; however, one is not a subset or superset of the other. In particular, extracting the contents of a program from a transport stream and creating a valid program stream is possible and is accomplished through the common interchange format of PES, but not all of the fields needed in a program stream are contained within the transport stream; some must be derived. The transport stream may be used to span a range of layers in a layers model and is designed for efficiency and ease of implementation in high-bandwidth applications.

The scope of syntactical and semantic rules set forth in the systems specification differ: the syntactical rules apply to systems-layer coding only and do not extend to the compression-layer coding of the video and

* To understand the full functionality of the PES, including the exact meaning of all fields, the reader is referred directly to the MPEG-2 standard.

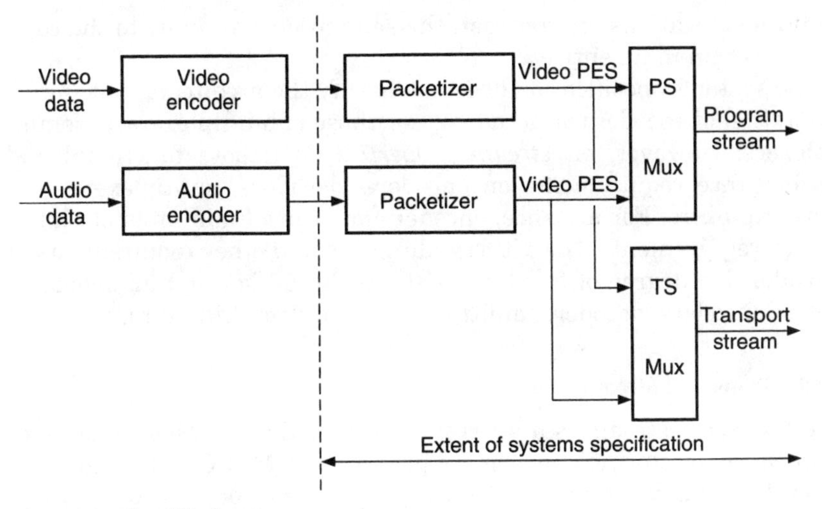

Figure 6.6 Simplified systems overview.

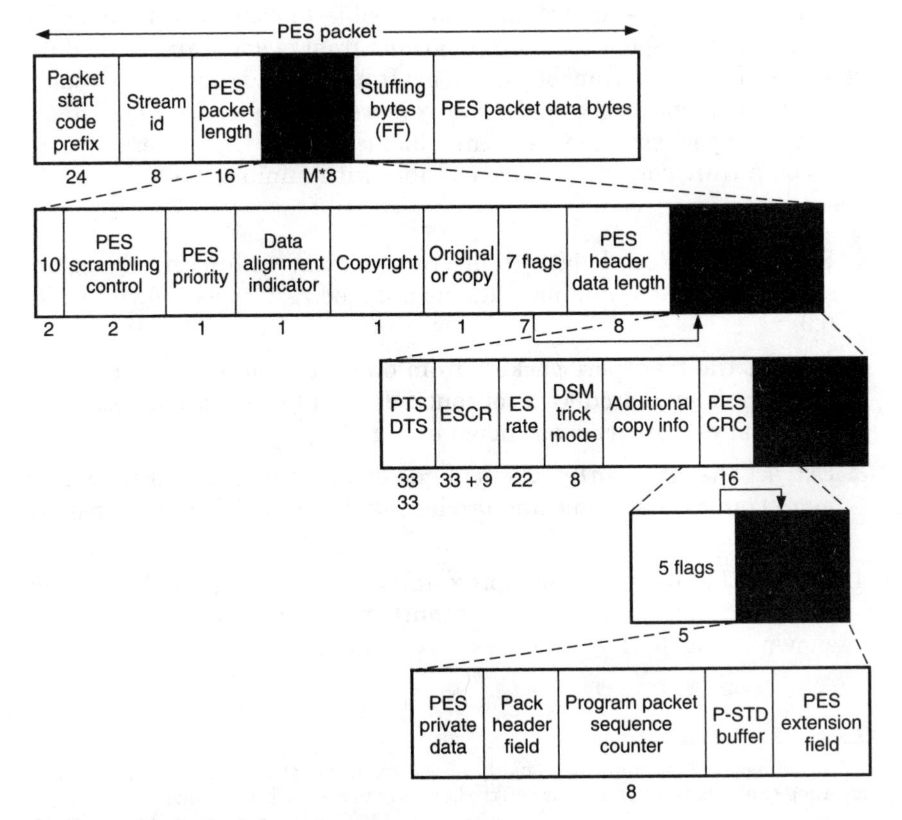

Figure 6.7 Packetized elementary stream.

audio specifications; by contrast, the semantic rules apply to the combined stream in its entirety.

The systems specification does not specify the architecture or implementation of encoders or decoders, nor those of multiplexers or demultiplexers. However, bit *stream properties* do impose functional and performance requirements on encoders, decoders, multiplexers, and demultiplexers. For instance, encoders must meet minimum clock tolerance requirements. Notwithstanding this and other requirements, a considerable degree of freedom exists in the design and implementation of encoders, decoders, multiplexers, and demultiplexers.

6.6.2 Transport stream

The transport stream is a stream definition that is tailored for communicating or storing one or more programs of MPEG coded data and other data in environments in which significant errors may occur. Such errors may be manifested as bit value errors or loss of packets.

The MPEG-2 transport stream may be constructed by any method that results in a valid stream. It is possible to construct a transport stream containing one or more programs from elementary coded data streams, from program streams, or from other transport streams which may themselves contain one or more programs. See Fig. 6.8*.

The transport stream is designed in such a way that several operations on a transport stream are possible with minimum effort. Among these are:

1. Retrieve the coded data from one program within the transport stream, decode it, and present the decoded results as shown in Fig. 6.9.
2. Extract the transport packets from one program within the transport stream and produce as output a different transport stream with only that one program as shown in Fig. 6.10.
3. Extract the transport packets of one or more programs from one or more transport streams and produce as output a different transport stream.
4. Extract the contents of one program from the transport stream and produce as output a program stream containing that one program as shown in Fig. 6.11.

---

* To understand the full functionality of the transport stream, including the exact meaning of all fields, the reader is referred directly to the MPEG-2 standard.

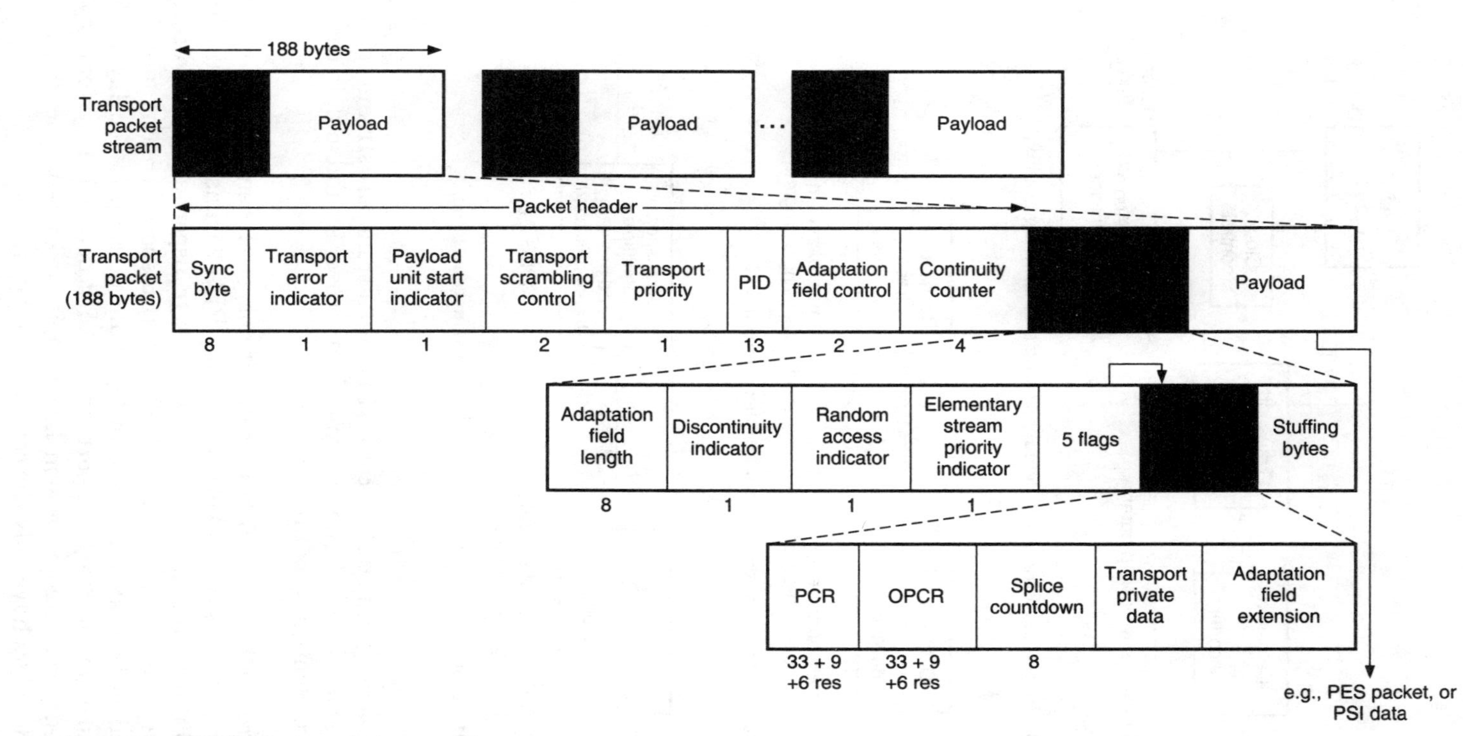

Figure 6.8 MPEG-2 transport stream syntax.

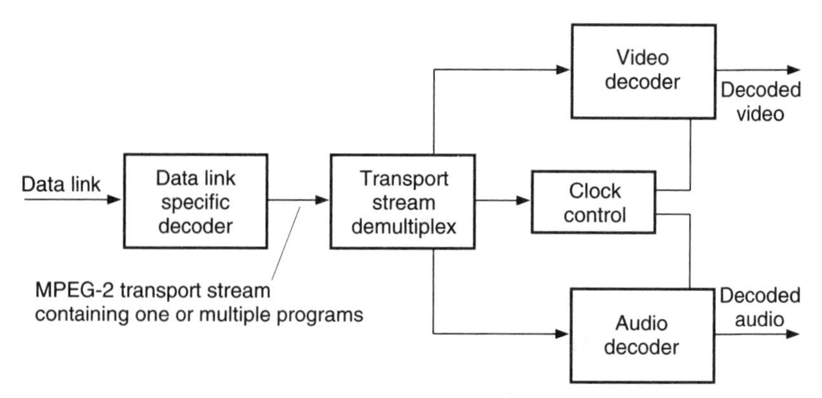

Figure 6.9 Prototypical transport demultiplexing and decoding example.

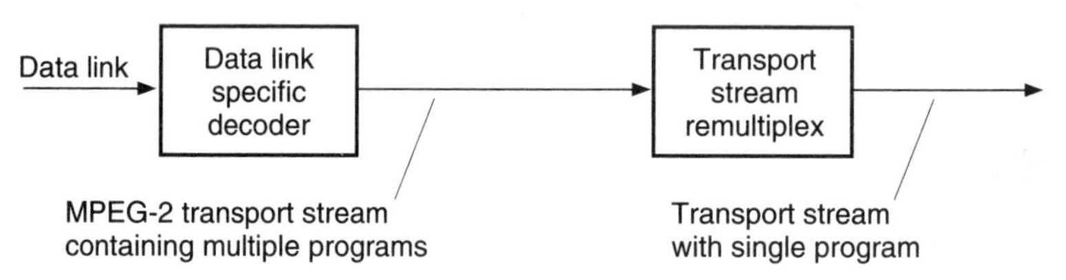

Figure 6.10 Prototypical transport multiplexing example.

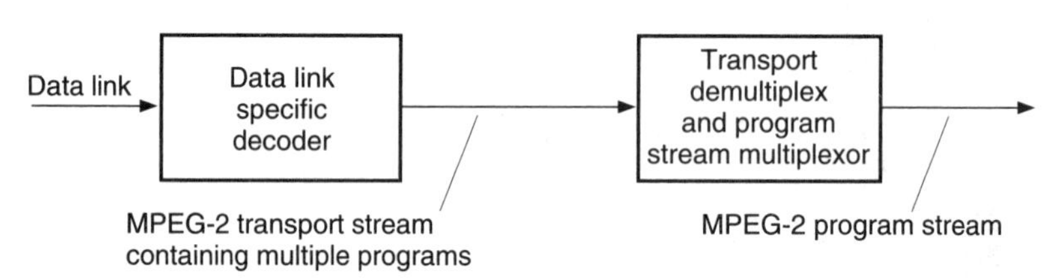

Figure 6.11 Prototypical transport to program stream conversion.

5. Take a program stream, convert it into a transport stream to carry it over a lossy environment, and then recover a valid and, in certain cases, identical program stream.

Figures 6.9 and 6.10 illustrate prototypical demultiplexing and decoding systems that take as input an MPEG-2 transport stream. Figure 6.9 illustrates the first case, where a transport stream is directly demultiplexed and decoded. MPEG-2 transport streams are constructed in two layers: a system layer and a compression layer. The input stream to the transport stream decoder has a system layer wrapped about a compression layer. Input streams to the video and audio decoders have only the compression layer.

Operations performed by the transport stream decoder either apply to the entire MPEG-2 transport stream (multiplex-wide operations), or to individual elementary streams (stream-specific operations). The MPEG-2 transport system layer is divided into two sublayers: one for multiplex-wide operations (the transport packet layer) and one for stream-specific operations (the PES packet layer).

A prototypical audio and video transport stream decoder is depicted in Fig. 6.9 to illustrate the function of a decoder. The architecture is not unique; some transport stream system decoder functions such as decoder timing control might equally well be distributed among elementary stream decoders and the data link specific decoder, but this figure is useful for discussion. Likewise, indication of errors detected by the data link specific decoder to the individual audio and video decoders may be performed in various ways, and such communication paths are not shown in the diagram. The prototypical decoder design does not imply any normative requirement for the design of an MPEG-2 transport stream decoder. Nonaudio and video data are also allowed but not shown.

Figure 6.10 illustrates the second case, where a transport stream containing multiple programs is converted into a transport stream containing a single program: in the case shown in Fig. 6.10, the remultiplexing operation will include the correction of PCR time stamps to account for the change in arrival of time stamps.

Figure 6.11 illustrates a case in which an MPEG-2 multiprogram transport stream is first demultiplexed and then converted into an MPEG-2 program stream.

Figures 6.10 and 6.11 indicate that it is possible and reasonable to convert between different types and configurations of MPEG-2 transport stream. There are specific fields defined in the transport stream and program stream syntaxes that facilitate the conversions illustrated. There is no requirement that specific implementations of demultiplexers or decoders include all of these functions.

6.6.2.1 Transport stream coding structure. The transport stream coding layer allows one or more groups of one or more elementary streams to be combined into a single stream. A group of elementary streams, with a common system_clock_frequency time base, is called a *program*. Data from each elementary stream are encoded and multiplexed together with information that allows elementary streams within a program to be replayed in synchronism.

An MPEG-2 transport stream consists of one or more programs, each containing one or more elementary streams and other streams multiplexed together. Each elementary stream consists of access units, which are the coded representation of presentation units. The presentation unit for a video elementary stream is a picture. The corresponding access unit includes all the coded data for the picture. The access unit containing the first coded picture of a group of pictures also includes any preceding data from the group of pictures, as defined in Part 2 of the MPEG-2 standard, starting the group_start_code. The access unit containing the first coded picture after a sequence header, as defined in Part 2 of MPEG-2, also includes a sequence header. The sequence_end_code is included in the access unit containing the last coded picture of a sequence. The presentation unit for an audio elementary stream is the set of samples that correspond to samples from an audio frame. (See Part 3 of the MPEG-2 standard for the definition of an audio frame.)

Elementary stream data are carried in PES packets. A PES packet consists of a PES packet header followed by packet data, as seen in Fig. 6.7. PES packets are inserted into transport packets, as seen in Fig. 6.8. The first byte of each PES packet header is located at the first available payload location of a transport packet.

The PES packet header begins with a 32-bit start code that also identifies the stream to which the packet data belong. The PES packet header may contain decoding and/or presentation time stamps (DTS and PTS) that refer to the first access unit that commences in the packet. The PES packet header also contains a number of flags opening up a range of optional fields. The packet data contain a variable number of contiguous bytes from one elementary stream.

Transport packets begin with a 4-byte prefix, which contains a 13-bit packet ID (PID). The PID identifies, via the program specific information (PSI) tables, the contents of the data contained in the transport packet (see below). Transport packets of one PID value carry data of one and only one elementary stream. The PSI tables are carried in the transport stream. There are four PSI tables:

- Program association table
- Program map table
- Network information table
- Conditional access table

These tables contain the necessary and sufficient information to demultiplexer and present programs. The program map table specifies, among other information, which PIDs and, therefore, which elementary streams are associated to form each program. This table indicates the PID of the transport packets that carry the PCR for each program.

Transport stream packets may be null packets. Null packets are intended for padding of the transport stream. They may be inserted or deleted by remultiplexing processes, and, therefore, the delivery of the payload of null packets to the decoder cannot be assumed.

PID. The payload_unit_start_unit_indicator contained in the transport stream is a 1-bit flag that has normative meaning for a transport packet that carries PES or PSI data. When the payload of the transport packet containes PES data, the payload_unit_start-indicator has the following signification: 1 indicates that the payload of this transport packet will commence with a PES packet header, and 0 indicates there is no PES header in this transport packet payload. When the payload of the transport packet contains PSI data, the payload_unit_start_indicator has the following significance: If the transport packet carries the first byte of a PSI section, the payload_unit_start_indicator value is 1, indicating that the first byte of the payload of this transport packet carries the pointer_field. If the transport packet does not carry the first byte of a PSI section, the payload_unit_start_indicator value is 0, indicating that there is no pointer_field in the payload (the meaning of this bit for private-data-carrying transport packets is not defined in the international standard).

6.6.3 Program stream

As in the case of the transport stream, a prototypical audio and video program stream decoder system is depicted in Fig. 6.12 to illustrate the function of a decoder. The architecture is not unique. System decoder functions, including decoder timing, might equally well be distributed among elementary stream decoders and the medium specific decoder, but this figure is useful for discussion. The prototypical decoder design does not imply any normative requirement for the design of an MPEG-2 program stream decoder. Nonaudio and video data are also allowed but not shown.

The prototypical MPEG-2 program stream decoder shown in Fig. 6.6 is composed of system, video, and audio decoders conforming to Parts 1, 2, and 3, respectively, of the MPEG-2 standard. In this decoder, the multiplexed coded representation of one or more audio and/or video streams is assumed to be stored on a *digital storage medium* (DSM), or network, in some medium-specific format. The medium-specific format is not governed by the MPEG-2 standard, nor is the medium-specific decoding part of the prototypical MPEG-2 program stream decoder.

Figure 6.13 depicts the program stream syntax*.

---

* To understand the full functionality of the program stream, including the exact meaning of all fields, the reader is referred directly to the MPEG-2 standard.

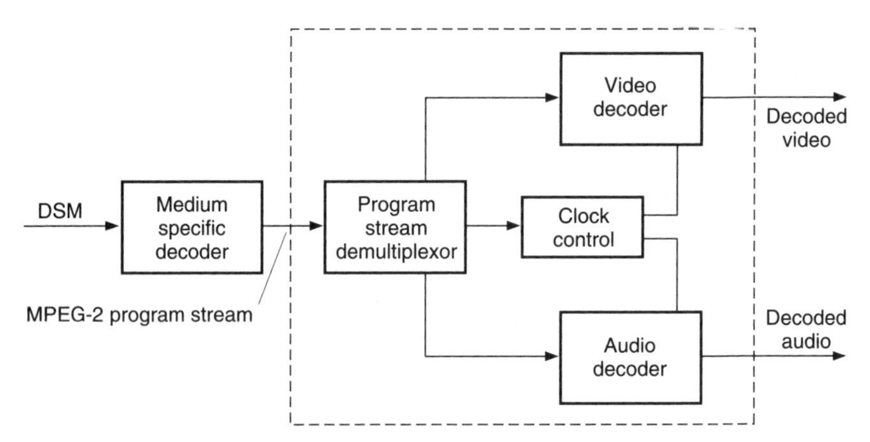

Figure 6.12 Prototypical program stream decoder.

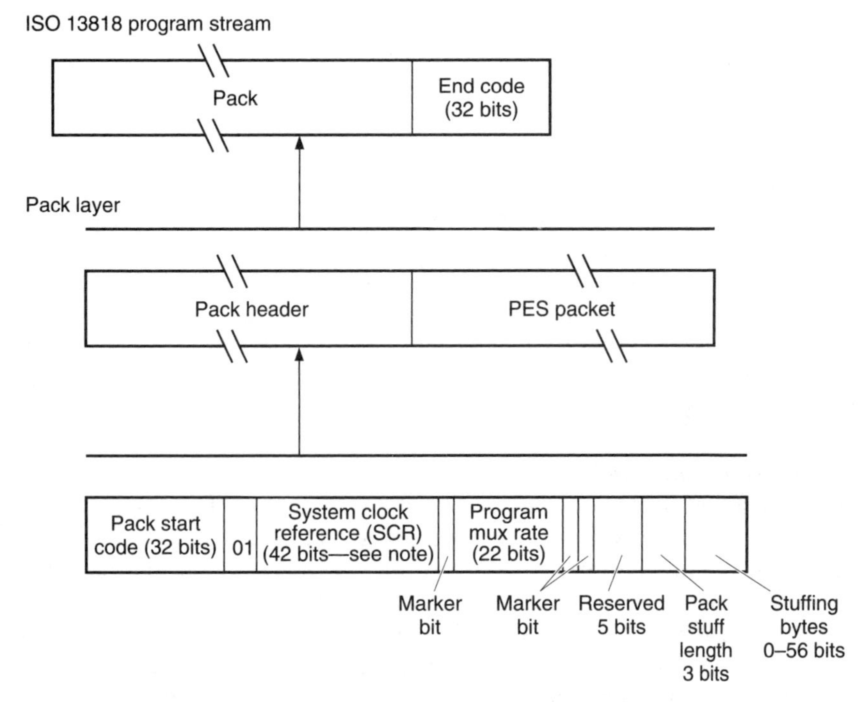

Note: SCR is a 42-bit number coded in four segments as follows:
3 bits (bits 32–30) system clock reference base
1 marker bit (additional to the 42 bits)
15 bits (bits 29–15) system clock reference base
1 marker bit (additional to the 42 bits)
15 bits (bits 14–0) system clock reference base
1 marker bit (additional to the 42 bits)
9 bits system clock reference extension

Figure 6.13 Program stream pack.

The prototypical decoder accepts as input an MPEG-2 program stream and relies on a program stream decoder to exact timing information from the stream. The program stream decoder demultiplexes the stream, and the elementary streams thus generated serve as inputs to video and audio decoders, whose outputs are decoder video and audio signals. Included in the design, but not shown in the figure, is the flow of timing information among the program stream decoder, the video and audio decoders, and the medium specific decoder. The video and audio decoders are synchronized with each other and with the DSM using this timing information.

MPEG-2 program streams are constructed in two layers: a system layer and a compression layer. The input stream to the program stream decoder has a system layer wrapped about a compression layer. Input streams to the video and audio decoders have only the compression layer.

Operations performed by the program stream decoder either apply to the entire MPEG-2 program stream (multiplexer-wide operations), or to individual elementary streams (stream-specific operations). The MPEG-2 program system layer is divided into two sublayers: one for multiplexer-wide operations (the pack layer), and one for stream-specific operations (the PES packet layer).

6.6.4 Conversion between transport stream and program stream

As noted, it is possible and reasonable to convert between transport stream and program streams by means of PESs. This results from the specification of transport stream and program stream. PESs may, with some constraints, be mapped directly from the payload of one multiplexed bit stream into the payload of another multiplexed bit stream. It is possible to identify the correct order of PES packets in a program to assist with this. Certain other information necessary for conversion, e.g., the relationship between elementary streams, is available in tables and headers in both streams. Such data must be available and correct in any stream before and after conversion. Not all transport streams will have been formed from program streams, but it is possible to create a valid program stream from a valid transport stream.

6.6.5 PES

Transport streams and program streams are each logically constructed from PES packets, as indicated in Fig. 6.8. PES packets are to be used to convert between transport streams and program streams; in some cases, the PES packets need not be modified when performing such conversions. PES packets may be much larger than the size of a transport packet.

A continuous sequence of PES packets of one elementary stream with one stream ID may be used to construct a PES. PES streams include *elementary stream clock reference* (ESCR) fields and elementary stream rate (ES_Rate) fields, and the elementary stream data must be contiguous bytes from the elementary stream in their original order. PES streams do not contain some necessary systems information that is contained in program streams and transport streams. Examples include the information in the pack header, program stream map, program stream directory, program map table, and elements of the transport packet syntax.

The PES stream is a logical construct that may be useful within implementations of this standard; however, it is not defined as a stream for interchange and interoperability. Applications requiring streams containing only one elementary stream can use program streams or transport streams that each contain only one elementary stream. These streams contain all of the necessary system information. Multiple program streams or transport streams, each containing a single elementary stream, can be constructed with a common time base and therefore carry a complete program, i.e., with audio and video.

6.6.6 Timing model

MPEG systems, video and audio, all have a timing in which the end-to-end delay from the signal input to an encoder to the signal output from a decoder is a constant. This delay is the sum of encoding, encoder buffering, multiplexing, communication or storage, demultiplexing, decoder buffering, decoding, and presentation. As part of this timing model, all video pictures and audio samples are presented exactly once, unless specifically coded to the contrary, and the interpicture interval and audio sample rate are the same at the decoder as at the encoder. The system stream coding contains timing information that can be used to implement systems that embody constant end-to-end delay. It is possible to implement decoders that do not follow this model exactly; however, in such cases, it is the decoder's responsibility to perform in an acceptable manner. The timing is embodied in the normative specifications of the MPEG-2 standard, which must be adhered to by all valid bit streams, regardless of the means of creating them.

All timing is defined in terms of a common system clock, referred to as a system time clock. In the program stream, this clock may have an exactly specified ratio to the video or audio sample clocks, or it may have an operating frequency that differs slightly from the exact ratio while still providing precise end-to-end timing and clock recovery. In the transport stream, the system clock frequency is constrained to have the exactly specified ratio to the audio and video sample clocks at all times; the effect of this constraint is to simplify sample rate recovery in decoders.

6.6.7 Conditional access

Encryption and scrambling for conditional access to programs encoded in the program and transport streams is supported by the system data stream definitions, but *conditional access* (CA) mechanisms are not specified in the MPEG-2 standard. The stream definitions are designed so that implementation of practical conditional access systems is reasonable, and there are some syntactical elements specified that provide specific support for such systems.

6.6.8 Multiplexer-wide operations

Multiplexer-wide operations include the coordination of data retrieval off the DSM or data link, the adjustment of clocks, and the management of buffers. The tasks are intimately related. If the rate of data delivery off the data link or DSM is controllable, then data delivery may be adjusted so that decoder buffers do not reach illegal states of overflow or underflow; but if the data rate is not controllable, then elementary stream decoders must slave their timing to the data source to avoid overflow or underflow.

Program streams are composed of packs whose headers facilitate the tasks listed earlier. Pack headers specify intended times at which each byte is to enter the program stream decoder and buffer management. The schedule need not be followed exactly by decoders, but they must compensate for deviations about it.

Similarly, transport streams are composed of transport packets with headers containing information that specifies the times at which each byte is intended to enter the program stream decoder from the data source. This schedule provides the same functions as that specified in the program stream.

An additional multiplex-wide operation is a decoder's ability to establish the resources that are required to decode a transport stream or program stream. The first pack of each program stream conveys parameters to assist decoders in this task. Included, for example, are the stream's maximum data rate and the highest number of simulations video channels. The transport stream likewise contains globally useful information.

The transport stream and program stream each contain information that identifies the pertinent characteristics of and between the elementary streams that constitute each program. Such information may include the language in audio channels, as well as the relationship between video streams when multilayer video coding is implemented.

6.6.9 Individual stream operations (PES packet layer)

The principal stream-specific operations are demultiplexing and synchronizing playback of multiple elementary streams.

6.6.9.1 Demultiplexing. On encoding, program streams are formed by multiplexing elementary streams, and transport streams are formed by multiplexing elementary streams, program streams, or the contents of other transport streams. Elementary streams may include private, reserved, and padding streams in addition to MPEG-2 audio and video streams. The streams are temporally subdivided into packets, and the packets are serialized. A PES packet contains codes from one and only one elementary stream.

In the program stream, both fixed and variable lengths are allowed. For transport streams, the packet length is 188 bytes. Both fixed and variable PES packet length are allowed but will be relatively long in most applications.

On decoding, demultiplexing is required to reconstitute elementary streams from the multiplexed program stream. Stream_id codes in program stream packet headers and packet ID codes in the transport stream make it possible.

6.6.9.2 Synchronization. Synchronization among multiple elementary streams is effected with *presentation time stamps* (PTS) in the program and transport bit stream. Time stamps are generally in units of 90 kHz, but the *system clock reference* (SCR), the *program clock reference* (PCR), and the optional elementary stream clock reference have extensions with a resolution of 27 MHz. Decoding of *N* elementary streams is synchronized by adjusting the decoding of streams to a common master time base rather than by adjusting the decoding of one stream to match that of another. The master time base may be one of the *N* decoder's clocks, the DSM or channel clock, or it may be some external clock. Each program in a transport stream that contains multiple programs has its own time base. The time bases of different programs with such a stream may be different.

Because presentation time stamps apply to the decoding of individual elementary streams, they reside in the PES layer of both the transport streams and program streams. End-to-end synchronization occurs when encoders have time stamps at capture time, when the time stamps propagate with associated coded data to decoders, and when decoders use those time stamps to schedule presentations.

Synchronization of a decoding system with a data source is achieved through the use of the SCR in the program stream and by the equivalent PCR in the transport stream. The SCR and PCR are time stamps encoding the timing of the bit stream itself in terms of the same time base as is used for the audio and video PTS values from the same program. Since each program may have its own time base, there are separate PCR fields for each program in a transport stream containing multiple programs. It is possible to have only one PCR for some or all programs in a multiprogram transport.

6.6.9.3 Relation to compression layer. The PES packet layer is dependent of the compression layer in some senses, but not in all senses. Specifically, it is independent in the sense that PES packets' payloads need not start at compression-layer start codes, as defined in Parts 2 and 3 of MPEG-2. For example, a video packet may start at any byte in the video stream. However, time stamps encoded in PES packet headers apply to presentation times of compression-layer construct (namely, presentation units).

References

1. D. Minoli, "Digital Video Compression: Getting Images Across a Net," *Network Computing,* July 1993, pp. 146 ff.
2. *Multimedia Week,* March 7, 1994, p. 3.
3. B. B. Mandelbrot, *The Fractal Geometry of Nature,* W. H. Freeman and Company, New York, 1983.
4. B. Doyle, "Crunch Time for Digital Video," *Newmedia,* March 1994, p. 47 ff.
5. L. W. Lockwood, "MPEG-2: A Wide Ranging Standard," *Communications Technology,* October 1993, p. 16 ff.
6. R. E. Chalfant, "MPEG-II with B-Frames: Video Compression Standard for the Decades to Come," *Communications Technology,* April 1993, p. 42 ff.
7. M. Barezzani et al., "Compression Codecs for Contribution Applications," *Electrical Communication,* 3rd Quarter, 1993, p. 220 ff.
8. T. Wechselberger, "Conditional Access and Encryption Options for Digital Systems," Communications Technology, November 1993, p. 20 ff.
9. D. Minoli and R. Keinath, *Distributed Multimedia Through Broadband Communications Services,* Artech House, Norwood, Mass., 1994.
10. O. Eldib and D. Minoli, *Telecommuting,* Artech House, Norwood, Mass., 1995.
11. P. E. Walker, "Squeezing the Picture: Video Compression," *Broadcast Engineer,* February 1994, p. 54 ff.
12. *Multimedia Week,* March 14, 1994, page 8.
13. G. K. Wallace, "The JPEG Still Picture Compression Standard," *Communications of the ACM,* vol. 34, no. 4, April 1991, p. 30 ff.
14. A. G. MacInnis, "The MPEG Systems Coding Specification," *Signal Processing: Image Communication,* Elsevier, New York, 4, 1992, pp. 153–159.
15. ISO/IEC JTC1/SC29/WG11 CD1-11172, *Coding of Moving Pictures and Associated Audio for Digital Storage Media at up to 1.5 Mbits/s, Part 1* (*Systems*), ISO, Geneva, Switzerland, November 1991.
16. P. Pancha and M. El Zarki, "MPEG Coding for Variable Rate Video Transmission," *IEEE Communications Magazine,* May 1994, p. 54 ff.
17. *Video Technology News,* December 20, 1993, p. 2.

Chapter

7

The Video Server

7.1 Introduction

Digital video servers will play a critical role in the advanced new video delivery services now being deployed, whether cable company provided or RBOC provided (i.e., VDT-based), particularly as their cost goes down from today's relatively high levels and the service penetration increases. VOD service is predicated on the principle of a library of stored movies, where, not only can the movies be selected in real time, but they can be controlled as if they were on a home VCR (i.e., they can be paused, rewound, fast-forwarded, etc.). Therefore, VOD services ideally require high-capacity, high-throughput, feature-rich digital video servers.

A video server comprises a storage subsystem* that supports a large number of simultaneous users, as well as control computer hardware and software to handle functions such as validation, authentication, billing, signaling support, consumer navigation, and communications. VIP or cable operators may choose a variety of storage technologies such as magnetic tape, magnetic disk†, and optical disk. They can also choose from a variety of platforms for the video server computer hardware. Massively parallel supercomputers are being positioned as the vehicle of choice for the control element of the video server for commercial-grade networks supporting hundreds of thousands of simultaneous viewers. Mainframe computers have also been offered as potential candidates. Finally, *reduced instruction set computing*

* In this description, the storage subsystem includes both the physical storage and the storage controller.

† Some use the term *disc;* except for the phrase *laser disc,* we use *disk* consistently.

(RISC)-based platforms are entering the market in some video server products*. More than just raw processing power, these servers need very high input/output (I/O) throughput. AT&T, IBM, Sun Microsystems, Hewlett-Packard, Digital Equipment Corporation, On-Demand Technologies, TRW, Oracle, Silicon Graphics, USA Video Corporation, Future Vision, and Bell Northern Research/Northern Telecom are all working on, or already have, video servers.[1]

Hardware vendors now tend to emphasize the technical capabilities of servers, and, indeed, there are no technology barriers to the commercial introduction of video servers because the technology is able to support up to 100,000 concurrent MPEG-2 video customers and store thousands of movies. Video operators, however, tend to focus on the cost of the new digital systems. Equipment vendors tend to talk extensively about VOD, while video operators talk about NVOD. Mainframe sales are going down at a rate of 5 to 10 percent a year, and supercomputer suppliers see a much-decreased military market. It is no surprise that vendors of this equipment are trying to generate new markets. These companies are retrofitting mainframes and/or adding new software to supercomputers to build video servers to support VOD, I-TV, and other services. Some observers see such business as the "basis for the revival of big iron."[2]

This chapter examines some technical and cost issues associated with video servers, as well as with optical storage technologies.

7.2 Server Technology and Approaches

7.2.1 Overview

Today many cable operators use tape recorders and automated tape libraries to support PPV and NVOD services. But maintaining a large number of videotape recorders leads to a "perpetual headache."[3] Evolving technology includes a laser disc player with four heads that allows a single laser disk to serve as many as four independent channels at once. Some operators have used a battery of stand-alone hard drives or PCs; while this solution begins to approach the digital era, these are not the ideal solution for large-scale VOD services. The video server is the next logical step in this process.

---

* RISC is now entering the high-end workstation market. RISC differs in approach from *complex instruction set* computing (CISC), now prevalent in the embedded base of PCs, as follows: in CISC, the microprocessor handles a variety of instructions (some multicycle, some microcoded); in RISC, only a subset of instructions (say 20 percent) are supported. The RISC microprocessors are faster (e.g., the Power-PC chip is claimed to be 40 percent faster than the Pentium chip), smaller, consume less power, and are less expensive to manufacture.

Computer servers have been around for many years; these servers support client access to databases. Hence, the basic concepts of server technology are well understood. However, the video server needs to support real-time output of multiple streams of video without variation in the time needed to transfer bits, frames, or cells of information (traditional data servers can occasionally slow down without a major degradation in the overall presentation of the data on the user's screen). To sustain the needed throughput, one requires: (1) a powerful processor, (2) a high-speed bus, and (3) high-capacity I/O devices and channels. Even then, there may be fluctuations in the information's streams. To smooth out these fluctuations, servers typically incorporate large output buffers. With this arrangement, data are retrieved very quickly from the hard drives, dumped into the buffer, and sent out over the transmission line at a preestablished transmission rate.

Figure 7.1 depicts a prototypical VIP/*Video Information Warehouse* (VIW). Figure 7.2 depicts a simplified view of IBM's ES/9000 video server. The IBM server has been adopted by the California Polytechnic Institute for distance-learning applications. Figure 7.3 shows a block view of the Digital Equipment Corporation video server, based on the Alpha APX platform. (NYNEX and Ameritech were expected to select Digital Equipment Corporation as the video server supplier;[4] previously, the company had also positioned itself as part of some U S WEST

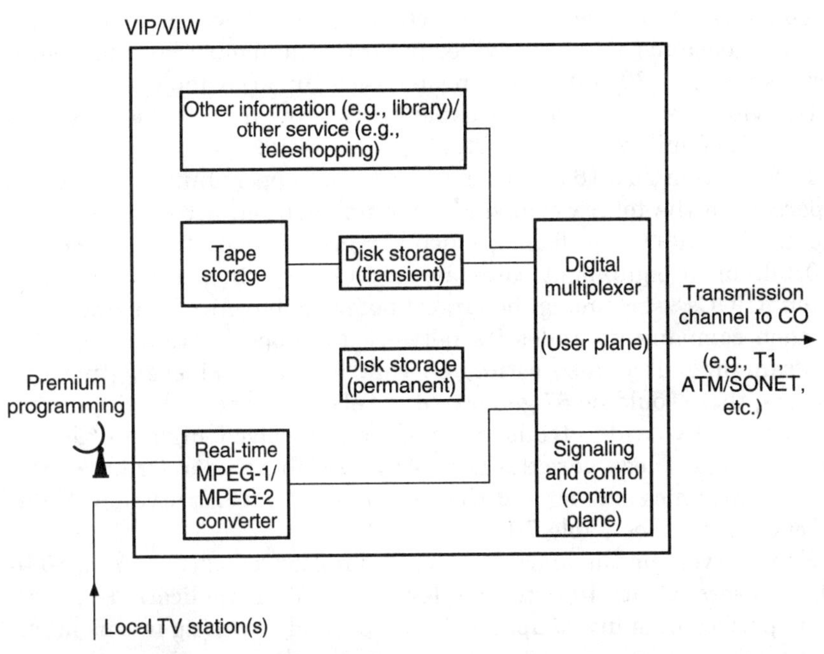

Figure 7.1 VIP/VIW.

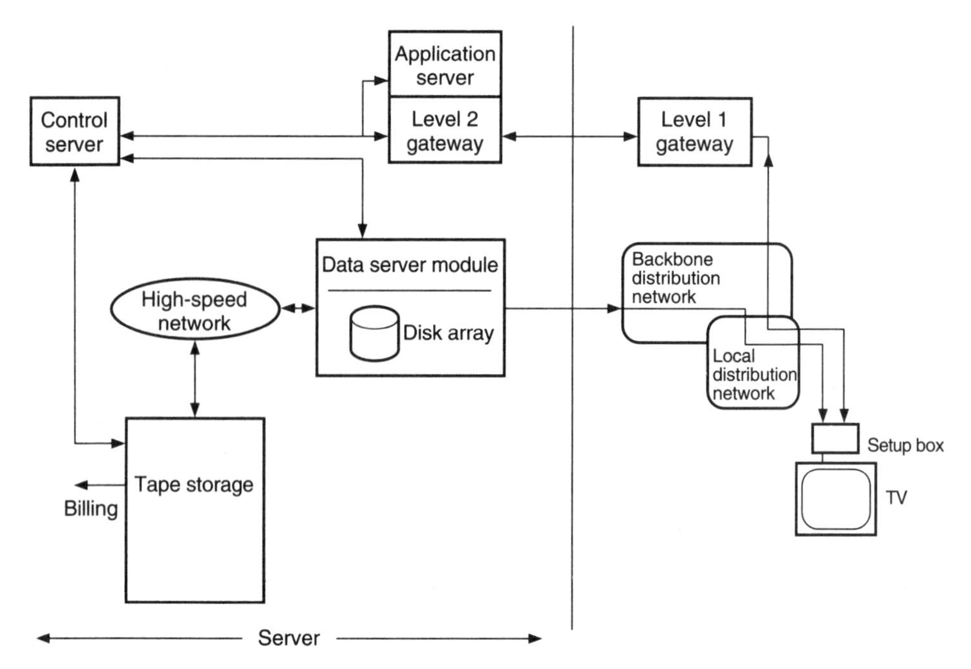

Figure 7.2 Simplified view of IBM's video server.

video trials.[5]) Video servers may include several thousand disk drives, tens of thousands of hours of video programs, and must be able to support as many as 30,000 simultaneous users (in large applications).

The video server market is expected to grow from $133 million in 1994 to $296 million in 1995, $1.327 billion in 1996, and $5.189 billion in 1997.[6] Although a $6 billion computer sales opportunity is just about 2 percent of the total worldwide computer market, accumulated over the 1994-to-1999 time frame, video servers are expected to generate $20 billion in equipment sales alone. Industry observers expect the seven RBOCs to be among the largest potential buyers of video servers, as each company upgrades its network to support interactive video programming. It is also estimated that for every $1 spent in video servers, there could be $7 generated in related services. At press time, there were two video trials planned that involved high-end digital video servers. These were the Time Warner Cable trial in Orlando, Fla. (Silicon Graphics server) and the Viacom trial in Castro Valley, Calif. (AT&T server). See Table 7.1.

Video servers on the market support simultaneous access from 80 to 30,000 users. Control of storage devices for VDT applications is still being perfected; many companies are reportedly working in this area.[3] Large video servers support 5000 or more simultaneous users; medium systems support 250 to 5000 users; small systems support less than

TABLE 7.1 Video Server Announcements by Mid-1994

Ameritech	Digital Equipment Corporation
Bell Atlantic	Oracle
NYNEX	Digital Equipment Corporation
Pacific Bell	Hewlett-Packard
Time Warner Cable	Silicon Graphics
U S WEST	Digital Equipment Corporation
Viacom	AT&T

250 users (e.g., a hotel). Many of the commercial systems in use today fit the "small" category and also use non-computer-based off-line storage (i.e., robotized analog video tapes with rugged video players). Video servers range in price from a relatively low end of a few hundred thousand dollars to $8 million or more. For example, a server provided by Oracle, Media Server, costs $7.5 million, plus a fee for software that is dependent on the number of users. The Hewlett-Packard HP 9000 server (in conjunction with the needed ATM switch) can cost in the range of $20 million. Servers are usually scalable; for example, the Oracle server can start out supporting 500 simultaneous users and grow to 30,000.

Servers capable of supporting 10,000 users cost around $6 million; hence, the per-*active* user cost is $600. Servers supporting 500 users cost around $1 million, and the per-*active* user cost is $2000. These numbers look attractive in a vacuum, but when considering, as covered in Chap. 2, that a typical consumer could order three PPV movies a year, the utilization could be small, and so the cost per viewing could be

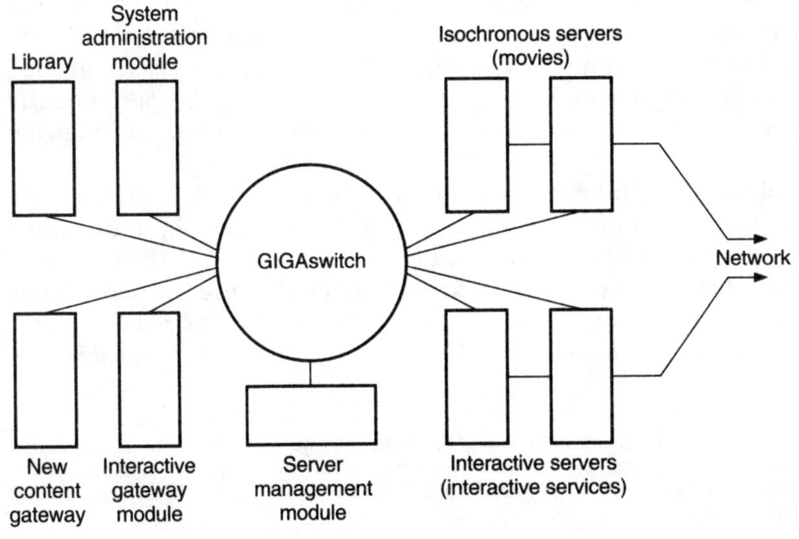

Figure 7.3 Digital Equipment Corporation's video server.

high, unless penetration increases. Additionally, this is hardware cost alone: these costs do not include program transmission costs, operations and maintenance costs, and program/program royalty costs (this topic is reexamined later).

The VIP may ultimately have to store thousands of movies to support VOD. A movie encoded in MPEG-1 requires about 1 GB of storage; MPEG-2 requires about 5 GB. Some of the video servers now being planned and under development (e.g., video jukeboxes) can support 10 to 15 TB, providing enough capacity for 10,000 MPEG-1 movies or 2000 MPEG-2 movies.

The VDT *network* of the future, supporting I-TV, VOD, teleshopping, and other services, will likely have the requirement of having to store the world's entire movie library, estimated at 65,000 films. Since each film requires 5 GB or more of storage when compressed* in MPEG-2 format, such a library needs about 325 TB. To that, one needs to add historical news footage and syndicated TV shows, electronic catalogs, interactive encyclopedias, library material, software, video games, etc. This material could be stored as an interconnected hierarchical system of VDT subnetworks and video servers, where, if such a user needs a movie not available locally, it could be obtained from another library, similar to the way a standard library works. See Fig. 7.4.

Much of the program material will be archived in the video server in near-line storage: batteries of automated jukeboxes that can mount tape cartridges or optical disks within a few seconds of a consumer's request. Upon such request, the server will likely copy the video material onto its local mass storage that has a lower access time than the tape or optical disks, possibly striping the data across arrays of hard disks for redundancy and expedited access. The server will buffer the data in *random access memory* (RAM) while transmitting it downstream to the consumer's setup box.[7] Frequently accessed material, such as the most popular movies and games, may be permanently cached in RAM, based on automatic statistical analysis of viewing habits.

RAM-based digital storage is still fairly expensive. The same is true for magnetic hard disks, as discussed later, even though, in historical terms, their costs have come down considerably. Some VIPs have contemplated the use of VCR tape for NVOD/VOD. Since the access time for tape is relatively slow, this solution is not the desired long-term approach. Besides access time, there is the issue of serial reusability: a

---

* A full-length box office feature would need about 1.5 terabits (1.5×10^{12} bits or 0.18×10^{12} bytes) of storage if it were stored uncompressed in video component format. If one were to store 65,000 movies, this would need 100×10^{15} bits (12×10^{15} bytes), that is, 100 pentabits (12 pentabytes).

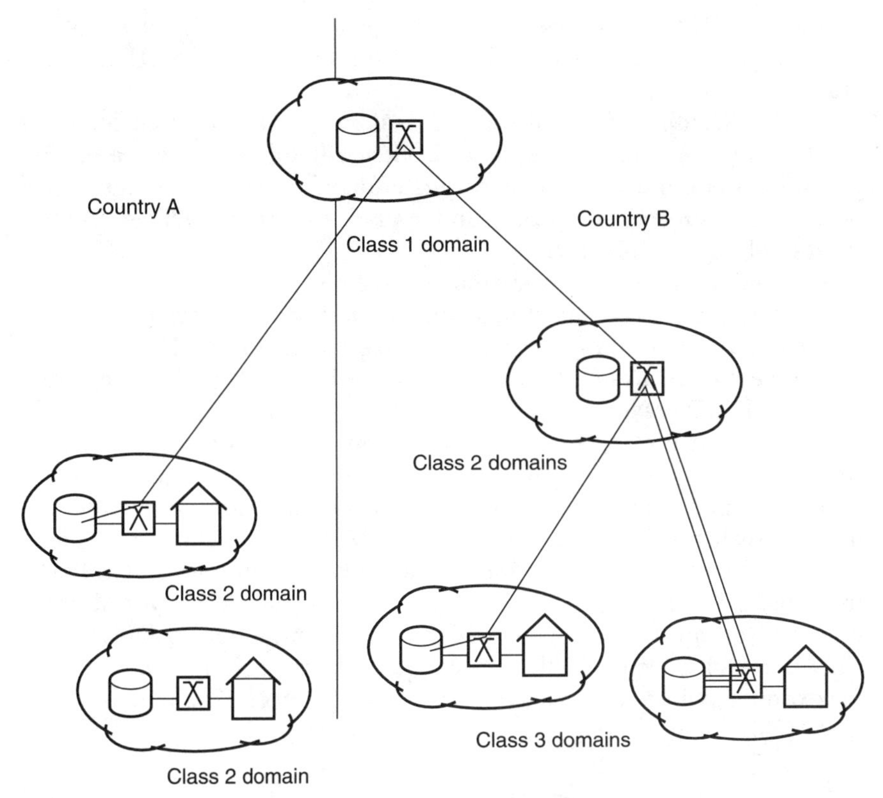

Figure 7.4 Hierarchical interconnection of video servers (future architecture).

VIP may have to play several versions of the movie at different times (say, within prime time) and for different customers. While this is easily accomplished when using disk-based storage (assuming no bus throughput bottlenecks), it would likely require multiple copies of a tape and multiple tape players in the predigital server arena. Also, VCR functions used with VOD are more difficult to implement when using tape. Figure 7.1 depicted the use of staging, where the movie that is stored on tape is sent, in small portions (typically 2 to 5 min of viewing time) to a disk drive, facilitating some functions (such as VCR functions). Tape libraries would most likely employ robotics.

Magnetic Disk Storage. Magnetic disk storage methods are better suited for real-time, on-line material than magnetic tape; access times are relatively fast. IBM, Sun Microsystems, Hewlett-Packard, Digital Equipment Corporation, On-Demand Technologies, USA Video Corporation, Future Vision, and Bell Northern Research/Northern Telecom are working on video servers based on magnetic storage devices. Current technology can store several hundred MPEG-1 or a few uncompressed NTSC movies. *Redundant arrays of inexpensive disks* (RAID) technology has been used (see Fig. 7.5 for an example of RAID Level 1 system).

Optical Storage. Optical storage technology can be used for digitized (compressed or uncompressed) video. Jukebox hardware on the market can store dozens or hundreds of disks; access to the disks can be by title, subject, or author, under processor control. This topic is treated at length in Sec. 7.3.

As noted, many first-generation servers now available are based on RAID; however, given the staggering amount of information involved, it is clear that second-generation servers appearing in 1996 to 1997 will have to be based on optical storage. For first-generation video servers, RAID results in high availability. This is accomplished through the use of one of six disk combinations called RAID levels (RAID Level 0 through RAID Level 5). The value of RAID is most important in systems that read and *write* information; one wants to protect freshly written information, and RAID keeps redundant copies (this must be done since the individual drives are not all that reliable in themselves). However, in movie playing, the data are never altered: it is written once and never needs to be instantaneously backed up. Therefore, for classical VOD (where the user interactions are minimal, for example, where one does not have a fully interactive movie where

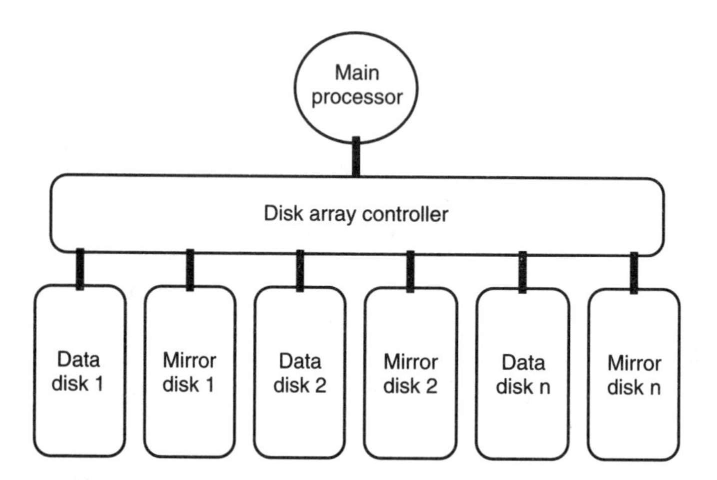

RAID Level 1 works by mirroring individual disk drives (data is stored onto a second disk every time a write instruction occurs). If a disk fails, the array controller automatically switches all read/write instructions to the alternate drive. This system is fairly expensive using a 1–1 backup rather than a 1–n backup.

Figure 7.5 RAID Level 1.

scenes may be written on the fly—something perhaps that might occur in the future), optical storage technology appears to be intrinsically better. Perhaps what is giving RAID a slight advantage now is the data access time: the access time is shorter on magnetic drives than in optical drives (this is based principally on the fact that the disk is larger, heavier, and the physical scan is longer). The gap, however, has been decreasing over time.

Figure 7.6 depicts some hierarchical memory arrangements that can be used in video servers (this concept is explored further in Sec. 7.3). Option (*b*) in Fig. 7.6 is more efficient than option (*c*) in terms of service features (particularly, interactivity) and memory. The case labeled (*c*) may use a VCR or laser disc as the off-line memory. Here a single customer may view a movie, unless multiple copies of the movie are available; alternatively, a group of viewers who are interested in the same movie at the same time can view such a program. This early approach, which has little or no interactivity, could be employed for NVOD services. The arrangement depicted in part (*b*) of Fig. 7.6 allows more sophistication. Movies to be played are first loaded from off-line or near-line storage to the magnetic disks. The memory controller provides the oversight function for the proper use of the magnetic disks. In this arrangement, it is possible to share one program among many subscribers. Of course, option (*a*) affords maximum flexibility. Here fast dynamic random access memory (DRAM) is added to the video server. DRAM is used to either store an entire movie or is used as a cache. When many subscribers (500 or more) are interested in some latest release to be played within some time window (say, prime time), then it pays to keep that movie in memory and play it out in a serially reusable manner. The cache architecture may be used in conjunction with some distribution approaches (particularly, ADSL) where the data may be burst across the network to remote buffering cards.

In some cases, it may be appropriate to cluster video servers to form a larger system. Clustering could be based on program content, hardware platforms (when more than one is present), or reliability considerations. Different vendors fit in different points of this clustered-to-loosely coupled systems continuum. Ultimately, the issue is one of performance: can a loosely coupled system meet the performance required to support VOD, or is a large symmetric multiprocessor system with a high-capacity back plane better able to meet those requirements?

Video servers usually include navigators, which are computer programs that categorize and present large amounts of data (e.g., program listings) in such a way that the user can quickly and conveniently get around the database. Navigation systems similar to Microsoft Windows and Apple Macintosh are being developed. Bell Atlantic, for

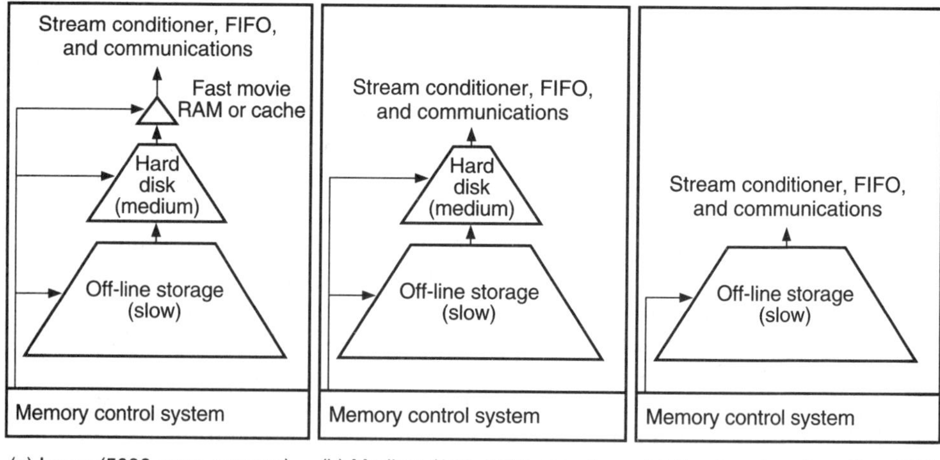

Figure 7.6 Server memory.

example, is experimenting with a concept called Stargazer, which aims at providing a full-motion, game-like navigation system that is simple to use. The screen could, for example, show an animated shopping mall; using a remote control pointer, a viewer would point to a store, enter the store, and receive information on that store's products.[3]

At a more macro level, looking inward to the video server from the network, an open (standard) interface at the VIW, rather than a proprietary interface, is desirable, so that existing and evolving (e.g., ATM-based) carrier transport services can be used. An immediate observation that emerges upon examination of the servers on the market is that there is no standardization in sight. Such standardization would, as per the classical argument, bring down the cost and enable interworking between video servers (for those operators that may expand by acquiring other systems) and between the video server and the setup box, which, if it occurred, would open the door for new services. The memory subsystem of a video server can be vendor-specific, where each vendor tries to outshine another vendor in terms of I/O performance and price, but looking toward the network, the transport element and the signaling element should be standardized to the extent possible. Vendors are already making excuses for divergent paths by stating that servers will have to vary in capacity to reflect different requirements of individual cable systems, that the various technology and market trials will point to different services, and that those differences need to be reflected in the servers. Although this is true, it should not have to imply lack of interoperability.

7.2.2 Video server components

Figure 7.7 depicts a somewhat more detailed view of the internals of a video server than the previous figures. The system-level components shown in the figure include*:

* The description that follows is representative of a video server such as Hewlett-Packard's.

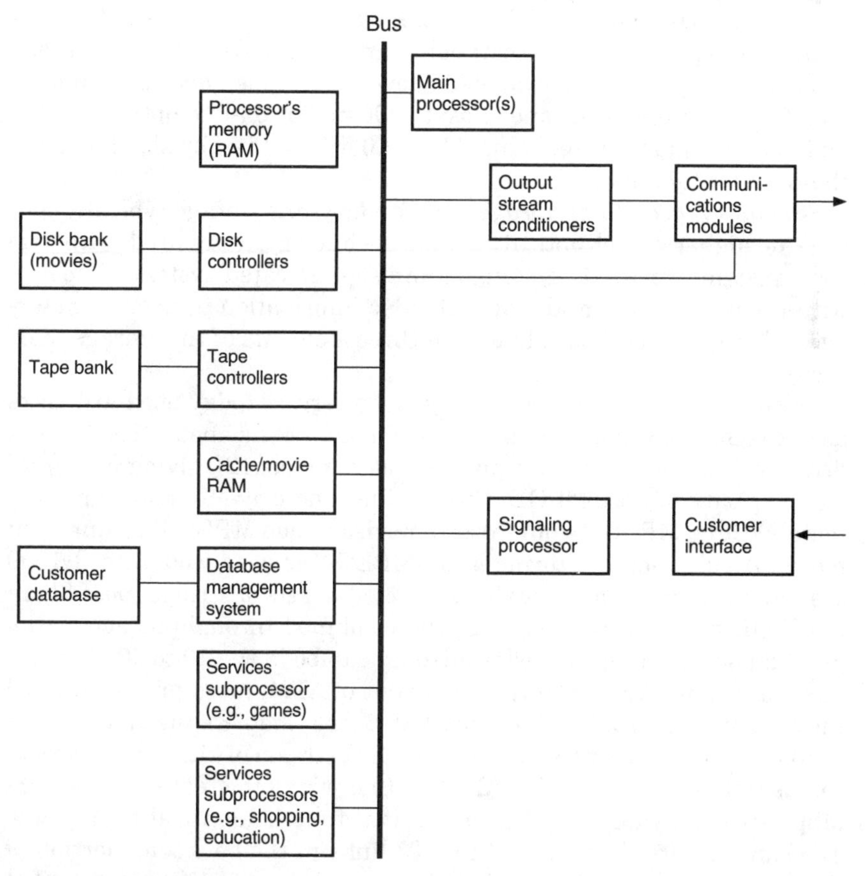

Figure 7.7 Video server—component level.

- Off-line
- Near-line storage
- On-line storage (e.g., hard disk)
- Storage controllers
- Steam controller
- System bus
- Movie RAM (DRAM cache)
- Main processor(s)
- Output stream conditioners
- Level 2 gateway-subscriber data

Tertiary (off-line) storage. Tertiary storage typically consists of VCRs, computer-based tapes (including DAT formats), optical disks in jukeboxes (CD-ROMs, WORMs, etc., see Sec. 7.3). Four characteristics of this storage are: (1) low cost, being around $5/GB; (2) large storage capacity, being in the TB range; (3) slow access time, measured in minutes for VCR tapes and tape silos, to 200 ms for on-line optical media; and (4) low transfer rate, being 0.2 to 2.0 MBps, thereby slowing down the transfer function.

Secondary (near-line) storage. Secondary-line storage typically consists of higher speed optical storage technology (compared with tertiary systems) or RAID systems. More sophisticated systems dispense with tertiary storage and store all video information in near-line storage; other systems choose to use all three elements of the storage hierarchy.

Primary (on-line) storage. Many video servers today use hard disks as the technology of choice. Magnetic memory costs about $1000/GB. A drive typically has many 5¼ platters (as many as 12), giving an aggregate capacity of about 3 GB; this enables one drive to store (approximately) three MPEG-1-compressed movies or one MPEG-2-compressed movie. A video server storing 1000 MPEG-2 movies would need 1000 systems. The drives are usually connected in sets of strings (possibly in the RAID configuration, or not); the combined throughput across the SCSI-2 ports feeding the disk controller can be in the 10 to 20 MBps.

Stream controllers. Until the advent of ATM, more precisely, until the advent of ATM video that uses AAL 5 (see Chap. 8), the video server needs to produce a continuous isochronous stream of bits for delivery to the user. For example, with ADSL-3, one wants to create a smooth 6-Mbps stream, where the interbit arrival time is identical for all bits. The same occurs if one uses T1 or T3 links in the backbone portion of the video distribution network. The same also occurs if one uses ATM

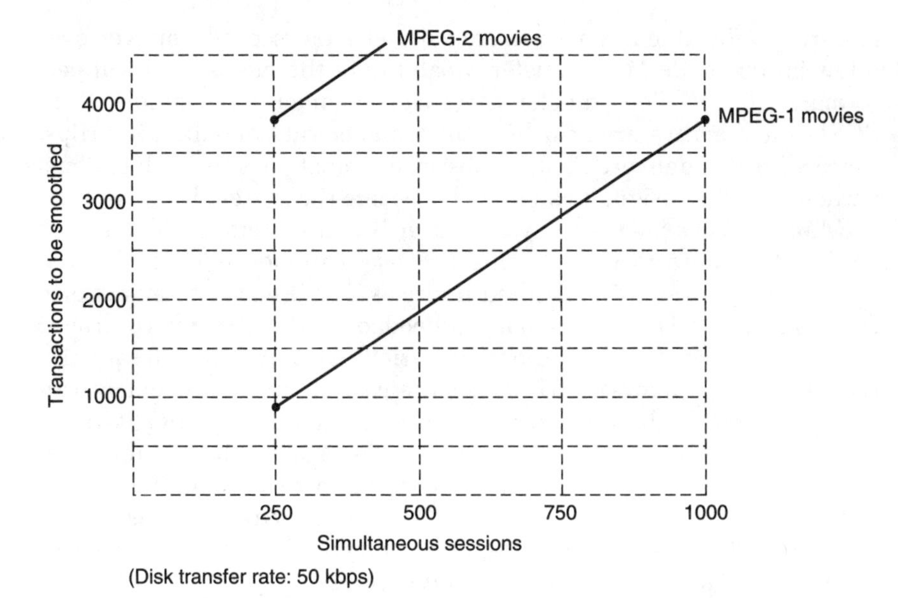

Figure 7.8 Throughput of stream controller.

but relies on circuit emulation using AAL1 (again, see Chap. 8)*. In these pre-ATM systems, the stream controller acts as the traffic manager for the video server. This device needs to support a high transaction rate. For example, if there are 1000 simultaneous streams (active customers), a movie information rate of 1.544 Mbps, and a disk data transfer segment size of 50 KB, then about 3860† transactions per second need to be handled by the controller (see Fig. 7.8).

Bus. Most computer processors use a bus-based architecture (massively parallel computers may not). Many video servers on the market use a high-speed bus. Since the information related to many movies comes across from the disk subsystem into the communication modules, and since video, in spite of some lulls‡, generates an isochronous stream

* Using AAL5, one is able to burst information at any instantaneous rate (up to some maximum peak rate) as it becomes available, without having to buffer or smooth it out.

† Calculation: (sessions)*(movie rate)/(throughput*8).

‡ Some engineers designing video compression systems, as well as others designing broadband communication systems, have taken the purely academic view of investigating the technical niceties of material with low-level video information output. At the consumer level, very few actual movies fit that category. Consumers gravitate to high-action movies and seem to prefer to be constantly bombarded with high-volume, high-entropy material such as music, as a trip to any local dentist, restaurant, or shopping mall will prove anecdotally. Therefore, we take the mental leap of questioning if unbuffered MPEG-2 streams over AAL5 really deliver, at the practical level, the bandwidth savings that designers are advertising.

of fairly predictable intensity averaged over a few seconds (or even over a few milliseconds), the transfer capability of the bus is a critical performance factor. The transfer rates can be huge; for example, if 10 SCSI-2 disk arrays are attached, the transfer rate may be 150 MBps. Since all traffic delivered to the subscribers must pass through the bus, without careful design, it can quickly become the system bottleneck.[8]

RAM and cache. Like any computer-based system, a video server needs RAM. More specifically, this discussion applies to RAM allocated to video single-cast, time-shifted multicast, or broadcast (this is also called *movieRAM*). This memory is needed by the streams controller to manage more efficiently the information associated with frequently requested video material. However, the cost of this arrangement can easily become high. For example, to store a 5-GB movie, it takes $100,000 in RAM memory ($20 per MB). Therefore, a RAM subsystem to store the 10 most popular movies would cost $1 million, in addition to the magnetic or optical storage. This expenditure becomes justifiable only when there is a large number of VOD users (as noted in Chap. 2, PPV services are not yet that popular).

Main processor(s). The main processor controls the data flow between the various components of the video server, as well as handling management and billing functions. New computer architectures (e.g., massively parallel systems such as those manufactured by nCube) may be required to support the number of simultaneous users in a large metropolitan area. Some systems use a mainframe processor or a RISC system.

Output streams conditioner. This subsystem adapts the video server to the transmission system at hand. Typically, this will be a traditional DS1 or DS3 link, and, even when ATM is used, many early systems (as well as hardware being made available by suppliers and public service offerings available from carriers), employ DS1 or DS3 circuit emulation. Therefore, buffering is needed to clock out the video stream at the appropriate constant rate. It is only with the availability of AAL5 ATM products and services that the controller will have to apply a more statistical type of conditioning; however, in this case, the conditioner will still have to buffer enough information to build an ATM cell, although the information cells do not have to be clocked out at a predetermined rate.

7.2.3 Parameters related to video servers

As implied by the previous discussion, there are some key parameters that characterize a video server.

Storage capacity. As discussed, this increases linearly with the number of movies, approaching a terabyte per 200 movies. See Table 7.2.

TABLE 7.2 Server Capacity

MPEG-2 movies	Total storage capacity (TB)
200	1
500	2.5
1,000	5
5,000	25
10,000	50
50,000	250

TABLE 7.3 Number of Drives per Server and Cost

MPEG-2 movies	Magnetic hard drives systems (@2.5 GB per system)	Optical hard drives (@7.5 GB per system)	Cost of storage (@$1000/GB, i.e., $1,000,000/TB)
200	400	133	$ 1,000,000
500	1,000	333	2,500,000
1,000	2,000	666	5,000,000
5,000	10,000	3,333	25,000,000
10,000	20,000	6,666	50,000,000
50,000	100,000	33,333	250,000,000

Disks per server. This parameter is important in terms of the number of physical ports that are required. The number depends on the capacity of the disk drive. Table 7.3 provides an estimate.

Bus capacity. For VOD, this increases linearly with the number of (simultaneous) streams/customers (in NVOD, PPV, or broadcast, this number would be an appropriate portion of the number of simultaneous users). Therefore, the bus capacity (for MPEG-2 movies) must be 6 Gbps per 1000 customers. As a comparison, small ATM switches now appearing support 2 Gbps; medium switches support 2 to 20 Gbps; large switches support 20 to 80 Gbps. Chapter 2 indicated that the movie rental business in the United States is $13.6 billion a year. That is $37.3 million per day. At $2.92 per movie (1994 Blockbuster Video rate), that equates to 12.77 million rentals per night in the United States. This is a rate of about one tape per day per eight households. Table 7.4 depicts what the demand for paid-movie watching is, in terms of city population size, along with the required bus capacity of a system serving that demand.

Costs of service. True VOD with full interactivity really costs.[8] The information previously provided can be used to estimate the cost of providing such service. Storing a movie takes $5000 of investment just for memory. A 36-month payback requires the recovery of about $5 per day per movie just for the storage. Since a video server is memory-intensive, one can assume a ratio such as 5:1 in storage memory to computing cost (in a PC, the ratio is the other way around, with 1:10

TABLE 7.4 Demand for Paid Movie Watching

City population	Estimated household	Total households watching paid movies per night	Required server/bus capacity (GBps) for 100% market share	Required server/bus capacity (GBps) for 10% market share
5,000	1,163	145	0.9	0.09
10,000	2,326	291	1.7	0.17
50,000	11,628	1,453	8.7	0.87
100,000	23,256	2,907	17.4	1.74
500,000	116,279	14,535	87.2	8.72
1,000,000	232,558	29,070	174.4	17.44
5,000,000	1,162,791	145,349	872.1	87.21
10,000,000	2,325,581	290,698	1,744.2	174.42

being more typical). This would mean a cost of $6 per movie per day for storage and computing. Additional assumptions are needed to complete the estimate. The model used here, based on the analysis of thousands of commercial narrowband networks, is as follows: hardware, when amortized, is usually 30 percent of a total network cost; transmission is another 30 percent; network management is 40 percent. Assuming that these numbers carry over to broadband*, a stored movie would have to fetch $20 per day, averaged across all the stored inventory, just in storage and networking costs (i.e., this does not include programming and royalties costs). Although this is easily accomplished for newly released movies, it may not be as easy to achieve as an average across all movies. Therefore, it is better to focus on the aggregate: a bank of 1000 movies would have to generate at least $600,000 per month to break even.

Tables 7.5 and 7.6 put these (admittedly simple) calculations into perspective. Table 7.5 allocates the cost across the population (this is not to imply that an average consumer would be charged the average fee shown whether the consumer used VOD or not), but it provides a sense of the per-capita costs. Table 7.6 translates the recovery cost in terms of movies watched, as previously derived. Both tables show that for a digital library of 1000 movies, the recovery cost makes the system viable, with today's costs, only for locations with 1 to 10 million people, and that the penetration must be fairly high. These numbers, which do not even include the programming cost and the cost of the setup box (communication costs were empirically accounted for), may be surprising at first.

* This would require a rethinking of the tariffs on the part of telecommunication carriers, since traditionally bandwidth has been sold by the weight rather than by the service.

TABLE 7.5 Equivalent per Capita Chargeable Monthly Rates to Recover Cost of 1000-Movie Library

City population	Estimated household	Total households watching paid movies per night	Equivalent hardware and communications *monthly* per household costs based on a 1000-movie library (50% VDT-service penetration)	Equivalent hardware and communications *monthly* per household costs based on a 1000-movie library (10% VDT-service penetration)
5,000	1,163	145	$8,275	$41,379
10,000	2,326	291	4,137	20,689
50,000	11,628	1,453	827	4,137
100,000	23,256	2,907	414	2,068
500,000	116,279	14,535	83	414
1,000,000	232,558	29,070	41	207
5,000,000	1,162,791	145,349	8	41
10,000,000	2,325,581	290,698	4	21

TABLE 7.6 Per Viewing Chargeable Rates to Recover Cost of 1000-Movie Library

City population	Estimated household	Total households watching paid movies per night	Movie events per month for 50% penetration	Cost per movie	Movie events per month for 10% penetration	Cost per movie for hardware alone
5,000	1,163	145	2,175	$275.86	435	$1,379.31
10,000	2,326	291	4,365	137.46	873	687.29
50,000	11,628	1,453	21,795	27.53	4,359	137.65
100,000	23,256	2,907	43,605	13.76	8,721	68.80
500,000	116,279	14,535	218,025	2.75	43,605	13.76
1,000,000	232,558	29,070	436,050	1.38	87,210	6.88
5,000,000	1,162,791	145,349	2,180,235	0.28	436,047	1.38
10,000,000	2,325,581	290,698	4,360,470	0.14	872,094	0.69

In vacuum, the hardware* cost of between $600 to $2000 per active user mentioned earlier looks small, but there is the issue of fill. This concept is well understood in the airline business: a $30 million plane may be cheap if it carries 300 passengers on every flight, but if it flies with only 3 passengers, the per actual passenger cost will be high. Table 7.7 takes another angle into this argument. Assume that a system supporting 10,000 simultaneous users needs to be amortized in 36 months. Each night the system theoretically supports a maximum of 20,000 viewings (say, by being used twice); also assume that it could be used for 10,000 other viewings during the day, for a total of 30,000 viewings a day. In 36 months this is 32,850,000 potential viewings. Just looking at the hardware cost, that would correspond to 6,000,000/32,850,000 or $0.18 per movie. However, as the utilization goes down,

* As noted, hardware costs alone do not include program transmission costs, operations and maintenance costs, and program-program royalty costs.

the cost per movie goes up. For example, if the utilization is only 1 percent, that is, only 300 VOD and NVOD requests per day are generated, then the hardware cost per movie would have to be $18; for a utilization of 10 percent (3000 viewing a day), the hardware cost per movie would be $1.8. Table 7.7 shows that for the cost to be reasonable, a base of about 2 to 10 million potential subscribers is needed for an RBOC example, and a base of 0.5 to 1 million potential subscribers is needed.

The aforementioned discussion used an empirical assessment of the cost of networking, extrapolated from narrowband. Reference 8 amplifies the discussion as follows. Each dedicated subscriber has a dedicated stream and dedicated channel. If 10,000 subscribers have access to 2000 server streams, then at least 2000 channels to the subscribers must be provided. In an FTTN architecture, five fibers could feed 10,000 homes with 2000 subscribers per home. With 400 streams per fiber, 2000 streams could be provided. As noted in Chap. 4, about six MPEG-2 streams per 6 MHz can be supported with advanced digital techniques, implying that 400 MHz of bandwidth must be allocated over the fiber. This system is complex and expensive: it requires 67 modulators per fiber to support the stream load. The conclusion is that true VOD is extremely expensive; even as prices fall in the future, VOD may not be feasible for small systems.[8] NVOD, on the other hand,

TABLE 7.7 Per-Viewing Hardware Chargeable Rates to Recover Cost of Video Server Supporting 10,000 Simultaneous Users

City population	Estimated household	Total households watching paid movies per night	Movie events per day for 10% penetration	Movie events per month for 10% penetration	Utilization (assumes ⅓ of movies occur simultaneously)	Cost per movie for hardware alone
RBOC example (low penetration)						
5,000	1,163	145	15	436	0.00	$382.22
10,000	2,326	291	29	872	0.00	191.11
50,000	11,628	1,453	145	4,360	0.00	38.22
100,000	23,256	2,907	291	8,721	0.01	19.11
500,000	116,279	14,535	1,453	43,605	0.05	3.82
1,000,000	232,558	29,070	2,907	87,209	0.10	1.91
5,000,000	1,162,791	145,349	14,535	436,047	0.48	0.38
10,000,000	2,325,581	290,698	29,070	872,093	0.97	0.19
Cable company (high penetration, but with possible wireless competition)						
5,000	1,163	145	109	3,270	0.00	50.96
10,000	2,326	291	218	6,541	0.01	25.48
50,000	11,628	1,453	1,090	32,703	0.04	5.10
100,000	23,256	2,907	2,180	65,407	0.07	2.55
500,000	116,279	14,535	10,901	327,035	0.36	0.51
1,000,000	232,558	29,070	21,802	654,070	0.73	0.25
5,000,000	1,162,791	145,349	109,012	3,270,349	Overload	Overload
10,000,000	2,325,581	290,698	218,023	6,540,698	Overload	Overload

would be more realistic, both in terms of transmission and of bus capacity; for example, if the start time is 1 min apart (so that the average wait would be 30 s), only 60 copies of the movie need to traverse the bus. As noted in Chap. 1, the expected transition in the delivery of digital video is from multiple-channel PPV, to NVOD in the immediate future, to VOD at some future point.

7.2.4 Some illustrative equipment examples

This section covers, for illustrative purposes, some examples of actual video servers. Vendor equipment is described only to give a sense of the technology; this information has not been counterverified with the vendors. No recommendation whatsoever is implied.

Figure 7.3 depicted Digital Equipment Corporation's video server based on the RISC processor (Alpha APX). Digital's architecture has been noted for the scalable design, enabling the operator to start small and grow as requirements increase. The system has a hierarchical storage architecture, as discussed in the previous sections, that keeps the most popular movies in RAM, with movies that are less in demand stored on hard disk arrays; a larger collection of films is also available in a tape library. The company was also in the process of developing a video version of the Alpha processor (called Rawhide) that is designed to handle up to 100,000 concurrent sessions. Unlike mainframe-based products, the APX-video server technology is based on a distributed client-server architecture that enables video servers to be installed either at a central site or elsewhere in the network.[9] The video server incorporates the company's StorageWorks disk storage arrays for hierarchical storage, digital linear tape library subsystems for archival or bulk video storage of movies requested less often, hard disks for active storage, and solid-state storage for buffering and video management. In addition, the system includes an interactive gateway unit, a server management unit, and the GIGAswitch high-speed networking switch (used initially in fiber distributed digital interface applications). The server management unit incorporates Unix-based video server management software that allows integrators and operators to add in billing systems, system administration software, and other specific applications. (In other servers, one is forced to use the software provided by the vendor, rather than any software of choice.) Reportedly, the APX-based server is able to operate with any vendor's setup box and with any networking technology, including ATM.

The Hewlett-Packard video server is built on as an integration of various hardware elements, including off-the-shelf hardware. The system follows the general architecture described in Sec. 7.2.2. It links two standard 800 series Precision Architecture computers with a specialized video server engine. One HP 800 server acts as a transaction processor, relaying the customers' requests over a proprietary link to the video server engine. The second HP 800 acts as a linked billing system. The server uses 2-Gbit SCSI-2 hard disks, 2 GB of RAM and ATM interfaces to communication switches.[10] As initially configured, the server could handle multiple VOD streams for up to 500 homes.

Silicon Graphics Inc. (SGI) is pursuing a software-based architecture using its scalable multiprocessing Challenger server. SGI's system of a ring of eight Challenger servers has been used in the Time Warner Cable trial of 4000 homes. In the trial, the customers will have access to 3000 video selections over an ATM network. The server uses digital tape, hard drives, and RAID in a hierarchical storage management scheme. These storage elements are tied together with a Session Manager, which is a proprietary software layer for VCR-like control over video data streams that run over the real-time extensions of SGI's Irix system software of the server.[10] The disks are 2-Gbit SCSI-2 drives (4-Gbit drives were expected to be used as soon as these become available). The total system storage capacity is 0.5 to 16 GB of main memory (movieRAM), 200 to 300 GB of video data on hard disk primary storage, 6 TB of RAID secondary storage (for 500 videos), and enough tertiary storage on digital tapes and magneto-optic disks to store 2500 videos. A navigator interface application has also been developed.

Oracle claims that it built its video server in 60 days, taking "the elite of the technology crop."[11] Using many processors connected in a massively parallel manner, the system achieves PC-level computing costs of $100 per MIPS. The vendor claims that the system will deliver VOD at $500 per video stream, this being one-tenth of the cost of a mainframe-based server.[12] The vendor also projects the cost to be $100 per video stream in 1995.

In addition to the mainframe approach, IBM has elected a clustered approach to video servers, using Groupe Bull's microprocessing technology to interconnect RS/6000s via Fibre Channel Standard facilities.

Sun Microsystems signed a software contract with Starlight Networks (see later discussion) to port Starlight's streaming video software to SparcCenters and SparcClusters.[10] The SparcCenter (Dragon) system (codeveloped by Sun and Xerox) can reportedly carry the backplane bandwidth that is typical of VOD services. SparcClusters (where SparcServer 10s are connected with ethernet links) can be upgraded to SparcCenters using the Fibre Channel Standard facilities or ATM.

In early 1994, Microsoft demonstrated a prototype server that featured massively parallel microprocessors.[13] This was followed by the formal introduction of the video server software, code-named *Tiger*, for

Windows NT-based systems. The software runs on NT Advanced Server and requires an Intel-based multiprocessing hardware platform.[14]

Sammons used video server technology developed by the Home Shopping Network to trial a system supporting four PPV movies in half-hour staggered starts.[15]

7.2.4.1 Corporate video applications. Although corporate video applications are not as demanding as those of carriers, usually employ LANs for distribution, and the quality of the video is lower, there are key similarities in approach. This section briefly surveys this tangential but related topic in terms of the type of equipment becoming available at press time to document the direction of the technology. Observers, however, note that the "corporate use of video servers is still limited."[16] As noted earlier, by 1997, the video server market may reach 2 percent of the total computer sales volume. Therefore, it is clear that the computer industry interest goes well beyond initial deployments of consumer video: selling services for VOD could help develop markets such as video-based business-information services, for corporate applications.[2] In particular, system integrators see opportunities to develop turnkey systems for corporate use, in addition to the carrier-cable company opportunities.[17] Specific applications include video database applications for advertising agencies and film production companies, training and education, store-and-forward video messaging, and videoconferencing. Video servers have the potential to replace desktop CD-ROM-based systems for training.

In 1994, The Network Connection Inc. introduced the Symmetric Multiprocessing M2V Multimedia Video Superserver and the Video Compression Station using Microsoft Corporation NT Advanced Server as the server's operating system. All files are compressed and run at 30 fps using MPEG-2 compression in the 1- to 5-Mbps range. The M2V supports Unix, Macintosh, and PCs; it can store and retrieve Intel's *Production Level Video* (PLV), *Digital Video Interactive* (DVI), JPEG, and Wavelet files with resolution up to 1024 × 768. The M2V is configured with 36 GB and can be extended to 168 GB, handling 120 simultaneous video users on a single system.[17]

In addition to providing video servers for the high-end market, Hewlett-Packard also introduced in late 1994 a corporate version of the video server. The server is targeted at videoconferencing, distance learning, and training applications.[18] The core of both server products is the video transfer engine that can manage the continuous and simultaneous flow of video material from disk drives to remote clients, over LANs or ATM-based WANs.

IBM enhanced its LAN-based Ultimedia server to provide greater on-line content storage and to support more users.[16] The server has a capacity of 100 GB and supports 40 simultaneous users. Multiple servers can be connected over a LAN to expand the base capacity. FDDI backbones are also supported.

Starlight Networks, who was an early entrant in the corporate video server market, upgraded its StarWorks video server software to support 40 users. The software runs on a 80486 *Extended Industry Standard Architecture* (EISA) server (with 25 MB of RAM) and supports DOS, Windows, and Macintosh clients.[19] The video storage was raised to 10 GB; video-audio streaming was upgraded to 50 Mbps, and bandwidth reservation is available to users with FDDI. The system also operates on Token Ring and Apple networks, delivering 6 to 8 Mbps.[20]

7.3 Storage Technology

The purpose of this section* is to survey a variety of optical video storage systems that can be used by VIPs†. This type of storage is also used in multimedia and imaging applications. CD-ROMs, WORMs, recordable CDs, and magneto-optic storage, are briefly examined.[21–24] Optical storage media are entering the market at rapid rate. To get a sense of the cost-effectiveness of optical storage at the business document level, note that the U.S. Congress Office of Technology Assessment reports that the annual cost of providing the *Congressional Record* to its repository libraries is $624 on paper, $84 on microfiche, and $10 on CD-ROM.[25] However, one of the challenges faced by the prospective users is that the proliferation of CD standards makes it difficult to decide which optical disk system to use.[26]

As noted earlier, for video applications, a typical movie encoded with MPEG-1 requires 1 GB and a movie encoded with MPEG-2 requires 5 GB. This requirement almost directly dictates the use of 12- or 14-in optical platters housed in jukeboxes. Table 7.8 depicts typical storage capacities for optical disks; for comparison, magnetic media are also included.[27]

7.3.1 A model of storage systems

Computer storage has traditionally been viewed as conforming to a hierarchy, implied in Sec. 7.2, modeled by a triangle (see, for example, Ref. 28). In this model, fast storage is found at the pinnacle, while

* This section is summarized from the book, *Imaging in Corporate Environments,* by D. Minoli, McGraw-Hill, 1994.

† As discussed in the previous section, a number of prototype or early-entry systems actually used magnetic storage; however, in the longer term, optical storage is the better approach.

TABLE 7.8 Typical Storage Capacities and Video Content

Medium	Size (in)	Min capacity (MB)	Max capacity (MB)*	Minutes of MPEG-1 video per disk†	Minutes of MPEG-2 video per disk†
CD-ROM‡	4.75	128	1,000	112	22
Hard drive	N/A§	20	800	89	18
Magnetic floppy	3.5	0.7	1.4	0.16	0.03
Magnetic floppy	5.25	0.35	1.2	0.14	0.03
Rewritable-erasable	3.5	128	256	28	5.6
Rewritable-erasable	5.25	512	1,024	113	23
WORM	5.25	600	1,280	142	28
WORM	12	4,400	9,000	1,000	200
WORM	14	6,800	10,200	1,133	227

* At press time.
† Based on maximum size.
‡ Typical capacity: 650 MB
§ N/A, not applicable.

larger, lower cost, lower performance storage is encountered as one descends the triangle toward the base. Figure 7.9 depicts this storage hierarchy, mapped to a traditional client-server model, including a VDT perspective. In the workstation or user device (e.g., setup box) a small amount of semiconductor memory (known as cache) is allocated to support fast, but expensive, access to the information. Cache retains data that are likely to be needed in the immediate future; accessing data across the network, as seen in Fig. 7.9 adds latency. In traditional client-server systems, some of the data can be retained in the local magnetic disk; this is effective, if the file is appropriately partitioned between the server and the client.

Figure 7.10 depicts the environment that one can expect in a video server system of the mid-1990s and beyond (in this figure of a VDT-like system, cache is shown as placed at the central office*; it can also be placed at the video server-VIP location). In this scenario, conventional hard drive disks have been replaced with RAIDs, and optical jukeboxes provide an intermediate level of storage, known as *near-line mode* that fits between the on-line mode and the off-line mode.

7.3.2 Optical storage developments

This section looks briefly at developments in optical storage in the past few years.

* An example where this may occur is if the central office acts as multicaster for PPV service.

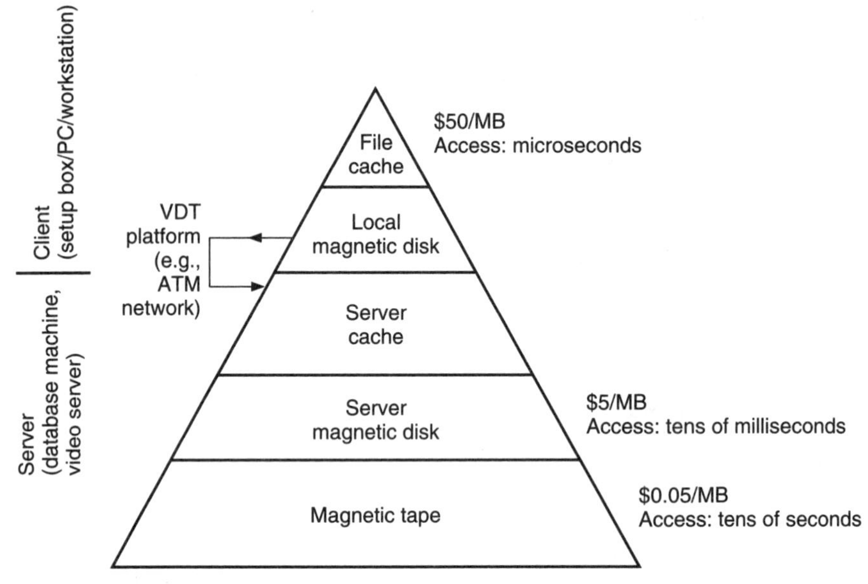

Figure 7.9 Storage hierarchy.

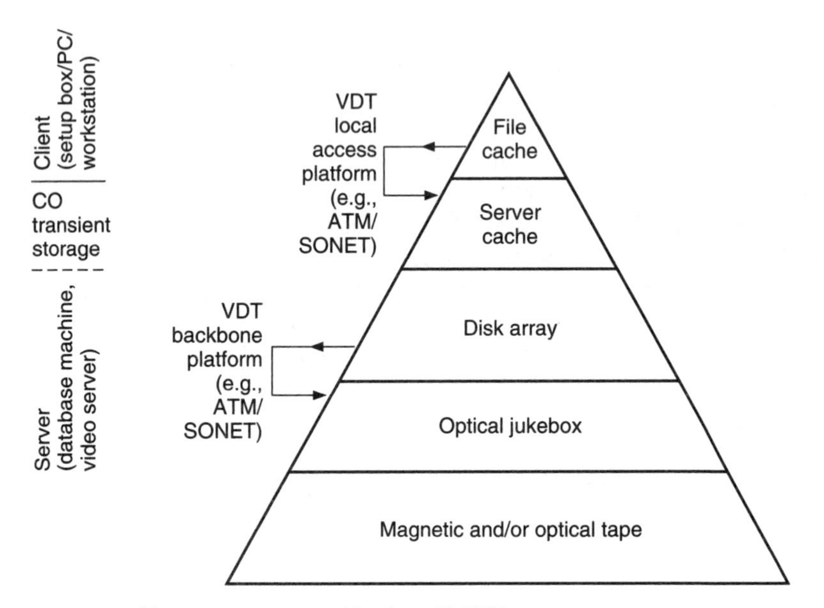

Figure 7.10 Memory management in the mid-1990s.

Laser video disks were introduced in the 1970s. About 30 min of analog video (54,000 frames) could be stored on one 12-in platter. The 12-cm *compact disk digital audio* (CD-DA) was introduced in 1982. The laser video disk technology was adapted to store up to 72 min of high-quality *digitally encoded* stereophonic audio.

The first optical drives for computer applications appeared in 1982; these were used for large library and database archival functions.[29] CD-ROM drives as PC peripherals appeared in 1984 to 1985. They provide storage for 600 to 680 MB of prestored information, including text, pictures, databases, and data. CD-ROM drives can now be purchased as PC peripherals for $300 to $600. These CD-ROMs are now faster and cheaper than the first-generation equipment that appeared in the mid-1980s. The cost of CD-ROM hardware and software has decreased to one-fourth the original cost in the past decade. CD-ROMs are the staple of desktop multimedia applications.

At about the same time, writable optical disks (*write once read many,* or WORM) became available. These 12-in platters initially stored 1 GB per side of video information; now they can store about 6 GB, and 10-GB systems are expected by 1995. The disk is written by a laser beam: when the beam is turned on, it records the data in the form of pits or bubbles within a specific layer of the disk. The data are read back, as is the case in all optical media, by detecting the variations in reflectivity of the disk surface. One of the limitations of this technology has been its transfer rate of 100 to 200 KBps. New systems, particularly at the high end, have transfer rates of 1 MBps.

In 1986, specifications for *compact disk interactive* (CD-I) were being developed for consumer market products supporting video, graphics, images, audio, data, text, and software. CD-I has been developed by Philips, Sony, and Microware. It is a *self-contained multimedia system* aimed principally at the consumer market. Products are now widely available.

In 1988 erasable (rewritable) optical disks became available. These disks store hundreds of Mbytes of information on removable media, enabling the ready (but manual) distribution of digitally encoded imaging information. The 5.25-in system appeared first, followed by 3.5-in systems.

Organizations that need to develop CD-ROM disks for archival material can already do so in a cost-effective manner, using *CD recordable* (CD-R) technology that had reached the market in the early 1990s. Machines to create CD-ROMs can cost as little as $4000, enabling production of disks for $50 each.[30] Blank writable CDs can be purchased for around $20 to $30.

Figure 7.11 depicts graphically some of the available optical storage formats and approaches.[27, 28]

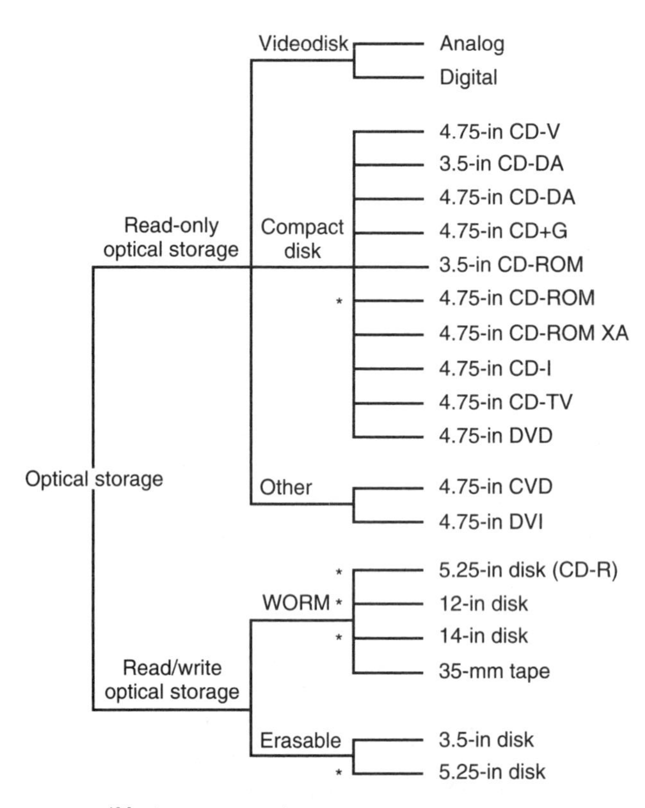

Technology	Description
CD+G	Enhancement to CD-DA, providing digital audio and still-frame graphics for television monitor display.
CD-DA	Compact Disk Digital Audio. Developed by Philips and Sony. Consumer stereo audio.
CD-I	Compact Disk Interactive. Multimedia extension of CD-DA aimed at the consumer market. Supports text, full-motion video, still video, graphics, and digital audio.
CD-ROM	Compact Disk Read Only Memory. Contains machine-readable audio, video, image, video, and textual information.
CD-ROM XA	CD-ROM Extended Architecture. Supports Philips/Sony/Microsoft format, particularly for consumer market. Stores data, text, graphics, and audio.
CD-V	Compact Disk Video. Philips/Sony technology for consumer market. Analog video and audio along with data (players can also utilize CD-DA).
DVD	DigitalVideo Disk. Toshiba's system introduced in 1995 which supports 270 minutes of full-motion video.
CD-TV	Commodore Dynamic Total Vision. Multimedia consumer product for Amiga computers.
CVD	Compact Video Disk. Developed by SOCS Research and Interactive Video Systems. Analog full-motion video and audio.
CD-R	CD Recordable. Low cost (less than $10,000) desktop presses for CD-ROMs. Used for prototyping titles for conventional presses, final production disks for limited distribution, and archiving.
DVI	Digital Video Interactive. Developed by Intel. Supports color digital video. Not necessarily specific to storage technology.
Optical Tape	Digital optical storage on polyester film.
RAID	Redundant Array of Inexpensive Disks. Combines multiple drives, typically based on magnetic media. A typical system with five 3.5-inch systems provides 8 GB for $20 K.

Figure 7.11 Optical storage technologies.

7.3.3 CD-ROMs

Although CD-ROMs currently may be a stretch technology for industrial quality video servers (because they currently can store only half a movie—1 hour of material—encoded in MPEG-1*), they are treated here because of their prevalence in the multimedia market. Digital data are stored in CD-ROM disks as microscopic pits and spaces. Stored data are read using a laser beam and interpreting the intensity of the light reflected from those pits and spaces as one or zero bits. The physical characteristics of CD-ROM disks were standardized in 1985 by the Sony/Philips *Yellow Book.* The logical file system used on CD-ROM disks follows the High-Sierra file system format (1986) and ISO 9660 (1988) standard (ISO 9660 is an update of the High-Sierra format).[31] These formats specify file management procedures that compensate for the slow access times of CD-ROM drives. File contents are not specified; several kinds of information (e.g., images, text, audio, video) can be stored on CD-ROM disks.

As seen in Fig. 7.11, CD-ROM disks are available in 4.75-in and 3.5-in forms. The 3.5-in CD-ROM disk can store about 180 MB of data, and the 4.75-in CD-ROM disk can store about 680 MB. Research is underway for using shorter wavelength blue-green lasers that permit the storage of an entire movie on a 4.75-in compact disk. The 3.5-in form is used in small portable systems such as laptop computers. CD-ROM drives come as 5.25-in peripheral devices and use *small computer system interface* (SCSI).

There is a variety of CD-ROM formats not only at the physical level, as just noted, but also in terms of the file layout: the *Red Book* specifies CD-DA; the *Yellow Book* specifies the CD-ROM; the *Green Book* specifies the CD-I; and the *Orange Book* describes the CD-R. ISO 9660 describes the logical structure and file format for a CD-ROM, while the *Yellow Book* specification describes the manner in which the data are organized on the disk. The other formats are used in the multimedia environment. Philips worked on the extension of the *Yellow Book,* to interleave text, sound, and video for CD-I. The resulting *Green Book* specifies not only the data format but also specific playback requirements. Sony and Microsoft now support a subset of CD-I called CD-ROM-XA, which provides standards for a number of compression modes, allowing continuous audio while textual data are read through, supporting interleaving of text and sound.

Video has also been compressed and stored using DVI methods. Intel's DVI is a de facto compression standard and supporting platform technology. DVI is a commercially successful technology of the late 1980s that has enabled video to be brought to the PC. A new system

---

* Evolving systems can store more video, as discussed later.

TABLE 7.9 Storage Capability of a CD-ROM with DVI Encoding

1 h full-screen, full-motion
or 4 h ¼ screen, full-motion
or 16 h ⅛ screen, 15 fps

called Indeo has since been introduced. Traditional CD-ROM can store only an hour of CD-quality audio, 500 high-resolution still images, or 30 s of uncompressed full-motion digital video. A DVI CD-ROM disk can, as an example, store 20 min of motion video* sequences, 1000 high-resolution still images, each accompanied by an average of 1 min of audio, and still have room for 50,000 pages of text. See Table 7.9.

DVI is an asymmetrical compression algorithm: specialized systems compress motion video in an off-line manner, while low-cost desktop hardware can play back the images in real time.

As already discussed, a CD-ROM drive is basically a CD-DA player enhanced with more elaborate data error detection and correction capabilities, data buffers, and SCSIs. CD-ROM applications search for data randomly while CD-DA applications are more sequential in nature, i.e., music in CD-DA players is played from beginning to end. Therefore, the internal components of CD-ROM drives must be more rugged and reliable than those used in CD-DA players. Moreover, an additional loading mechanism is needed in CD-ROM drives to load the caddy that is used to protect the CD-ROM disk. There are two CD-ROM drive designs at the disk-feeder level: Sony-style CD-ROM caddy case and traditional tray- or drawer-configured players based on the CD-DA design. Caddies protect the disks from hand-transferred oils and dust; however, disk loading is more cumbersome, particularly for cases of high-exchange activity. Tray designs facilitate CD-ROM handling but may allow dirt and dust to enter the drive mechanism. Self-cleaning lenses and double doors may be used to minimize the influx of foreign particles.

As noted, CD-ROMs can store as much as 680 MB of information† in a compact, portable optical medium (for comparison, a WORM disk can store 6.55 GB—3.27 GB per side; to read the opposite side, the platter

* In early 1995, two new schemes were advanced for the storage of movies on CDs. Toshiba, four major electronics companies, and four major film studios announced DigitalVideo Disk, which allows playback of 270 minutes of full-motion video. This scheme can also be used for multimedia and I-TV. Sony and Philips Electronics are pursuing a different scheme called CD-V. At press time it was not clear which approach will ultimately prevail.

† In 1994, IBM announced a technology to store information on CD-ROMs in layers; this technology, if developed and implemented, could increase storage capacity by an order of magnitude. Observers expect products to be developed by 1997 to 1999.

needs to be flipped robotically or manually; otherwise, systems with dual-sided read heads can be used). The current generation of CD-ROM drives can deliver data at a (maximum) rate of 150 KBps. The access time of most drives is between 200 and 400 ms, compared with 16 ms on a magnetic hard drive. Improvements are being made: some models now reaching the market have access time in the 200-ms range. While for applications involving only text transfer or low-resolution images, the user can use a slower and, hence, less expensive unit, multimedia applications involving video, graphics, and imaging applications require the faster units. These relatively slow access times in CD-ROM drives are the result of having to move the read head continuously in search of data usually requested in random order and located in noncontiguous areas of a disk's surface. Optical drives with new lighter read head designs are starting to appear for improved performance characteristics.

CD-ROM data files are laid down sequentially; this implies that sustained transfer rates are generally more important than average access time, particularly when dealing with nontext data. A slow drive can severely impact a video application. Applications are designed to minimize the distances the read head must travel to access related data. Some new CD-ROM drives come with cache memory, typically 64 KB of RAM, to reduce the access times of frequently requested data even further. These buffers are used also to smooth out the fluctuations in the data transfer rate for better quality and performance when running applications. The standard data transfer rate for CD audio files is 150 KBps. This data transfer rate is relatively slow for handling high-resolution imaging files (and also for video-multimedia applications where 30 frames per second need to be fetched). If processor overhead is considered, the average data transfer rate for most CD-ROM drives is about 90 to 100 KBps.

Some CD-ROM drives support different data transfer rates for different information types. For instance, some drives support 300-KBps data transfer rates for video-multimedia files and support the 150-KBps data transfer rate for audio files.

SCSI is the channel interface found on PC disk drives, as well as on a variety of peripherals such as tape drives, optical disk drives, and image scanners. SCSI provides a high-level message-based protocol for communication between initiators and targets.[32] SCSI (ANSI X3.131-1986) was developed starting in 1982 and was completed with the publication of the standard in 1986. This SCSI protocol (now known as SCSI-1) has been revised to achieve higher performance. The SCSI-1 uses a 10-MHz clock rate supporting transfers at 5 MBps. The new SCSI interface, SCSI-2 (also known as *fast SCSI*) uses a 20-MHz clock rate to support 10 MBps. Newly announced high-performance disk systems support SCSI-2. There is also a wide SCSI supporting transfer of 16-bit words (32 bits when using a second cable) rather than 8-bit words as is customary in SCSI-1; this interface supports transfer rates at 20 MBps. SCSI-2 is an expansion and enhancement of SCSI-1; consequently, a high degree of compatibility exists between the two standards.

Examples of commands are: Read, Write, Track Select, Rewind, Request Sense, Read Block Limits, Write File Marks, Recover Buffer Data, Release Unit, Copy, Erase, Send Diagnostic, Locate, Read Position, Compare, Copy and Verify, and Read Buffer. Work on SCSI-2 started in 1985, as some vendors sought to increase the functionality of SCSI-1. SCSI-2 added many new options, made several others mandatory, and eliminated obsolete SCSI-1 features. New features and improvements include (in addition to those listed in the previous paragraph): command queueing (up to 256 commands per initiator), high-density connector to improve signal quality, asynchronous event notification, addition of new commands to support CD-ROMs, scanners, WORMs, medium changer, and communication devices.[33] There is also a SCSI-3 effort to allow operating systems to support both SCSI-1 and SCSI-2 command sets.

As noted earlier, at press time, the major consumer electronic firms, led by an alliance of Sony Corp. and Philips Electronics, were cooperating on the creation of a new digital video player (CD-V) that uses disks capable of storing 2 h or more of full-motion video. Separately, Time Warner and Toshiba have codeveloped a similar digital video disk (DVD) system and are attempting to make it the de facto world standard.[34] The new CD player would be based on advances in digital compression, enabling it to hold an entire movie. The alliance between Sony and Philips is viewed by some in the industry as being strong because: (1) Philips already has the CD-I player that offers full-screen, full-motion digital video (several studios are delivering movies for the CD-I format) and (2) Sony, which owns Columbia Pictures and TriStar, is a major software and consumer products supplier.

7.3.3.1 Server aspects. Multiworkstation access to CD-ROM drives over the network is not well supported because of the lack of standard client-server computing models for CD-ROM in particular and optical disks in general. A CD-ROM drive is usually connected to a file server or workstation host that provides network access to one user or workstation at a time. Additional hardware and/or software is required to support shared multiworkstation access. A typical product used for this purpose is basically a stack of CD-ROM drives arranged in a pedestal configuration.

7.3.4 CD recordable

CD-Rs are desktop recordable systems (drives and media) that provide low-cost "printing presses" for three kinds of applications: (1) prototyping titles that eventually may go to a conventional press; (2) final production of disks that have limited distribution (possibly even only one copy); and (3) archiving.[35] The CD-ROM drive is attached to a PC or workstation. CD-R techniques allow a user to produce CD-ROM replicas in low-volume quantities. Once a CD-R disk is produced, it can be used on a traditional CD-ROM drive; data are indexed and organized the same way as they would be on a CD-ROM. The software needed for formatting the data for CD-ROM production is often provided with the CD-R drive. In contrast with the mastering technique employed with CD-ROMs, CD-R blanks can be recorded locally in a single step. The CD-R is immediately usable. In VDT, these could be used by the VIP to "burn" advertisements, announcements, standard clips, etc.

CD-R technology provides the benefits of CD-ROM approaches, while improving the economics and rapidity of production and distribution. Users are able to eliminate the mastering costs and expedite the turnaround time compared with traditional CD-ROM production. CD-Rs have been available for a number of years, but were expensive (e.g., $60,000). First-generation systems cost in the $30,000 range. The newest CD-Rs allow multiple write sessions. Although, once a section is written it cannot be altered, these drives allow the user to append information to already (partially) recorded disks. These systems can be purchased in the $3,000 to $10,000 range. Blank media currently sell for $20 to $30 each, and the price is projected to fall to $10 by 1996.

Blank CD-R media are available in either 540 or 630 MB; therefore, the CD-R media are not produced at the outer limits of the *Red Book* or *Yellow Book* specifications. Only the best CD-R drives can record the 630-MB media, particularly at twice real time. CD-R media are the same size as CD-ROM and, once installed into a CD-ROM drive, are indistinguishable from a traditional CD-ROM. A CD-R disk has a gold reflective layer, in contrast to the silver layer used in CD-ROMs. Table 7.10 provides some highlights of the features of CD-R compared to competitive technologies such as WORM and MO.

The first generation of CD-R drives was introduced in the United States in 1989. This generation of equipment was capable of writing the entire disk in one session (once a section of a CD-R disk has been written, it cannot be altered). These devices recorded the disk in real time, requiring over an hour of recording time to write one disk. The second generation of equipment can record at two speeds: real time and twice real time.

TABLE 7.10 A Comparison of CD-R and Competitive Technologies

CD-R	Standardized format: archived data playable on any CD-ROM drive. Incremental, file-oriented write or update capability not yet implemented (ECMA* 168 specification); user must write a large amount of data at a time. Not fast enough for interactive use. More expensive (≈ $10 K) compared to WORM (≈ $4 K) and MO (≈ $4.5 K).
WORM	Interchangeability problems. Market being eroded by MO technology (sales of 5¼ systems: 40,000 units in 1992 to 1,000 in 1997). 12-in systems with 6.5-GB capacity (and packageable in jukeboxes[21]) not threatened by CD-R.
MO	Proprietary aspects (Bernoulli, Syquest). Increasing market (sales of 5¼ systems: 170,000 units in 1992 to 400,000 in 1997). Multifunction MO drives (e.g., HPs, Sony, Hitachi): standard MO media as well as WORM support.

* European Computer Manufacturer's Association.

With first-generation systems, once the recording of a track on a CD-R disk has started, the entire track must be recorded in one session; hence, the computer system driving the CD-R drive must be able to sustain the CD-R I/O rate. This rate is 150 KBps for real-time systems and 300 KBps for twice real-time systems. If the PC, its magnetic media, the PC software-operating system, or the network is not able to sustain the throughput, the CD-R drive will abort the recording, and the disk will be useless (clearly, this is more of a consideration in the twice real-time situations). Some vendors may not guarantee a high rate of success in the recording operation; others precheck the layout of files on magnetic disk and estimate the success rate; still others fully buffer the CD-ROM image in a magnetic disk dedicated to the CD-R operations.[36]

7.3.5 Rewritable or erasable systems

High-end writable videodisk recorders can cost in the $40,000 range. Some systems aimed at the broadcast industry can store 58,000 frames per side or 32 min of full-motion video. The disk can be re-recorded more than 1 million times without signal degradation; the disks, however, are fairly expensive, costing approximately $1300. Much more cost-effective systems are now appearing.

MO is based on a combination of magnetic and optical recording techniques and support erasable storage. The disk is composed of a material that becomes more sensitive to magnetic fields at high temperatures. A laser beam is used to heat up the appropriate section of the disk surface; once heated, a magnetic field is used to record on the surface. Optical techniques are used for reading the information off the disk, by detecting how the laser beam is deflected by different magnetizations of the disk surface.[31]

Rewritable optical disk drives appeared first in the 5¼-in configuration; second-generation 3½-in rewritable optical disks appeared more recently. Many or these products now sell for less than $1500. The 5¼-in systems can store as much as 1 GB, with 500 MB per side (650 MB—325 MB per side—is more typical, however), while 3½-in systems can store 256 MB. They can be rewritten over 100,000 times. Most erasable optical drives use magneto-optical mechanisms; however, systems that do not require magnetization, e.g., phase-change methods, are appearing. Rewritable optical storage can be very cost-effective (e.g., for archiving): $0.13 per MB compared with $1.25 per MB for removable hard drives. About 400,000 3½-in drives were expected to be sold in 1994 worldwide, equating to $260 million, and 700,000 in 1997, equating to $500 million.

The 3½-in drive systems have a much faster access time than the larger MO drives; the access time is comparable to that of magnetic hard drives. For example, the 256-MB systems have average seek time of 35 ms and an average access time of 45 ms. The recording process is as follows.[29] The disk used in MO recording has a thin film of magnetic material that responds to a biased magnetic field when heated to its Curie point (200°C) by a focused laser beam. Recording is performed in two stages: erasing an area and writing information into that area. Erasing of data occurs when the bias magnet is turned on in the erase direction, and a continuous laser beam heats the magnetic layer of the media to its Curie point. Writing into an erased area occurs when the bias magnet is polarized from the erase direction to the write direction. A data bit is formed when the laser emits a high-energy burst at a selected spot within the erased area. As the laser heats the spot, the magnetic properties held within the thin film of the magnetic material are altered; these alterations create a different light reflection when read by the laser and are interpreted as 1 bit (unchanged selected spots are read as 0 bits). Once the data have been written, they stay there permanently until a strong magnetic field and high localized heat is applied. Retrieval of data in MO recording is similar to the process of reading data from a CD-ROM.

One drawback of this technology is read/write speed. For example, it typically takes 19 to 60 s to write a 3-MB file and 16 to 33 s to read the same file (depending on model and/or product).[37] This relatively slow speed depends on three factors:

1. Magneto-optical methods require a three-pass procedure to write data: (*a*) erasure of existing data; (*b*) writing of new data; and (*c*) verification of integrity of the operation.

2. Since laser-based assemblies are expensive, optical drives contain only one assembly, in contrast with hard disk drives that have multiple read/write assemblies to operate on multiple platters simultaneously.
3. Laser-based assemblies are larger and heavier than hard drive read/write heads, slowing down physical movement.

Techniques used to speed up the process include: (1) turning off the verify cycle; (2) using RAM caches; (3) using new split-head optics (a lighter head where only one prism, the laser and the sensor, are on the actual head); (4) increase the spindle speed, from 2400 to 3600 rpm and as high as 5400 rpm. Throughput as high as 2 MBps is being claimed on some units.[37]

MO disk drives and media are now used in a variety of data-storage applications, including video applications. They have a burst data rate of 1.25 MBps, sustained rates of 0.5 MBps, and average access time of around 50 ms (including latency); see Table 7.11. They are characterized as providing removability, high capacity, and random access capabilities.[36] MOs are being used for direct access secondary storage. The drives have a small footprint. The cost is around 30 cents per MB (this compares with $3 per MB for primary magnetic storage: fixed disks). The information transfer rate is similar to those of conventional magnetic disks, while the access time is still slower than that of a magnetic disk (this is driven by the relatively bulky read/write mechanism).

There is an expectation that the density of MO media will increase in the future while the access time will decrease. The performance characteristics are a function of the following four facts:

1. Optics of drive
2. Mechanics of drive
3. Electronics of drive
4. Media substrate configuration

All of these will see technical improvements in the next few years.

TABLE 7.11 Typical Parameters* of Rewritable Magneto-Optical Media

Rotation speed	3000 to 3600 rpm
Average access time	35 to 45 ms
Data buffer memory	64 to 256 KBps
Sustained data transfer rate	0.6 to 1.0 MBps
Maximum data transfer rate	1.5–2.5 MBps
Disk diameter	90 to 130 mm (3.5 to 5.25 in)
Disk cost	$100 to $270
Drive cost	$2500 to $5000

* Press time data.

7.3.6 Optical jukeboxes

Systems are now being deployed for generic archival, imaging archival, and video archival, that provide relatively rapid access to large amounts of data stored on removable optical and magnetic media. Many CD-ROMs and/or WORMs can be housed in a jukebox. Systems already support 240 CD-ROMs (using multiple drives—disk access time is 8 to 14 s).[37] A CD-ROM jukebox supporting 60 disks can be purchased for about $8000; a CD-ROM jukebox supporting 240 disks can be purchased for about $20,000.[38] Practitioners note that 5¼-in jukeboxes are not as reliable as the 12-in ones. Large jukeboxes can cost around $150,000.

This type of archival is accomplished by storing the media on shelves that can be accessed by robotic handlers. When a file needs to be retrieved, the file management system identifies where it can be found within the tape or optical library; the robotic handler exchanges the currently loaded media with the one containing the file in question. The completely automated procedure typically takes a few seconds. High-end systems (e.g., Kodak's Optical Disk System 6800 Automated Disk Library) can support up to 150 14-in platters and can therefore store 1000 GB (1 TB); this equates to 3 GB for each side of the 150 platters.

7.3.7 Holographic memories for video servers and other applications

Optical memories based on holographic principles are now emerging. A holographic storage device is a page-oriented device that writes and reads data in an optical form. The data are two-dimensional arrays of spots called a *page*. Multiple pages are multiplexed holographically to create a stack of pages, all in the space normally required to store a single two-dimensional image. Holographic storage is a generic storage technology in the same way as today's magnetic recording storage technology. According to some in the computer storage industry "holographic storage products could exceed $30 billion worldwide in 1995, including applications in multimedia, computing and imaging/records storage."[39] (Such figures are clearly optimistic.) Holographic technology is currently under development by companies such as IBM, Sony, and Tamarack Storage Devices Inc.

Holographic storage systems (i.e., three-dimensional storage systems) under development at Tamarack Storage Devices Inc., Austin, Texas, use beam-steering retrieval, thereby eliminating motors and moving arms and, consequently, greatly improving the information access time of optical storage systems. Table 7.12 compares the access time of various storage technologies, including holographic technology. By using new laser and semiconductor technologies, these companies can bring to the market the first cost-effective systems. For example, by press time, Tamarack Storage Devices was planning to introduce a holographic jukebox holding 30 2½-in disks, each capable of storing 640 MB of data. Tamarack was also planning a 1-GB removable storage drive designed to fit in the PCMCIA slots of notebook computers. In the view of some, however, in their rush to market, some companies have compromised too much: Some use "all of the bad aspects of the current storage technology and do not take full advantage of the good aspects of the new holographic technology—namely, the very large data bandwidth and the short access time."[40]

Commercial magnetic and optical media are inherently limited to one-dimensional arrays of bits, organized in spirals on a disk. A very efficient way of storing data in three dimensions is in holographic fashion, using the depth of the media. This technique also provides robustness and error insensitivity, since any point defect only lowers the signal-to-noise ratio of the readout, rather than obliterating it. In the three-dimensional storage scheme, such as the one used by Tamarack, high bandwidth can be achieved by handling data as an array of bits instead of the current single-bit approach (one can think of this as reading and writing data in parallel versus serial). This feature is provided by optics and has a considerable advantage over electronics in that each pixel of the image does not have to be isolated from its neighbors by separate channels or insulators.

Figures 7.12 and 7.13 show how a holographic storage device is a page-oriented device that writes and reads data as a page of two-dimensional arrays of spots. The storage medium can be a photorefractive crystal for rewritable applications, or a photopolymer for write-once applications. The information is stored as a hologram that is reconstructed for the readout. The hologram is stored as the Fourier transform of the two-dimensional array of spots; this approach is chosen in such a way to get a small recording size and to provide immunity to material defects. Pages are placed in the storage media as a two-dimensional array of stacks. Pages composed by a two-dimensional array of light valves called the *spatial light modulator* are stored in the media, with a stack of up to 30 to 40 pages stored in a 1-mm^2 spot size.

TABLE 7.12 Portable System Storage Tradeoffs
(Courtesy Tamarack Storage Devices Inc.)

Technology	Capacity (bytes)	Access time (ms)	Drive cost per MByte	Media cost per MByte
Magnetic fixed disk (7200 rpm)	2.1 G	12	$1.04	Nonremovable
CD-ROM (triple speed)	650 M	195	$0.84	$0.002
Flash (PCMCIA)	40 M	0.0001	$50.00	$50.00
1.8-in hard disk (PCMCIA)	105 M	18	$4.70	$4.70
Holographic	1 G	30	$0.50	$0.005

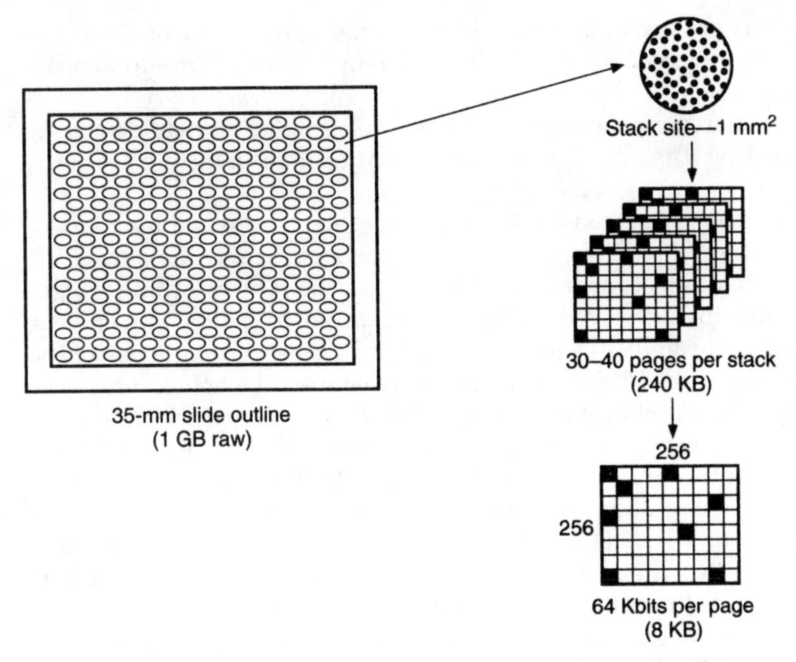

Figure 7.12 Stacks and pages of data stored in a holographic storage tile.

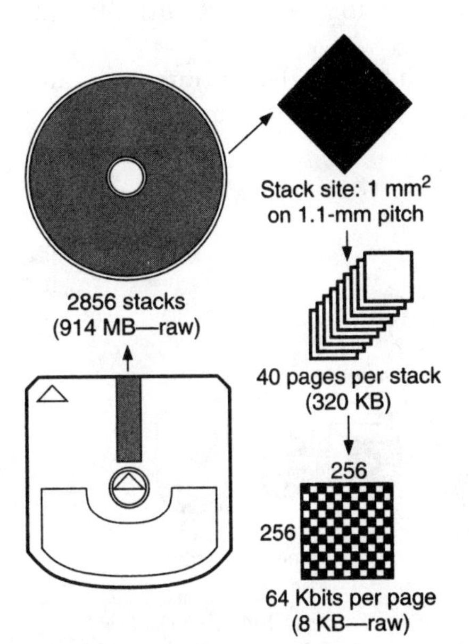

Figure 7.13 Multimedia floppy organization.

Multiple pages are superimposed by changing the angle of the reference beam by a fraction of a degree or through other proprietary encoding means.

Four major components constitute a holographic storage device, as developed by Tamarack: tiles, lasers, optics, and detectors (see Fig. 7.14).[41]

Tiles. Each tile consists of a substrate (e.g., glass) and a *photosensitive polymer film* (PPF); CAT scan technology currently employs PPF. The PPF reacts to laser transmission focused on it by recording images of the data being stored. A tile is organized by pages and stacks; approximately 30 to 40 vertically arranged pages constitute one of the 4096 stacks in the PPF. This architecture enables the holographic storage device to store data across the three dimensions of width, length, and depth. One page contains 8000 characters (bytes). Presently, the tiles will contain 983,040,000 bytes, nearly 1 GB. Five pages per stack are already possible in the laboratory; this allows for 1.6 GB of information in one tile. Tile durability, when exposed to a number of variables, indicates that normal daylight does not affect an empty tile; more work on durability is underway.

Lasers. A laser writes and reads information to/from a tile. The laser's wavelength inversely factors into the amount of data information that can be written to a tile. Hence, reductions in wavelength increase tile capacity. In terms of colors, blue lasers transmit in shorter wavelength than red. For example, Tamarack presently works with green-blue lasers. These lasers' wavelength equal 690 nm. Further developments should promote a move to blue lasers with 500-nm wavelength.

Optics. The optics system splits the laser beam into two and focuses both onto a tile. The assembly contains two moving mirrors

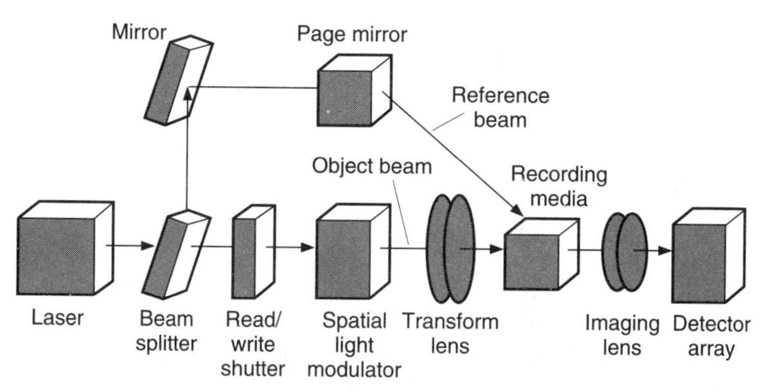

Figure 7.14 Tamarack's holographic storage system. (*Courtesy Tamarack Devices Inc.*)

and several lenses. Careful alignment of the optical components is crucial to drive performance. The beam splitter divides the laser beam into reference and object beams. Then a spatial light modulator imposes a data pattern of a page on the object beam, creating numerous (64,000 in Tamarack's first product) independent pencil beams that represent the data to be stored. Next, a transform lens focuses the beams onto the media where the image is written by recording the interference pattern of the reference and object beams. To record multiple pages in the stack, the page mirror changes the angle at which the reference beam intersects the media and the object beam by a few degrees. Data are retrieved from the media by shutting off the object beam and passing the reference beam through the media. The angle of the reference beam set by the page mirror selects the page to be read, which is then projected on a charge-couple device (CCD) detector array by the imaging lens (see subsequent paragraph). Because of the spatial invariance properties of plane-wave holograms, x-y misalignment tolerances as loose as hundreds of a micrometer are allowed for the read alignment of the drive, compared with fractions of a micrometer tolerances of magneto-optic and magnetic disk drives. The lower precision requirements of this technology enable cost-effective holographic storage products.

Detectors. A CCD reads the information from the laser when it focuses on a stack. Many of the video camcorders and automatic focusing cameras on the market today employ CCD technology, making it inexpensive and reliable. The detector array converts the optical display to electrical signals, which are then decoded and transmitted from the device.

In conclusion, the areal density chart shown in Fig. 7.15, contributed by Tamarack Storage Devices, illustrates the tradeoffs in areal density

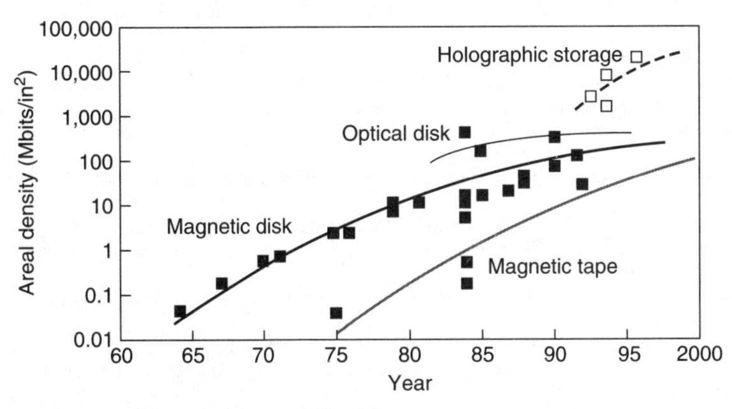

Figure 7.15 Comparative areal densities.

of the major storage technologies. Magnetic and optical technologies, both disk and tape, are becoming track-bound and are approaching fundamental limits in the technology. However, holographic storage technology, which does not use tracks or continuously moving media, offers much higher densities than the other technologies and has the capability to increase dramatically in capacity as laser, encoding, and drive technologies improve.[41]

References

1. *Communications Technology,* December 1993, p. 10.
2. G. McWilliams, "They Can't Wait to Serve You," *Business Week,* January 24, 1994.
3. S. Scully, "From Videotape to Video Servers, Technology Drives PPV," *Broadcasting and Cable,* November 29, 1993, p. 66 ff.
4. "NYNEX, Ameritech Are Said to Choose Digital Equipment Corporation to Supply Video-Server Systems," *Wall Street Journal,* 2/25/1994, p. B8.
5. "DEC Finds More Success With Video Server," *Network Week,* March 4, 1994, p. 1.
6. *Multimedia Week,* February 21, 1994, p. 5.
7. A. Reinhardt, "Building the Data Highway," *Byte,* March 1994, p. 46 ff.
8. L. W. Lockwood, "Video Servers," *Communications Technology,* December 1993, p. 18 ff.
9. "NT-Based Video Server is Announced," *Network Week,* January 31, 1994, p. 1.
10. "SGI Ready Opposing Video-Server Architectures: Video On Demand Battle," *Electronic Engineering Times,* October 4, 1993, p. 1.
11. M. Hennon, "Oracle Says It Has a Video Server," *Video Store,* September 5, 1993, p. 26.
12. "Video Servers: nCube and Oracle Challenge IBM," *Computergram International,* October 29, 1993, p. 17.
13. D. Bank, "Microsoft Counters Oracle with Plans for Video Server," *San Jose Mercury News,* February 19, 1994, p. 14D.
14. M. Semilof, "Microsoft Primes Video-Server Ware," *Communications Week,* May 16, 1994, p. 1 ff.
15. "Video Server Technology for Sammons," *Communications Daily,* December 3, 1993.
16. S. Pollili, "IBM Beefs Up Features and Capacity of Its Video Server," *Infoworld,* February 7, 1994, p. 5.
17. "Integrators Target Niches in Video-Server Market," *Computer Reseller News,* December 13, 1993, p. 2.
18. B. Francis, "HP Develops Video Server for Business," *Infoworld,* February 7, 1994, p. 5.
19. "Starlight to Upgrade Video Server," *Infoworld,* February 28, 1994, p. 4.
20. R. Karen, "Full-Motion Video Won't Wait for the Data Superhighway: Starlight Works with Existing Systems," *InfoWorld,* March 7, 1994, p. 43.
21. S. Ranade, *Jukebox and Robotic Libraries for Computer Mass Storage,* Meckler, Westport, Conn., 1992.
22. J. P. Roth, *Case Studies of Optical Storage Applications,* Meckler, Westport, Conn., 1991.
23. S. Ranade, *Mass Storage Technologies,* Meckler, Westport, Conn., 1991.
24. W. Saffady, *Optical Storage Technology 1992—A State of the Art Review,* Meckler, Westport, Conn., 1992.
25. T. J. Thiel, "Costs of CD-ROM Production—What They Are and How to Overcome Them," *CD-ROM Professional,* March 1993, pp. 43 ff.
26. T. Halfhill, "Buying a CD-ROM Drive," *Byte,* February 1993, p. 120.
27. B. Wiggins, "Document Image Processing—An Overview," *Document Image Automation,* vol. 12, no. 3, Fall 1992.
28. F. Richardson, "Hierarchical Storage Management: New Model for Network Storage," *Computer Technology Review,* February 1993, pp. 75 ff.

29. L. Payne, "Increased Capacities and Speeds Spur Growth of 3.5-in Optical," *Computer Technology Review,* February 1993, pp. 97 ff.
30. *MACWEEK,* 3/1/1993, p. 7.
31. H. Nabil, "CD-ROM Supports Multimedia Applications on the Notebook," *Computer Technology Review,* February 1993, pp. 91 ff.
32. R. H. Katz, "High-Performance Network and Channel Based Storage," *Proceedings of the IEEE,* vol. 80, no. 8, August 1992, pp. 1238 ff.
33. B. A. McFalls, "SCSI-2 Command Sets Address New Device Type Requirements," *Computer Technology Review,* February 1993, pp. 67 ff.
34. J. A. Trachyenberg, "Sony, Philips Forge Alliance to Create Digital Disk Player for 2 Hours of Video," *The Wall Street Journal,* May 20, 1994.
35. J. Udell, "Start the Presses," *Byte,* February 1993, pp. 116 ff.
36. A. Anderson, "Rewritable Magneto-Optical Disk Technology: The Best is Yet to Come," *Document Image Automation,* Sept/Oct 1991, pp. 281 ff.
37. A. Streeter, "Kubik Jukebox Spins Up 240 CDs on Four Drives," *MACWEEK,* 3/8/1993, p. 10.
38. C. Piller, "Optical Update," *MacWorld,* November 1992, pp. 124 ff.
39. J. Wyckoff, "Industry News," *Smart Electronics,* June/July 1994, p. 12.
40. W. W. Gibbs, "Ready or Not," *Scientific American,* October 1994, pp. 128 ff.
41. J. F. Stockton, "Portable Electronic Storage Systems," *IEEE Micro,* February 1994, pp. 69 ff.

Chapter

8

ATM and ATM Switches

8.1 Introduction

*Asynchronous Transfer Mode** (ATM) is a high-bandwidth, low-delay switching and multiplexing technology now becoming available for both public and private networks. ATM principles and ATM-based platforms form the foundation for the delivery of a variety of high-speed digital communication services aimed at corporate users for high-speed data,[2] LAN interconnection,[3] imaging,[4] and multimedia[5] applications. Residential applications, such as video distribution (VDT), videotelephony, and other information-based services, are also planned. Video applications are of particular interest in this context. ATM is the technology of choice for evolving broadband integrated services digital network (B-ISDN) public networks, for next-generation LANs, and for high-speed seamless interconnection of LANs and WANs. ATM supports transmission speeds of 155 Mbps and 622 Mbps and can support speeds as high as 10 Gbps in the future. As an option, ATM will operate at the DS3 (45 Mbps) rates; some proponents are also looking at operating at the DS1 (1.544 Mbps) rates. While ATM in the strict sense is simply a data link layer protocol, ATM and the many supporting standards, specifications, and agreements constitute a technology platform supporting the integrated delivery of a variety of switched high-speed digital services. Figure 8.1 depicts a physical view of an ATM network. Table 8.1 depicts some highlights of the technology. In summary:

* The early portions of this chapter are a short summary of material contained in the book *ATM and Cell Relay Service in Corporate Environments,* McGraw-Hill, 1994, by D. Minoli and M. Vitella.[1]

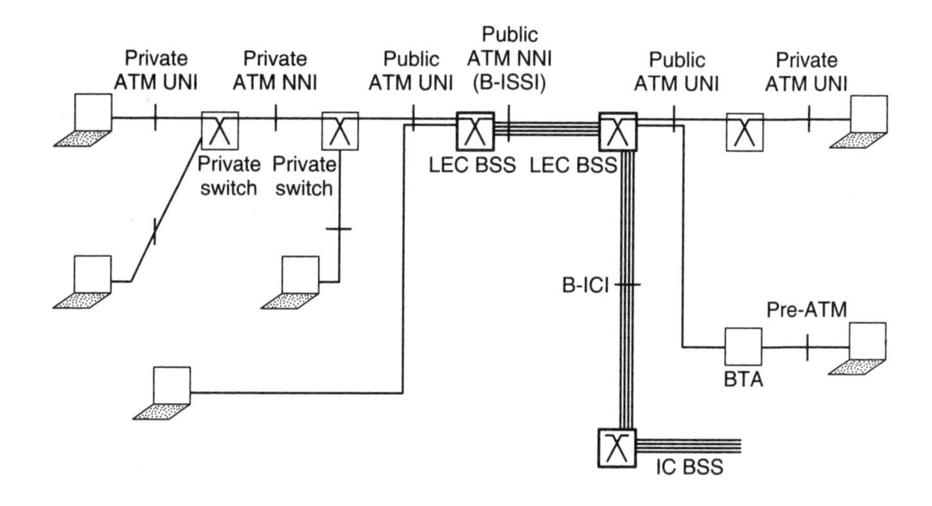

BSS = Broadband Switching System (BISDN switch)
BTA = Broadband Terminal Adapter
B-ISSI = Broadband InterSwitching System Interface
B-ICI = Broadband InterCarrier Interface
LEC = Local Exchange Carrier
IC = Interexchange Carrier

Figure 8.1 A physical view of ATM/CRS private and public network.

An ATM network supporting *cell-relay service* (CRS*) accepts user data units, called *cells,* formatted according to a certain layout and sends the data units in a connection-oriented manner (i.e., via a fixed established path), with sequentiality of delivery, to a remote recipient (or recipients). Every so often a cell may be dropped by the network to deal with network congestion; however, this is a very rare event. The user needs a signaling mechanism to tell the network what he/she needs. The signaling mechanism consists of a data link layer capability (where the data link layer has been partitioned into a number of sublayers) and an application-level, call-control layer. ATM switches and other network elements supporting cell-relay service can also support other "fastpacket" services. If the user wishes to use ATM to achieve a circuit-emulated service, e.g., for constant bit rate video, certain adaptation protocols in the user equipment will be required. Other adaptation protocols in the user equipment are also needed to obtain fastpacket services over an ATM platform. ATM supports certain operations and maintenance procedures, enabling both the user and the provider to monitor the "health" of the network.

---

* In this chapter, the term *cell relay* is used synonymously with *cell-relay service.* The term *ATM* used by itself refers to the underlying technology, platform, and principles. B-ISDN refers to the overall blueprint for the evolution of public networks. ATM protocol refers to the data link layer protocol (ITU-T I.361).

TABLE 8.1 ATM Highlights

- International standard for segmenting, multiplexing, and switching customer's information using fixed-length 53-byte packets (cells).
- Supports both private and public networks, LANs and WANs.
- Supports integrated transport for multimedia applications, including video.
- ATM cells are typically mapped into SONET frames for transport (SONET provides the conduit through the public network; private ATM networks may not use SONET).

Cell-relay service is one of the key new services enabled by ATM. CRS can be used for enterprise data communications networks that use completely private communication facilities, completely public communication facilities, or a hybrid arrangement. Cell-relay service can also support a variety of evolving nondata corporate applications such as the following: (see also Table 8.2): video distribution-VDT, desk-to-desk videoconferencing of remote parties, access to remote multimedia video servers (e.g., in network-based, client-server video systems), multimedia conferencing, multimedia messaging, distance learning, business imaging (including CAD/CAM), animation, and cooperative work (e.g., joint-document editing). CRS is one of three fastpacket* technologies that have entered the scene in the 1990s (the other two are frame-relay service and *switched multimegabit data service,* or SMDS). A generic ATM platform supports all of these fast-packet services (namely, it can support cell-relay service, frame-relay service, SMDS), as well as circuit emulation service. Circuit emulation enables the user to obtain the equivalent of a dedicated DS1 or DS3 line over an ATM network.[6,7]

1994 saw the culmination of 10 years of ATM standard-making efforts. Work started in early 1995 and experienced an acceleration in the late 1980s and early 1990s. With the *International Telecommunications Union-Telecommunication* (ITU-T) standards and The ATM Forum implementers' agreements, both sets of which were finalized in 1993 (some related specifications and/or extensions were completed in 1994),† the technology is ready for introduction. In particular, a user-network interface (UNI) specification that supports switched cell-relay service as well as the critical point-to-multipoint connectivity, important for new applications, has been finalized (multiservice UNIs are also being specified). In 1993, The ATM Forum also published an initial *broadband intercarrier interface* (B-ICI) specification; this specification is equally critical for *wide area network* (WAN) interLATA service. A variety of vendors (hubs, routers, switches, PC-workstation cards,

* Some also use *fast packet* (two words).

† Phase 2 follow-up standards are now underway.

TABLE 8.2 Possible Early Applications of ATM in Real Environments (partial list)

Application	Advantages of ATM use	Associated business issues
WAN interconnection of existing enterprise network	High bandwidth; switched service	Unknown cost; geographic availability; equipment availability
WAN interconnection of existing LAN [especially *fiber distributed data interface* (FDDI) LANs]	High bandwidth; switched service	Unknown cost; geographic availability
WAN interconnection of mainframe and super-computer channel[2]	High bandwidth; only service that support re-quired throughput (200 Mbps); switched service	Unknown cost; geographic availability; equipment availability
WAN interconnection of ATM-based LANs	High bandwidth; switched service; multipoint connectivity	New application, not widely deployed; un-proven business need; unknown cost; geographic availability
Support of distributed multimedia	High bandwidth; switched service; multipoint connectivity	New application, not widely deployed; un-proven business need; unknown cost; geographic availability
Support of statewide distance learning with two-way video	High bandwidth; switched service; multipoint connectivity	New application, not widely deployed; unproven market; other solutions exist; unknown cost; geo-graphic availability
Support of video-conferencing (including desktop video)	High bandwidth; switched service; multipoint connectivity	Not widely deployed; unproven market; other solutions exist, particu-larly at lower end (e.g., 384 kbps H.xxx video); unknown cost; geographic availability
Residential distribution of video (VDT)	High bandwidth; switched service; multipoint connectivity	Unproven market; other solutions exist, particularly coaxial-based cable TV; expensive for this market; needs MPEG-2 hardware; geographic availability

and some video equipment vendors, etc.) now have ATM products on the market.

A key aspect of B-ISDN in general, and ATM in particular, is the support of a wide range of data, video, and voice applications in the same public network. An important element of service integration is the provision of a range of services using a limited number of connection types and multipurpose user-network interfaces. ATM supports both non-switched *permanent virtual connections* (PVCs) as well as *switched virtual connections* (SVCs). In a PVC service, virtual connections between endpoints in a customer's network are established at service subscription time through a provisioning process; these connections or paths can be changed via a subsequent provisioning process or via a customer network management (CNM) application. In SVC, the virtual connections are established as needed (that is, in real time) through a signaling capability. Both the PVC and SVC services are being contemplated for VDT, as discussed in Chap. 5 and elsewhere. ATM supports services requiring both circuit-mode and packet-mode information transfer capabilities. ATM can be used to support both connection-oriented (e.g., frame-relay service) and connectionless services (e.g., SMDS).[8] For the purpose of VDT, frame-relay service and SMDS are not as important as CRS and circuit emulation.

A glossary of some of the key ATM and related concepts is contained in Table 8.3, based on a variety of ATM standards and documents.

Table 8.4, based on data from Ref. 9, depicts for illustrative purposes typical frames-per-cell statistics for several movie sequences.

8.2 Basic ATM Concepts

8.2.1 ATM protocol model: an overview

ATM's functionality corresponds to the physical layer and *part* of the data link layer of the *open systems interconnection reference model* (OSIRM). This protocol functionality must be implemented in appropriate user equipment, for example, routers, hubs, multiplexers, video servers, and setup boxes, and in appropriate network elements, for example, switches and service multiplexers. A *cell* is a block of information of short fixed length (53 octets) that is comprised of an overhead section and a payload section (5 of the 53 octets are for overhead and 48 are for user information), as shown in Fig. 8.2. Effectively, the cell corresponds to the data link layer frame that is taken as the atomic building block of the cell-relay service. The term *cell relay* is used since ATM transports user cells reliably and expeditiously across the network to the destination. *ATM* is a transfer mode in which the information is organized into cells; it is asynchronous in the sense that the recurrence of cells containing information from an individual customer is not necessarily periodic.

The ATM architecture uses a logical protocol model to describe the functionality it supports. The ATM logical model is composed of a user plane, a control plane, and a management plane. The *user plane,* with its layered structure, supports user information transfer. Above the

TABLE 8.3 Glossary of Key ATM Terms

AAL	An adaptation layer of functionality that sits above the ATM layer and prepares the protocol data units of the upper-layer protocols (e.g., TCP/IP, or the output of a video coder) for carriage over ATM.
AAL connection	Association established by the AAL between two or more next higher layer entities.
Asynchronous time division multiplexing	A multiplexing technique in which a transmission capability is organized in a priori unassigned time slots. The time slots are assigned to cells upon request of each application's instantaneous real need.
Asynchronous transfer mode	A transfer mode in which the information is organized into cells. It is asynchronous in the sense that the recurrence of cells containing information from an individual user is not necessarily periodic.
ATM layer connection	An association established by the ATM layer to support communication between two or more ATM service users (i.e., two or more next higher layer entities or two or more ATM management entities). The communication over an ATM layer connection may be either bidirectional or unidirectional. When bidirectional, two *virtual channel connections* (VCCs) are used. When unidirectional, only one VCC is used.
ATM layer link	A section of an ATM layer connection between two adjacent active ATM layer entities (ATM-entities).
ATM link	A *virtual path link* (VPL) or a *virtual channel link* (VCL).
ATM peer-to-peer connection	A VCC or a *virtual path connection* (VPC).
ATM traffic descriptor	A generic list of traffic parameters that can be used to capture the intrinsic traffic characteristics of a requested ATM connection.
ATM user-user connection	An association established by the ATM layer to support communication between two or more ATM service users (i.e., between two or more next higher layer entities or between two or more ATM management entities). The communication over an ATM layer connection may be either bidirectional or unidirectional. When bidirectional, two VCCs are used. When unidirectional, only one VCC is used.
Broadband	A service or system requiring transmission channels capable of supporting rates greater than the *integrated service digital network* (ISDN) primary rate.
Call	An association between two or more users or between a user and a network entity that is established by the use of network capabilities. This association may have zero or more connections.
Cell	ATM layer protocol data unit.

TABLE 8.3 Glossary of Key ATM Terms (*Continued*)

Cell-delay variation	A quantification of variability in cell delay for an ATM layer connection.
Cell header	ATM layer protocol control information.
Cell-loss ratio	In a network, cell ratio is $(1 - x/y)$, where y is the number of cells that arrive in an interval at the ingress of the network, and x is the number of these y cells that leave at the egress of the network. In a network element, cell loss ratio is $(1 - x/y)$, where y is the number of cells that arrive in an interval at the ingress of the network element, and x is the number of these y cells that leave at the egress of the network element.
Cell-transfer delay	The transit delay of an ATM cell successfully passed between two designated boundaries.
Connection	An ATM connection consists of the concatenation of ATM layer links to provide an end-to-end information transfer capability to access points.
Connection admission control (CAC)	The procedure used to decide if a request for an ATM connection can be accepted based on the attributes of both the requested connection and the existing connections.
Corresponding entities	Peer entities with a lower layer connection among them.
Header	Protocol control information located at the beginning of a protocol data unit.
Layer connection	A capability that enables two remote peers at the same layer to exchange information.
Layer entity	An active element within a layer.
Layer function	A part of the activity of the layer entities.
Layer service	A capability of a layer and the layers beneath it that is provided to the upper-layer entities at the boundary between that layer and the next higher layer.
Layer user data	Data transferred between corresponding entities on behalf of the upper layer or layer management entities for which they are providing services.
Multipoint access	User access in which more than one *terminal equipment* (TE) is supported by a single network termination.
Multipoint-to-multipoint connection	A multipoint-to-multipoint connection is a collection of associated ATM VC or VP links and their associated endpoint nodes, with the following properties: (1) all N nodes in the connection, called *endpoints,* serve as a root node in a point-to-multipoint connection to all of the $(N - 1)$ remaining endpoints; (2) each of the endpoints on the connection can send information directly to any other endpoint, but the

TABLE 8.3 Glossary of Key ATM Terms (*Continued*)

	receiving endpoint cannot distinguish which of the endpoints is sending information without additional (e.g., higher layer) information.
Multipoint-to-point connection	A point-to-multipoint connection where the bandwidth from the root node to the leaf node is zero, but the return bandwidth from the leaf node to the root node is nonzero.
Network node interface (NNI)	The interface between two network nodes.
Operation and maintenance (OAM) cell	A cell that contains ATM LM information. It does not form part of the upper-layer information transfer.
Peer entities	Entities within the same layer.
Physical layer (PHY) connection	An association established by the PHY between two or more ATM entities. A PHY connection consists of the concatenation of PHY links to provide an end-to-end transfer capability to PHY SAPs.
Point-to-multipoint connection	A collection of associated ATM VC or VP links, with associated endpoint nodes, with the following properties: (1) one ATM link, called the *root link,* serves as the root in a simple tree topology. When the root node sends information, all of the remaining nodes on the connection, called *leaf nodes,* receive copies of the information; (2) each of the leaf nodes on the connection can send information directly to the root node. The root node cannot distinguish which leaf is sending information without additional (higher layer) information; (3) the leaf nodes cannot communicate directly to each other with this connection type.
Point-to-point connection	A connection with only two endpoints.
Primitive	An abstract, implementation-independent interaction between a layer service user and a layer service provider or between a layer and the management plane.
Protocol	A set of rules and formats (semantic and syntactic) that determines the communication behavior of layer entities in the performance of the layer functions.
Protocol control information (PCI)	Information exchanged between corresponding entities, using a lower layer connection, to coordinate their joint operation.
Protocol data unit (PDU)	A unit of data specified in a layer protocol and consisting of protocol control information and layer user data.
Relaying	A function of a layer by means of which a layer entity receives data from a corresponding entity and transmits it to another corresponding entity.
Service access point (SAP)	The point at which an entity of a layer provides services to its layer management entity or to an entity of the next higher layer.

TABLE 8.3 Glossary of Key ATM Terms (*Continued*)

Service data unit (SDU)	A unit of interface information whose identity is preserved from one end of a layer connection to the other.
Source traffic descriptor	A set of traffic parameters belonging to the ATM traffic descriptor used during the connection set up to capture the intrinsic traffic characteristics of the connection requested by the source.
Structured data transfer	The transfer of AAL-user information supported by the CBR AAL when the AAL-user data transferred by the AAL is organized into data blocks with a fixed length corresponding to an integral number of octets.
Sublayer	A logical subdivision of a layer.
Switched connection	A connection established via signaling.
Symmetric connection	A connection with the same bandwidth value specified for both directions.
Traffic parameter	A parameter for specifying a particular traffic aspect of a connection.
Trailer	Protocol control information located at the end of a PDU.
Transit delay	The time difference between the instant at which the first bit of a PDU crosses one designated boundary and the instant at which the last bit of the same PDU crosses a second designated boundary.
Unstructured data transfer	The transfer of AAL-user information supported by the CBR AAL when the AAL-user data transferred by the AAL is not organized into data blocks.
Virtual channel (VC)	A communication channel that provides for the sequential unidirectional transport of ATM cells.
Virtual channel connection (VCC)	A concatenation of VCLs that extends between the points where the ATM service users access the ATM layer. The points at which the ATM cell payload is passed to, or received from, the user of the ATM layer (i.e., a higher layer or ATM management entity) for processing signify the endpoints of a VCC. VCCs are unidirectional.
Virtual channel link (VCL)	A means of unidirectional transport of ATM cells between the point where a VCI value is assigned and the point where that value is translated or removed.
Virtual path (VP)	A unidirectional logical association or bundle of VCs.
Virtual path connection (VPC)	A concatenation of VPLs between virtual path terminators. VPCs are unidirectional.
Virtual path link (VPL)	A means of unidirectional transport of ATM cells between the point where a VPI value is assigned and the point where that value is translated or removed.

TABLE 8.4 Cell Counts for ATM-Based Video

Movie	Peak cells per frame	Mean cells per frame	Standard deviation cells per frame	Cells in movie (based on 7000 s, using mean)	Cells in movie (based on 7000 s, using mean for 6000 s and peak for 1000 s)
Taxi Driver 1	743	242	143	50,820,000	65,850,000
Last Emperor 2	809	352	110	73,920,000	87,630,000
News show	842	248	113	52,080,000	69,900,000
Indiana Jones 1	931	405	112	85,050,000	100,830,000
Taxi Driver 2	1,001	356	161	74,760,000	94,110,000
Last Emperor 1	1,095	604	222	126,840,000	141,570,000
League of Their Own	1,252	537	172	112,770,000	134,220,000
Boxing show	1,337	780	187	163,800,000	180,510,000
Indiana Jones 2	1,452	500	204	105,000,000	133,560,000

physical layer, the ATM layer provides information transfer for all applications the user may contemplate; the *ATM adaptation layer* (AAL), along with associated services and protocols, provides service-dependent functions to the layer above the AAL. In approximate terms, the AAL supplies the balance of the data link layer not included in the ATM layer. AAL supports error checking, multiplexing, segmentation, and reassembly. It is generally implemented in user's equipment or may occasionally be implemented in the network at an interworking (i.e., protocol conversion) point. The *control plane* also has a layered architecture and supports the call-control and connection functions. The control plane uses AAL capabilities as seen in Fig. 8.3; the layer above the AAL in the control plane provides call control and connection control. It deals with the signaling necessary to set up, supervise, and release connections. The *management plane* provides network supervision functions. It provides two types of functions: layer management and plane management. Plane management performs management functions related to a system as a whole and provides coordination among all planes. Layer management performs management functions relating to resources and parameters residing in its protocol entities.

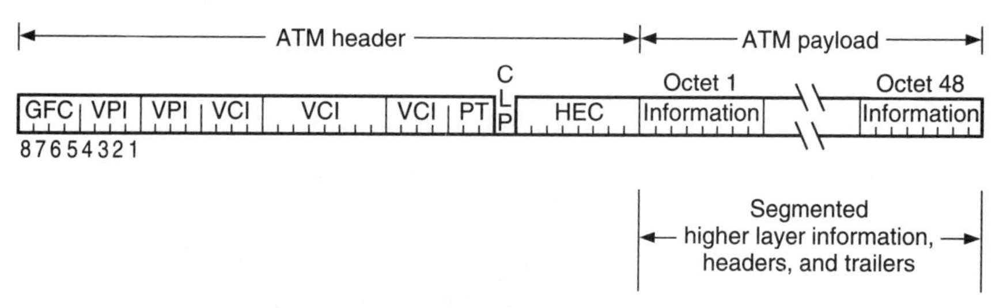

Figure 8.2 ATM cell layout.

As noted in this description, four user plane protocol layers are needed to undertake communication in an ATM-based environment:

1. A layer below the ATM layer, corresponding to the physical layer. The function of the physical layer is to manage the actual medium-dependent transmission. SONET is the technology of choice for speeds greater than 45 Mbps.
2. The ATM layer (equating approximately, for comparison, to the upper part of a LAN's medium access control layer) that has been found to meet specified objectives of throughput, scalability, interworking, and consistency with international standards. The function of the ATM layer is to provide efficient multiplexing and switching, using cell relay mechanisms.
3. The layer above the ATM layer, that is, the AAL. The function of the AAL is to insulate the upper layers of the user's application protocols, e.g., Transmission Control Protocol/Internet Protocol (TCP/IP), from the details of the ATM mechanism.
4. Upper layers, as needed. These include TCP/IP, IBM APPN, OSI TP, etc.

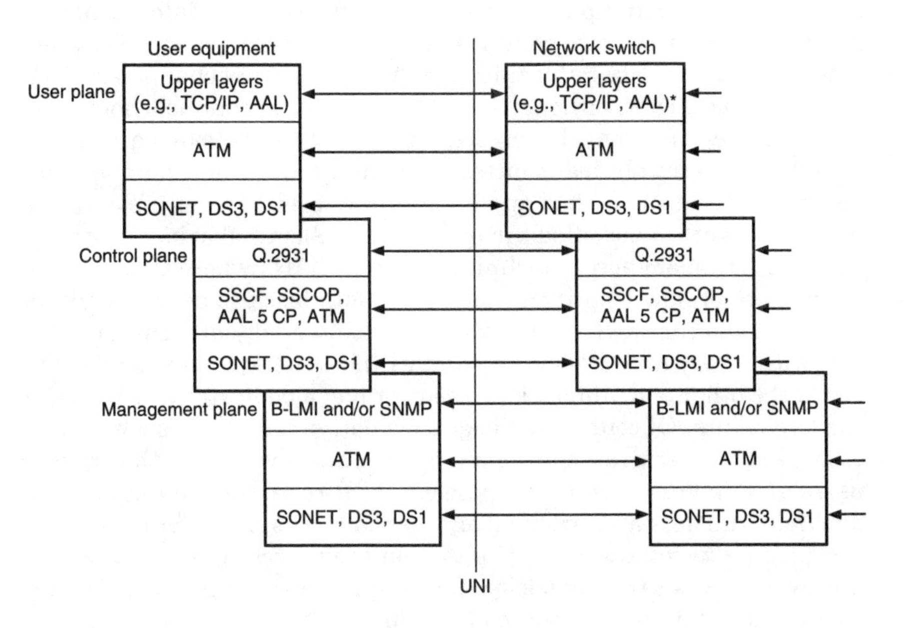

Figure 8.3 Planes constituting the ATM protocol model.

Several layers are needed in the control plane. Early PVC service users do not need the signaling stack in the control plane. SVC service needs both an information transfer protocol stack and a companion signaling protocol stack.

ATM is aimed at supporting a variety of user needs, including high-speed data, video, and multimedia applications. These applications have varying quality of service (QOS) requirements. For example, video-based services have stringent delay, delay variation, and cell-loss goals, while other applications have different QOS requirements. Carriers are proposing to support a number of service classes to tailor cell relay to a variety of business applications. In particular, there have been proposals to support a *guaranteed* and a *best effort* class.

8.2.2 Classes of ATM applications

Two main service categories for ATM have been identified (from the network point of view): (1) interactive broadband services and (2) distributive broadband services. See Table 8.5.[8]

8.2.3 Virtual connections

Just as in traditional packet switching or frame relay, information in ATM is sent between two points *not* over a dedicated, physically owned facility, but over a shared facility comprised of virtual channels.* Each user is assured that, although other users or other channels belonging to the same user may be present, the user's data can be reliably, rapidly, and securely transmitted over the network in a manner consistent with the subscribed quality of service class. The user's data are associated with a specified virtual channel. ATM's sharing is not the same as a random access technique used in LANs where there are no guarantees of how long it can take for a data block to be transmitted: in ATM, cells coming from the user at a stipulated (subscription) rate are, with a very high probability and with low delay, guaranteed delivery at the other end, almost as if the user had a dedicated line between the two points. Of course, the user does not, in fact, have such a dedicated—and expensive—end-to-end facility, but it will seem that way to users and applications on the network. Cell-relay service allows for a dynamic transfer rate, specified on a per call basis. Transfer capacity is assigned by negotiation and is based on the source requirements and the available network capacity. Cell sequence integrity on a virtual channel connection is preserved by ATM.

Cells are identified and switched by means of the label in the header, as seen in Fig. 8.2. In ATM, a *virtual channel* (VC) is used to describe

---

* The access lines are owned by the user, but the WAN facilities are shared.

TABLE 8.5 Broadband Services Supported by ATM

Interactive services
Conversational services. Provide the means for bidirectional communication with real-time, end-to-end information transfer between users or between users and servers. Information flow may be bidirectional symmetric or bidirectional asymmetric. Examples: high-speed data transmission, image transmission, videotelephony, and videoconferencing.
Messaging services. Provide user-to-user communication between individual users via storage units with store-and-forward, mailbox, and/or message handling (e.g., information editing, processing, and conversion) functions. Examples: message handling services, and mail services for moving pictures (films), store-and-forward image, and audio information.
Retrieval services. Allow users to retrieve information stored in information repositories (information is sent to the user on demand only). The time at which an information sequence is to start is under the control of the user. Examples: film, high-resolution images, information on CD-ROMs, and audio information.
Distributive services
Distribution services without user individual presentation control. These broadcast services provide a continuous flow of information that is distributed from a central source to an unlimited number of authorized receivers connected to the network. User can access this flow of information without the need to determine at which instant the distribution of a string of information will be started. User cannot control the start and order of the presentation of the broadcast information, so that depending on the point of time of the user's access, the information will not be presented from its beginning. Examples: broadcast of television and audio programs.
Distribution services with user individual presentation control. Provide information distribution from a central source to a large number of users. Information is rendered as a sequence of information entities with cyclical repetition. The user has the ability of individual access to the cyclically distributed information and can control the start and order of presentation. Example: broadcast videography.

unidirectional transport of ATM cells associated by a common unique identifier value, called the *virtual channel identifier* (VCI). Even though a channel is unidirectional, the channel identifiers are assigned bidirectionally. The bandwidth in the return direction may be assigned symmetrically, asymmetrically, or it could be zero. A *virtual path* (VP) is used to describe unidirectional transport of ATM cells belonging to virtual channels that are associated by a common identifier value, called *virtual path identifier* (VPI). The bandwidth in the return direction may be assigned symmetrically, asymmetrically, or it could be zero.

VPIs are viewed by some as a mechanism for hierarchical addressing. The VPI-VCI address space allows in theory up to 16 million virtual connections over a single interface; however, most vendors are building equipment supporting (a minimum of) 4096 channels on the user's interface. Note that these labels are only locally significant (at a given interface). They may undergo remapping in the network; how-

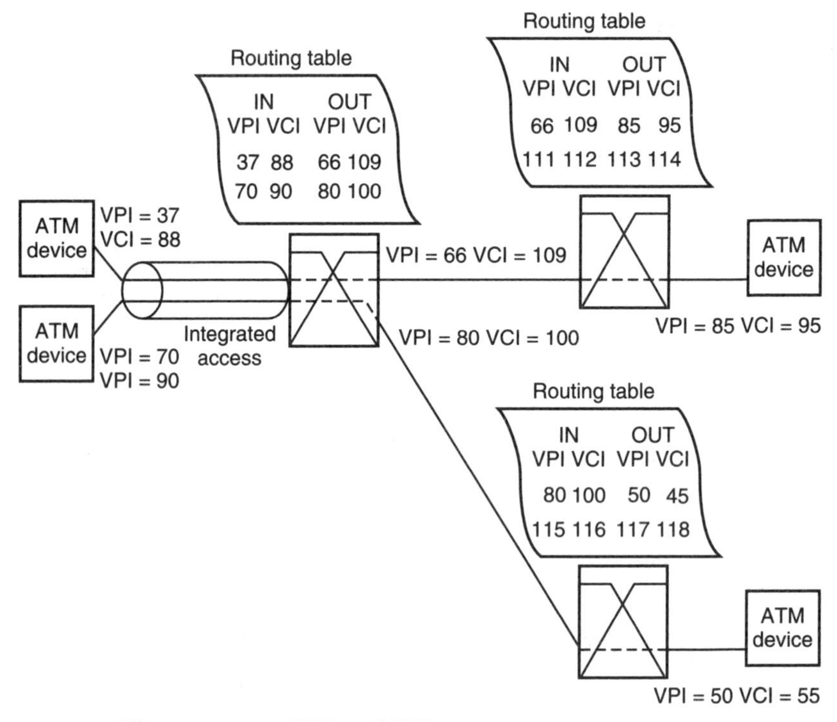

Figure 8.4 Illustrative use of VPIs and VCIs.

ever, there is an end-to-end identification of the user's stream so that data can flow reliably. Also, note that on the network trunk side more than 4096 channels per interface are supported.

Figure 8.4 illustrates how the VPI-VCI field is used in an ATM WAN. Figure 8.5 depicts the relationship of VPs and VCs as may be used in an enterprise network.

8.3 ATM Protocols: An Introductory Overview

Figure 8.6 depicts the cell-relay protocol environment, which is a particularization of the more general B-ISDN protocol model described earlier. The user's equipment must implement these protocols, as must the network elements to which the user connects.* Some of the key functions of each layer are described next.

* In the network (as well as in the user's equipment), two pieces of equipment may implement the stack. For example, the channel service unit may implement the network-bound physical layer, while the ATM router may implement the network-bound ATM layer. (A special physical layer protocol may be implemented between the channel service unit and the router.)

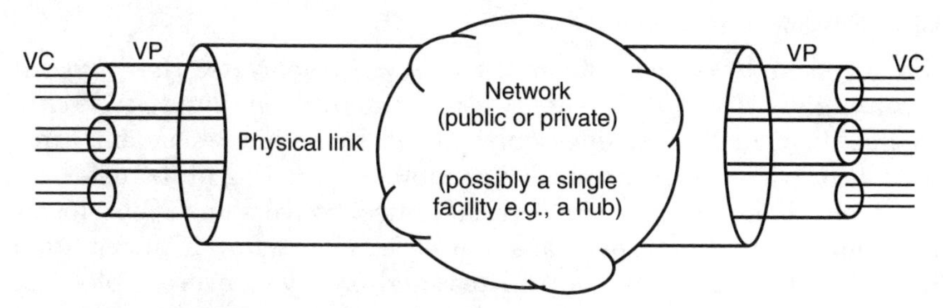

Figure 8.5 Relationship of VCs and VPs.

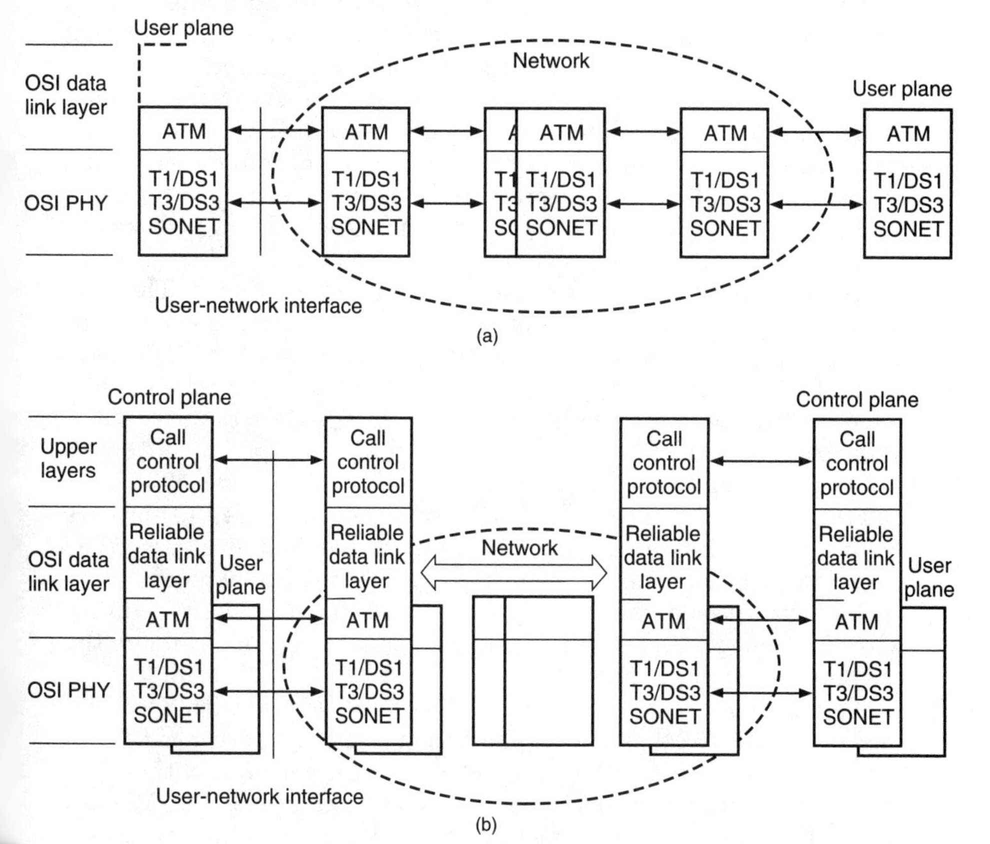

Figure 8.6 CRS environment—protocol view: (*a*) user plane (information flow); (*b*) control plane (signaling).

8.3.1 Physical layer functions

The physical layer consists of two *logical* sublayers: the *physical medium-dependent* (PMD) sublayer and the *transmission convergence* (TC) sublayer. PMD includes only physical-medium-dependent functions. It provides bit transmission capability, including bit transfer, bit alignment, line coding, and electrical-optical conversion. TC performs functions required to transform a flow of cells into a flow of information (i.e., bits) that can be transmitted and received over a physical medium. TC functions include: (1) transmission frame generation and recovery; (2) transmission frame adaptation; (3) cell delineation; (4) *header error control* (HEC) sequence generation and cell header verification; and (5) cell rate decoupling.

The transmission frame adaptation function performs the actions that are necessary to structure the cell flow according to the payload structure of the transmission frame (transmit direction) and to extract this cell flow out of the transmission frame (receive direction). In the United States, the transmission frame requires SONET envelopes above 45 Mbps. Cell delineation prepares the cell flow to enable the receiving side to recover cell boundaries. In the transmit direction, the payload of the ATM cell is scrambled. In the receive direction, cell boundaries are identified and confirmed, and the cell flow is descrambled. The HEC mechanism covers the entire cell header, which is available to this layer by the time the cell is passed down to it. The code used for this function is capable of either single-bit correction or multiple-bit error detection. The transmitting side computes the HEC field value. Cell-rate decoupling includes insertion and suppression of idle cells to adapt the rate of valid ATM cells to the payload capacity of the transmission system.

The service data units crossing the boundary between the ATM layer and the physical layer constitute a flow of valid cells. The ATM layer is unique, that is, it is independent of the underlying physical layer. The data flow inserted in the transmission system payload is independent of the physical medium; the physical layer merges the ATM cell flow with the appropriate information for cell delineation, according to the cell delineation mechanism.

The transfer capacity at the UNI is 155.52 Mbps, with a cell-fill capacity of 149.76 Mbps, due to physical layer framing overhead. Since the ATM cell has 5 octets of overhead, the 48-octet information field equates to a maximum of 135.631 Mbps of actual user information. A second UNI interface is defined at 622.08 Mbps, with the service bit rate of approximately 600 Mbps. Access at these rates requires a fiber-based loop. The first UNI makes sense for a VIU; the first and second UNI make sense for a VIP. Other UNIs are also being contemplated in the United States at the DS3 rate and perhaps at the DS1 rate. The

DS1 UNI is discussed in the context of an electrical interface (T1); so is the DS3 UNI.

8.3.2 ATM layer functions

ATM supports a flexible transfer capability common to all services. The transport functions of the ATM layer are independent of the physical layer implementation. As noted, connection identifiers are assigned to each link of a connection when required and are released when no longer needed. The label in each ATM cell is used to explicitly identify the VC to which the cells belong. The label consists of two parts: the VCI and the VPI. A VCI identifies a particular VC link for a given virtual path connection. A specific value of VCI is assigned each time a VC is switched in the network. With this in mind, a VC can be defined as a unidirectional capability for the transport of ATM cells between two consecutive ATM entities where the VCI value is translated. A VC link is originated or terminated by the assignment or removal of the VCI value.

The functions of ATM include:

Cell multiplexing and demultiplexing. In the transmit direction, the cell-multiplexing function combines cells from individual VPs and VCs into a noncontinuous composite cell flow. In the receive direction, the cell-demultiplexing function directs individual cells from a noncontinuous composite cell flow to the appropriate VP or VC.

Virtual path identifier and virtual channel identifier translation. This function occurs at ATM switching points and/or cross-connect nodes. The value of the VPI and/or VCI fields of each incoming ATM cell is mapped into a new VPI and/or VCI value (this mapping function could be null).

Cell header generation / extraction. These functions apply at points where the ATM layer is terminated (e.g., user's equipment). The HEC field is used for error management of the header. In the transmit direction, the cell header generation function receives cell-payload information from a higher layer and generates an appropriate ATM cell header except for the HEC sequence (which is considered a physical-layer function). In the receive direction, the cell header extraction function removes the ATM cell header and passes the cell-information field to a higher layer.

For the UNI, as can be seen in Fig. 8.2, 24 bits are available for routing: 8 bits for the VPI and 16 bits for the VCI. Three bits are available for payload type identification; this is used to provide an indication of whether the cell payload contains user information or network information. In user information cells, the payload consists of user information and, optionally, service adaptation function information. In network information cells, the payload does not form part of the user's information transfer. The HEC field consists of 8 bits.

The initial thinking was that if the *cell-loss priority* (CLP) is set by the user (CLP value is 1), the cell is subject to discard, depending on the network (congestion) conditions. If the CLP is not set (CLP value is 0), the cell has higher priority. More recent thinking proposes not to make use of this bit (i.e., it must always be set at 0 by the user).

8.3.3 ATM adaptation layer

Additional functionality on top of the ATM layer (i.e., in the ATM adaptation layer) may have to be provided by the user (or interworking) equipment to accommodate various services. The ATM adaptation layer enhances the services provided by the ATM layer to support the functions required by the next higher layer. The AAL function is typically implemented in the user's equipment, and the protocol fields it requires are nested within the cells' payload.

The AAL performs functions required by the user, control, and management planes and supports the mapping between the ATM layer and the next higher layer. Note that a different instance of the AAL functionality is required in each plane. The AAL supports multiple protocols to fit the needs of the different users; hence, the AAL is service-dependent (namely, the functions performed in the AAL depend on the higher layer requirements.) The AAL isolates the higher layers from the specific characteristics of the ATM layer by mapping the higher layer protocol data units into the information field of the ATM cell and vice versa. The AAL entities exchange information with the peer AAL entities to support the AAL functions.

The AAL functions are organized in two logical sublayers: the convergence sublayer (CS) and the segmentation and reassembly sublayer (SAR). See Fig. 8.7.[1] The function of CS is to provide the AAL service to the layer above it; this sublayer is service-dependent. The functions of SAR are segmentation of higher layer information into a size suitable for the information field of an ATM cell and reassembly of the contents of ATM cell information fields into higher layer information. The SAR and CS sublayers provide different service access points to the layer above the AAL. The CS depends on the particular service that the ATM transport is aiming at supporting and may undertake clock recovery and data structure recovery. In some applications, the SAR is responsible for segmentation and reassembly operations. Different combinations of SAR and CS sublayers provide different service access points to the layer above the AAL. In some applications, the SAR and/or CS

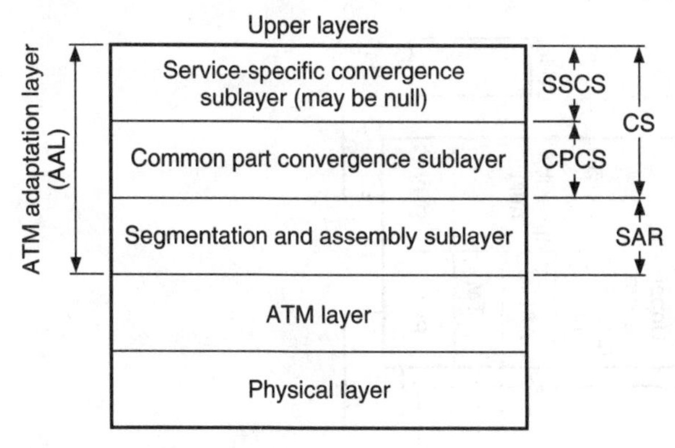

Figure 8.7 ATM layer structure.

may be empty. Fig. 8.8 depicts the role played by AAL in end-to-end video transport (note that the AAL resides in the VIU-VIP equipment, only in the case of service interworking would it also reside in the network).

Connections in an ATM network support both circuit-mode and packet-mode (connection-oriented and connectionless) services of a single medium and/or mixed media and multimedia. ATM supports two types of traffic: *constant bit rate* (CBR) and *variable bit rate* (VBR). CBR transfer rate parameters for on-demand services are negotiated at call setup time. (Changes to traffic rates during the call may eventually be negotiated through the signaling mechanism; however, initial deployments will not support renegotiation of bit rates.) CBR transfer rate parameters for permanent services are agreed with the carrier from which the user obtains service. This service could be used, for example, to transmit real-time video. VBR services are described by a number of traffic-related parameters (minimum capacity, maximum capacity, etc.). It supports packet-like traffic (e.g., variable rate video or LAN interconnection). The AAL protocols are used to support these different connection types.

Given the discussion of the preceding paragraph, to minimize the number of AAL protocols, a service classification is defined based on the following three parameters: (1) timing relation between source and destination (required or not required), (2) bit rate (constant or variable, already discussed), and (3) connection mode (connection-oriented or connectionless). Other parameters, such as assurance of the communication, are treated as quality of service parameters and, therefore, do not lead to different service classes for the AAL. The five classes of application are (see Fig. 8.9):

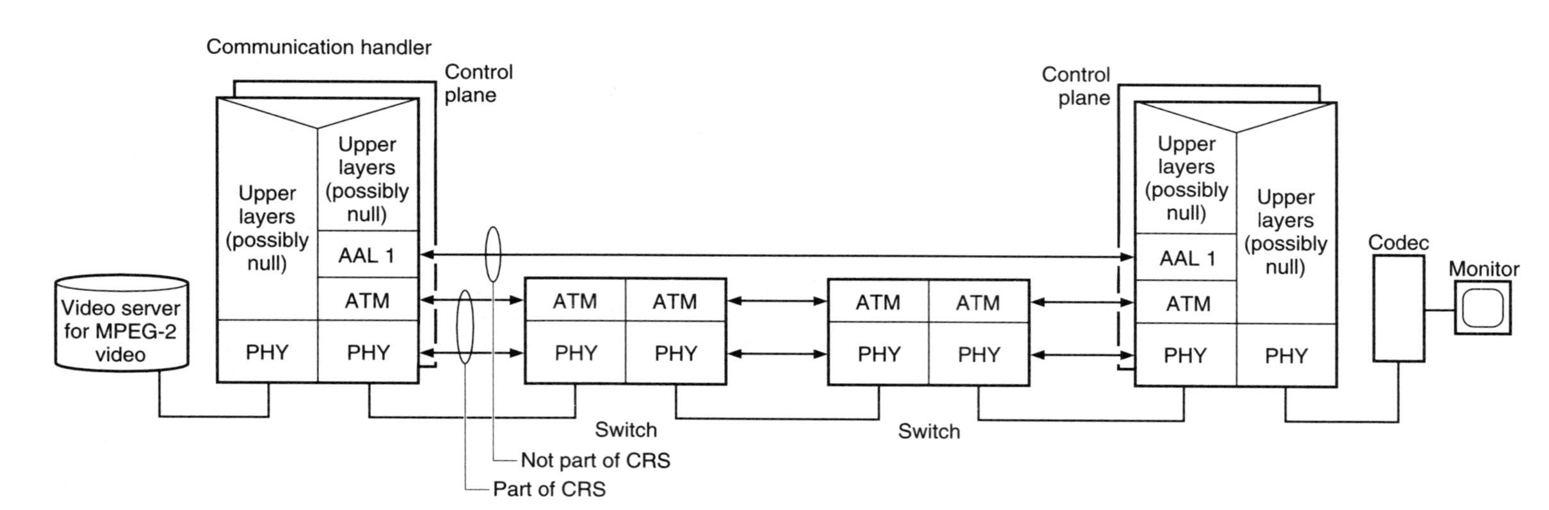

Figure 8.8 Use of AAL functionality over an ATM network.

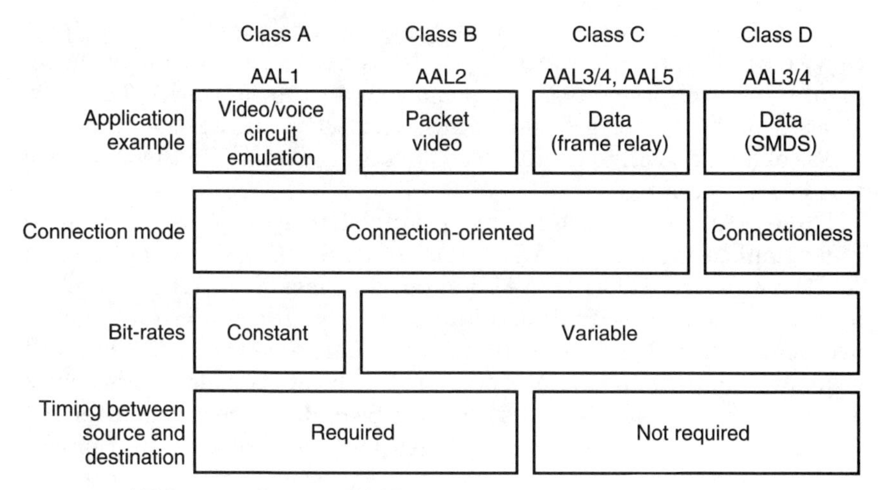

Figure 8.9 ATM service classes and AALs.

Class A: Timing required, bit rate constant, connection-oriented

Class B: Timing required, bit rate variable, connection-oriented

Class C: Timing not required, bit rate variable, connection-oriented

Class D: Timing not required, bit rate variable, connectionless

Class X: Unrestricted (bit rate variable, connection-oriented or connectionless)

Class A service is an on-demand connection oriented, constant-bit-rate ATM transport service. Class A service has end-to-end timing requirements. Class A service may require stringent cell loss, cell delay, and cell-delay variation performance. The user chooses the desired bandwidth and the appropriate QOS during the signaling phase of an SVC call to establish a Class A connection (in the PVC case this is prenegotiated). This service can provide the equivalent of a traditional dedicated line and may be used for VDT, videoconferencing, multimedia, etc.

Class B service is not currently defined by formal agreements.

Class C service is an on-demand, connection-oriented, variable-bit-rate ATM transport service. Class C service has no end-to-end timing requirements. The user chooses the desired bandwidth and QOS during the signaling phase of an SVC call to establish a Class C connection.

Class D service is a connectionless service. It has no end-to-end timing requirements. The user supplies independent data units that are delivered by the network to the destination specified in the data unit. SMDS is an example of a Class D service.

Class X service is an on-demand, connection-oriented ATM transport service where the AAL, traffic type (VBR or CBR), and timing requirements are user-defined (i.e., transparent to the network). The user chooses only the desired bandwidth and QOS during the signaling phase of an SVC call to establish a Class X connection (in the PVC case this is prenegotiated).

Three AAL protocols have been defined in support of these user plane applications: AAL1, AAL3/4, and AAL5. AAL1 supports Class A; AAL3/4 supports Class D; AAL5 supports Class X. AAL1 is intended for constant bit rate circuit emulation services. It contains one octet of overhead with fields for *synchronous residual time stamp* (SRTS), cell sequence number, and error protection for the above fields. This leaves 47 bytes for payload. AAL2 has not yet been developed; it is intended for variable rate video. AAL3/4 SAR, intended for connectionless SMDS, contains two octets of header and two octets of trailer per cell; the trailer includes length indication and *cyclic redundancy check* (CRC). AAL5 requires no per-cell overhead, but the CS contains eight octets of trailer per user packet that include length indicator and CRC. One of the issues being debated is which AAL should be used to support video over ATM and, further, which AAL should be used to support video in the most effective manner over ATM. It appears that the computer communication community (e.g., LAN and multiplexing equipment) will use AAL5. There is now work under way to rethink the transport of video over ATM and to use AAL5 rather than AAL1. Additionally, the ATM service likely to be available first (and the one supported by evolving computer equipment vendors) is the one known as Class X (i.e., cell-relay service).

Note that two stacks must be implemented in the user's equipment to obtain VCs on demand (i.e., SVC service) from the network. With this capability, the user can set up and take down multiple connections at will. The control plane needs its own AAL; there has been agreement to use AAL5 in the control plane. Initially, only PVC service will be available. In this mode, the control plane stack is not required, and the desired connections are established at service initiation time and remain active for the duration of the service contract. Also, note that AAL functions (SAR and CS) must be provided by the user equipment (except in the case where the network provides interworking functions). Additionally, the user equipment must be able to assemble and disassemble cells (i.e., run the ATM protocol).

8.4 Multiservice ATM Platforms

SMDS and frame-relay PVC are currently available fastpacket services. SMDS is a high performance, packet-switched, public data service being deployed by the RBOCs, GTE, SNET, and MCI in the

United States. SMDS is also being deployed in Europe. Frame-relay PVC is a public data service widely available today and is expected to be deployed by all RBOCs and most interexchange carriers by the end of 1994. Frame-relay SVC should be available in the 1995 time frame (see Chap. 14). ATM is a switching and multiplexing technology that is being embraced worldwide by a wide spectrum of carriers and suppliers. This new technology can switch and transport voice, data, and video at very high speeds in a local or wide area. It can also support multiple data services. See Fig. 8.10.

SMDS and frame relay are carrier service while ATM is, by lingua franca, a technology, as indicated at the beginning of this chapter. ATM will be used by carriers to provide SMDS, frame relay, and other services including cell-relay service (a fastpacket service based on the native ATM bearer service capabilities). Customers who deploy SMDS or frame relay now will be able to take advantage of the benefits of ATM technology without changing the services they use as the carriers upgrade their networks to ATM. The customer's investment in SMDS or frame-relay equipment and applications is thus preserved.[10]

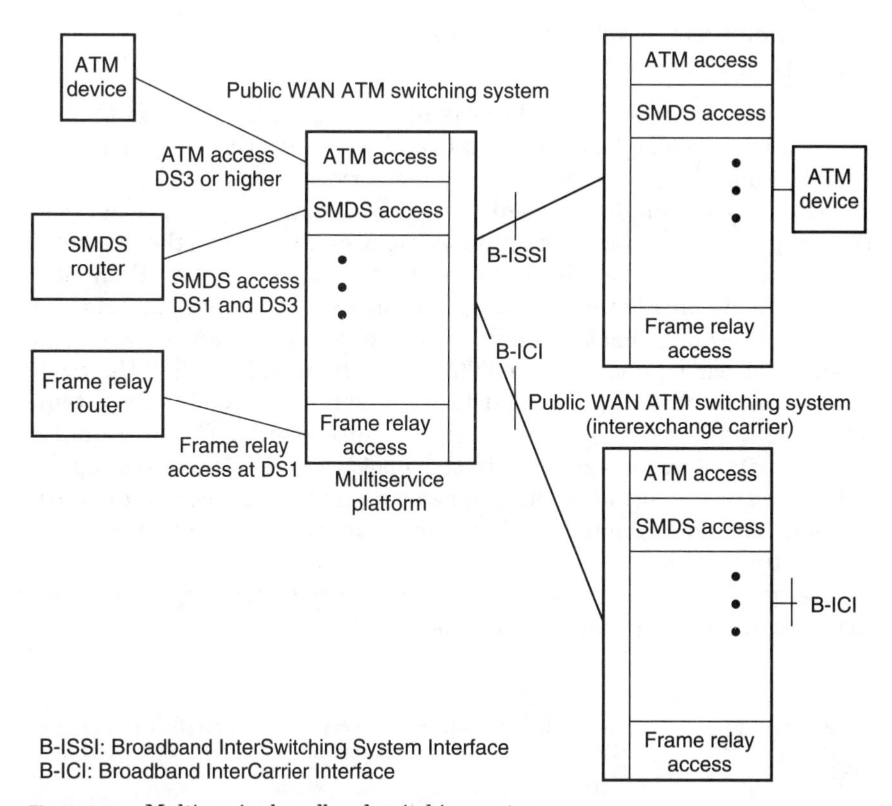

Figure 8.10 Multiservice broadband switching system.

For the same reasons that carriers are choosing ATM technology (i.e., speed, flexibility), workstation, computer, hub, and LAN manufacturers are turning to ATM for their next-generation networking needs. This is happening because current networks based on Ethernet, FDDI, etc. have limitations when handling the multimedia communications (video, voice, and data) that will flow among future workstations in a network. These manufacturers see global, multimedia communications among devices as essential. To meet these networking needs, future workstations and computers will transport user information in ATM cells. Public carriers will offer cell-relay service that will transport ATM cells across MANs, WANs, and internationally as networks evolve. Cell-relay service is targeted initially toward high-end users with multimedia needs to transport video, voice, and data across their WANs. When ATM technology extends from the desktop and throughout the network, cell-relay service will join SMDS and frame relay as another service that can be put to use by data communication managers to support evolving high-bandwidth corporate applications.

8.5 Commercial Availability of ATM Equipment and Network Services

As with any other service, there is at least a triumvirate at play to make this technology a commercial reality (if any other of these three parties fails to support the service, the service will not see any measurable commercial deployment): (1) carriers must deploy the service, (2) equipment manufacturers must bring user products to the market,* and (3) users must be willing to incorporate the service in their networks. (Some observers add two more forces: (1) agencies supporting R&D and standardization and (2) trade press to educate the end users.) The early phases of the ATM research, including all of the work already accomplished in standards organizations, cover the first item. The industry activity is covered briefly subsequently. The user analysis that will follow (not covered in this book), where users assess applicability, cost, support of embedded base, manageability (all of it in situ, in their own environment, rather than in a multicolor brochure), covers the third item.

A large set of vendors have developed or are developing ATM products, particularly for data applications, including

* For item 1 to occur, some vendors must bring out network products; this point refers to user products (see ISDN switches versus availability of cost-effective terminal adapters).

- Large switching systems targeted toward shared public networks
- Enterprise network switches, targeted toward high-end private networks
- LAN switches targeted toward high-end client-server users
- LAN switches targeted toward use as campus backbone hubs
- *Digital service units* (DSUs) which, through the *data exchange interface* (DXI), allow existing internetworking equipment (e.g., routers) to access ATM networks
- Interface cards that allow high-end workstations to access ATM LANs
- Chipsets that are marketed toward manufacturers of all of these

The paragraphs to follow describe industry activities that indicate encouraging signs of the acceptance of cell relay as a commercially viable networking technology. However, as with all new technologies, there are a number of potential hurdles and roadblocks that can delay or deter the success of cell relay. History has shown that, in spite of industry standards, interoperability problems can exist if different manufacturers implement subsets (or supersets) of the required networking features. Networking hardware may preceed the availability of software applications designed to exploit the networking power of ATM, this may slow user acceptance of cell relay. In addition, advances in existing technologies (e.g., the emergence of "fast" Ethernet) may extend the life cycle of existing products and slow the acceptance of new technologies. These challenges must be met to make ATM cell relay a long-term commercial success.

Vendors are in the process of bringing products to the market. By press time, there were several vendors of ATM hubs and a dozen vendors of ATM workstation plug-ins. Some equipment vendors are building stand-alone switches; others are adding switching capabilities to their hubs and at the same time are developing ATM adapter cards for workstations to allow them to connect to the hub. Some are also working on bridge-router cards for ATM hubs that enable Ethernet LANs to connect to the ATM. About 36 vendors had announced firm equipment plans by publication time. Over 320 companies have joined The ATM Forum, which is an organization whose goal is to expedite and facilitate the introduction of ATM-based services. PC-workstation cards are expected to become available for about $1000 per port, although the initial cost was in the $2800 to $5000 range.

Carriers are deploying *broadband switching systems* (BSSs) based on ATM *technology* to support a variety of *services.* As noted earlier, ATM is designed to be a multiservice platform. For example, frame relay and SMDS will be early services supported on these platforms; another early service is cell-relay service, which provides a network service allowing users to connect their ATM equipment using the native ATM bearer service.

Early entrants in the user-products arena included, among others, Adaptive, Cabletron, Digital Equipment Corporation, Fore Systems, GDC, Hughes, Newbridge, Stratacom, Sun, SynOptics (now Bay Networks), and Wellfleet; in the carrier switch arena, early entrants included ADC, Alcatel, AT&T Network Systems, DSC, Fujitsu Systems, Northern Telecom, and Siemens. In 1994, these vendors were demonstrating ready or near-ready products for a variety of networking needs. Products appearing were first targeted to the local connectivity environment, but WAN products are also expected soon.

Hubs and switches to support the bandwidth-intensive applications listed earlier, such as video, have already appeared. Typical switches now support 8 to 16 155-Mbps ports over shielded twisted-pair or multimode fibers (lower speeds—45 or 100 Mbps—are also supported). Some systems can grow to 100 ports. Typical backplane throughput ranges from 1 or 2 Gbps up to 10 Gbps. A number of these products support not only PVC but also SVC; some also support multipoint SVC service. Products already on the market (e.g., Hughes LAN, Synoptics/Bay Networks, Newbridge, Adaptive, and Fore Systems) are priced as low as $1200 per port.

Interface cards for high-end workstations (e.g., SPARCstation) are also appearing (e.g., Bay Networks and Adaptive). These typically support 45 Mbps (DS3) on twisted-pair cable and 100 or 155 Mbps on multimode fiber, consistent with The ATM Forum specification. Some even support prototype 155-Mbps connectivity on shielded twisted pair. These boards are already available for as little as $1250.

One major push now is in the network management arena. Users need capabilities to integrate the support of ATM products into the overall enterprise network, specifically the corporate management system. Some typical features recently introduced include automatic reconfiguration of virtual connections in case of failure, loopback support, performance and configuration management, and Simple Network Management Protocol (SNMP) functionality (with private *management information base,* MIB, extensions). Some of the hubs also act as multiprotocol routers, either (1) accepting ATM devices internally for WAN interconnection over SMDS and frame-relay networks, or (2) accepting ATM devices internally for WAN interconnection over a cell-relay network, or (3) accepting traditional devices internally for WAN interconnection over a cell-relay network (these are stand-alone ATM multiprotocol routers).

As can be inferred from the products reaching the market, ATM has been aimed at the data (more specifically, LAN interconnection) fast-packet market in the enterprise network context. Only more recently have video and VDT been given proper consideration.

Specifically, for WAN cell-relay service, AT&T, Sprint, and WilTel have already announced plans for ATM-based services, while MCI is pursuing SMDS initially, and ATM at a later date.[10,11] As early as 1993, MSF Datanet announced the availability of commercial ATM service in 14 U.S. cities. A three-phase approach has been announced publicly by Sprint. Phase 1 (1993) entails frame-relay interconnectivity with local exchange carriers. Phase 2 (1993 to 1994) supports PVC cell-relay service at DS3 rate. Phase 3 (1994 to 1995) enhances the cell-relay service to 155 Mbps. Other carriers are expected to follow suit in the next 6 to 18 months. All seven RBOCs have also announced deployment plans for ATM platforms and for cell-relay service. 1995 is the year users can expect public cell-relay service in a number of key metropolitan areas. Chapter 14 provides additional details on this topic.

In addition to the international and domestic standards, one needs additional details and clarifications to enable the deployment of the technology. To this end, the RBOCs have published generic requirements needed for suppliers to start building ATM equipment that will enable PVC cell-relay services to be offered (see Table 8.6).

The technical advisories and technical requirements can be used by (1) *local exchange carriers* (LECs) interested in providing nationally consistent cell-relay PVC exchange service to their customers, (2) suppliers of ATM equipment in the local customer environment (e.g., ATM LANs, ATM routers, ATM DSUs, ATM switches), and (3) suppliers of ATM equipment in LEC networks.

The development of nationally consistent LEC cell-relay PVC (as well as an exchange access cell-relay PVC service) is important for a consistent set of service features and service operations for customers who will want to use the service on a national basis. However, for VDT services, a nationally consistent service is less critical.

The following phases of nationally consistent service have been advanced. It is possible that LECs may be offering prenationally consistent cell-relay PVC to meet near-term demand for the service by customers in the 1995 time frame. These carriers are expected to support

TABLE 8.6 Technical Advisories and Requirements for Cell-Relay Service

TA-TSV-001408: generic requirements for exchange PVC cell-relay service
TA-TSV-001409: generic requirements for exchange access PVC cell-relay service
TA-TSV-001501: generic requirements for exchange SVC cell-relay service

a nationally consistent cell-relay PVC exchange service in the second half of 1995.

- Phase 1.0: Nationally consistent cell-relay PVC exchange service by the end of 1994 based on a core set of service features.
- Phase 2.0: Nationally consistent cell-relay PVC exchange service by the middle of 1995. Phase 2.0 builds on the capabilities of Phase 1.0 and supports expanded capabilities in some areas, such as traffic management, congestion management, and customer network management.
- Phase 3.0: Phase 3.0 will see the initial support of a cell-relay SVC exchange service in mid to late 1995.

8.6 ATM Equipment

A number of cable companies, Time Warner Cable in particular, have announced the intention to use ATM in the delivery of video. AT&T has a family of ATM equipment for video delivery; IBM recently announced it was developing a family of ATM products and states that it will spend $100 million per year developing the technology; other suppliers have or are readying products.[12] Although the cost of ATM is now high and is expected to remain high for the immediate future, there are efforts underway to reduce that figure to $500. The 1994 cost per desktop connection is about $1500 to $2500.

8.6.1 General equipment

A scan of product literature provided by the manufacturers of ATM equipment shows that there are a number of distinct switching technologies being used for ATM. Indeed, there are arguably a confusing number of terms manufacturers use to describe their core-switching technology, including

- Crosspoint switch fabric
- Output-buffered fabric
- Self-routing switch fabric
- Time-memory switch fabric
- Time-space-time switch fabric

8.6.2 Broadband switching architectures

There are several basic choices a switch designer can make in designing a broadband switching fabric, including architecture, control, and physical topology. These switches will support, among other devices, the video servers discussed in Chap. 7. Daddis and Torng developed a tree-based taxonomy of broadband switching systems design (Fig. 8.11).[13] Their goal in classifying switch designs in this way was to group together switches with fundamentally similar designs and to separate those based on radically different philosophies in architecture. In their tree structure, the end nodes represent particular switch designs and each inner node specifies a design decision. This collection of design points classifies broadband switches through their fundamental properties. Five design points were chosen. In order of descending levels on the classification tree, they are:

1. Dedicated versus shared links: A link is dedicated if, at most, one switch element may transmit onto it; otherwise the link is shared.
2. The transport mode is the discipline governing the transmission of data units onto links; may be statistical or assignment-based (this corresponds roughly to circuit versus packet switching).
3. Centralized versus decentralized routing control: Routing decisions can be made at one point in the switch or distributed throughout the switch.
4. Buffered versus nonbuffered switch elements: A buffered switch element stores data units within the element to resolve output link contention. Unbuffered switch elements incur a constant delay on data units relative to congestion in the system.
5. The switching technique is the method of directing the data units from one switch element to another. The four techniques are space switching, time switching, frequency switching, and address filtration.

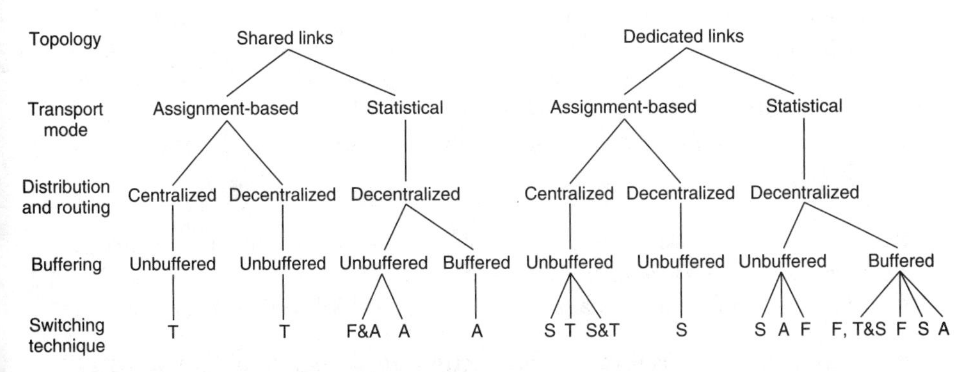

Figure 8.11 Taxonomy of broadband switches.

8.6.2.1 Dedicated versus shared links. A look at Fig. 8.11 indicates that we are mainly interested in the right-hand side of the tree, namely, dedicated links. The majority of ATM products on the market or in development today use dedicated links, although a few products employ a shared link (e.g., bus) switching approach.

8.6.2.2 Transport mode. A transport mode is the method the switch uses to transmit data over links in the interconnection fabric. This mode can be categorized into links whose capacities may be dedicated to each call, or they may be made statistically available to all calls through a need-based scheme. Dedicated links form the basis of circuit switching that has dominated the telecommunications business to the present day.

8.6.2.3 Distribution of routing intelligence. With centralized routing, a single logical entity controls the routing of data over the switch elements and interconnection fabric at a given level. In distributed (decentralized) routing, data routing is executed via processing that is localized to each switch element, possibly with the aid of global state information replicated among each of the elements. A self-routing switch fabric would be considered to have distributed control. An example of distributed routing is a Batcher/Banyan switch fabric, where the routing control is decided on an element-by-element basis as the data unit traverses the switch. Distributed routing is, in general, more complex than centralized routing. This complexity includes routing control for multiple services, deciding between the arbitration of access of cells to a link, and the switching of data. Output link contention occurs when data from one or more switch elements contend for the transmission capacity of a link.

8.6.2.4 Buffering in switch elements. A buffered switch element stores cells within the element to resolve output link contention. Data may be buffered before switching (input buffering), after switching (output buffering), within the switch elements (shared buffering), or not at all (no buffers). The choice of buffering method can have a dramatic effect on the characteristics of a switch (e.g., performance under heavy traffic loads).

8.6.2.5 Switching techniques. The switching technique influences or determines the physical structure, distribution, and execution of the switch intelligence. There are four primary switching techniques:

- *Space switching* refers to the technique whereby a transmitting switch element chooses among multiple physical routes to send data. An example of space switching is a simple crossbar switch, in which a crosspoint either routes a data unit along a vertical output bus or ignores it and allows it to continue along the horizontal input bus.

- *Time switching* refers to a technique whereby the transmitting switch element delays a data unit for an amount of time, depending on the routing.
- *Frequency switching* refers to when the transmitting switch element chooses among two or more carrier frequencies on which to route data units.
- *Address filtration* refers to a technique whereby the receiving switch element selects labeled data units from among multidestination traffic. Fujitsu's FETEX 150 ATM switch architecture is an example of such an arrangement.[14] In the FETEX 150, buffered intelligent crosspoints select cells through addressing, and cells are injected onto horizontal input buses that later transmit the cells onto the attached vertical output buses. The switch elements do not route data units among physical links but, instead, select cells from the broadcast input bus and retransmit onto a single output bus.

These four switching techniques may be used alone or, more commonly, in combinations.

8.7 ATM Carriage of Video

8.7.1 Overview

There is now a lot of discussion about video transport over ATM. As an indication, in 1993 the president of Cable Television Labs, Louisville, Colorado, gave the keynote speech at an ATM Forum meeting, where he stated[12]: "I come here as an emissary of the cable industry, suggesting that the deliberations of The ATM Forum can and should play an important role in the cable industry's efforts to participate in the digitization of communications in America . . . I want you to know that the cable industry pledges to participate with you in the effort to refine and coordinate interindustry technical standards."

In particular, ATM signaling capabilities for the establishment of SVC connections make the technology ideally suited to an environment where there are bandwidth limitations; hence, rather than simultaneously bringing 100 channels to a consumer when he or she only wants to watch one, ATM switches in, on demand, only the one of interest. Even individuals within the cable industry note that no other method "compares with ATM in the potential to supply the functionality required to support long-term evolution of the cable network."[12] Some cable systems are investigating the technology. For example, the Time Warner Cable trial of the Full Service Network (see Chap. 10) uses ATM. In that trial, ATM provides circuit emulation and switches DS3 links. (As noted in Chap. 4, Fig. 4.12, this system carries 20 DS3 links to each 400-house service area.) U S WEST, Southwestern Bell, Southern New England Telephone, and other telephone companies are reportedly considering similar technology.[12]

In the layered model of the CRS protocol model, the physical and ATM layers represent the network transmission and switching fabric associated with a CRS PVC or SVC connection. The ATM adaptation layer is associated with points at which the ATM layer transport is terminated to recover the service. This capability plays a key role in enabling an ATM network to support video distribution, videoconferencing, and/or multimedia. The AAL functionality can reside in either the network or the user's equipment; however, for the discussion at hand, the functionality is located at the endpoints, as seen in Fig. 8.8. The AAL is responsible for adapting services utilizing ATM transport to the ATM network environment. The AAL is defined in terms of protocol elements available for the support of different classes of service in an ATM environment (i.e., variable bit rate connectionless data, constant bit rate).

A discussion of video carriage (MPEG-1 or MPEG-2) over ATM transport, based on current proposals for cell-relay service, is provided herewith. Two approaches have evolved:

1. Use of AAL1 over ATM to provide a circuit-emulated link over which digital video information will flow
2. Use of AAL5 over ATM

Key issues of interest relate to traffic characteristics based on coder parameters, to ascertain that the AAL-ATM service is adequate. See Ref. 9 for a discussion of this topic.

8.7.2 Discussion of issues

For the transport of MPEG-2 video and audio information over ATM, several issues related to MPEG-2 systems and ATM need to be taken under consideration. In particular, a number of transmission factors need to be examined to ensure that the carriage of MPEG-2 streams meets the required performance goals. These issues include the mapping of MPEG-2 bit streams into ATM cells, cell-loss detection and replacement (ultimately related to bit error detection and correction), the use of time stamps, and jitter. In summary, from an ATM service point of view, two key issues come into play:

When using circuit emulation (AAL1): cell loss and transmission channel efficiency

When using CRS (AAL5): cell loss and jitter

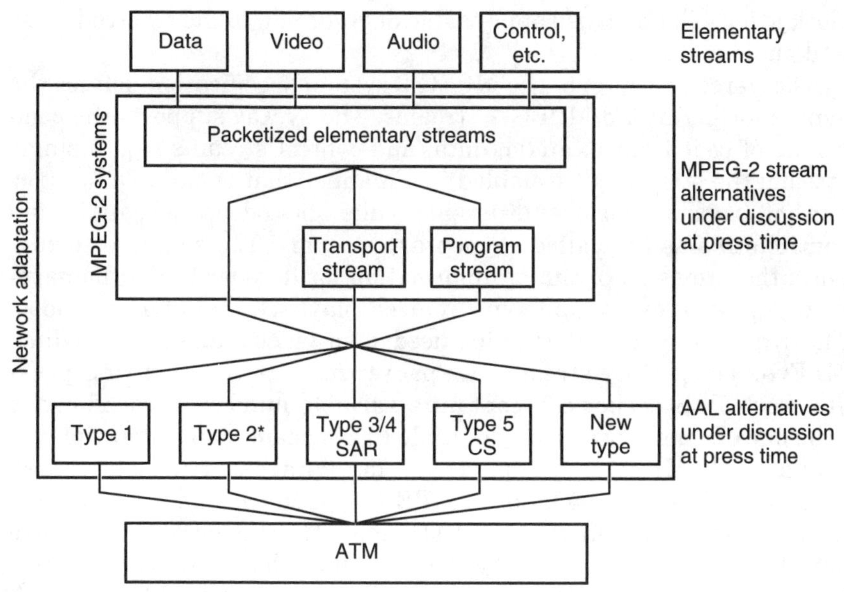

Figure 8.12 Carrying MPEG-2 over ATM.

At press time, these issues were being actively discussed in various organizations, including ANSI T1S1, ITU-T SG15, ISO MPEG, and The ATM Forum.* Clearly, it is desirable to arrive at a standard solution to these issues to minimize the cost of the encoder/decoder and to support equipment interoperability. A diagram of some of the options under study by ITU-T SG15 is shown in Fig. 8.12. These issues are discussed in the subsections that follow. (There also are proposals to map MPEG-2 to *synchronous transfer mode,* i.e. SONET; see Sec. 8.7.3.) To focus the ATM discussion, however, some highlights from Chap. 6 are provided in the next few paragraphs.

As covered in Chap. 6, the MPEG-2 system specification uses time stamps and a constant delay timing model for the recovery of the decoder clock. As implied by the discussion of the previous sections, in ATM networks, this MPEG-2 model is not automatically valid because of the cell-delay jitter. In particular, the cell-delay jitter in the network will introduce jitter in the recovered decoder system clock. Therefore, it is critical to the viability of ATM-based video that the decoder system

* The ATM Forum established in early 1994 a subgroup of the Service Aspects & Applications group, called the Audiovisual Multimedia Service subgroup. The charter of this group is to develop a specification to carry MPEG-2 video (TS) over AAL5 as the transport protocol. The specification was slated for completion by the end of 1995.

clock jitter will not cause appreciable degradation to the received video and audio material.

As covered in Chap. 6, the MPEG-2 system specification defines the syntax for permitted MPEG-2 streams. The syntax supports the combining of coded video, audio, data, and control streams into a single stream; the stream is assembled in a manner that is suitable for storage and transmission over different media. The coded video, audio, and control streams are called elementary streams. The syntax also supports the provision of timing information for decoder buffer, management, clock recovery, and synchronized playback of audio and video. The syntax provides information needed for video and audio decoding. MPEG-2 elementary streams are packetized into series of PES packets. Each PES packet can contain a variable number of coded octets from one elementary stream; multiple PES streams can be multiplexed into a single MPEG-2 system stream for storage or transmission. As covered in Chap. 6, two kinds of MPEG-2 system streams are defined: program streams and transport streams. The PS packets can have variable length and can be long. They are intended for relatively error-free environments (or where errors are being corrected, either through a forward error correction method, or some medium-dependent method, as in a CD-ROM). The TS is suitable for relatively error-susceptible environments. The TS packets have a fixed length of 188 octets (4 octets of packet header and 184 octets of payload). The PS and TS have been designed for different applications; however, it is possible to convert from one to the other. This discussion focuses on the TS.

The MPEG-2 systems timing model, which comes into play when talking about network services such as ATM, is shown in Fig. 8.13. Since the video and audio encoding produce variable bit rates, the delays through each of the encoder and decoder buffers (items A and B in Fig. 8.13) are variable. Timing information is needed for the decoder to read the data out of the buffers at the correct time to prevent the decoder buffers from overflow or underflow. Time stamps called *pro-*

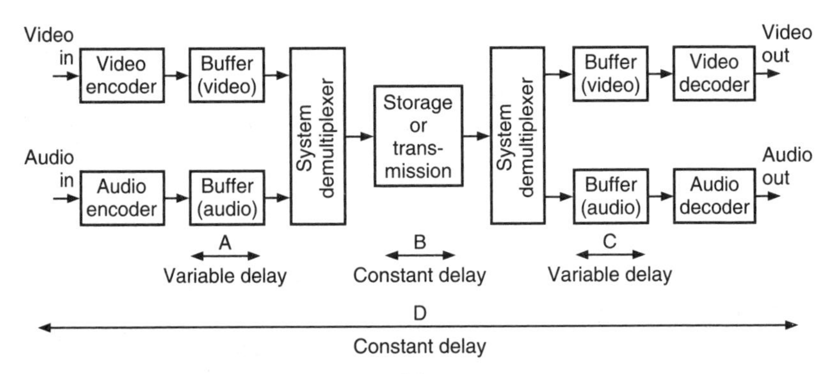

Figure 8.13 MPEG-2 system timing model.

gram clock references (PCRs), which specify the encoder system clock time, are routinely added to the bit stream. Observe that the delay from the input of the encoder to the output of the decoder (item D in Fig. 8.13) is constant. Based on this observation, time stamps that indicate the correct decoding and presentation timing are added into the bit stream in the encoder, enabling the decoder to read the data out at the appropriate time. As discussed in Chap. 6, the time stamps that indicate the presentation time of video and audio are called *presentation time stamps* (PTS); the time stamps that indicate the decoding time are called *decoding time stamps* (DTS). In addition to supporting decoder buffer management, these time stamps are also used to synchronize the playback of video and audio.

Another function of the MPEG-2 system is to provide the timing information for decoder timing recovery. In the MPEG-2 system model, both the encoder and the decoder have a 27-MHz system clock that is used to derive other clocks (such as the video and the audio sampling clocks). The decoder system clock must be synchronized to the encoder system clock. To accomplish this, the MPEG-2 timing model assumes the transmission channel has constant delay. This is an issue of relevance: MPEG-2 as it exists, has been developed with a synchronous transfer mode discipline in mind, not an asynchronous transfer mode discipline; there lies the industry debate in the current proceedings on how to carry MPEG-2 effectively over ATM. PCR time stamps need to arrive at the decoder with a fixed delay. At the decoder, they are compared with the stamps generated locally, based on the decoder's system clock. The differences between the time stamps from the encoder and those from the decoder are used to control decoder phase lock loop to synchronize the decoder system clock to the encoder system clock. Therefore, MPEG-2 assumes constant channel delay to properly receive the PCRs for decoder timing recovery.

As previously discussed, MPEG-2 systems assume constant channel delay for decoder timing recovery; hence, when MPEG-2 bit streams are carried over an ATM network, there are concerns about the effect of the ATM cell-delay jitter on the feasibility and complexity of clock recovery schemes using the MPEG-2 PCRs. The cell-delay jitter will cause the PCRs to arrive at the decoder with an unpredictable and variable delay. Given that the arrival time of the PCRs is used in generating the error signal that controls the phase lock loop, PCR jitter results in jitter in the recovered decoder system clocks. In turn, since all the clocks in the decoder are derived from the system clock, the jitter in the system clock will reflect itself in jitter in the derived clocks, with ensuing picture degradation.*

---

* For quality transmission of TV signals, the color subcarrier is expected to have a frequency jitter less than 10 Hz and a frequency drift of less than 0.1 Hz/s.

With typical delay cell jitter, it could turn out to be difficult to meet the specifications required for quality video using the PCRs and an inexpensive phase lock loop. One proposal, therefore, is to use AAL1 and *synchronous residual time stamp* (SRTS) (see App. A). AAL1 provides a jitter-free clock at the receiver, when the VDT network uses the same primary clock.* Issues related to SRTS are covered in ITU-T I.363[15] and ANSI T1.631.[16] Since the CBR service's clock may not be locked directly to a clock available from the ATM network, the CBR AAL must provide a method for recovering asynchronous† clocks. Clock recovery, as described in ANSI T1.631, is used in support of DS1 and/or DS3 circuit emulation for CBR traffic (including traditional MPEG-2 streams) that needs to be carried over an ATM network. DS1 and DS3 services that have been used to transmit video in the past have stringent jitter requirements (as specified in ANSI T1.403 and ANSI T1.404.[17,18]) SRTS encodes the frequency offset between the encoder system clock and a common network clock that is, by assumption, available to both the encoder and the decoder. (The reference clock is assumed to be derived from network clocks available to the CBR AAL transmitter and receiver, e.g., a SONET 155.52-MHz clock on a STS-3c physical layer access.) From this frequency offset and the common network clock, the receiver can generate a jitter-free clock, synchronized to the encoder clock even in the presence of cell-delay jitter. As noted, the concern in using SRTS is that a common network clock may not be available in some situations. Figure 8.14 depicts the operation of the recovery mechanism at a macro level.[19]

There are other issues besides jitter that need to be taken into consideration. These are discussed next.

Bit error considerations. The *bit error rate* (BER) depends on the physical media. In general, ATM can be carried over different physical media, including coaxial cable, fiber-optic cable, and twisted pair. Usually, it is higher for coaxial cable and twisted-pair cable than in a fiber-optic cable. The BER is important in MPEG because a bit error usually causes the decoding to lose synchronization and propagate errors; this is due to the characteristics of the variable-length codes used in the video coding. For example, for a channel with BER of 10^{-9}, an MPEG-2 bit stream will encounter a bit error every 4 min or so. The MPEG-2 bit stream syntax uses several start codes for synchronization purposes.‡

---

* In a nationwide network consisting of several carriers and possibly several unsynchronized clocks, the issue of the feasibility of SRTS is still under study.

† A CBR service clock not locked to a network clock.

‡ A start code cannot be emulated by concatenations of other code words, making it suitable for synchronization.

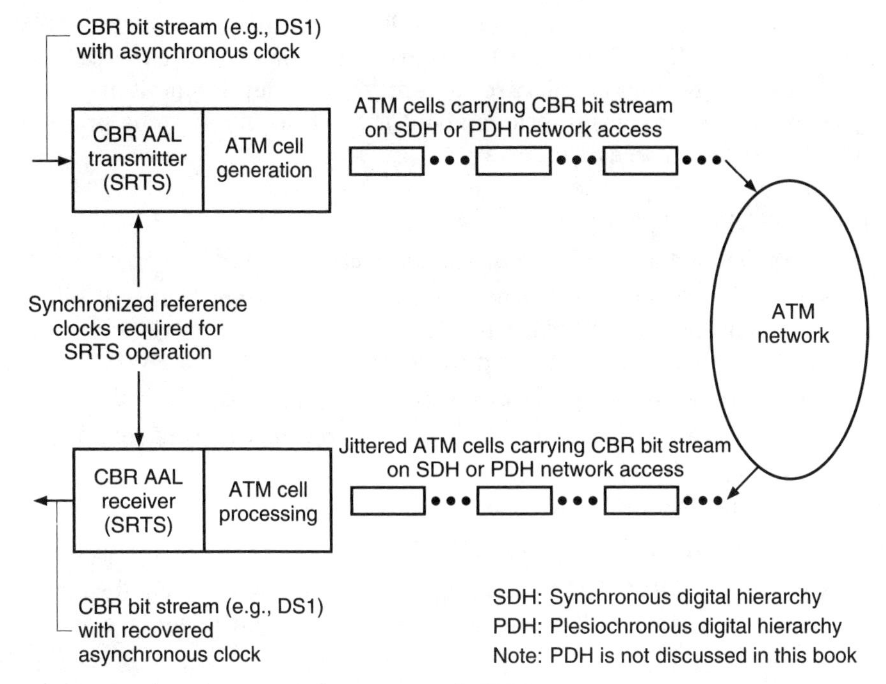

Figure 8.14 Asynchronous clock recovery with SRTS.

Synchronization codes can come no sooner than a slice, which may contain up to a horizontal strip of macroblocks in the picture.* Therefore, a bit error typically results in error propagation until the next video slice. Compounding this, since MPEG-2 uses interframe coding, the error in a slice may also propagate until the next intraframe-coded picture, which typically appears once in every 15 frames. Error concealment techniques may be applied to make the degradation less conspicuous; however, further research is needed to determine if results are acceptable to the viewing public.

Cell loss considerations. The *cell-loss ratio* (CLR) depends on the physical media, the switching technique, the amount of switch buffering, the number of switches traversed, the QOS class used for the service, and whether the application is CBR or VBR. For VBR applications, it is more likely that the network will encounter congestion and drop cells. A cell loss causes 48 octets of data to be lost, and perhaps more if the cell is part of a long frame. The effect on video quality is similar to a bit error because a cell loss is likely to also cause error

* At a 4-Mbps video, the average number of octets per slice is 462 (based on 720×576 picture at 30 fps, and noting that a slice is one horizontal strip of macroblock).

propagation until the next video slice. For a variable-rate MPEG-2 bit stream and a CLR of 10^{-7}, a cell loss every 17 min or so is expected.

While the specifics of CRS are currently under development and subject to change, key assumptions about the service likely to be present in a CRS offering are as follows:

1. Private virtual connection CRS
   - provisioned as semipermanent connection
   - provides access to a broadband switching system that will allow for partitioning of ATM capacity among PVC connections associated with virtual paths and the virtual channels of a PVC
   - point-to-point and point-to-multipoint connections
   - provides a wide range of rates (i.e., 64 kbps to a current maximum around 132 Mbps has been proposed)
   - assumes DS3, STS-3c, or STS-12c access*
2. Switched virtual connection CRS
   - upgrade of PVC CRS when signaling agreements needed to support real-time CRS connection setup and teardown are made available

Current descriptions for CRS propose preliminary performance objectives for ATM network impairments. The proposed CRS objectives recognize two classes of connection performance identified as QOS Class I and QOS Class II. Class I differs from Class II by offering stricter cell-loss and cell-delay variation objectives. The objectives currently proposed for QOS Class I and Class II are:

Accuracy

Cell-loss ratio

Class I: less than 8.5×10^{-10}

Class II: less than 5×10^{-7}

Delay

Cell-transfer delay

Classes I and II: 4-ms propagation delay (99 percent objective)

* DS1 access is also under consideration; this access option is important for videoconferencing applications; otherwise, the use of a DS3 physical access facility is needed at all (multipoint) ends, making the cost of communication outside the reach of desktop applications.

Cell-delay variation (CDV)

Classes I and II: mean CDV of 0 within 0.01 ms

Classes I and II: CDV less than 0.5 ms (99 percent objective)

Class I: no more than 1 in 10^9 cells should experience CDV exceeding 1.0 ms

(One should distinguish between the ATM QOS for a single switch and those for the service, measured end-to-end.)

CRS currently supports two classes of connections from a performance viewpoint. Class I has more stringent cell-loss and cell-delay variation objectives than Class II. An examination of performance indicates that:

- Class I could support video distribution with an acceptable grade of service under either scenario: shared CBR pipe or dedicated CRS connections.
- Class II would not provide an acceptable level or support the shared CBR pipe (i.e., DS1) scenario.
- Class II may provide an acceptable level of performance for the dedicated CRS connection scenario, but enhanced user-equipment error control, such as forward error correction, may be needed to achieve a satisfactory level of service.

Error correction. Aspects of error correction that were being debated at press time (and which may take a couple of years to solve by way of consensus) include the following: what are acceptable levels for BER and CLR for the transport of MPEG-2 streams over an ATM network? Does error concealment in the video decoder provide acceptable quality so that bit-error detection or correction is not required? What is a suitable error detection/correction and cell-loss detection/correction code if this functionality is needed? Fortunately, because of the use of self-synchronized scrambler and the line codes in the physical layer, a single bit error typically produces multiple correlated errors. Given this correlation, efficient Reed-Solomon codes can be employed. When cell-loss correction via forward error correction methods is needed, a group of cells may be octet-interleaved and then covered by Reed-Solomon codes (several interleaving structures are being studied by the ITU-T); however, this requires additional memory, affecting not only cost, but also delay.

Mappings to ATM. Aspects of mapping that were being debated at press time (and which may take a couple of years to solve by way of consensus) include: Should the TS, PS, or PES be used in the transport of MPEG-2 over ATM? Which AAL should be used? Considerations being used in evaluating the various proposals include delay, packing efficiency, error robustness, timing recovery, and implementation cost. There are proposals to use PES or PS in mapping MPEG-2 streams over ATM, instead of the TS. Such streams would be more desirable in terms of packing efficiency and protocol processing, because TS and ATM transport contain duplicate functions.* Furthermore, TS requires large PES packets for efficiency. (This arises from the fact that MPEG-2 requires that the PES packet headers must be aligned with the TS packet header.)

In terms of error robustness, a video slice is an intrinsically atomic unit for error control in MPEG-2. Hence, alignment of the start of each video slice with the start of an ATM cell payload provides error localization.† It follows that it appears desirable to want to pack a video slice into a PES packet and then map the PES packet directly into multiple ATM cells. When this approach is used, cells subject to transmission impairments impact only a video slice. Given this argument, observe that, since a slice is about 460 octets as noted earlier, the efficiency of mapping such a short PES packet to TS packets and then to ATM cells is relatively low; hence, this reinforces the premise of not mapping TSs, but other streams. However, in the context of MPEG-2 systems, PES is not intended for transmission over networks. Because of VDT time-to-market considerations, the consensus-building effort needed to agree on how to transport MPEG-2 PES over ATM may sidetrack the potential technical benefits.

Many industry groups, however, argue in favor of using TSs over ATM. This originates from the fact that most video services under discussion—VOD and HDTV services, whether over satellite, coaxial cable, or fiber—are expected to use MPEG-2 TS. Hence, using TS facilitates interoperability, abrogating the need for format transcoding. Upon analysis, it is clear that the overlap of functionality between ATM and TS originates in the 4-octet TS header, and since a TS packet is 188 octets long, the overhead of a 4-octet header may not be ultimately significant.

AALs. The driving force behind the decision as to which AAL to employ impinges on the mapping efficiency, the required functionality (timing recovery support, bit-error, and cell-loss detection or correction, etc.), and implementation cost and time. AAL1 can carry the TS

* Examples include packet delineation, multiplexing, and prioritization.

† As a counterargument, the probability of a transmission impairment affecting two consecutive video slices may not be high enough to dictate the alignment of a video slice with a PES packet. Furthermore, if bit- and cell-loss correction are employed, the error rate could be below a threshold of concern.

packets fairly efficiently (mapping a TS packet into four ATM cells) but has no capability for carrying forward error correction codes. Also, although it supports SRTS as discussed in the next section, the number of bits assigned for SRTS may be insufficient for VBR applications. AAL2 is intended for video applications, but, as noted earlier, it is currently undefined. AAL3/4 is fairly inefficient, while AAL5 is very efficient. However, AAL3/4 and AAL5 would require extensions to provide timing recovery; furthermore, neither supports the mapping of a 188-octet TS packet directly and neatly into four cells. The status at press time was to either use AAL1, map TS into AAL5 and derive five cells, or develop AAL2. These proposals were actively discussed in the standards bodies. These proceedings come at a time when chips for AAL5 (and also AAL3/4 from an SMDS genesis) are becoming generally available. Given the hardware availability, one can expect to see the short-term emergence of video over AAL1/ATM and/or over AAL5/ATM.

8.7.3 Video transport over ATM through circuit emulation-AAL1

This section looks at the use of AAL1 in greater detail. AAL1 has been the assumed method for transport of video over ATM until recently. All of the ATM video trials at press time (both for entertainment and for distance-learning applications) use AAL1 (also called *CBR AAL*).

8.7.3.1 General observations. As discussed, since existing codec technology and standards are based on the assumption of a constant bit-rate service, it is assumed for near-term deployment scenarios that the CBR AAL is used with VDT services. AAL1 segments incoming protocol data units into blocks of 47 octets, plus 1 octet for AAL1 overhead. Notice that a 188 MPEG-2 transport stream packet fits exactly into four AAL1 service data units ($188 = 4 \times 47$), which are then sent in four cells.

Two scenarios have been proposed for the transport of the video information channel over ATM as a constant bit-rate service*:

The channel shares a CRS connection supporting a CBR channel (e.g., DS1) with other applications, e.g., telephony

The channel uses a CRS connection dedicated to support the video service

Before discussing specifics of the CBR AAL, it is necessary to establish the basic role of the CBR AAL in the CBR video service offered over CRS. If the codecs, video servers, setup boxes, etc., are assumed to be

* This discussion is based on Ref. 5, which also includes original references on which the treatment draws.

non-ATM terminal equipment, then the codec was not designed for compatibility with ATM. In this context, the basic functionality of the CBR AAL can be viewed as a *terminal adapter* (TA) that permits the video equipment and an ATM network to interwork. Two scenarios for introduction of the TA with the CBR AAL functionality are possible. In the first scenario, the TA is not exclusively serving the needs of the video channel but is supporting a generic CBR service, such as DS1. In the second scenario, the TA is dedicated to the support of the video equipment which provides direct access to an ATM CRS connection dedicated to the video channel. This scenario treats the TA functionality as an add-on to the encoder/decoder. The sending TA maps the video channel from the transmitting encoder into the ATM layer transport (i.e., ATM cell-information payloads) and recovers the channel when the ATM layer transport is terminated at the TA associated with the receiving decoder.

The CBR AAL adapts video channels to the ATM transport environment. The AAL's functions and options are defined in ITU-T Recommendation I.363 for AAL1 and the ANSI T1.631 CBR AAL standards. Table 8.7 summarizes the functions of the AAL. Figure 8.15 depicts the format of AAL1 protocol data units.

8.7.3.2 AAL1. This section provides additional information on the use of AAL1. Appendix A at the end of the chapter provides more technical details for the AAL1 protocol.

The existing agreements in ITU-T Recommendation I.363 and the ANSI T1.631 CBR AAL standard for AAL1 provide two basic modes of operation to the CBR AAL:

Unstructured data transfer (UDT)

Structured data transfer (SDT)

When the UDT mode is operational, the AAL protocol assumes the incoming data from the AAL user is *a bit stream* with an associated bit

TABLE 8.7 Functions of the AAL in the Video Context

Mapping the video channel information into and out of the cell payloads; specifically, it provides a circuit-emulated link at a requested bit rate, even though the access channel may be of higher speed (e.g., 384 kbps over a DS3 access facility)
Clock recovery
Error detection and control (cell loss and CDV problems)
8-kHz integrity or other octet-block structures—SDT mode of operation

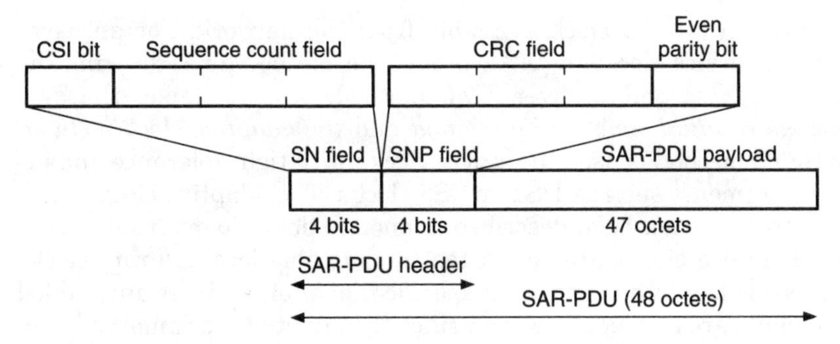

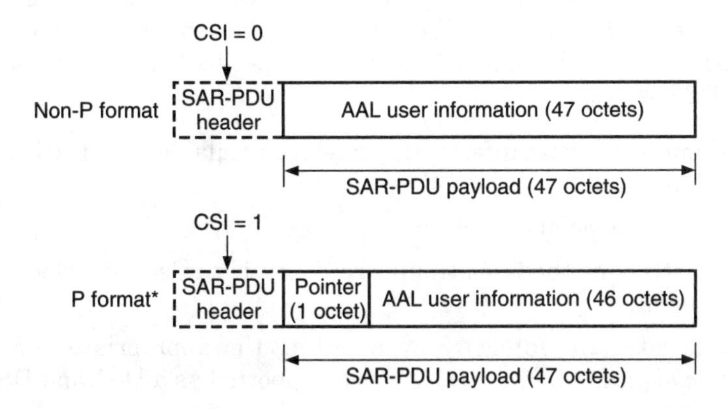

Figure 8.15 AAL1 formats: *top:* AAL1 frame layout; *bottom:* Non-P and P formats.

clock. When the SDT mode is operational, the AAL protocol assumes the incoming information is *octet-blocks* of a fixed length (e.g., such as an $n \times 64$ kbps channel with 8-kHz integrity) with an associated clock. The other basic CBR service attribute that determines the AAL functionality required to support a service is the status of the CBR service clock:

Synchronous
Asynchronous

Since the service clock is assumed to be frequency-locked to a network clock in the synchronous case, the recovery of the service clock is done directly with a clock available from the network. For an asynchronous service clock, the AAL provides a method for recovering the source clock at the receiver. Two methods are available: the *synchronous residual time stamp method* and the *adaptive clock method.* The SRTS method is used to recover clocks with tight tolerance and jitter requirements such as DS1 or DS3 clocks. The adaptive clock recovery method has not been described in enough detail to determine what type of service clocks are supported (presumably less accurate clocks with looser low-frequency jitter specifications) or what, if any, added agreements are needed. However, since the adaptive clock method is in common use in user equipment, this method is assumed to be available.

For the first scenario outlined earlier (shared CRS PVC or SVC), the most probable CBR signals to be carried over a CRS ATM connection are those associated with DS1 and DS3 service. Since the CBR AAL supports performance objectives and error control techniques determined by the more generic needs of DS1 or DS3 service, the applications must use the DS1 and DS3 service as specified. The support of DS1 and DS3 service:

uses the entire 47-octet information payload available with the basic CBR AAL protocol

uses the UDT mode of operation

uses the SRTS methods of timing recovery if the service clock is asynchronous

maintains bit-count integrity by inserting the appropriate *alarm indication signal* (AIS)* for the service supported as a DS1 and DS3 error control measure

The discussion will examine the issue from the perspective of the two CBR AAL modes of operation: UDT or SDT. In this scenario, the TA functionality is most likely to be a part of the user's equipment.

Dedicated CRS PVC or SVC-UDT mode. For the case of UDT operation, the support of video connections is analogous to the support provided for DS1 and DS3 service. The entire 47-octet information payload available with the CBR AAL protocol is available to carry video channel information, unless additional AAL protocol procedure(s) requiring overhead, such as forward error correction, are implemented. The video equipment output is treated as a bit stream with an associated bit clock. If the user's equipment is not synchronous with the network clock, then an asynchronous method of timing recovery for the bit clock is required of the AAL. Although other methods are avail-

* The reader interested in DS1 or DS3 formats, including AIS, may see Ref. 2.

able, typically, SRTS is expected to be used in the case of asynchronous clock recovery. The AAL aids in the control of the following error conditions at the receiving TA servicing the video decoder:

detecting cell loss conditions

monitoring for cell-delay variation problems that could result in buffer starvation conditions or a loss of information due to buffer overflows

In the event of either detected cell loss or a CDV problem, the AAL is assumed to be maintaining bit count to the video decoder through the substitution of appropriate dummy information as with DS1 and DS3 service. For DS1 and DS3, network interworking requirements dictated that the bit pattern substitution be consistent with their AIS.

Dedicated CRS PVC or SVC-SDT mode. The prior discussion covering UDT remains largely applicable for SDT. The AAL retains the function of supporting clock recovery for the bit clock and checks for cell-loss or buffer underflow/overflow problems at the receiving TA. Once again, the issue of maintaining bit-count integrity is important.

The SDT mode of operation provides the CBR AAL with a new capability at the expense of added protocol implementation and information payload overhead. With the SDT protocol implemented, the AAL can convey information about octet-block boundaries of fixed size between the transmitting AAL entity and the receiving AAL entity. The SDT protocol could also provide information about fixed-size octet-block structures in the video channel itself. However, in the context of existing technology, the video equipment is assumed to be responsible for recognizing and processing video channel data structures. Therefore, it can be assumed that the SDT capability is used only to support generic CBR service attributes, such as 8-kHz integrity structure. If user's equipment needs 8-kHz integrity structure, then the SDT mechanism provides a verification of this structure that is independent of the bit clock or framing patterns internal to the video channel. This capability is important for $n \times 64$ kbps systems, e.g., videoconferencing and videotelephony. Its use beyond $n \times 64$ kbps remains to be explored.

8.7.4 Video transport over ATM through AAL5

There has been a lot of work in ANSI Committee T1 and in The ATM Forum to define the approach to carry video over a pure ATM transport. However, at press time, there was no technical consensus, as described earlier in the general comments. Therefore, the treatment on AAL5 limits itself to a technical discussion of the capabilities of AAL5. This is done in App. B, to which the interested reader is referred.

8.7.5 Proposed STM transport of MPEG-2

This section briefly describes the carriage of MPEG-2 over a synchronous transport, as proposed by Scientific Atlanta,[20] for illustrative purposes. Scientific Atlanta proposes a *unidirectional synchronous optical network* (UNISON-1) which provides a point-to-point optical network that uses SONET, without requiring the complete conformance to SONET specifications (see Chap. 4 for some information on SONET). UNISON-1 is capable of transporting a variety of information types in an integrated manner. STS-3c is the basic rate used in the *interactive video services network* (IVSN) which is developed by Scientific Atlanta.

The 270 × 9-octet SONET frame of 2430 octets was shown in Fig. 4.14. The definition of the subset of functionality to be supported in the overhead section of the frame (the 3 × 9-octet section)[20] is not covered here because it is not the thrust of the discussion. Figure 8.16 shows how the MPEG-2 TS packets are planned to be transported. R represents reserved octets (R = 00000000). P is the Reed-Solomon parity. The Reed-Solomon class of coding is part of a large class of powerful random error-correcting cyclic codes known as the *Bose-Chaudhuri-Hocquenghem* (BCH) codes (Reed-Solomon codes are a subclass of the nonbinary BCH codes). The code is implemented on the preceding MPEG-2 TS packet.

8.7.6 Audiovisual multimedia service

In 1994, The ATM Forum initiated work to facilitate the definition of multimedia conferencing and video distribution solutions that are optimized for ATM networks. This work is carried out under the banner of AudioVisual Multimedia Service (AMS). The goal is to define these services to take full advantage of the bandwidth capabilities while accommodating the statistical nature of cell-relay service. That is, to enable new applications* with high bandwidth and/or stringent QOS requirements to operate over ATM LANs and WANs. The work addresses two issues (see Fig. 8.17):

1. Which AAL type is appropriate
2. Which sublayer of MPEG-2 should be used

This work takes I.211 ("B-ISDN Service Aspects"), I.374 ("Network Capabilities to Support Multimedia Services"), and about 50 contributions on these topics made to The ATM Forum up to 1994, as the point of departure. The work seeks to establish, among other things, syn-

* The ATM Forum identified the following applications: broadcast video, videoconferencing, desktop multimedia, desktop audiovisual services, audio and data, video on demand, interactive video, near-video on demand, emergency broadcast, distance learning, interactive gaming.

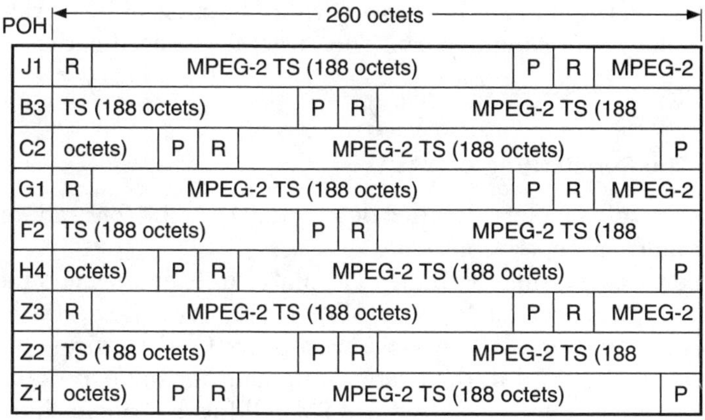

POH	260 octets				
J1	R	MPEG-2 TS (188 octets)	P	R	MPEG-2
B3	TS (188 octets)	P	R	MPEG-2 TS (188	
C2	octets)	P	R	MPEG-2 TS (188 octets)	P
G1	R	MPEG-2 TS (188 octets)	P	R	MPEG-2
F2	TS (188 octets)	P	R	MPEG-2 TS (188	
H4	octets)	P	R	MPEG-2 TS (188 octets)	P
Z3	R	MPEG-2 TS (188 octets)	P	R	MPEG-2
Z2	TS (188 octets)	P	R	MPEG-2 TS (188	
Z1	octets)	P	R	MPEG-2 TS (188 octets)	P

Figure 8.16 Scientific Atlanta proposed SONET mapping of MPEG-2 TSs.

chronization-decoder timing recovery functions. There clearly are tradeoffs between the end-user video display device (e.g., setup box or desktop video workstation) and the cost of the bandwidth on the transmission line, or the cost of additional hardware in the ATM switching nodes and associated line card. The Forum is also looking to see if there can be a unique solution. Less demanding applications may be burdened with costs of technology to support more demanding applications (e.g., HDTV). A single solution may bring a gamut of applications

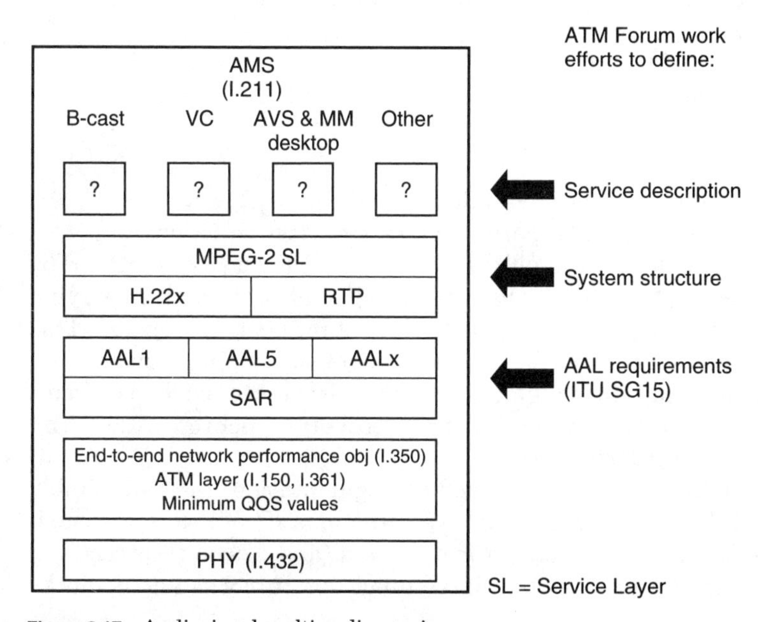

Figure 8.17 Audiovisual multimedia service.

to the market sooner and may lower costs overall because of large volumes for technology components parts. The target for a specification is late 1995.

Appendix 1 The Functioning of AAL1

This appendix contains a more detailed description of AAL1 for readers who require a more in-depth treatment.

One of the services possible with an ATM platform is emulation of a dedicated line (typically at 1.544 or 45 Mbps). This type of service is also known as Class A or CBR service. To support CBR services, an adaptation layer is required in the user's equipment for the necessary functions that cannot be provided by the ATM cell header. Some characteristics and functions that may be needed for an efficient and reliable transport of CBR services are identified subsequently. Ideally, CBR services carried over an ATM-based network should appear to the user as equivalent to CBR services provided by the circuit-switched or dedicated network. Some characteristics of these CBR services are:

1. Maintenance of timing information
2. Reliable transmission with negligible reframes
3. Path performance monitoring capability

CBR services with these characteristics can be provided by assigning the following functions for the CBR adaptation layer:

1. Lost cell detection
2. Synchronization
3. Performance monitoring

(These functions may not be required by all the CBR services.)

Therefore, the CBR AAL performs the necessary functions to match the service provided by the ATM layer to the CBR services required by its service user. It provides for the transfer of AAL_SDUs carrying information of an AAL user supporting constant bit-rate services. This layer is service-specific, with the main goal of supporting services that have specific delay, jitter, and timing requirements, such as circuit emulation. This layer provides timing recovery, synchronization, and indication of lost information.

The AAL1 functions are grouped in segmentation and reassembly functions and in convergence sublayer functions. The SAR is responsible for the transport and bit-error detection (and possibly correction). The CS performs a number of related functions: it blocks and deblocks

AAL_SDUs, counting the blocks (modulo 8) as it generates or receives them; also, it maintains bit-count integrity, generates bit-timing information when needed, recovers timing, generates and recovers data structure information (if required), and detects and generates indications of error conditions and signal loss (to the appropriate management layer). The CS may receive reference clock information from the appropriate management layer.

AAL1 services and functions

The CBR AAL functions are grouped into two sublayers: the SAR sublayer and the convergence sublayer. The SAR is responsible for the transport and bit-error detection (and possibly correction) of CS protocol control information. The CS performs a set of service-related functions. It blocks and deblocks AAL_SDUs, counting the blocks, modulo 8, as it generates or receives them. Also, it maintains bit-count integrity, generates timing information (if required), recovers timing, generates and recovers data structure information (if required), and detects and generates indications to the *AAL management* (AALM) entity of error conditions or signal loss. The CS may receive reference clock information from the AALM entity that is responsible for managing the AAL resources and parameters used by the AAL entity. The services provided by AAL1 to the AAL user are:

- Transfer of service data units with a constant source bit rate and the delivery of them with the same bit rate
- Transfer of timing information between the source and the destination
- Transfer of structure information between the source and the destination
- Indication of lost or errored information that is not recovered by AAL1, if needed

Specifically, the functions are:

Segmentation and reassembly of user information

Handling of cell-delay variation

Handling of cell payload assembly delay

Handling of lost and misinserted cells

Source clock recovery at the receiver

Recovery of the source data structure at the receiver

Monitoring of AAL-PCI for bit errors

Handling of AAL-PCI for bit errors

Monitoring of user information field for bit errors and possible corrective actions

SAR functions

The SAR functions are:

Mapping between the CS_PDU and the SAR_PDU The SAR sublayer at the transmitting end accepts a 47-octet block of data from the CS and then prepends a 1-octet SAR_PDU header to each block to form the SAR_PDU.

Indicating the existence of CS function The SAR can indicate the existence of CS function; the use of the indication mechanism is optional.

Sequence numbering For each SAR_PDU payload, the SAR sublayer receives a sequence number value from the CS.

Error protection The sequence number and the CSI bits are protected.

A buffer is used to handle cell-delay variation. When cells are lost, it may be necessary to insert an appropriate number of dummy SAR_PDUs. Figure 8.15 depicts the AAL1 frame layout.

Convergence sublayer functions

The functions of CS are:

Handling of cell-delay variation for delivery of AAL_SDUs to the AAL user at a constant bit rate. (The CS layer may need a clock derived at the S_B or T_B interface to support this function.)

Processing the sequence count to detect cell loss and misinsertion.

Providing the mechanism for timing information transfer for AAL users requiring recovery of source clock frequency at the destination end.

Providing the transfer of STRUCTURE information between source and destination for some AAL users.

Possibly supporting forward error correction (particularly for video).

For those AAL users who require transfer of structured data (e.g., 8-kHz structured data for circuit-mode bearer services for 64-kbps-based ISDN), the STRUCTURE parameter is used.

The STRUCTURE parameter can be used when the user data stream to be transferred to the peer AAL entity is organized into groups of bits. The length of the structured block is fixed for each instance of the AAL service. The length is an integer multiple of 8 bits. An example of the use of this parameter is to support circuit-mode services of the 64-kbps-based ISDN. The two values of the STRUCTURE parameter are:

Start. Used when the DATA is the first part of a structured block that can be composed of consecutive data segments

Continuation. Used when the value *Start* is not applicable

The use of the STRUCTURE parameter depends on the type of AAL service provided; the use of the parameter is agreed upon prior to or at the connection establishment between the AAL user and the AAL.

I.363 notes that "for certain applications such as speech [or video], some SAR functions may not be needed." For example, I.363 provides the following guidance for CS for voice-band signal transport (which is a specific example of CBR service):

- *Handling of AAL user information:* The length of the AAL_SDU (i.e., the information provided to the AAL by the upper-layer protocols) is 1 octet; for comparison, the SAR_PDU is 47 octets.
- *Handling of cell-delay variation:* A buffer of appropriate size is used to support this function.
- *Handling of lost and misinserted cells:* The detection of lost and inserted cells, if needed, may be provided by processing the sequence count values. The monitoring of the buffer fill level can also provide an indication of lost and misinserted cells. Detected misinserted cells are discarded.

P and non-P formats

The 47-octet SAR_PDU payload used by CS has two formats called non-P and P formats, as seen in Fig. 8.15. These are used to support transfer of information with STRUCTURE.

Note that in the non-P format, the entire CS_PDU is filled with user information.

Partially filled cells

I.363 notes that in some situations the SAR_PDU payload may be filled only partially with user data to reduce the cell-payload assembly delay. In this case, the number of leading octets utilized for user information in each SAR_PDU payload is a constant that is determined by the allowable cell-payload assembly delay. The remainder of the SAR_PDU payload consists of dummy octets.

Clocking issues

This section describes SRTS, based on Ref. 19.

The synchronous residual time stamp[21] method is used to support asynchronous clock recovery for CBR services with stringent clock specifications, such as those associated with the 1.544- and 44.736-MHz hierarchies. The SRTS method uses a measure that conveys timing information related to the frequency difference between a reference clock and the clock to be recovered, which is referred to as the source clock. It requires that SRTS at the CBR AAL transmitter and SRTS at the CBR AAL receiver use synchronized reference clocks. This requirement is necessary to control the jitter in the recovered clock introduced by the ATM transport (particularly, low-frequency jitter known as *wander*).

Since the timing information generation process is related to the *time stamp* (TS) approach of clock recovery, SRTS supports a range of source clock frequencies with a single reference clock frequency. If the source clocks to be recovered are stable and accurate, then the most significant bits of a TS value convey information that does not impact the clock adjustment at the receiver. Therefore, the SRTS method transmits the equivalent of the least significant portion of a TS value referred to as the *residual time stamp* (RTS).

The SRTS method measures a time interval defined to be an integral number of source clock cycles in terms of a reference clock derived from a network clock. The measurement generates the equivalent of a truncated time stamp value. The following notation is used in the discussion:

f_{nx} = reference clock derived from network clock

f_s = source clock frequency

N = period of RTS in source clock (f_s) cycles

ε = Source clock tolerance in parts per million (ppm)

M(max, nom, min) = maximum, nominal, or minimum, number of f_{nx} cycles in an RTS sampling period

Figures 8.18 and 8.19 from ANSI contributions[19] illustrate the concept and generation of RTS. The sampling period is measured within the accuracy permitted by the f_{nx} clock. An RTS with a size P bits and the f_s clock variation bounds (i.e., M(min) and M(max) as measured in f_{nx} clock cycles) must obey the relationship:

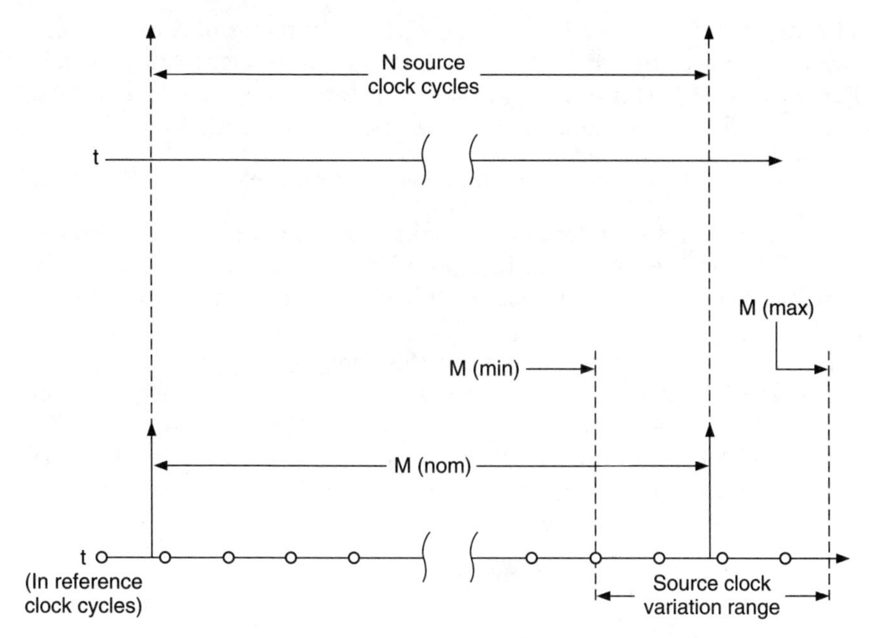

Figure 8.18 Residual time stamp concept.

$$M(\max) - M(\min) < 2^P \tag{A.1}$$

if the RTS values are to provide a valid measure of source clock variation. The RTS size bounds the magnitude of the range of source clock variation that the SRTS method can tolerate and still provide a valid measure of source clock variation. The maximum magnitude of the f_s clock variation, M(max) – M(min), which can be accommodated by SRTS, is bounded by the RTS sampling interval, N, the ratio of f_{nx}/f_s, and the clock tolerance ε:

$$M(\max) - M(\min) \leq N * \left(\frac{f_{nx}}{f_s}\right) * \varepsilon \tag{A.2}$$

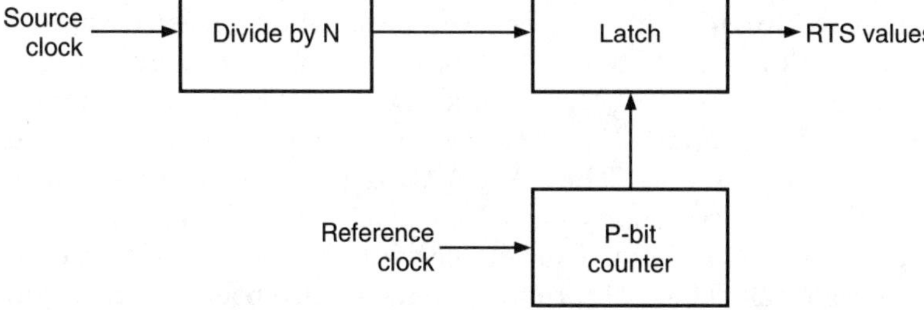

Figure 8.19 Generation of RTS values.

The value of N is fixed to coincide with the number of f_s clock cycles associated with the CBR AAL-user information in eight-cell payloads. For the case of DS1 and DS3 service, 47 octets of a cell payload is AAL-user data. The RTS sampling interval then corresponds to:

$$N = 8*8*47 = 3008 \text{ source clock cycles} \tag{A.3}$$

Selecting ε to have a maximum value of 200 ppm and fixing the maximum value of the ratio f_{nx}/f_s to 2, and RTS of 4 bits satisfies the relationship in Eq. (A.1) for the magnitude of variation bound given by Eq. (A.2).

DS1 and DS3 clock tolerances are 130 ppm and 20 ppm, respectively; hence, RST can accommodate both of these services. The use of a common SRTS reference clock frequency at both the transmitter's SRTS entity and the receiver's SRTS entity is necessary for interoperability. The reference clock frequencies used with SONET-based physical-layer transport of ATM are:

$$f_{nx} = \frac{(155.52 \text{ MHz})}{2^k} = \frac{[(8*9*270) \times f_8]}{2^k} \tag{A.4}$$

where $k = 0, 1, \ldots, 11$
$f_8 = 8$ kHz

for supporting f_s clock rates from 64 kbps to 132.8 Mbps (132.8 Mbps is the available ATM cell-payload bandwidth on an STS-3c with a 1-octet AAL). For the case of DS3, the highest reference clock frequency needed is 38.88 MHz. The f_{nx}/f_s ratio constraint:

$$1 \leq \frac{f_{nx}}{f_s} < 2 \tag{A.5}$$

determines the appropriate reference clock to use. For example, the reference clock used for the recovery of a DS1's 1.544-MHz clock is 2.43 MHz.

CBR AAL mechanism to support functions

The CBR AAL provides its service over preestablished AAL connections. The establishment and initialization of an AAL connection is performed through the AALM. The transfer capacity of each connection and other connection characteristics are negotiated prior to or at connection establishment. (The CBR AAL is not directly involved in the negotiation process, which may be performed by management or signaling.) The AAL receives from its service user a constant-rate bit stream with a clock. The AAL provides to its service user this constant rate bit stream with the same clock. The CBR service clock can be either synchronous or asynchronous relative to the network clock. The CBR service is called synchronous if its service clock is frequency locked to the network clock. Otherwise, the CBR service is called asynchronous.

The service provided by the AAL consists of its own capability plus the capability of the ATM layer and the physical layer. This service is provided to the AAL-user (e.g., an entity in an upper layer or in the management plane). The service definition is based on a set of service primitives that describe in an abstract manner the logical exchange of information and control. Functions that are performed by the CBR AAL entities are shown in Table 8.8.

The logical exchange of information between the AAL and the AAL user is represented by two primitives, as shown in Table 8.9.

Description of parameters

AAL_SDU: This parameter contains one bit of AAL-user data to be transferred by the AAL between two communication AAL-user peer entities.

STRUCTURE: As discussed, this parameter is used to indicate the beginning or continuation of a block of AAL_SDUs when providing

TABLE 8.8 Functions Performed by CBR AAL

Detection and reporting of lost SAR_PDUs	Detects discontinuity in the sequence count values of the SAR_PDUs and senses buffer underflow and overflow conditions.
Detection and correction of SAR_PDU header error	Detects bit errors in the SAR_PDU header and possibly corrects a 1-bit error.
Bit-count integrity	Generates dummy information units to replace lost AAL_SDUs to be passed to the AAL user in an AAL-DATA.indication.
Residual time stamp (RTS) generation	Encodes source service clock timing information for transport to the receiving AAL entity.
Source clock recovery	Recovers the CBR service source clock.
Blocking	Maps AAL_SDUs into the payload of a CS_PDU.
Deblocking	Reconstructs the AAL_SDU from the received SAR_PDUs and generates the AAL-DATA.indication primitive.
Structure pointer generation and extraction	Encodes in a 1-octet structure pointer field at the sending AAL entity the information about periodic octet-based block structures present in AAL-DATA request primitives. The receiving AAL entity extracts the structure pointer received in the CS_PDU header field to verify locally generated block structure.

TABLE 8.9 Primitives for CBR AAL

AAL-DATA.request (AAL_SDU, STRUCTURE)	This primitive is issued by an AAL-user entity to request the transfer of an AAL_SDU to its peer entity over an existing AAL connection. The time interval between two consecutive AAL-DATA.request primitives is constant and a function of the specific AAL service provided to the AAL user.
AAL-DATA.indication (AAL_SDU, STRUCTURE, status)	This primitive is issued to an AAL-user entity to notify the arrival of an AAL_SDU over an existing AAL connection. In the absence of error, the AAL_SDU is the same as the AAL_SDU sent by the peer AAL-user entity in the corresponding AAL-DATA.request. The time interval between two consecutive AAL-DATA.indication primitives is constant and a function of the specific AAL service provided to the AAL user.

for the transfer of a structured bit stream between communicating AAL-user peer entities (structured data transfer service). The length of the blocks is constant for each instance of the AAL service and a multiple of 8 bits. This parameter takes one of the following two values: start and continuation. It is set to start whenever the AAL_SDU being passed in the same primitive is the first bit of a block of a structured bit stream; otherwise, it is set to continuation. This parameter is used only when SD service is supported.

Status: This parameter indicates that the AAL_SDU being passed in the same indication primitive is judged to be nonerrored or errored. It takes one of the following two values: valid or invalid. The invalid value may also indicate that the AAL_SDU being passed is a dummy value. The use of this parameter and the choice of the dummy value depends on the specific service provided.

Service Expected from the ATM Layer

The AAL expects the ATM layer to provide for the transparent and sequential transfer of AAL data units, each 48 octets long, between communicating AAL entities, over an ATM layer connection, at a negotiated bandwidth and QOS. The ATM layer transfers the information in the order in which it was delivered to the ATM layer and provides no retransmission of lost or corrupted information.

Interactions between the SAR and the convergence sublayer

The logical exchange of information between the SAR and the convergence sublayer is represented by the primitives of Table 8.10.

TABLE 8.10 SAR Primitives

SAR-DATA.invoke (CSDATA, SCVAL, CSIVAL)	This primitive is issued by the sending CS entity to the sending SAR entity for requesting the transfer of a CSDATA to its peer entity.
SAR-DATA.signal (CSDATA, SNCK, SCVAL, CSIVAL)	This primitive is issued by the receiving SAR entity to the receiving CS entity for notifying of the arrival of a CSDATA from its peer CS entity.

Description of the parameters

CSDATA: This parameter represents the interface data unit exchanged between the SAR entity and the CS entity. It contains the 47-octet CS_PDU.

SCVAL: This 3-bit parameter contains the value of the sequence count associated with the CS_PDU contained in the CSDATA parameter.

CSIVAL: This 1-bit parameter contains the value of the CSI bit.

SNCK: This parameter is generated by the receiving SAR-entity. It represents the results of the sequence number protection error check over the SAR_PDU header. It can assume the value of SN-valid or SN-invalid.

Appendix 2 The Functioning of AAL5

This appendix contains a more detailed description of AAL5 for readers who require a more in-depth treatment.

The goal of AAL5 is to support, in the most streamlined fashion, those capabilities that are required to meet upper-layer data transfer over an ATM platform. The *Common Part of AAL5* (CPAAL5) protocol provides for the transport of variable-length frames (1 to 65,535 octets) with error detection (the frame is padded to align the resulting PDU with an integral number of ATM cells). A length field is used to extract the frame and detect additional errors not detected with the CRC-32 mechanism.

The convergence sublayer has been subdivided into the *Common Part CS* (CPCS) and the *Service-Specific CS* (SSCS) as shown in Fig. 8.20. Different SSCS protocols, to support specific AAL user services, or groups of services, may be defined. The SSCS may also be null, in the sense that it only provides for the mapping of the equivalent primitives of the AAL to CPCS and vice versa. SSCS protocols are specified in separate recommendations than, say, ITU-T I.363. This discussion, therefore, focuses on CPCS and SAR. Notice that CPAAL5 = SAR + CPCS. Also see Fig. 8.21.

Service provided by the CPAAL5

The Common Part of AAL5 provides the capabilities to transfer the CPAAL5_SDU from one CPAAL5 user to another CPAAL5 user through the ATM network. During this process, CPAAL5_SDUs may be corrupted or lost (in this case, an indication of the error is provided). Corrupted or lost CPAAL5_SDUs are not recovered by CPAAL5. CPAAL5 supports a message mode and a streaming mode.

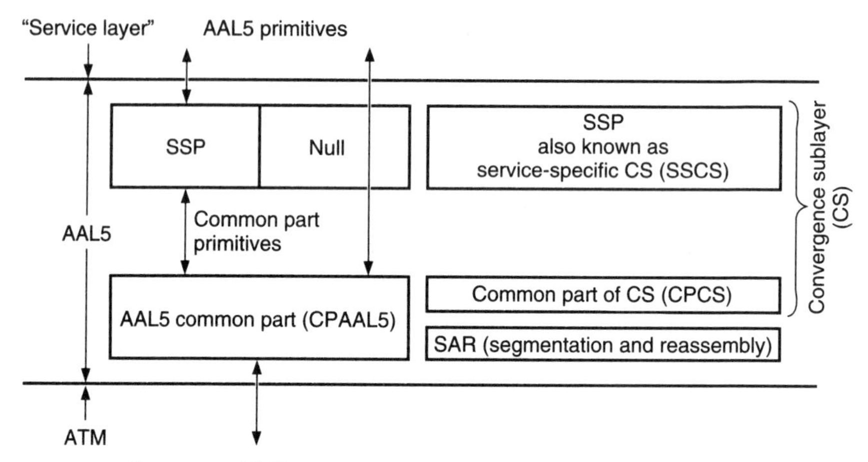

Figure 8.20 Structure of AAL5.

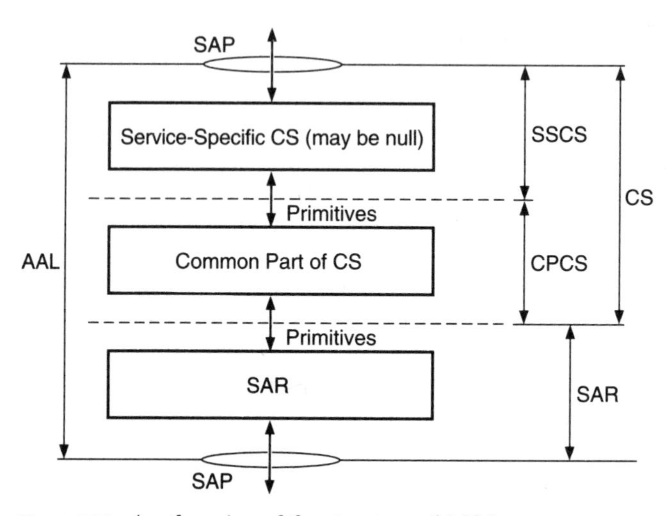

Figure 8.21 Another view of the structure of AAL5.

1. *Message-mode service:* The CPAAL5_SDU is passed across the CPAAL5 interface in exactly one *Common Part AAL Interface Data Unit* (CPAAL5_IDU). This service provides the transport of fixed-size or variable-length CPAAL5_SDUs.
   a. In case of small fixed-size CPAAL5_SDUs, an internal blocking/deblocking function in the SSCS may be applied; it provides the transport of one or more fixed-size CPAAL5_SDUs in one SSCS_PDU.
   b. In case of variable-length CPAAL5_SDUs, an internal CPAAL5_SDU message segmentation/reassembling function in the SSCS may be applied. In this case, a single CPAAL5_SDU is transferred in one or more SSCS_PDUs.
   c. Where the preceding options are not used, a single CPAAL5_SDU is transferred in one SSCS_PDU. When the SSCS is null, the CPAAL5_SDU is mapped to one CPCS_SDU.
2. *Streaming-mode service:* The CPAAL5_SDU is passed across the CPAAL5 interface in one or more CPAAL5_IDUs. The transfer of these CPAAL5_IDUs across the CPAAL5 interface may occur separated in time. This service provides the transport of variable-length CPAAL5_SDUs. The streaming-mode service includes an abort service by which the discarding of an CPAAL5_SDU partially transferred across the AAL interface can be requested.
   a. An internal CPAAL5_SDU message segmentation/reassembling function in the SSCS may be applied. In this case, all the CPAAL5_IDUs belonging to a single CPAAL5_SDU are transferred in one or more SSCS_PDU.
   b. An internal pipelining function may be applied. It provides the means by which the sending CPAAL5 entity initiates the transfer to the receiving CPAAL5 entity before it has the complete CPAAL5_SDU available.
   c. Where option *a* is not used, all the CPAAL5_IDUs belonging to a single CPAAL5_SDU are transferred in one SSCS_PDU. When the SSCS is null, the CPAAL5_IDUs belonging to a single CPAAL5_SDU are mapped to one CPCS_SDU.

Both modes of service may offer the following peer-to-peer operational procedures:

- *Assured operations:* Every assured CPAAL5_SDU is delivered with exactly the data content that the user sent. The assured service is provided by retransmission of missing or corrupted SSCS_PDUs. Flow control is provided as a mandatory feature. The assured operation may be restricted to point-to-point AAL connections.

- *Nonassured operations:* Integral CPAAL5_SDUs may be lost or corrupted. Lost and corrupted CPAAL5_SDUs will not be corrected by retransmission. An optional feature may be provided to allow corrupted CPAAL5_SDUs to be delivered to the user (i.e., optional error discard). Flow control may be provided as an option.

Description of AAL connections

The CPAAL5 provides the capabilities to transfer the CPAAL5_SDU from one AAL5-SAP to one other AAL5-SAP through the ATM network. The CPAAL5 users have the capability to select a given AAL5-SAP associated with the QOS required to transport that CPAAL5_SDU (e.g., delay- and loss-sensitive QOS).

The CPAAL5 in nonassured operation also provides the capability to transfer the CPAAL5_SDUs from one AAL5-SAP to more than one AAL3-SAP through the ATM network.

CPAAL5 makes use of the service provided by the underlying ATM layer. Multiple AAL connections may be associated with a single ATM layer connection, allowing multiplexing at the AAL; however, if multiplexing is used in the AAL, it occurs in the SSCS. The AAL user selects the QOS provided by the AAL through the choice of the AAL5-SAP used for data transfer.

Primitives for the AAL

These primitives are service-specific and are contained in separate recommendations on SSCS protocols.

The SSCS may be null, in the sense that it only provides for the mapping of the equivalent primitives of the AAL to CPCS and vice versa. In this case, the primitives for the AAL are equivalent to those for the CPCS but identified as CPAAL5-UNITDATA.request, CPAAL5-UNITDATA.indication, CPAAL5-U-Abort.request, CPAAL5-U-Abort.indication, and CPAAL5-P-Abort.indication, consistent with the primitive naming convention at an SAP.

Primitives for the CPCS of the AAL

Because there exists no SAP between the sublayers of the AAL5, the primitives are called *.invoke* and *.signal* instead of the conventional *.request* and *.indication* to highlight the absence of the SAP.

```
CPCS-UNITDATA.invoke and the CPCS-UNITDATA.signal
```

These primitives are used for the data transfer. The following parameters are defined:

- *Interface data (ID).* This parameter specifies the interface data unit exchanged between the CPCS and the SSCS entity. The interface data unit is an integral multiple of one octet. If the CPCS entity is operating in the message-mode service, the interface data represents a complete CPCS_SDU; when operating in the streaming-mode service, the interface data does not necessarily represent a complete CPCS_SDU.
- *More (M).* In the message-mode service, this parameter is not used. In the streaming-mode service, this parameter specifies whether the interface data communicated contains a beginning/continuation of a CPCS_SDU or the end of/complete CPCS_SDU.
- *CPCS-loss priority (CPCS-LP).* This parameter indicates the loss priority for the associated CPCS_SDU. It can take only two values, one for high priority and the other for low priority. The use of this parameter in streaming mode is for further study. This parameter is mapped to and from the SAR-LP parameter.
- *CPCS congestion indication (CPCS-CI).* This parameter indicates that the associated CPCS_SDU has experienced congestion. The use of this parameter in streaming mode is for further study. This parameter is mapped to and from the SAR-CI parameter.
- *CPCS user-to-user indication (CPCS-UU).* This parameter is transparently transported by the CPCS between peer CPCS users.
- *Reception status (RS).* This parameter indicates that the associated CPCS_SDU delivered may be corrupted. This parameter is used only if the corrupted data delivery option is used.

Depending on the service mode (message or streaming, discarding or delivery of errored information), not all parameters are required.

`CPCS-U-Abort.invoke and CPCS-U-Abort.signal`

These primitives are used by the CPCS user to invoke the abort service. They are also used to signal to the CPCS user that a partially delivered CPCS_SDU is to be discarded by instruction from its peer entity. No parameters are defined. These primitives are not used in message mode.

`CPCS-P-Abort.signal`

This primitive is used by the CPCS entity to signal to its user that a partially delivered CPCS_SDU is to be discarded because of the occurrence of some error in the CPCS or below. No parameters are defined. This primitive is not used in message mode.

Primitives for the SAR sublayer of the AAL

These primitives model the exchange of information between the SAR sublayer and the CPCS.

Because there exists no SAP between the sublayers of the AAL5, the primitives are called *.invoke* and *.signal* instead of the conventional *.request* and *.indication* to highlight the absence of the SAP.

```
SAR-UNITDATA.invoke and the SAR-UNITDATA.signal
```

These primitives are used for the data transfer. The following parameters are defined:

- *Interface data (ID).* This parameter specifies the interface data unit exchanged between the SAR and the CPCS entity. The interface data unit is an integral multiple of 48 octets. The interface data does not necessarily represent a complete SAR_SDU.
- *More (M).* This parameter specifies whether the interface data communicated contains the end of the SAR_SDU.
- *SAR-loss priority (SAR-LP).* This parameter indicates the loss priority for the associated SAR interface data. It can take on two values, one for high priority, and the other for low priority. This parameter is mapped to the ATM layer's submitted loss priority parameter and from the ATM layer's received loss priority parameter.
- *SAR-congestion indication (SAR-CI).* This parameter indicates if the associated SAR interface data has experienced congestion. This parameter is mapped to and from the ATM layer's congestion indication parameter.

Functions of the SAR sublayer

The SAR sublayer functions are performed on an SAR_PDU basis. The SAR sublayer accepts variable length SAR_SDUs that are integral multiples of 48 octets from the CPCS and generates SAR_PDUs containing 48 octets of SAR_SDU data. It supports the preservation of SAR_SDU by providing for an "end of SAR_SDU" indication.

SAR_PDU structure and coding

The SAR sublayer function uses the *ATM-layer-user to ATM-layer-user* (AUU) parameter of the ATM layer primitives to indicate that a SAR_PDU contains the end of an SAR_SDU. A SAR_PDU where the value of the AUU parameter is 1 indicates the end of a SAR_SDU; the value of 0 indicates the beginning or continuation of a SAR_SDU. The structure of the SAR_PDU is shown in Fig. 8.22.

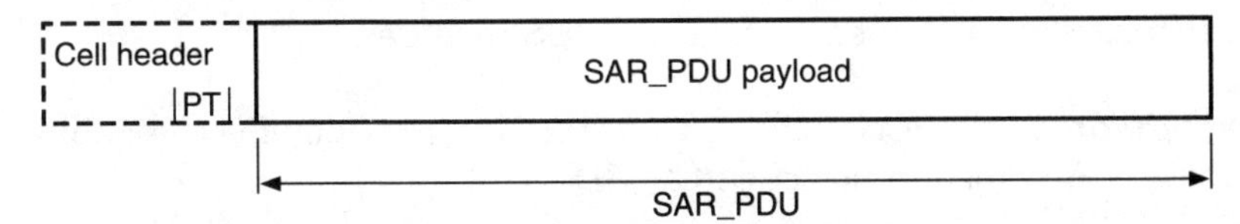

Note: The payload type (PT) field belongs to the ATM header. It conveys the value of the AUU parameter end-to-end.

Figure 8.22 SAR_PDU format for AAL5.

Convergence sublayer

The CPCS has the following service characteristics.

- Nonassured data transfer of user data frames with any length measured in octets from 1 to 65,535 octets
- CPCS connection established by management or by the control plane
- Error detection and indication (bit error and cell loss or gain)
- CPCS_SDU sequence integrity on each CPCS connection

Functions of the CPCS

The CPCS functions are performed per CPCS_PDU. The CPCS provides several functions in support of the CPCS service user. The functions provided depend on whether the CPCS service user is operating in message or streaming mode.

1. *Message-mode service:* The CPCS_SDU is passed across the CPCS interface in exactly one CPCS-IDU. This service provides the transport of a single CPCS_SDU in one CPCS_PDU.
2. *Streaming-mode service:* The CPCS_SDU is passed across the CPCS interface in one or more CPCS-IDUs. The transfer of these CPCS-IDUs across the CPCS interface may occur separated in time. This service provides the transport of all the CPCS-IDUs belonging to a single CPCS_SDU into one CPCS_PDU. An internal pipelining function in the CPCS may be applied which provides the means by which the sending CPCS entity initiates the transfer to the receiving CPCS entity before it has the complete CPCS_SDU available. The streaming-mode service includes an abort service by which the discarding of a CPCS_SDU partially transferred across the interface can be requested.

Note: At the sending side, parts of the CPCS_PDU may have to be buffered if the restriction ("interface data are a multiple of 48 octets") cannot be satisfied.

The functions implemented by the CPCS include:

1. *Preservation of CPCS_SDU.* This function provides for the delineation and transparency of CPCS_SDUs.
2. *Preservation of CPCS user to user information.* This function provides for the transparent transfer of CPCS user to user information.
3. *Error detection and handling.* This function provides for the detection and handling of CPCS_PDU corruption. Corrupted CPCS_SDUs are either discarded or are optionally delivered to the SSCS. The procedures for delivery of corrupted CPCS_SDUs are for further study. When delivering errored information to the CPCS user, an error indication is associated with the delivery. Examples of detected errors would include received length and CPCS_PDU length field mismatch including buffer overflow and improperly formatted CPCS_PDU, and CPCS CRC errors.
4. *Abort.* This function provides for the means to abort a partially transmitted CPCS_SDU. This function is indicated in the length field.
5. *Padding.* A padding function provides for 48 octet alignment of the CPCS_PDU trailer.

CPCS structure and coding

The CPCS functions require an 8-octet CPCS_PDU trailer. The CPCS_PDU trailer is always located in the last 8 octets of the last SAR_PDU of the CPCS_PDU. Therefore, a padding field provides for a 48-octet alignment of the CPCS_PDU. The CPCS_PDU trailer, together with the padding field and the CPCS_PDU payload, comprise the CPCS_PDU.

The coding of the CPCS_PDU conforms to the coding conventions specified in 2.1 of Recommendation I.361. See Figs. 8.23 and 8.24.

1. *CPCS_PDU payload.* The CPCS_PDU payload is the CPCS_SDU.
2. *Padding (PAD) field.* Between the end of the CPCS_PDU payload and the CPCS_PDU trailer, there will be from 0 to 47 unused octets. These unused octets are called the padding (PAD) field; they are strictly used as filler octets and do not convey any information. Any coding is acceptable. This padding field complements the CPCS_PDU (including CPCS_PDU payload, padding field, and CPCS_PDU trailer) to an integral multiple of 48 octets.
3. *CPCS user-to-user indication (CPCS-UU) field.* The CPCS-UU field is used to transparently transfer CPCS user to user information.

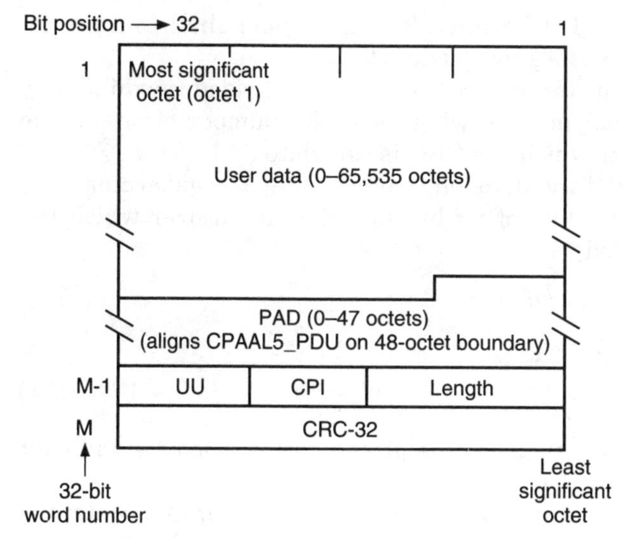

Figure 8.23 CPAAL5_PDU.

4. *Common part indicator (CPI) field.* One of the functions of the CPI field is to align the CPCS_PDU trailer to 64 bits. Other functions are for further study. Possible additional functions may include identification of layer management messages. When only the 64-bit alignment function is used, this field is coded as zero.

5. *Length field.* The length field is used to encode the length of the CPCS_PDU payload field. The length field value is also used by the receiver to detect the loss or gain of information. The length is binary encoded as number of octets. A length field coded as zero is used for the abort function.

6. *CRC field.* The CRC-32 is used to detect bit errors in the CPCS_PDU. The CRC field is filled with the value of a CRC calculation which is performed over the entire contents of the CPCS_PDU, including the CPCS_PDU payload, the PAD field, and the first four

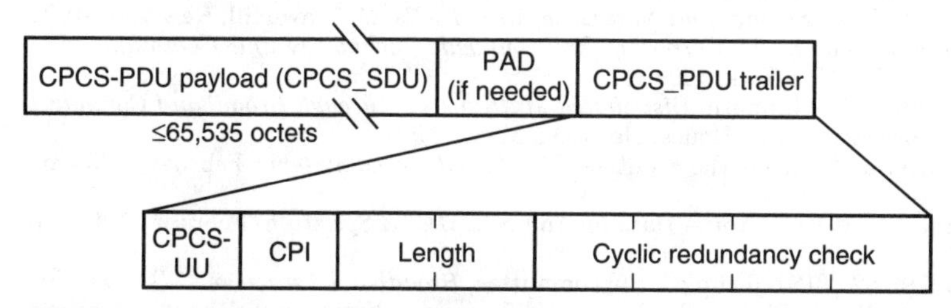

Figure 8.24 Another view of CPAAL5_PDU.

octets of the CPCS_PDU trailer. The CRC field shall contain the ones complement of the sum (modulo 2) of:

a. The remainder of $xk^*(x^{31} + x3^0 + \ldots x + 1)$ divided (modulo 2) by the generator polynomial, where k is the number of bits of the information over which the CRC is calculated

b. The remainder of the division (modulo 2) by the generator polynomial of the product of x^{32} by the information over which the CRC is calculated

The CRC-32 generator polynomial is:

$$G(x) = x^{32} + x^{26} + x^{23} + x^{22} + x^{16} + x^{12} + x^{11} + x^{10} + x^{8} + x^{7} + x^{5} + x^{4} + x^{2} + x + 1 \qquad \text{(B.1)}$$

The result of the CRC calculation is placed with the least significant bit right justified in the CRC field.

As a typical implementation at the transmitter, the initial content of the register of the device computing the remainder of the division is preset to all 1s and is then modified by division by the generator polynomial (as previously described) on the information over which the CRC is to be calculated; the ones complement of the resulting remainder is put into the CRC field.

As a typical implementation at the receiver, the initial content of the register of the device computing the remainder of the division is preset to all 1s. The final remainder, after multiplication by x^{32} and then division (modulo 2) by the generator polynomial of the serial incoming CPCS_PDU, will be (in the absence of errors):

$$C(x) = x^{31} + x^{30} + x^{26} + x^{25} + x^{24} + x^{18} + x^{15} + x^{14} + x^{12} + x^{11} + x^{10} + x^{8} + x^{6} + x^{5} + x^{4} + x^{3} + x + 1 \qquad \text{(B.2)}$$

References

1. D. Minoli and M. Vitella, *Cell Relay Service and ATM in Corporate Environments,* McGraw-Hill, New York, 1994.
2. D. Minoli, *Enterprise Networking—Fractional T1 to SONET, Frame Relay to BISDN,* Artech House, Norwood, Mass., 1994.
3. D. Minoli, *First, Second, and Next Generation LANs,* McGraw-Hill, New York, 1993.
4. D. Minoli, *Imaging in Corporate Environments Technology and Communication,* McGraw-Hill, New York, 1994.
5. D. Minoli and B. Keinath, *Distributed Multimedia: Through Broadband Communication Services,* Artech House, Norwood, Mass., 1994.
6. S. Chatterjee, "Fencing the Frontier of ATM," *Advanced Systems,* February 1994, pp. 17 ff.
7. J. Lane, "ATM Knits Voice, Data on Any Net," *IEEE Spectrum,* February 1994, pp. 42 ff.
8. T1S1.5/93-52, T1S1 Technical Subcommittee, *Broadband Aspects of ISDN Baseline Document,* Rajeev Sinha (ed.), August 1990; reissued Erwin Fandrich (ed.), February 1993.

9. P. Pancha and M. El Zarki, "MPEG Coding for Variable Rate Video Transmission," *IEEE Communications Magazine,* May 1994, pp. 54 ff.
10. D. Minoli, *ATM Technology, Products, Services & Competitive Markets,* Probe Research Corp., Cedar Knolls, N.J., January 1995.
11. R. Karpinski, "MCI Pulls SMDS Trigger, Explains ATM Caution," *Telephony,* January 24, 1994, pp. 7 ff.
12. G. Lawton, "Interoperating Cable into the Great Opportunity," *Communications Technology,* November 1993, pp. 26 ff.
13. G. E. Daddis, Jr. and H. C. Torng, "A Taxonomy of Broadband Integrated Switching Architectures," *IEEE Communications Magazine,* 27(s): 32–42 (1989).
14. S. Nojima, E. Tsutsui, H. Kukuda, and M. Hashimoto, "Integrated Services Packet Network Using Bus Matrix Switch," *IEEE J. on Sel. Areas in Commun.* (October 1987), vol. SAC5, no. 8, pp. 1284–1292.
15. ITU-T Recommendation I.363, "B-ISDN ATM Adaptation Layer (AAL) Specification," ISO, Geneva, CH, 1992.
16. ANSI T1.631, *B-ISDN–ATM Adaptation Layer Constant Bit Rate Services Functionality and Specification,* New York, 1993.
17. ANSI T1.403, *Carrier-to-Customer Installation—DS1 Metallic Interface,* New York, 1991.
18. ANSI T1.404, *Carrier-to-Customer Installation—DS3 Metallic Interface,* New York, 1991.
19. B. Kittams, "Proposal for a Liaison to T1X1," Contribution to T1 Committee T1S1.5/94-027, January 10 to 14, 1994, Dallas.
20. Scientific Atlanta, *Broadband Transport for Interactive Video Services,* doc. no. 69L007Z, Atlanta, 1993.
21. R. C. Lau, "Synchronous Frequency Encoding Technique for Circuit Emulation," *SPIE Conference on Visual Processes and Image Communication,* September 1989.

Chapter

9

Terrestrial-Guided Distribution Systems: ADSL

ADSL is being viewed by some as a quick and dirty way to deliver digital video over the existing copper plant. With an embedded base of over 125 million local loops in the United States, it makes sense to find a way to reuse this infrastructure, at least for a subset of VDT services. This chapter examines briefly a number of technical issues associated with this type of mechanism for video delivery.

9.1 Overview

ADSL specifications in general, and line modulation formats in particular, are being addressed by the *Alliance for Telecommunications Industry Solutions* (ATIS), formerly ESCA. Working Group T1E1.4 was planning at press time to publish an interface standard for ADSL. In addition to defining the interface and the line transmission format, there are several architectural alternatives under consideration. In conjunction with such ANSI standards, the RBOCs are pursuing the formulation of ADSL systems requirements.[1] As noted elsewhere, Bell Atlantic, as well as other East Coast RBOCs, have expressed interest in this technology.

The desired transmission scope of ADSL is the coverage of loops that meet *carrier serving area* (CSA) criteria, whether the loops originate at a central office or at an outside plant multiplexer terminal, such as a digital loop carrier system. The various ADSL systems can support 1, 2, or 4 simultaneous sessions of MPEG-1 video over existing copper telephone lines; ADSL-3 supports 1 MPEG-2 channel and/or combinations of MPEG-1 channels. Figure 9.1 depicts at a high level various combinations of ADSL usage.

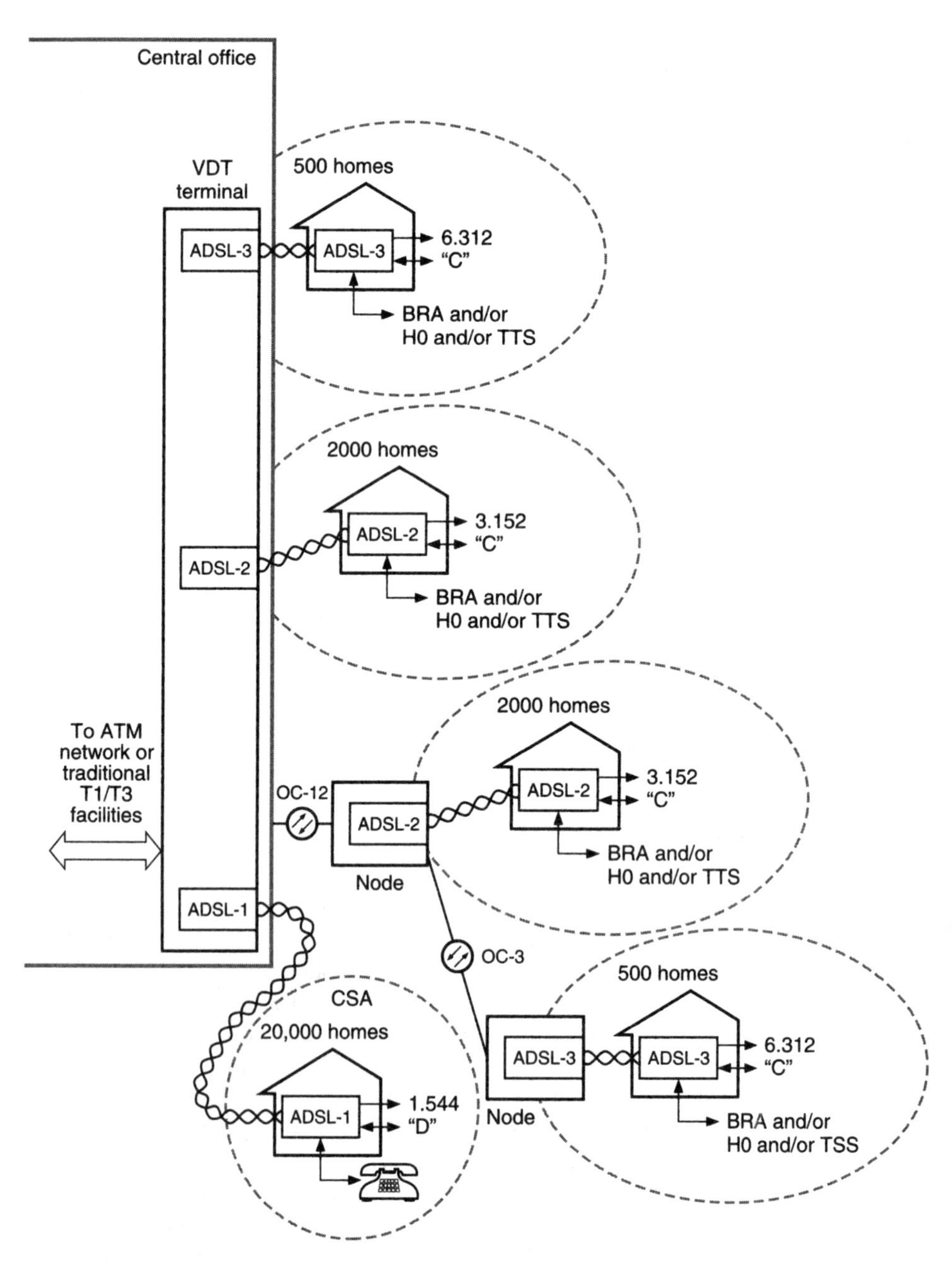

BRA: Basic rate access
TTS: Traditional telephone service (also called POTS)

Figure 9.1 Typical ADSL arrangements.

ADSL is based on advanced *digital signal processing* (DSP) principles. ADSL technology has arisen out of the research and development on *digital subscriber lines* (DSLs) undertaken in the past 10 years. ADSL employs adaptive signal processing for the purpose of automatically adjusting echo cancellers and digital equalizers to compensate, in a precise manner, for signal distortions generated by a subscriber loop. Recent advances in VLSI have allowed complex, high-speed adaptive digital filters to be placed on a single chip, which in turn can be set on a line card of small surface area. Traditional metallic loops required manual treatment techniques, where human intervention was needed to adjust equalizer and echo canceller settings (however, the channel characteristics change over time because of, for example, changes in ambient temperature). ADSL does not require such manual intervention: DSLs are installed, and soon after they are powered, the link is active. After that, continuously adapting filters in the DSL monitor the channel for any changes in characteristics. Figure 9.2 depicts an ADSL system at the component level, depicting the asymmetric echo canceling elements.

The twisted-pair plant only supports limited bandwidths at a given loop length, say 12,000 ft, even after removing the loading coil that is employed to optimize the performance of the circuit from a voice point of view. Besides the length issue, the loop may employ wire of different gauges (e.g., 24 and 26 gauge); this gives rise to signal reflections that, in practical terms, further limit the transmission spectrum. ADSL systems deal with these plant issues and enable the carrier to derive a certain digital bandwidth at a certain loop length. As seen in Figs. 9.1 and 9.2, electronics have to be added at both ends of the twisted-pair component of the loop; this could be from the *central office* (CO) but more likely is from the *optical network unit* (ONU—a device similar to a traditional digital loop carrier system), closer to the user. See Fig. 9.3, which also shows the *host digital terminal* (HDT). Figure 9.4 provides additional details for the home's equipment that is typically needed.

Two modulation schemes are available for ADSL. *Carrierless amplitude-phase* (CAP) is a modulation technique used in early ADSL systems; it was developed by AT&T Paradyne (it is also used to support HDSL*). *Discrete multitone* (DMT) techniques are used to obtain the

---

* High-bit-rate digital subscriber line (HDSL) is similar to ADSL by using copper twisted pair to provide repeaterless T1 service. HDSL uses two pairs of wire to divide two-way 1.5 Mbps transport. CSA guidelines are applicable, and distance is limited to 12 kft. Both non-ISDN-based and ISDN-based HDSL equipment is available; suppliers include Adtran, Artel, PairGain, Tellabs, ADC, Westell, AT&T Paradyne, and Performance Technologies. HDSL is currently targeted at business customers for data and videoconferencing applications where the use of traditional T1-DS1 is not cost-effective.

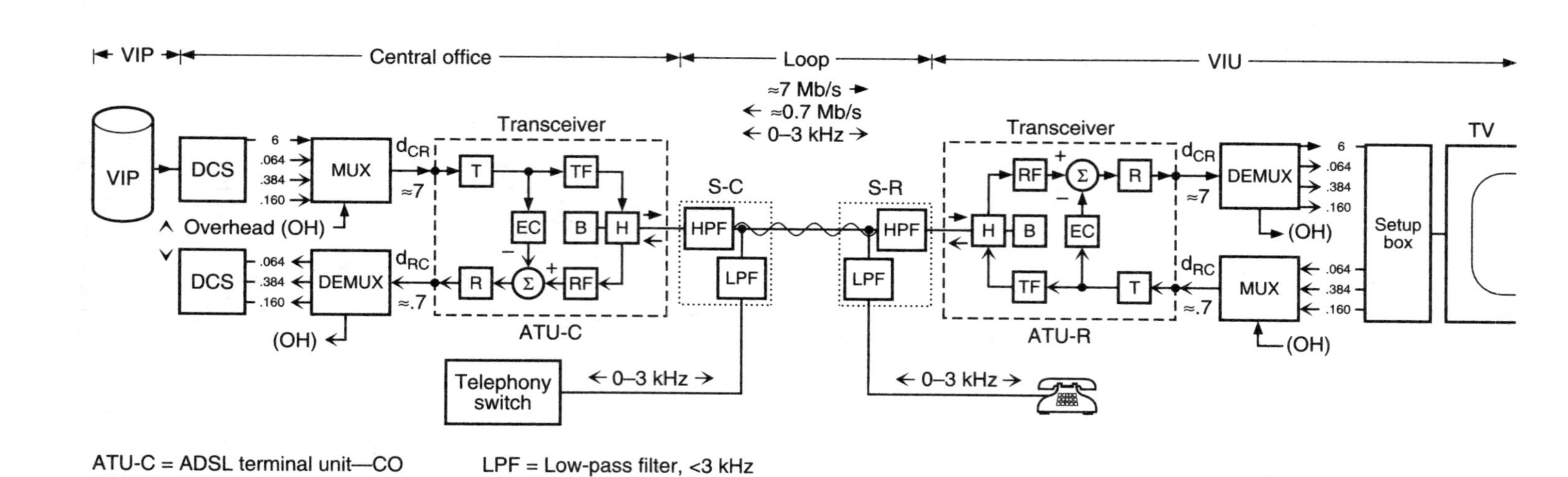

ATU-C = ADSL terminal unit—CO
ATU-R = ADSL terminal unit—remote
B = Balance impedance
d_{CR} = Data C-to-R, ≈7.280 Mb/s
d_{RC} = Data R-to-C, ≈652 kb/s
EC = Echo canceller
H = Hybrid
HPF = High-pass filter, >10 kHz
LPF = Low-pass filter, <3 kHz
RF = Receiving filter
Σ = Summer
T = Transmitter
TF = Transmitting filter
S-C = Splitter-C
S-R = Splitter-R

Figure 9.2 ADSL with asymmetric echo cancelers.

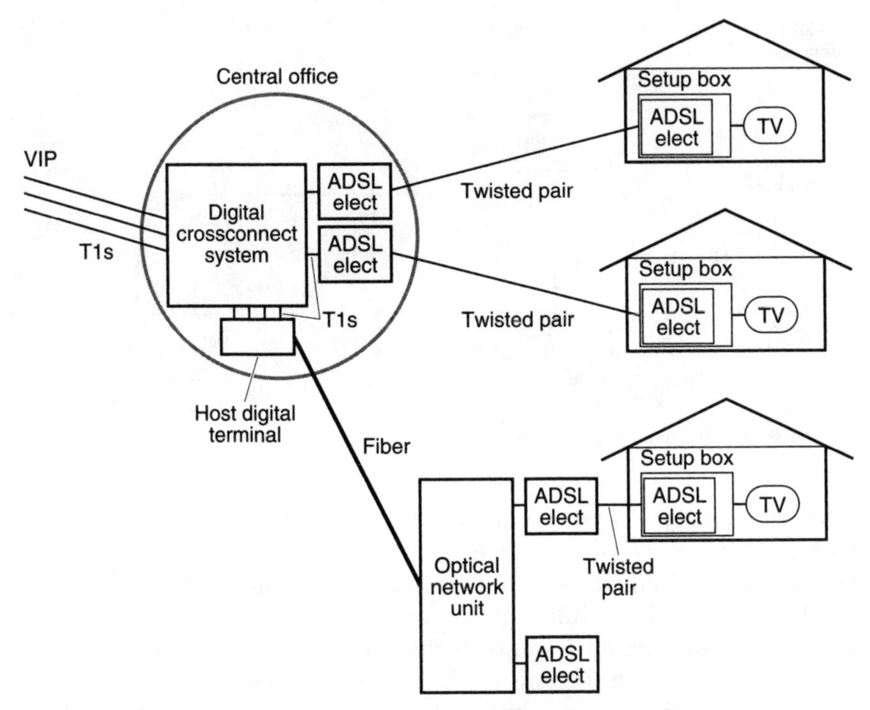

Figure 9.3 An example of an ADSL-based video distribution network.

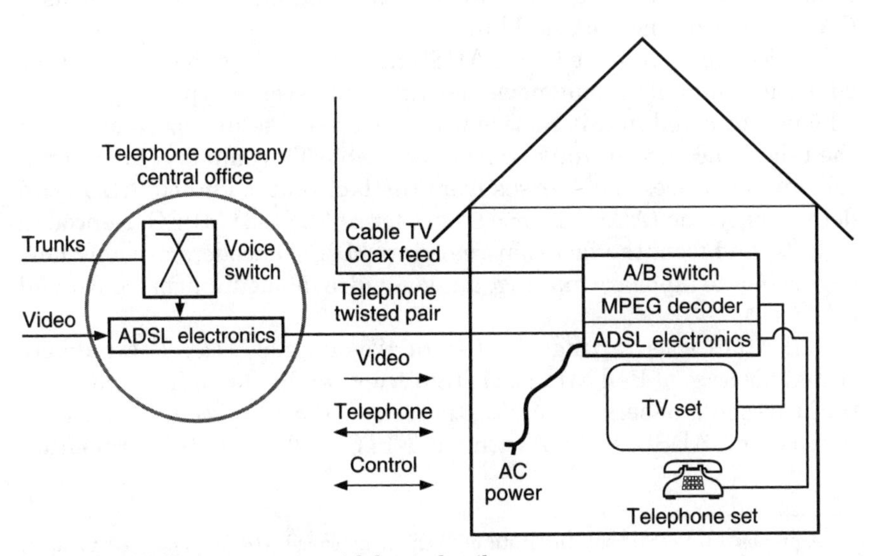

Figure 9.4 ADSL arrangement, with home detail.

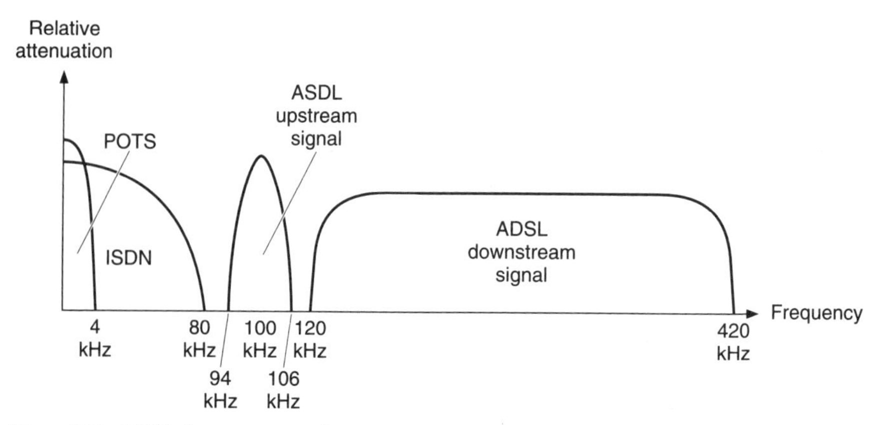

Figure 9.5 ADSL frequency spectrum.

6-Mbps capacity. Figure 9.5 depicts the frequency spectrum employed by ADSL on two-wire twisted pair. Each signal medium uses a separate portion of the frequency domain available on the twisted-pair cable. Traditional telephony information occupies the lower end of the spectrum; the upstream data (control) channel occupies the middle portion of the spectrum, and the downstream channel occupies the upper part of the spectrum. *Multitone* refers to the use of many subchannels to transport the information. Depending on the loss and interference detected on each subchannel, the system allocates up to 15 bits of data among each of 512 subchannels. The channels with the better signal-to-noise ratio are assigned more information; each subchannel uses QAM to encode the allocated bits.

Services to be provided with ADSL include VOD, games, interactive education, and electronic news delivery. However, a typical consumer who is interested in this service would need two setup boxes (one from the telephone carrier and one from the cable TV company) to obtain a full set of services. This arises from the fact that, as noted, ADSL can deliver only one (ADSL-1) or at most four (ADSL-3) MPEG-1-encoded signals, and from the fact that real-time MPEG encoders are only now beginning to appear, making analog video a medium of continued interest.*

Bell Atlantic is testing ADSL-1 in Washington, D.C. The publicly stated position of Bell Atlantic is that they would like to first see a certain threshold of service penetration, say 15 to 20 percent before they replace the ADSL system with an FTTC system. If this transition

* A proposal from Stevens Institute of Technology called *ADSL-4* would carry more channels.

strategy works, Bell Atlantic is able to reuse the ADSL electronics in another market, until the appropriate penetration in the new market is also achieved.

ADSL-1 pricing at press time was about $1500 per unit; providers forecast $300 per unit pricing by 1996, based on large quantities. Suppliers of ADSL-1 equipment include AT&T Paradyne, PairGain, Performance Technologies, and Westell.[2]

ADSL-3 is a more recent technology that has been standardized by the ATIS T1E1 committee. It is similar to ADSL-1 in that a twisted-wire pair is the transmission medium, but it uses DMT modulation to achieve approximately one 6-Mbps downstream channel, as well as a data channel and digital voice channel. CSA guidelines apply, and distance is limited to 6 to 9 kft. Clearly, in the context of video, ADSL-3 allows the transmission of four 1.5-Mbps MPEG-1 channels to the customer; it can also be used for delivery of one 6-Mbps MPEG-2 channel. ADSL-3 equipment is more expensive than ADSL-1 equipment. Critical issues for ADSL-3 include acceptable system cost, timely availability of equipment, physical size, and power consumption.

Northern Telecom was an early supplier of ADSL-3 equipment (Northern Telecom is attempting to license ADSL-3-DMT technology to other manufacturers who wish to develop such equipment). Production-grade equipment supporting 6-Mbps video (as well as telephone and data services) was expected to be available by 1995 (production-grade, real-time compression hardware was also expected to be available by 1995).

9.2 Technical Details

9.2.1 General considerations

A number of modulation schemes and bandwidth allocation mechanisms for ADSL have been studied. In summary, the tradeoffs are multidimensional, including issues of spectral compatibility with other services, high-channel loss for long loops, insertion of impulse noise at the residence, coupling of signaling transients from regular telephony into the digital receivers, and radio frequency interference. The telecommunications industry has worked to resolve this complex set of engineering issues. At the March 1993 meeting of the T1E1.4 ATIS working group, there was consensus to define such interface based on the DMT technology. Table 9.1 depicts some of the features of the ADSL family (the ADSL standard being developed by T1E1 is designed to accommodate ADSL-1, ADSL-2, and ADSL-3 system rates and payloads).

A proposed ADSL architecture that does not require echo cancellation employs FDM techniques to separate the narrowband upstream transmission from the wideband downstream transmission, as seen in

TABLE 9.1 ADSL Family

ADSL-1	1.544-Mbps downstream payload over loops conforming to CSA guidelines; 16-kbps, full-duplex control channel and analog telephony support
ADSL-2	3.152-Mbps downstream payload over loops conforming to CSA guidelines; with a 64-kbps, full-duplex control channel and telephony support; analog telephony and/or ISDN basic access and/or H0
ADSL-3	6.312-Mbps downstream payload over distribution area loops less than 6 kft in length; with a 64-kbps, full-duplex control channel and telephony support; analog telephony and/or ISDN basic access and/or H0*

* There has been research to enhance the performance of ADSL-3 sufficient to cover a CSA.

Figs. 9.5 and 9.6 part *a*. This approach also uses TDM: one direction of the narrowband channels is time-interleaved with the wideband channel in the downstream direction. FEC will probably be used to deal with impulse noise. Both of these factors result in signal latency of the order of several milliseconds. Latency is acceptable for the one-way wideband downstream channel, but it is undesirable in the narrowband channels because of its adverse effects on voice.* A more efficient architecture uses asymmetric echo cancellation (Fig. 9.6 part *b*), an advanced technique that is a spin-off of the echo cancellation method used in ISDN.

The introduction of ADSL in the telephone plant brings a number of challenges that telephone companies need to address. The issue of spectral compatibility with the older T-carrier digital system becomes important; this compatibility is desired since restrictions on the combination of services coexisting in a cable would be detrimental. Consequently, ADSL systems need to operate even under the worst case signal scenario of services such as ISDN, HDSL, and T-carrier systems. Furthermore, ADSL systems are not allowed to adversely affect these systems. This issue of spectral compatibility requires one to look at worst case crosstalk interference and at the maximum permissible ADSL transmit power; optimal distribution of the transmit power spectrum is sought.

9.2.2 ADSL-1

ADSL-1 allows the transmission, up to 18 kft, of a downstream 1.5-Mbps signal, a low-speed data channel, and digitized voice over a twisted copper pair.[3] Traditional line repeaters found on T1 systems are not necessary in ADSL systems, reducing equipment cost and provisioning time. Therefore, ADSL-1 expands the utility of the copper

* A method proposed by Amati Communications selectively applies coding to part of the payload to decrease latency.

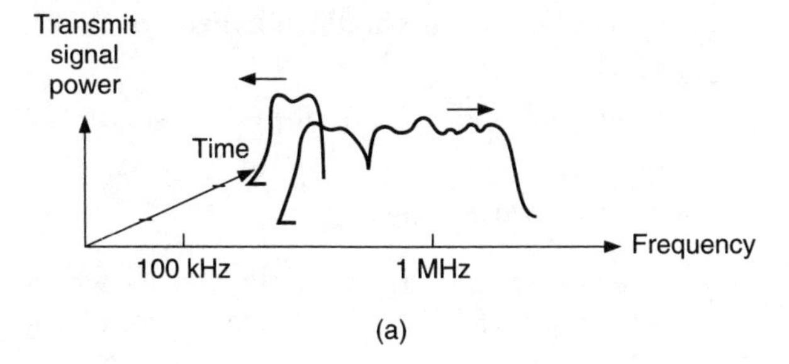

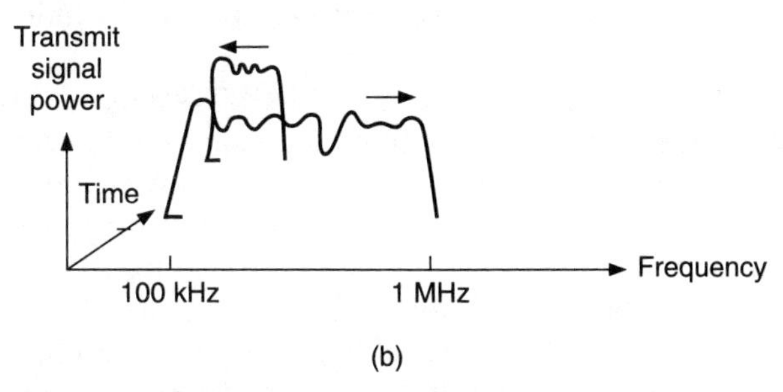

Figure 9.6 ADSL-3 spectra: (*a*) frequency division multiplexing; (*b*) asymmetric echo cancellation.

plant for services with asymmetric flows (i.e., broadcast, multicast, and distribution). ADSL-1 uses CAP modulation technology. ADSL-1 results in three signal components appearing simultaneously on one pair. Their spectra are separated in frequency with FDM and are recovered by line coupling and filtering circuitry.

ADSL-1 supports VDT VOD services, where the VIUs can access an inventory of movies and other entertainment video compressed in MPEG-1; as noted elsewhere, MPEG-1 provides VCR* picture quality. Although only one MPEG-1 channel can be provided over ADSL-1, the inventory of available movies can be large. This approach to video and VOD may be satisfactory to a (small) percentage of VIUs.

Carrier serving area guidelines are followed in the ADSL system. These guidelines are:

- Use only nonloaded cable.
- Use 26-AWG cable (used alone or in combination with other gauge cables), not to exceed 9 kft including bridged taps.

* VCR quality is a picture quality that is comparable to a VHS tape played in a conventional VCR when subjectively tested with a group of viewers. It consists of 352 × 240 pixels (broadcast NTSC consists of 512 × 480 pixels).

- Single-gauge 19-, 22-, or 24-AWG cable should not exceed 12 kft including bridged tap.
- Total bridged tap length should not exceed 2.5 kft and no single tap is to be greater than 2 kft.
- Multigauge cable is restricted to two gauges.

Hence, ADSL-1 systems operate over a single nonloaded loop conforming to *revised resistance design* (RRD) rules (these account for about 98 percent of nonloaded loops in the North America plant). ADSL-1 supports a 16-kbps, full-duplex control channel and analog telephony.

Figure 9.7 depicts the typical components of an ADSL system. ADSL circuits can be provisioned like residential telephone service. Service activation may be performed via a flow-through approach with minimum human intervention. ADSL is designed to be transparent to the telephony service. The traditional ringing and signaling voltages pass through the ADSL units and are delivered to the voice station sets. The full-duplex data channel supports a bandwidth of 9.6 to 16 kbps. This bandwidth is sufficient to transmit basic (nongraphic) menu information from the VIP to the VIU and to transmit keystrokes and mouse movements into the network for service feature control. The duplex channel is digitally encoded and placed above the voice band, enabling telephony to operate at baseband (see Fig. 9.3). Notice that if local power fails, the digital service fails, but telephony still operates normally.

9.2.3 ADSL-2 and ADSL-3 systems

Studies and early prototyping activity by companies such as Amati Communications demonstrated that throughputs up to 7 Mbps can be achieved over twisted pairs at a (maximum) length of 12 kft using

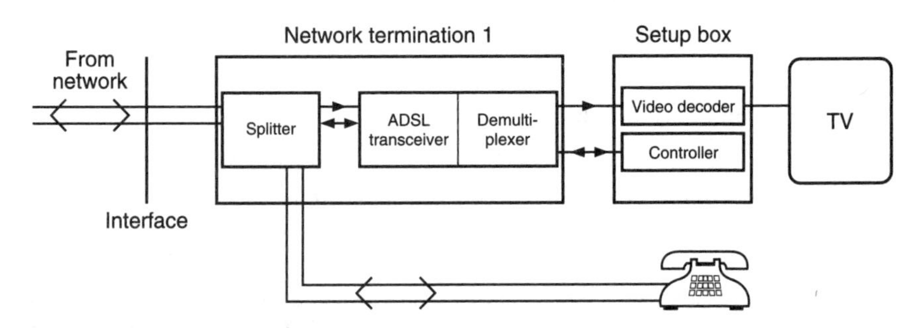

Figure 9.7 Termination electronics for ADSL circuits.

ADSL-based techniques. This higher rate allows for simultaneous viewing of multiple MPEG-1 channels or of one MPEG-2 channel. Beyond higher capacity for the downstream payload, there has been interest in achieving increased throughput for the low-speed channel: the control channel has been extended from 16 to 64 kbps. Also, support of narrowband telephony is provided by way of ISDN basic access transport; full-duplex H0 transport (384 kbps) is also of interest for video telephony. The higher rate systems have been called ADSL-2 and ADSL-3. ADSL-2 has a downstream payload of 3.152 Mbps, and ADSL-3 has a downstream payload of 6.312 Mbps. It does not appear that ADSL-2 will be developed commercially.

ADSL-3 systems need to be accommodated in telephone networks in a number of ways. Although ADSL-3 conforms with the DS2 digital hierarchy, there has not been a major deployment of that unit of bandwidth in the North American PSTN, since DS1 and DS3 systems have been the prevalent transmission mechanisms. Hence, the ADSL-3 interfaces need to be physically defined and implemented for terminating both kinds of channels (high-speed and low-speed channels). These terminations need to fit into the network hierarchy, whether traditional or SONET-based. At the VIU's location, these interfaces need to interwork easily and inexpensively with the VIU's devices. There is also the issue of transmission of the signal over the VIU's own (in-house) wiring.

Because of the work on ISDN in the past two decades, the integration of narrowband telephony signals from a digital access subnetwork into the rest of the network is well understood; however, integration of the one-way wideband downstream channels, called *AH*1, *AH*2, *AH*3, and *AH*4 (*A* stands for asymmetric, and *H* is the ISDN H channel), into the network remains a new transport requirement; developers suggest the use of SONET tributary mechanisms to accomplish such integration, although this technology has seen almost no deployment to date.

From one to four synchronous streams are multiplexed by the ADSL-3 terminal unit at the CO (ATU-C), or possibly in the feeder node. The ADSL-3 terminal unit at the remote site (ATU-R) demultiplexes these signals down to the constituent elements (as noted, other signals are also multiplexed). These interfaces accept synchronous streams at the DS1, E1, DS1C, and DS2 levels (also, two E1s could be premultiplexed, or three DS1s could be premultiplexed). See Table 9.2. The required ADSL-3 overhead brings the total system throughput up to about 7 Mbps; see Table 9.3.

Carrying these bandwidths can be challenging. The problem has two dimensions: the cable construction will dictate the amount of crosstalk coupling between pairs, but also the topology of the particular feeder and distribution cables in question comes into play.[4] Distribu-

TABLE 9.2 Signal Combinations for ADSL-3

Option	Layout			
1	DS2 signal			
2	DS1c	DS1c		
3	E1	E1	E1	
4	DS1	DS1	DS1	DS1
5	Two premultiplexed E1s	E1		
6	Three premultiplexed DS1s	DS1		

TABLE 9.3 Bandwidth Associated with an ADSL-3 System

	Payload							
	Video		Telephony			Overhead		
Channel	Video	Control	H0	DSL	Voice	OAMP	FEC	Total rate
Downstream	6312 kbps	64 kbps	384 kbps	160 kbps	3 kHz	24 kbps	336 kbps	7280 kbps 3 kHz
Upstream	—	64 kbps	384 kbps	160 kbps	3 kHz	24 kbps	20 kbps	642 kbps 3 kHz

DSL = Digital Subscriber Line (same signal format as 2B + D ISDN)

tion plant topology is the most difficult element to model accurately because topologies vary considerably, are constantly changing, and detailed topological information is relatively difficult to obtain. The interested reader should consult Ref. 4 for a thorough discussion.

Table 9.4 shows three ways that the ADSL-3 signals can be integrated into the network. In the first case, a DCS provides multiple output channels that are fed to the ATU-C. In the second case, an ATM switch supports circuit emulation, generating streams that are fed to the ATU-C. In the third case, the ATU-C is located in the feeder, and SONET transmission equipment is used between the CO and the ATU-C.

As discussed elsewhere in this text, there is interest in transporting ATM cells directly to the customer (cell demultiplexing is accomplished in the VIU's terminating equipment). In the ADSL context, this can be realized by embedding cells within a 6.312-Mbps synchronous stream; the stream is then transported through the ADSL-3 system.

TABLE 9.4 Accommodating ADSL-3 Streams

Example	Component	Component	Component	Component	Component	Component	Component
1	DCS	ATU-C	Twisted pair	ATU-R			
2	ATM switch	ATU-C	Twisted pair	ATU-R			
3	ATM switch	SONET mux	Fiber	SONET demux	ATU-C	Twisted pair	ATU-R

References

1. "Notice to the Industry: ADSL Equipment," *Bellcore Digest of Technical Information,* June 1993, p. 40.
2. R. Karpinski, "ADSL: Alive and (Seemingly) Well," *Telephony,* March 14, 1994, pp. 26 ff.
3. Bellcore, *Framework Generic Requirements for Asymmetric Digital Subscriber Lines,* FA-NWT-001397, no. 1, December 1992.
4. W. Y. Chen, D. L. Laring, "Applicability of ADSL to Support Video Dial Tone in the Copper Loop," *IEEE Communications Magazine,* May 1994, pp. 102 ff.

Chapter

10

Terrestrial-Guided Distribution Systems: HFC

10.1 Introduction

A variety of systems exists today for delivery of broadband signals to the customer. This chapter looks at a number of hybrid fiber-coaxial systems that can be deployed today and some that are on the horizon. Ameritech, Bell Atlantic, BellSouth, GTE, NYNEX, Pacific Bell, U S WEST, and several of the major cable TV companies have all announced commitments to HFC architectures. The second part of the chapter contains a detailed description of Time Warner's Full-Service Network, based on the publicly available RFP, as an example of a high-end HFC system under development.

Currently, most of the cable industry is working on upgrading the traditional, all-coaxial network to fiber-coaxial hybrids. In 1992, 3000 fiber-optic nodes were installed, and the number reached approximately 6000 in 1993. By the end of 1994, 16,000 fiber nodes had been installed, covering 35 percent of the industry household base.[1]

In many cable TV networks today, fiber is being used at least in the trunk portions of the network; as time goes by, fiber will make even more inroads. The single-mode fiber is certified for operation at the traditional 1310 nm, as well as the 1550-nm region. As part of this conversion to fiber, both the physical and logical topologies of the cable TV network have been substantially altered. HFC networks are star-based with dedicated, optically passive connections between the headend and the optical receiver that supports a small cluster of homes. Star-based networks are more reliable than bus networks with dozens of cascaded amplifiers; in the latter, any amplifier failure can create an outage for *all the customers* beyond the point of failure; in the former, not only is the reliability increased by the fact that passive optical components are used in the trunk portion but also by the fact that a segment failure only impacts a smaller population of customers (in the so-called *fiber-serving area*)—say, 500 rather than 5000. It is estimated that 26 percent of all outages are caused by the failure of active network devices, principally amplifiers and power supplies.[2] Replacing a traditional cable TV system with an HFC system has the effect of reducing the total number of active devices by 15 to 20 percent. The downtime for a customer on a traditional network is about 2.5 hours per year* (for a well-maintained cable system); on an HFC system, the downtime can be reduced to 25 minutes per year (99.995 availability), by keeping only four or five distribution amplifiers between the optical receiver and any customer.[2] Not only does this equate to improved customer satisfaction, but the operational cost of running the cable network can be decreased by reducing the labor and infrastructure needed to maintain the analog amplifiers.

In addition to improved reliability, transition to an HFC plant also decreases power consumption (by about 12 to 14 percent). Moving to an FTTC or FTTH architecture would further improve reliability and power consumption. The use of fiber also provides major improvements in picture quality by eliminating electrical analog amplifiers and makes the use of the upstream channel more practical.

Another advantage of HFC systems is that each optical node serves a smaller population of customers, thereby offering the opportunity of frequency reuse. This cellularization can work to the advantage of services such as PCS. While in a tree and branch arrangement, the 5- to 30-MHz return bandwidth is shared by all customers on the trunk line (say 4000 customers), in an HFC arrangement, the return bandwidth (usually 5 to 42 MHz) is available to each fiber-serving area of 200 to 500 homes, giving the equivalent of a fourfold to twentyfold increase in the upstream bandwidth.†

As noted in Chap. 4, proposals are being made to extend and/or modify the traditional cable frequency plan to increase the bandwidth (i.e., extend the use to 900 to 1000 MHz) and/or add new services, including two-way PCS. A bandwidth of 12 MHz (6 MHz in each direction) could serve about 2000 customers by providing throughput for 400 lines—

---

* This number is higher for a customer far from the headend; the number is smaller for a customer close to the headend.

† This works as described when each optical node is connected directly to the headend—if several optical nodes share a fiber at the same optical frequency, then to make this work, the optical nodes need to shift their individual return paths to a different frequency band or multiplex the signal in a different set of time slots (this would not be required if the optical nodes used a different optical frequency, e.g., 1310 and 1550 nm).

some blocking would be experienced, but it should be small (0.01 to 0.05). (A system serving 500 homes can provide throughput for 1000 lines or two lines per served home.)

Figure 4.7 depicted an example of a (typical) fiber upgraded to an existing cable network: in this figure (except for the digital headend-to-headend system) the fiber is analog so as to interwork easily with the coaxial system; the various nodes only handle transport-media conversion (thereby being physical-layer relays in the open system reference model terminology), rather than handling other types of conversion. The fiber-optic backbone overlay with analog fiber is an approach that has seen increased deployment since the early 1990s. Figures 10.1*a* to 10.1*d* depict some of the possible transitions, using the type of commercially available fiber-coaxial equipment illustrated in Fig. 4.7. In the future, it is likely that digitally modulated fiber (SONET-based) will be employed more extensively, particularly for FTTH and FTTC architectures, especially to support ATM. A straight HFC system is not meant to support VDT in regard to multiple VIPs and Level 1/Level 2 gateways; however, such architecture can be modified to accommodate it.

Fiber offers several well-documented advantages, including:

- Elimination of various types of analog amplifiers
- Improved signal quality due to elimination of analog amplifiers, reduced crosstalk, and extraneous signal injection
- More reliable, cheaper, and less power-intensive components

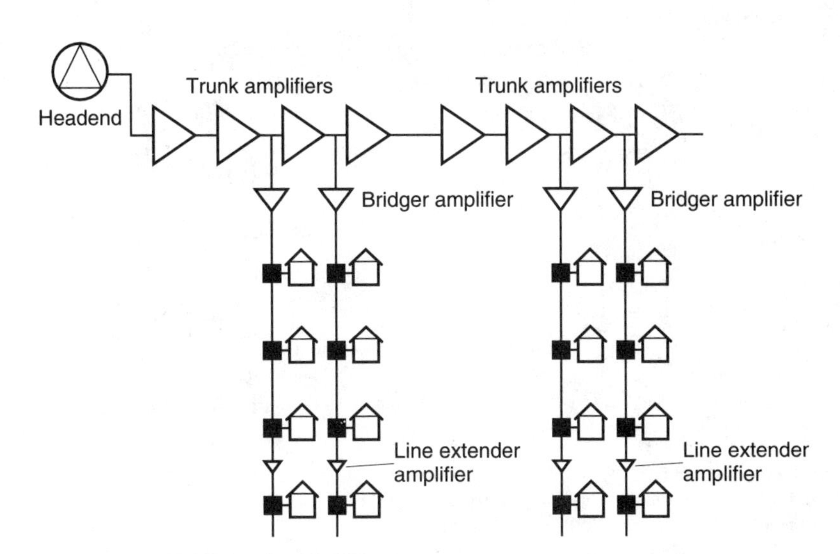

Figure 10.1*a* Traditional cable TV system.

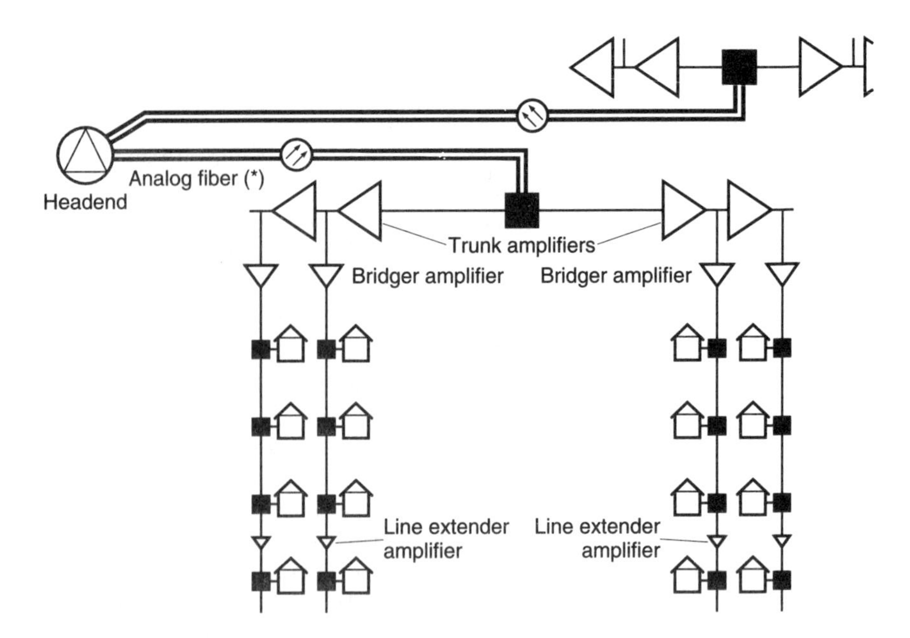

Figure 10.1*b* Fiber-coaxial upgrade: backbone upgrade.

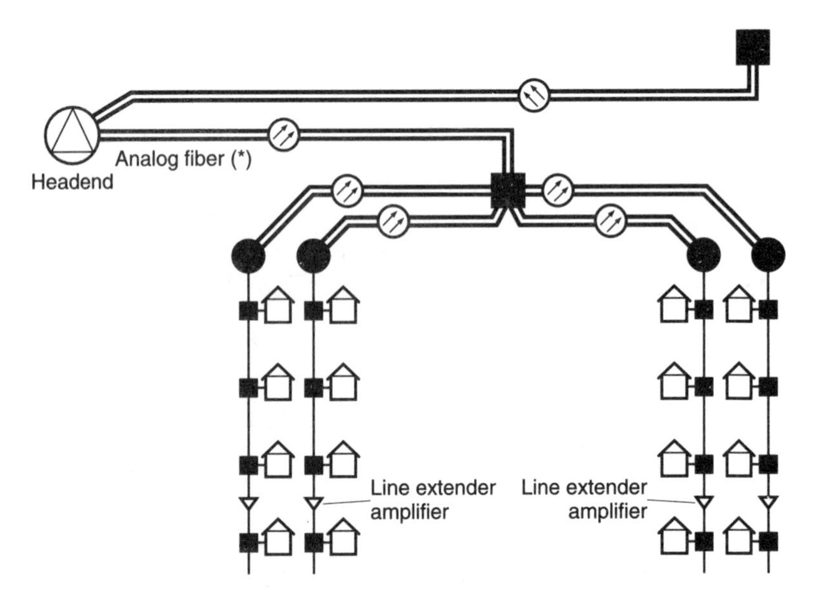

Figure 10.1*c* Fiber-coaxial upgrade through the old bridger amplifier.

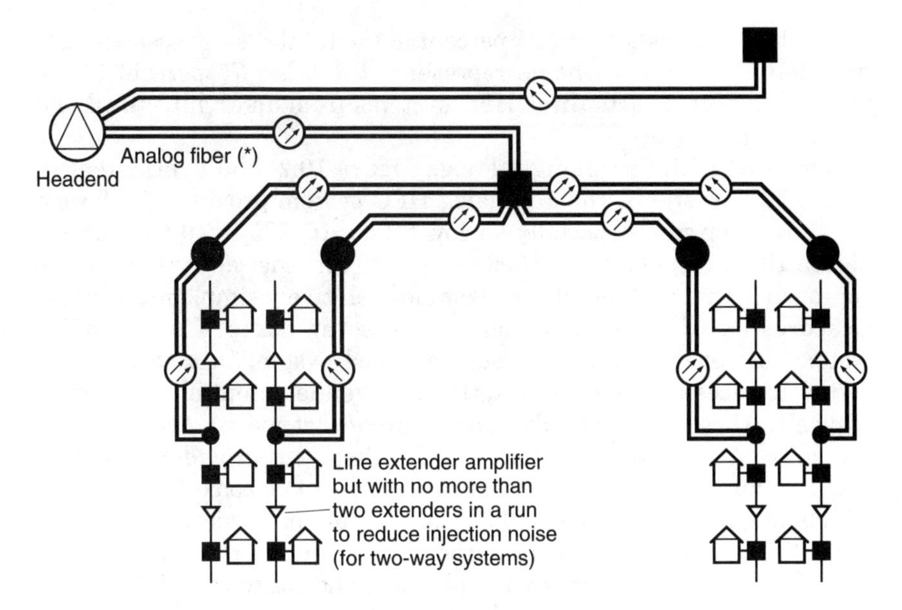

Figure 10.1*d* Fiber-coaxial upgrade to limit a run to require at most two-line extenders.

- Higher capacity, digital format
- Lower signal spillage
- Scalable design, enabling addition of new services at a later date
- Ability to support two-way communication (when two fibers or two WDM signals are used)
- Ability to undertake a partial conversion of the plant, using HFC techniques

Therefore, as seen in Figs. 10.1*a* to 10.1*d*, fiber is being introduced to replace part of the cable system, coexisting with a remaining portion of coaxial facilities. Terms such as *Coax To The Curb* (CTTC) and *Coax To The Home* (CTTH) are used in the industry in the context of these HFC variations. There are relatively few feeder components in the backbone; therefore, these can be replaced cost-effectively. On the other hand, there are thousands of subscribers, and to replace the infrastructure to every home would be expensive. HFC provides a compromise by immediately bringing in the advantages of fiber for a portion of the cable network, while allowing the distribution plan to evolve separately. Studies have shown that the infrastructure from the server to the feeder node costs about 25 percent of the total system cost; the distribution plant to each home represents the other 75 percent of the cost. Table 10.1 depicts three HFC systems from an architectural and service point of view.

Compared with the all-digital (e.g., Figure 10.2) and hybrid analog-digital HFC systems, the all-analog HFC system provides the lowest per-subscriber cost, especially for low take rates (25 to 40 percent) (in Figure 10.2, cost includes a $500 setup box). Another advantage of this approach is that the operator can employ relatively simple interdiction measures based on directly addressable setup boxes. The all-analog system, however, is the least sophisticated system in terms of supported services and, hence, in terms of revenue potential for carriers (typically, these systems only support broadcast video). Also, the cost per subscriber increases considerably when the providers want to deliver a number of channels that necessitates two coaxial cables into each home. In a VDT context, the need for more bandwidth may be dictated by the equal-access requirements to support multiple VIPs, rather than just the desire to provide more choices to the VIUs.

TABLE 10.1 Some HFC Architectural Alternatives

All-analog fiber-coaxial	Fiber from the headend or CO to a node in the consumer's proximity handling optoelectronic conversion. Beyond the node, a coaxial bus distribution network is employed. Such a system typically provides a combination of analog broadcast video channels, in addition to a number of premium pay channels, selectable on a PPV or subscription basis. This access method does not support the full complement of services of interest, e.g., VOD. However, it can be used as a benchmark for economic comparisons because it is very similar to the cable TV networks currently deployed.
Hybrid analog and digital fiber-coaxial	Fiber from the headend or CO, followed by a coaxial tree-and-branch distribution network beyond the node handling optoelectronic conversion. Such a system typically provides a selection of both analog channels (basic and extended programming) and digital pay channels (premium, PPV, and VOD). At the home, customers access pay channels through digital setup boxes that demultiplex and decode the video signals and convert them to the analog format needed by the entrenched TV set.
All-digital fiber-coaxial	This system entails fiber from the headend or CO, followed by a coaxial tree-and-branch distribution network beyond the node handling optoelectronic conversion. The system supports a combination of broadcast and switched video channels in digital form to the subscribers' homes. At home, subscribers use digital setup boxes to access all the video channels.

10.2 Coaxial Systems That Form Part of an HFC Distribution System

This section briefly recapitulates the various coaxial-based systems now in place that, at the pragmatic level, are likely to be part of HFC architectures that will be increasingly introduced in the next couple of years.

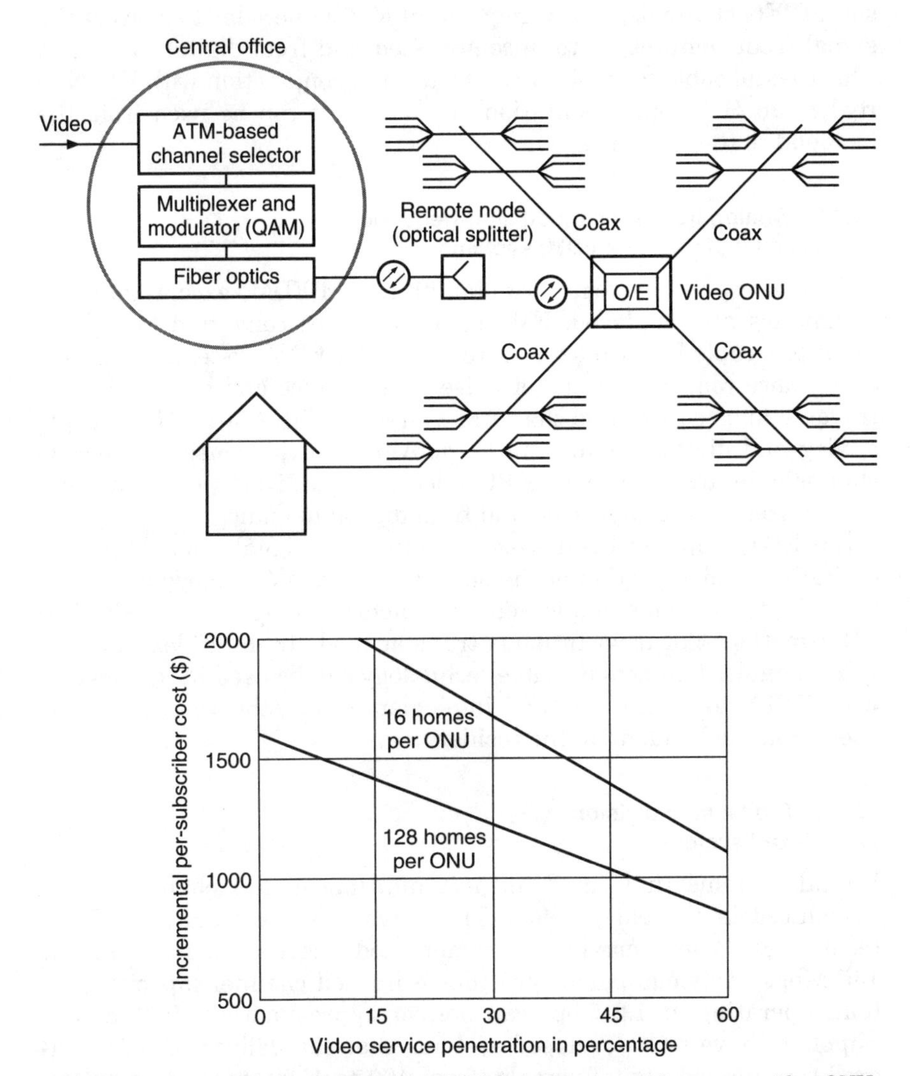

Figure 10.2 *Top:* example of an all-digital HFC supporting 64-home per video ONU; *bottom,* incremental cost to upgrade existing narrowband FTTC system to support all-digital HFC system.

10.2.1 Analog amplitude modulated-frequency division multiplexed (AM-FDM) systems

AM-FDM coaxial systems are used extensively today by cable TV operators. A 550-MHz system can deliver 60 television channels. A 750-MHz system can be used to deliver 60 analog TV channels in the lower portion of the bandwidth and 125 MPEG-2 (6-Mbps) digital signals in the upper portion of the bandwidth. With an AM system, the analog channels are in the proper format to feed directly into a cable-ready TV set. MPEG channels, if any, require an MPEG decoder to convert the signal from compressed to uncompressed and from digital to analog. This coaxial cable technology can be used in conjunction with FTTN to realize an HFC implementation; otherwise, it can be used from the headend to the residence.

10.2.2 Analog frequency modulated-frequency division multiplexed (FM-FDM) systems

FM-FDM coaxial systems are similar to AM-FDM, except that FM techniques are employed. FM signals must be converted to AM for input to a cable-ready television receiver. FM-FDM systems are inherently more robust in terms of noise performance and can, therefore, traverse longer distances than AM systems. With AM-FDM, multiple analog and digital signals can be delivered to the consumer. MPEG channels, if any, require an MPEG decoder to convert the signal from compressed to uncompressed and from digital to analog.

AM-FDM and FM-FDM systems multiplex signals (including the digitally encoded portion of the spectrum) in the frequency domain—pure digital systems employ either synchronous transfer mode (i.e., TDM methods) or asynchronous transfer mode to multiplex multiple video signals. This coaxial cable technology can be used in conjunction with FTTN to realize an HFC implementation; otherwise, it can be used from the headend to the residence.

10.2.3 Digital time division multiplexed systems

Digital systems that can transport multiple digital channels were introduced in the early 1990s. These systems use traditional TDM techniques. When carrying noncompressed television signals (say at 135 Mbps), this equipment provides a limited channel capacity; systems operating at 1.2 Gbps can deliver approximately 10 channels. Suppliers have now developed systems that can deliver several hundred compressed digital signals (from 400 to 1500, depending on the system). Some systems carry 200 MPEG-2 signals on a 1.2-Gbps transport; others carry 1000 MPEG-1 over the same digital bandwidth.

This coaxial cable technology can be used in conjunction with FTTN to realize an HFC implementation; otherwise, it can be used from the headend to the residence. Until digital-ready television receivers become available, a decoder is required in the setup box to decompress and undertake the digital-to-analog conversion.

10.2.4 Digital ATM-based systems

Systems that use ATM techniques to carry the various video signals are now being introduced. These systems are either completely digital, using SONET or DS3 as the underlying transport, or use multilevel encoding (e.g., quadrature amplitude modulation) to encode the digital signal over an analog transport (e.g., a segment of bandwidth in the coaxial spectrum). Hence, this technology can be used in conjunction with FTTN to realize an HFC implementation. See Fig. 10.3.

10.3 Some Examples of Recently Proposed Approaches

This section briefly reviews some HFC systems that are planned for trial or deployment, in order to appreciate where the technology may be headed. All information is based on the open literature.

10.3.1 Time Warner

Time Warner proposed a 1-GHz system, extending the current generation of coaxial systems that operate at 750 MHz. This system requires

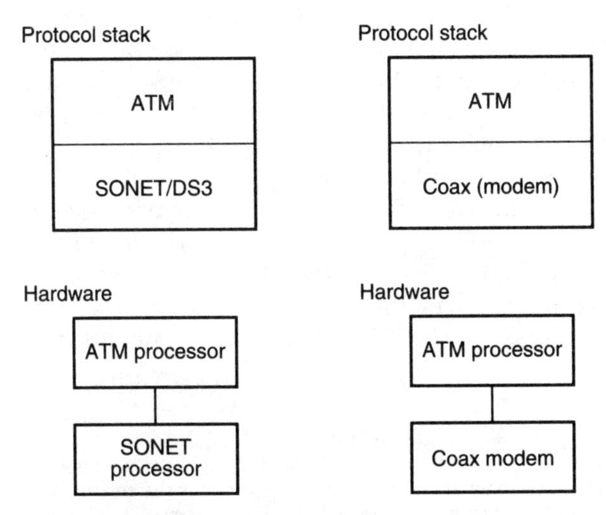

Figure 10.3 Digital video transmission systems.

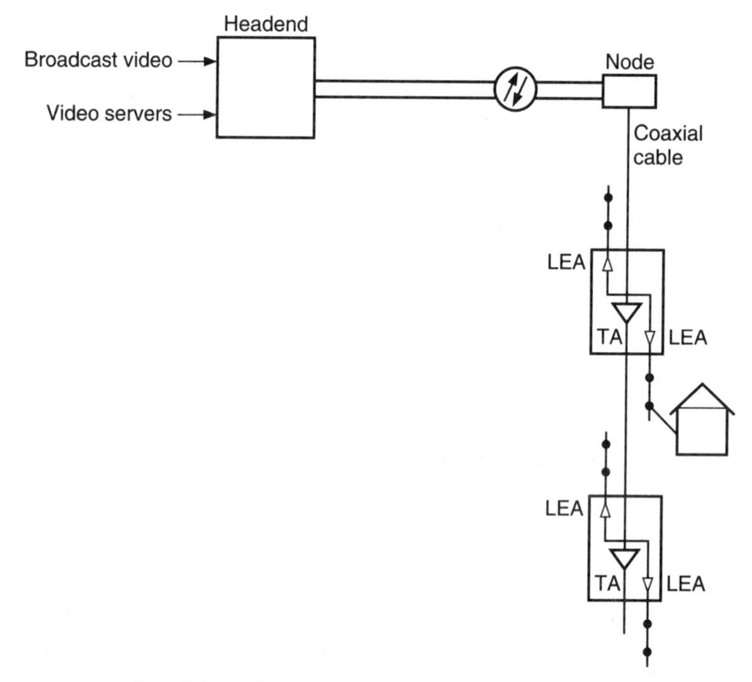

Figure 10.4 Time Warner's super distribution.

newly developed amplifiers. The 1-GHz system is employed to support 96 traditional programming channels (some analog and some digital) and 54 PPV channels. The proposed bandwidth allocation on the analog transport portion of the system (the cable component) was shown in Fig. 4.12.* The amplifier box actually includes three amplifiers: a trunk amplifier and two line extender amplifiers; this arrangement eliminates the need to pass power to the taps. See Fig. 10.4. The line extenders support two-way communication, enabling interactive services and telephony (the handoff of the telephone traffic occurs at the hub, by connecting the hub to a telephone carrier's switch).

Time Warner's system entails four fibers to each fiber node (three downstream and one upstream). One analog fiber carries the broadcast analog channel; the second analog fiber carries the digitally encoded channels; the other two fibers carry the downstream and upstream telecommunications services. At the fiber node, these signals are

---

* As indicated in the figure, there are several proposals being circulated, including the one by Time Warner and one by Cable Television Labs. Some of these proposals allocate bandwidth to 1.3 GHz, but these systems are not currently available at the commercial level, although work is underway to increase the higher limit of operation.

appropriately multiplexed to be delivered over a single coaxial cable to the home.* Each fiber node will serve 600 to 1000 homes. The system employs ATM techniques and uses a $5000 setup box developed by Scientific Atlanta and Toshiba. The video server is provided by Silicon Graphics, while the ATM engine is a 20-Gbps GCNS-2000 switch provided by AT&T. All cells are broadcast down into the cluster of homes connected over the coaxial cable to the fiber node; the setup box is given the responsibility of selecting only the cells that belong to a specific home. See Fig. 10.5.

The Time Warner system is not a VDT system in the exact sense of the word because it only has one video programmer and, therefore, does not support equal access. Also, it should be noted that the system allows for 46 distinct channels (at T1 rate) that are shared among the population of homes on the fiber node; one concern is the blocking probability for these channels at busy hour or busy season (this may force the designer to break down the serving area into smaller and smaller clusters). The trial of this system started in December 1994.

Section 10.4 provides a more detailed discussion of this system for illustrative and pedagogical purposes.

10.3.2 Scientific Atlanta

Figure 10.6 depicts, for illustrative purposes, Scientific Atlanta's HFC system that provides fiber to the serving area.

10.3.3 Cablevision Systems

Figure 10.7 depicts an HFC system envisioned by Cablevision Systems that supports PCS capabilities that use the landline network for trunking. Here the dedicated optical feeder node can be used to transport PCS signals between remote antennas and the headend-located switch.

10.3.4 Some RBOCs' HFC plans

Some of the published plans by the RBOCs are summarized here. In each case, note that the intent of the RBOCs is to support VDT (e.g.,

* Sixty analog channels are provided in the 50- to 450-MHz portion of the bandwidth; the 450- to 695-MHz band supports 20 digitally encoded channels on 12-MHz slots (each of the resulting 45-Mbps channels can support 7 MPEG-2 channels); the 695- to 735-MHz band supports PCS services; the bandwidth between 735 and 900 MHz is not used because of possible radiation and interference with land mobile communication and to provide a guard band between the downstream and the upstream signal; the next 70 MHz are used to support upstream communication; the last 30 MHz are used for PCS. The first of the three fibers carries the signal that fits into the 50 to 450 MHz; the second fiber carries the signal that fits into the 450 to 695 MHz; the last two fibers carry the downstream and upstream PCS-telephony services.

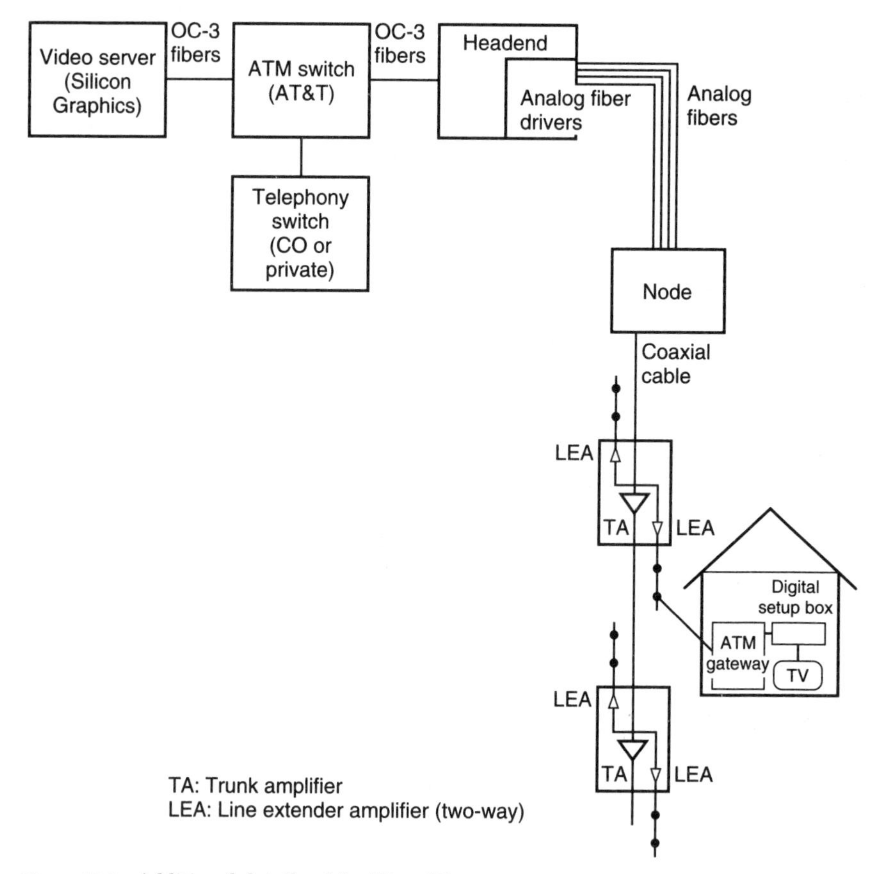

Figure 10.5 Additional details of the Time Warner system.

multiple VIPs), not just to deploy an upgraded video distribution system (i.e., upgraded compared with a traditional 40-channel system).

Ameritech (also see Chap. 14) is planning to deploy a 750-MHz HFC overlay network for video and use the existing voice network to support traditional services. See Fig. 10.8. A single fiber carries the video signal from the end office to a four-way passive splitter. Each node supports 2000 homes over four 500-home segments. A coaxial cable mechanism delivers the video from the node to the customer. Each node supports four coaxial cables, each supporting 125 homes, permitting future upgradability. The system supports 70 channels of analog video, 240 channels of compressed digital broadcast video, and 80 channels of compressed digital video to support VOD services for a 500-home segment. Bandwidth of 5 to 30 MHz is used for upstream communications. The video server is manufactured by DEC, and the ATM switch by AT&T.

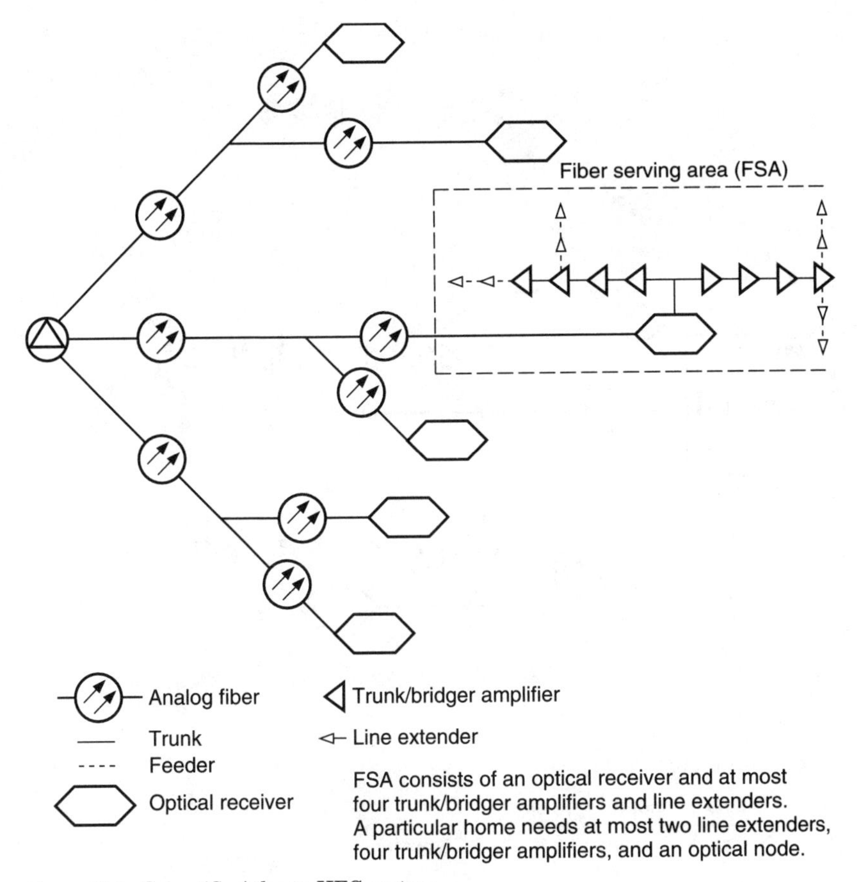

Figure 10.6 Scientific Atlanta HFC system.

Pacific Bell (see Chap. 14) is looking at an integrated (rather than overlay) network that carries voice, video, and data. See Fig. 10.9. The 750-MHz system to be supplied by AT&T carries analog video up to 450 MHz (60 channels), and then uses QAM (six bits per Hertz) to carry video and QPSK (two bits per Hertz, more noise resilience) to carry upstream bandwidth. Each bandwidth supports 450 homes. The server is manufactured by Hewlett-Packard.

U S WEST (see Chap. 14) is planning to use an overlay video system based on HFC technology, while the voice is supported with FTTC technology. See Fig. 10.10. The Omaha trial announced at press time consisted of five fibers into each feeder node and two coaxial cables to each home. Coaxial cable #1 is mapped to Fiber #1, Fiber #2, and Fiber #3. Fiber #1 carries a 5- to 30-MHz upstream signal; Fiber #2 carries 75 channels of analog video that are positioned on the 50- to 550-MHz portion of the coaxial spectrum; Fiber #3 carries 175 shared channels of

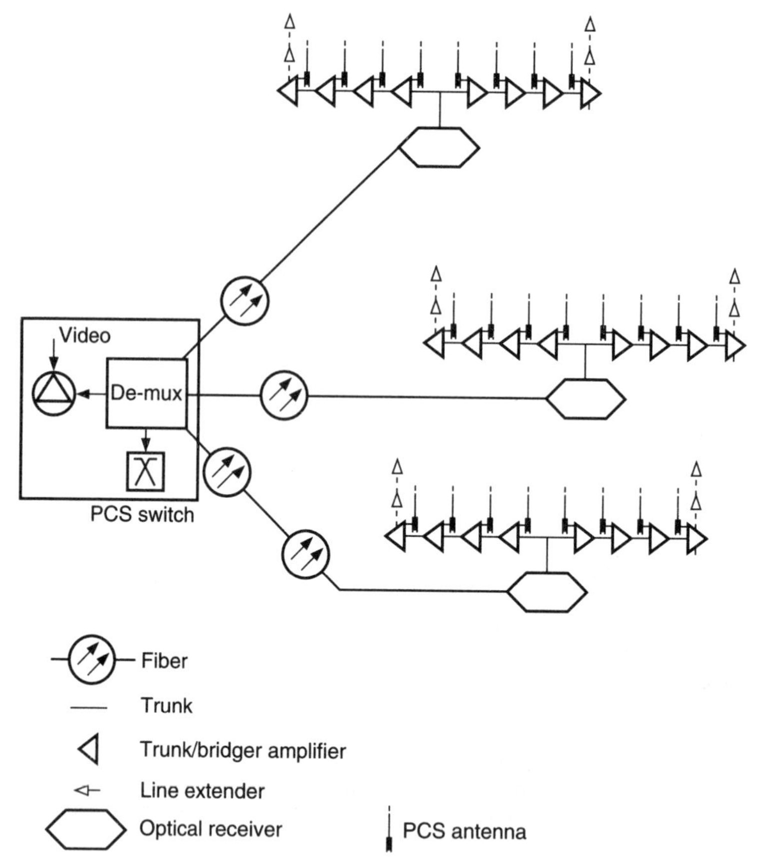

Figure 10.7 Cablevision PCS and video support over HFC systems.

digital video that are positioned on the 550- to 650-MHz portion of the coaxial spectrum. Coaxial cable #2 is mapped to Fiber #4 and Fiber #5. Fiber #4 carries a 5- to 130-MHz upstream signal; Fiber #5 carries 800 shared channels of digital video that are positioned on the 150- to 600-MHz portion of the coaxial spectrum. (Coaxial cable #2 also provides power for the ONU.) Each node supports 400 homes.

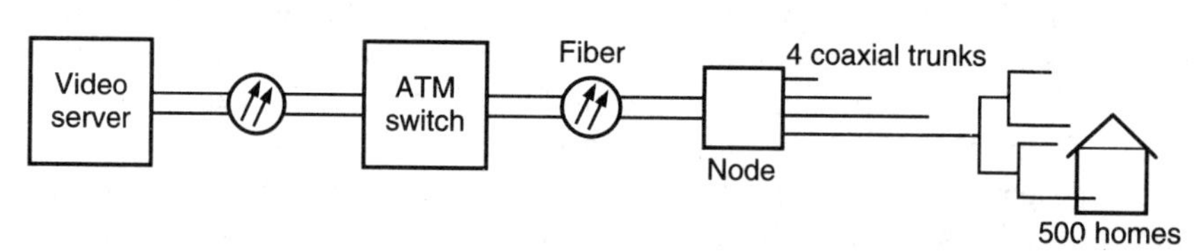

Figure 10.8 Ameritech's HFC.

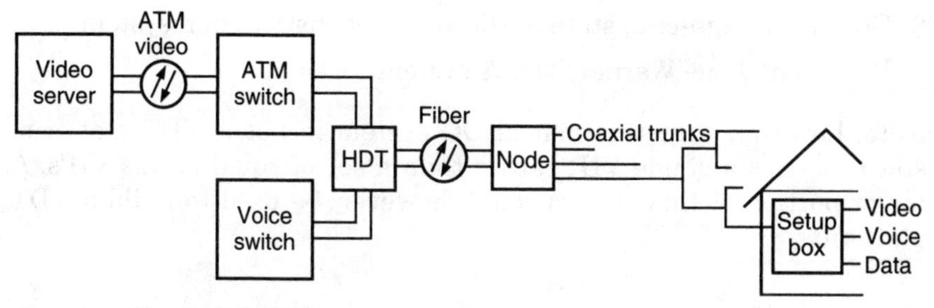

Figure 10.9 Pacific Bell VDT system.

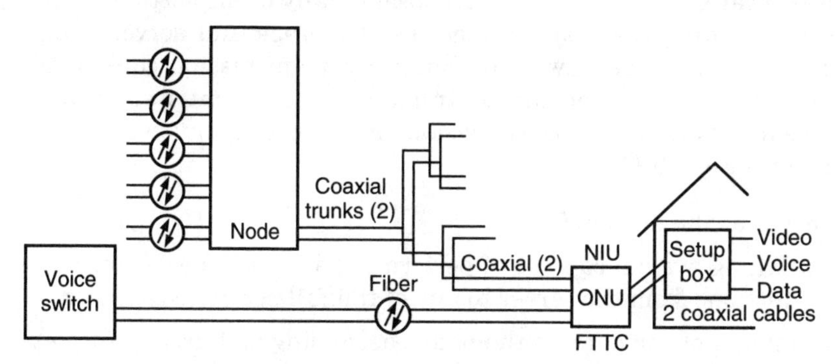

Figure 10.10 U S WEST VDT trial.

This system does not currently use switching. Broadcast video, as well as video on demand, are sent to all users on a segment. The addressable network interface unit, which performs an ONU function outside the home, is employed to selectively enable the channels to which a particular home is entitled. The plan is to ultimately use ATM.

Close to press time, NYNEX announced an overlay 750-MHz HFC system to carry video; integrated voice and video over a single system were planned for a future date.

10.4 Time Warner Full-Service Network

This section reviews for pedagogical and illustrative purposes the HFC system that was recently proposed by Time Warner, expanding the short synopsis provided earlier. (This discussion is based on the RFP the company issued.[3]) The material that follows builds on many of the concepts covered in the preceding nine chapters. Therefore, the motivation in including this material was to:

1. Illustrate with an actual example the concepts covered in the book to this point

2. Describe a high-end, state-of-the art video distribution system
3. Document Time Warner Cable's system

Note, however, that this non-RBOC system is not a VDT platform, since there is a single VIP, rather than a set of equal-access VIPs. A major portion of this system could, however, be used to build a VDT platform.

10.4.1 Overview

Time Warner Cable requested a proposal on any or all portions of the project, including the major categories of storage and server equipment, headend devices, switching, network transmission equipment, customer terminal equipment, control and management systems, and systems integration service. The vision for this project involves the following (see Fig. 10.11).

- A server at the headend
- A switching system capable of performing all of the necessary routing functions from the server to any of the full-service customers
- Assembling of the data on the node basis using add-drop multiplexing or equivalent technologies

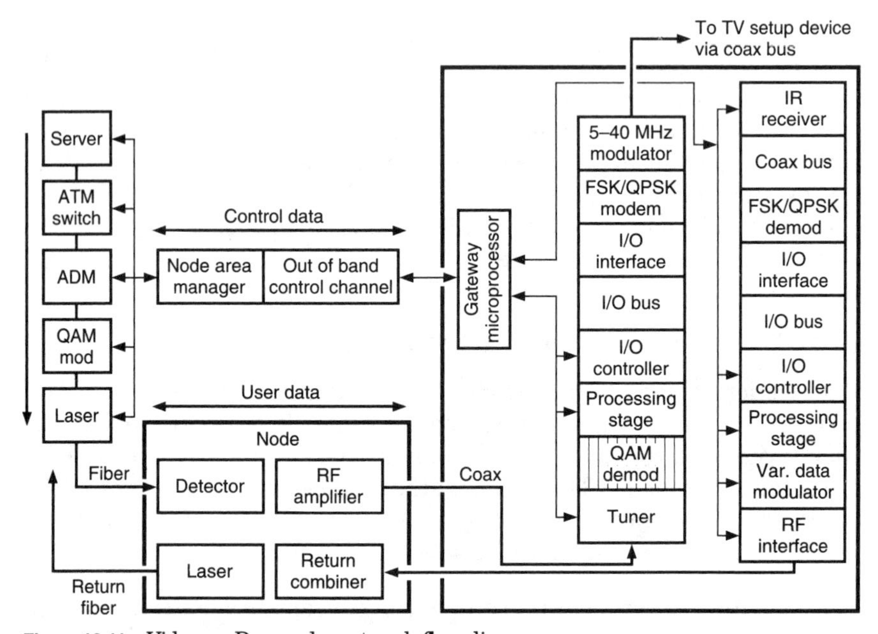

Figure 10.11 Video on Demand—network flow diagram.

- Digital modulation devices using QAM modulation or equipment techniques
- Optical links from the headend to the node locations fed by single or multiple laser configurations
- Coaxial plant to the customers homes
- Appropriate customer terminal equipment to translate the bit streams into usable services

In addition, various types of security and error correction are necessary at several points within the network.

10.4.2 Network configuration and general capabilities

Time Warner Cable anticipated that, in the project area, approximately 6000 customers will subscribe to traditional cable services, and 4000 of those customers will make use of the new service to some extent. Time Warner Cable will make extensive use of fiber-optics transmission to divide the project area into nodes that serve approximately 500 to 600 *passings* (potential customers passed by the network). The Orlando architecture will result in a total of 16 to 20 nodes installed from a central origination point and approximately 100 miles of coaxial plant with a bandwidth of 1 GHz. The bandwidth from 50 to 450 MHz will be used to provide traditional cable services in an AM-VSB format. The spectrum from 450 to 750 MHz will be used for digital services such as video on demand. Time Warner Cable plans to use a high-split configuration to allow for return information, from the home to the headend, in the spectrum from 850 to 1000 MHz.

Time Warner Cable anticipates that they will experience a 25 percent peak use demand of high-data-rate digital service (i.e., full-motion video) across the customer base. It is likely that the requests will be unevenly distributed among node areas, resulting in up to 100 simultaneous full-service accesses in any combination of node areas. Time Warner Cable is assuming that the majority of accesses will be for VOD services operating at a rate of up to 4 Mbps.* Such use will result in a potential bit-rate demand of up to 400 Mbps in the 450- to 750-MHz spectrum. It will also be necessary to maintain downstream control communications links, error correction, and other maintenance overhead. This will result in an even greater required bit rate.

* For the quality and profiles of interest, the buffered MPEG-2 rate can vary in the 4- to 6-Mbps range, as discussed in Chap. 6.

10.4.3 Server

The server (Fig. 10.12) delivers ATM cells to the ATM switch for distribution to the home. ATM cells must be delivered at *optical carrier* (OC) rates of OC-12 (622 Mbps) or with an interim rate of OC-3 (155 Mbps). The server must provide for random, simultaneous access to top hits of movies, music, and software. The server must also provide for sequential, batched access to on-line media, such as digitally stored movies on disk or tape. It must be possible to distribute the inventory in the server on appropriate storage devices, memory or physical media, so as to maximize the number of viewers or revenue flow. The server must be designed for the graceful recovery from failure of application software or failure of peripheral storage devices. The server must also allow for the introduction of additional servers without network or software changes.

The server must have a feedback mechanism to correct erroneous data or downloaded software. Since a major aspect of the project is to deliver multimedia services, the server must provide constant bit-rate services, such as voice, music, and video, in a jitter-free stream. The server should be capable of providing up to 1000 simultaneous accesses to accommodate the 25 percent peak loading over the customer base of 4000. For the purpose of this project, Time Warner Cable is assuming that video will be transmitted at a rate of 4 Mbps, requiring the server to have a throughput of up to 4 Gbps for downstream video, in addition to any overhead and maintenance information that originates within the server.

The server should provide redundancy to the extent that it is possible to recover and restore the server to full operation without signifi-

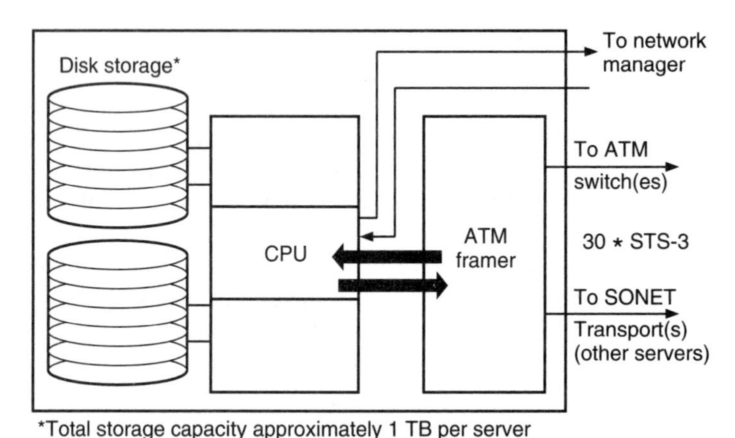

Figure 10.12 Time Warner Cable server—general structure.

cant downstream on the network. This may or may not involve physical duplication of the server in the headend.

Time Warner Cable plans for the server to be feeding an ATM switch that accepts data in the form of STS-3c signals. It will be necessary for the server to accommodate whatever type of data input format is required by the switching equipment, and vice versa. This will require a high degree of communication and cooperation between the server manufacturer and the switch manufacturer.

10.4.4 ATM switch

The ATM switch (Figs. 10.13 and 10.14) provides synchronous switching and routing at the speed of the physical medium and the demand rate of the user. The ATM switch architecture must support enveloped services of differing bandwidths from many access ports.

The ATM switch provides switching and routing of ATM cells as delivered to it by the server using *self-routing modules* (SRMs) based on the address in the ATM cell header. It provides the electric and/or optical interfaces necessary to accept signals from the server and distribute signals into the full-service network. The ATM switch provides direct connection to telephone interexchange carriers for long distance service via an external ATM to a DS-3 conversion multiplexer.

The ATM switch:

- Must provide *constant bit-rate* (CBR) service over a single-access line.
- Must support both real-time connection-oriented and non-real-time connectionless data services in one system.

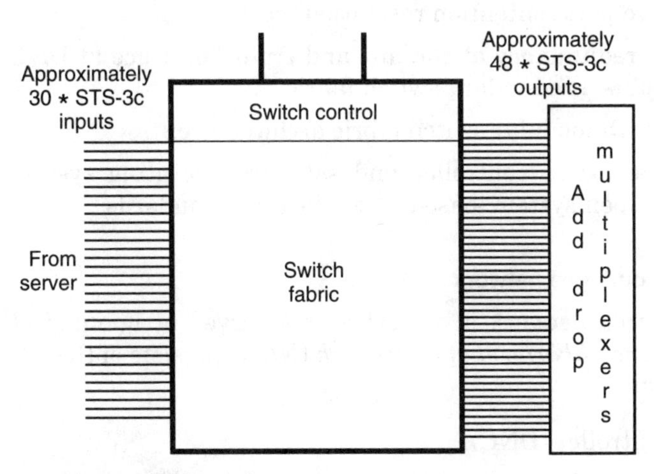

Figure 10.13 ATM switch.

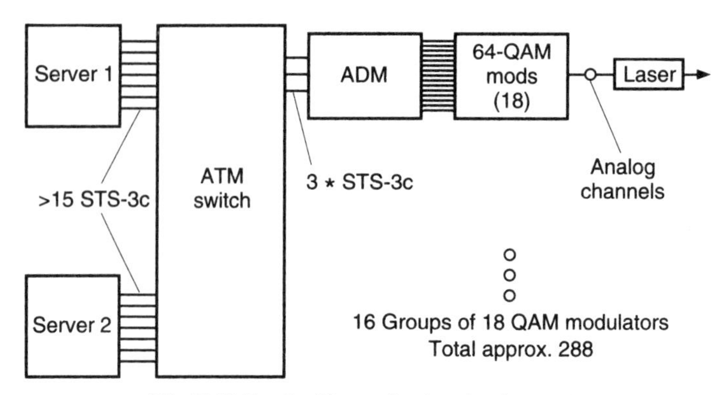

Figure 10.14 The Full-Service Network—headend overview.

- Must meet the 53-byte cell format standard as defined by the ITU-T and ANSI for ATM.
- Must have physical interfaces at DS-1, DS-3, and STS-3c (or equivalent OC-3).
- Must have throughput capacity to switch and route the equivalent of 30 STS-3c input ports and 48 STS-3c output ports.
- Must have networking capabilities based on the establishment of associations or connections between Layer 2 ATM entities. At a minimum, these connections are *preprovisioned* (PVC). The ATM switch must have the ability to provision PVCs into VPs. The VP is a group of PVCs carried between two points and may also involve many ATM links.
- Must have a self-routing switch fabric, provide for cell buffering, and have an effective port contention resolution mechanism.
- Must have full redundancy at the line and trunk interface at DS-3 rate or above, plus a redundant switch fabric.
- Must support with modular switch fabric architecture.
- Must include a switch controller and software operating system designed as an open system, based on application standards.

10.4.5 Neighborhood area network

A neighborhood area network is a service area served by one digital node controller. A *neighborhood area network* (NAN) consists of the following subsections:

Digital node controller (DNC)

Neighborhood optical node (Node)

Fiber-optic transport

RF amplifiers

Home interfaces

A typical NAN (see Fig. 10.15 as well as Fig. 1.7 in Chap. 1) will encompass a physical area of approximately five coaxial plant miles and will pass 300 to 1000 potential customers. The distance from the headend to the Node can be up to 10 miles. There are no more than four RF amplifiers in series, and there is a maximum of 1.5 miles of coaxial cable separating the RF amplifier series. There is a minimum of four fibers terminating at each Node. Optical branching of the broadband analog signals may be used in some cases.

The NAN manages traffic into and out of the area that it serves. The NAN receives data from the ATM switch and routes this information to the home in the service area via RF carriers. The NAN is managed by the NAN manager. The NAN manages all terminals in the service area. Management responsibilities include bandwidth allocations, frequency allocations, information packaging, and diagnostics within service area.

10.4.6 Digital node controller

The DNC is the interface between the digital and the analog portions of the network. See Fig. 10.16. The DNC receives STS-x information from the ATM switch and converts it into a signal that can be carried

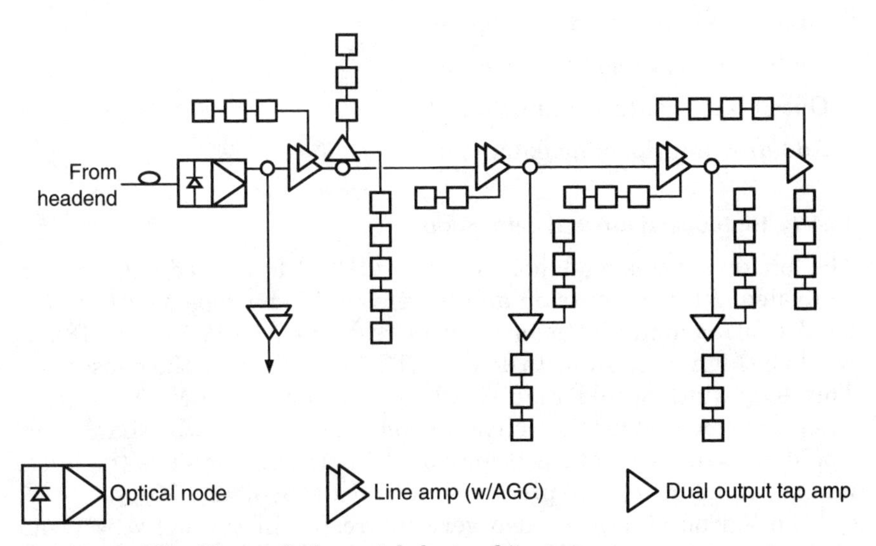

Figure 10.15 Time Warner Cable coaxial plant architecture.

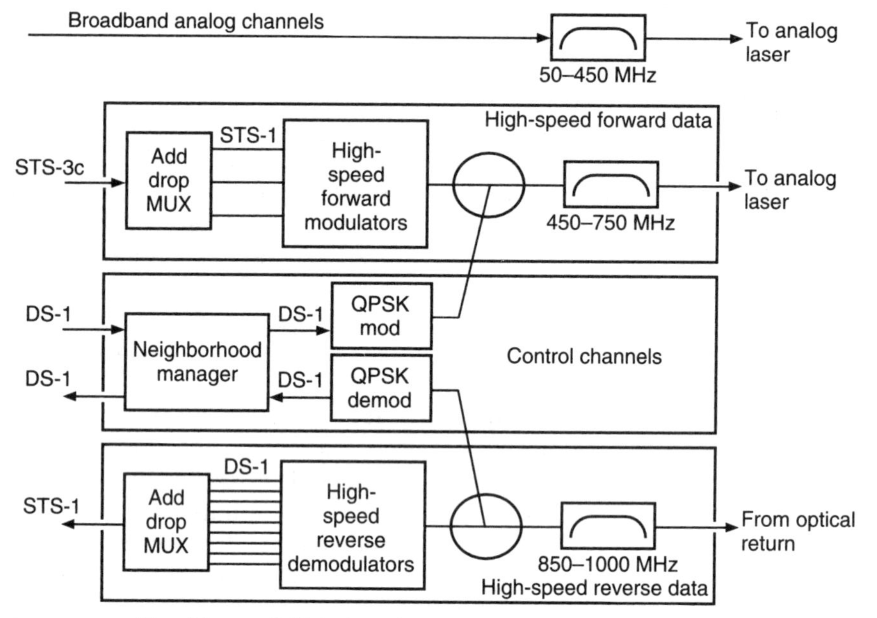

Figure 10.16 Time Warner Cable's digital node controller.

on the analog fiber and coaxial portion of the network. The DNC also receives the reverse signals from subscriber and multiplexes these signals into an STS-x format signal.

The DNC consists of the following sections:

High-speed forward data module

High-speed reverse data module

Out-of-band control channels

Add-drop multiplexing equipment

10.4.7 High-speed forward data module

The function of this module is to convert STS-1 formatted data into an equivalent RF transmission and to deliver RF envelope over a broadband cable communications network (see Figs. 10.17 and 10.18). It would be highly desirable to deliver STS-1 data rates to the subscriber. This bandwidth would require 12 MHz using 64 QAM. This is an acceptable method for the delivery bandwidth. On the other hand, this would make the cost of electronics at the home higher since the information would have to be processed at STS-1 rate of 51.8 Mbps.

Time Warner Cable is also very interested in the delivery of an STS-3c over the cable. The minimum bandwidth delivery method that

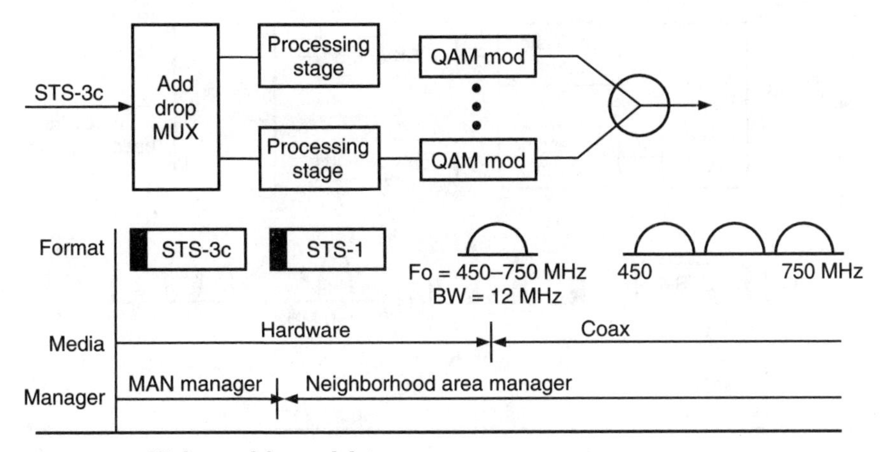

Figure 10.17 High-speed forward data.

will be considered is 4 Mbps. This is driven by the need to deliver video on demand, which is assumed to be 4 Mbps.

10.4.8 High-speed reverse data module

The high-speed reverse data unit receives RF carries in the 850- to 1000-MHz area and converts these signals into an STS-1. It would be highly desirable for this unit to receive ATM packets that are TDMA into DS-1 rate RF channels. The system must be able to support 100 simultaneous users at an average rate of 64 kbps per user (see Fig. 10.19).

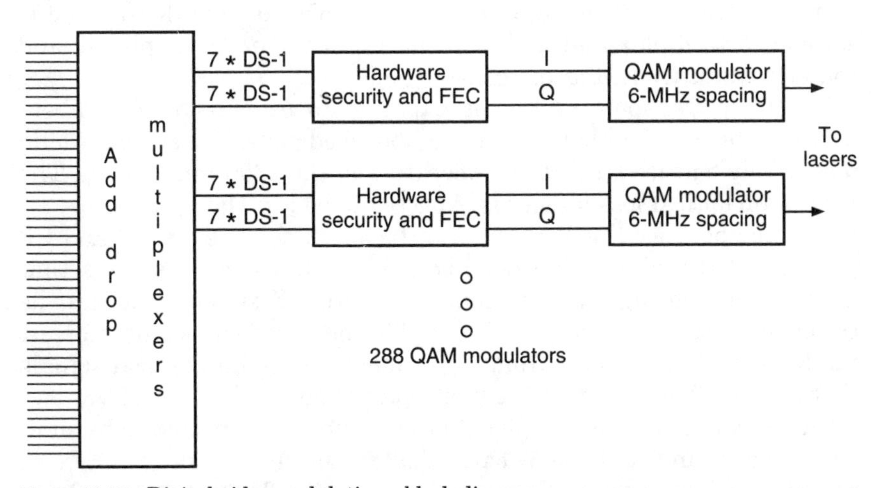

Figure 10.18 Digital video modulation—block diagram.

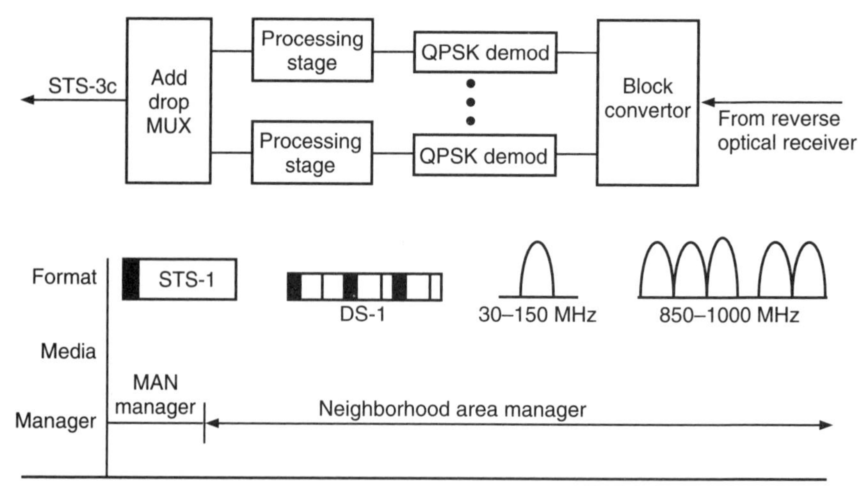

Figure 10.19 High-speed reverse data.

10.4.9 Fiber-optic transport

The purpose of the broadband fiber-optic system (see Fig. 10.20) is to transport the 50- to 750-MHz forward and 850- to 1000-MHz return spectrum between the headend and the coaxial service area. In the forward, or downstream, direction the 50- to 450-MHz analog NTSC signals and the 450- to 750-MHz RF modulated digital carriers can be directly applied to separate lasers for each respective band (see Fig. 10.21). As an alternative, the two forward RF bands can be multiplexed together into a 50- to 750-MHz band and applied directly to a single AM laser transmitter (see Fig. 10.22).

In the return or upstream direction, the 850- to 1000-MHz band is selected with diplex and high-pass filters in the node, amplified, and routed to a return AM laser transmitter.

The fiber transport system is expected to use up to three fibers between the headband and each neighborhood node. The system would most likely feature DFB lasers operating in the 1300-nm window. The maximum optical loss budget is expected to be less than 10 dB.

The neighborhood node is the interface between the optical and RF portions of the plant. The neighborhood node converts signals that arrive from the digital node controller into RF signals that can be transported on coaxial plant. The neighborhood node also must convert the RF signals that come from subscribers' homes into optical signals that can be transported on the fiber-optic plant (see Fig. 10.23).

The coaxial transport, or distribution system, is expected to be similar to that found in Time Warner Cable's Queens or Rochester, N.Y., cable television plants, with one major exception. The exception relates to use of the upper end of the spectrum, between 850 and 1000 MHz, to

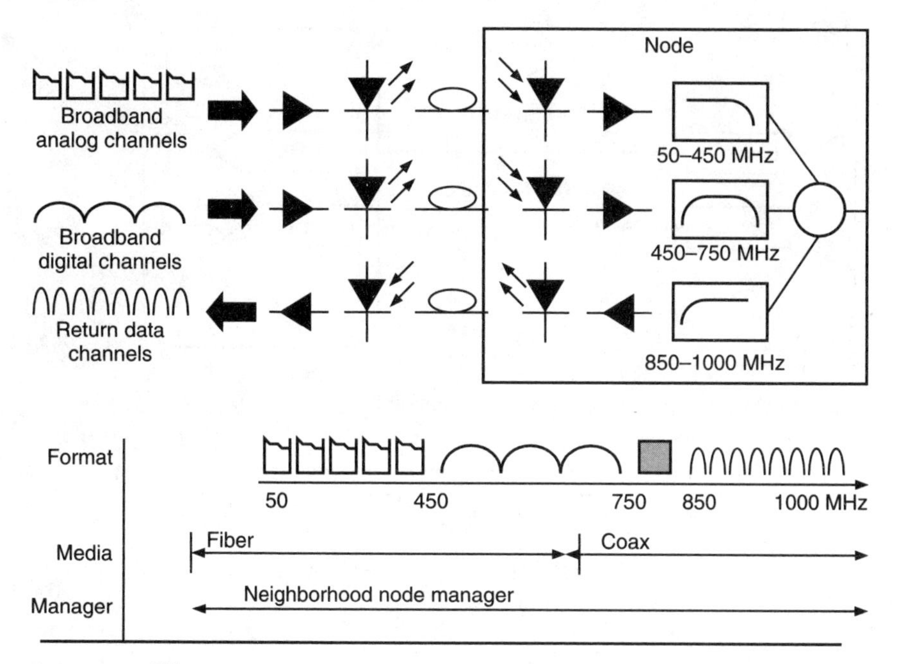

Figure 10.20 Fiber transport.

transport the return signals from the customers' homes back to the neighborhood node. In the Queens, N.Y., and Rochester, N.Y., systems, the return path is located between 5 and 30 MHz.

The only portion of the coaxial network affected by the relocation of the return path from the lower to the upper end of the spectrum is the RF amplifiers. Since the attenuation of the passive elements, such as cable and directional couplers, is significantly higher at the upper end of the frequency band, the gain of the return amplifier modules must

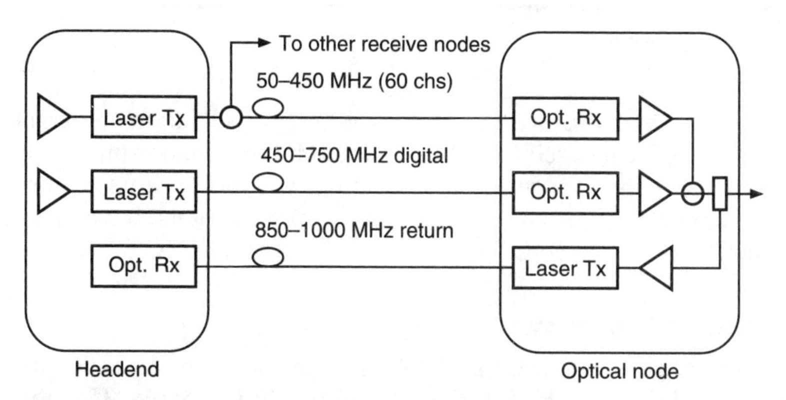

Figure 10.21 Fiber transmission scheme—Option 1.

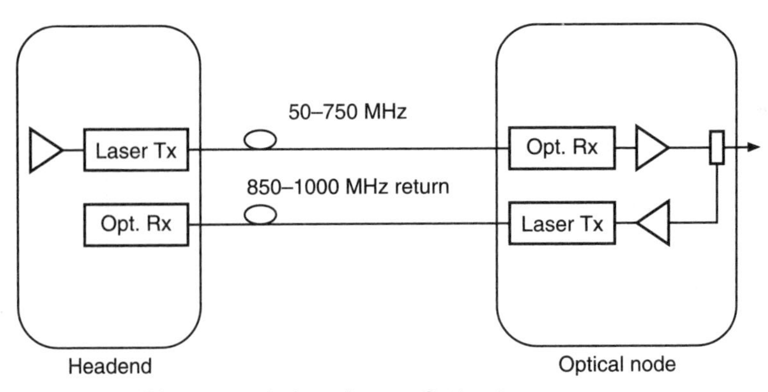

Figure 10.22 Fiber transmission scheme—Option 2.

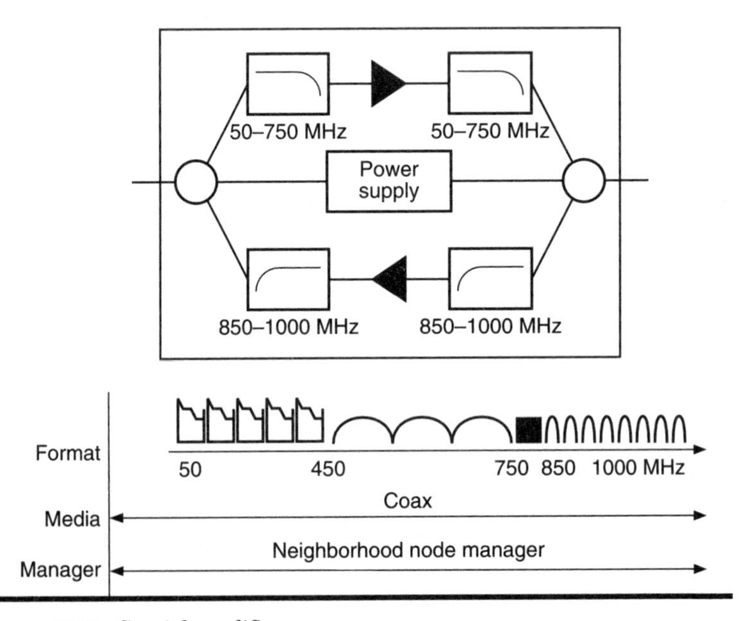

Figure 10.23 Coaxial amplifier.

be increased. Internal filtering in the amplifier, as well as the match of external devices, are crucial to obtain stable bidirectional amplifiers. The coaxial transport network architecture will be optimized once practical amplifier gain and performance specifications have been determined.

10.4.10 Residential gateway

The residential gateway (see Fig. 10.24) is a device to decouple the RF and high-speed digital circuitry from terminals in the home. The phi-

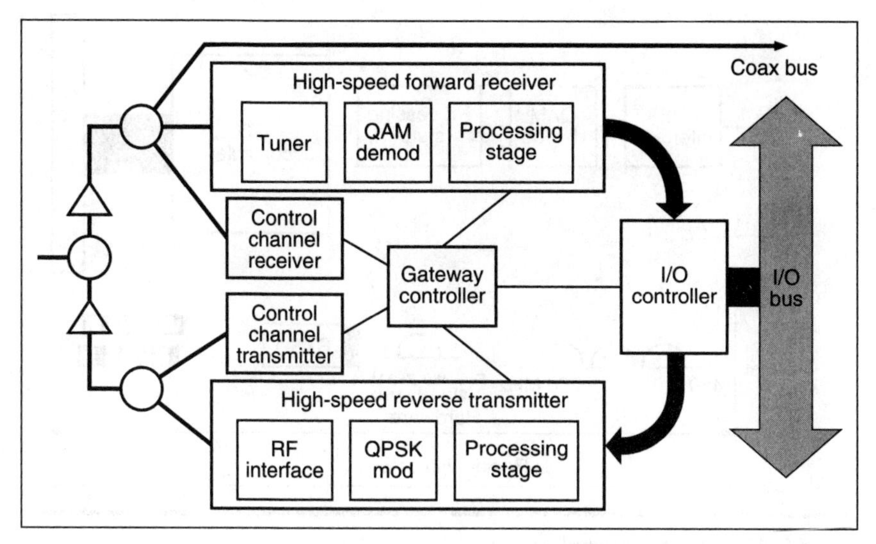

Figure 10.24 Residential gateway.

losophy behind the residential gateway is that there are a variety of devices that need to communicate over the cable communications network. Each of these devices communicates using some defined interface standard. If an interface is created for these devices that is driven off the gateway, then a variety of present and future devices could communicate over the cable communications network.

The residential gateway is the interface to all devices and services that need bandwidth from the cable supplier. The gateway receives high-speed forward data and converts the data into the format specified by the I/O controller. The gateway collects data from devices in the home though the I/O controller. These data are then transmitted via the high-speed reverse data modulator to the neighborhood node. The residential gateway also receives information via the forward control channel received and transmits request commands via the control channel transmitter.

The minimum capability for the residential gateway is:

- Receive one 4-Mbps data stream from the neighborhood node
- Transmit one 1.544-Mbps data stream to the neighborhood node
- Exchange appropriate control information with the neighborhood with the neighborhood node

Also see Figs. 10.25 and 10.26.

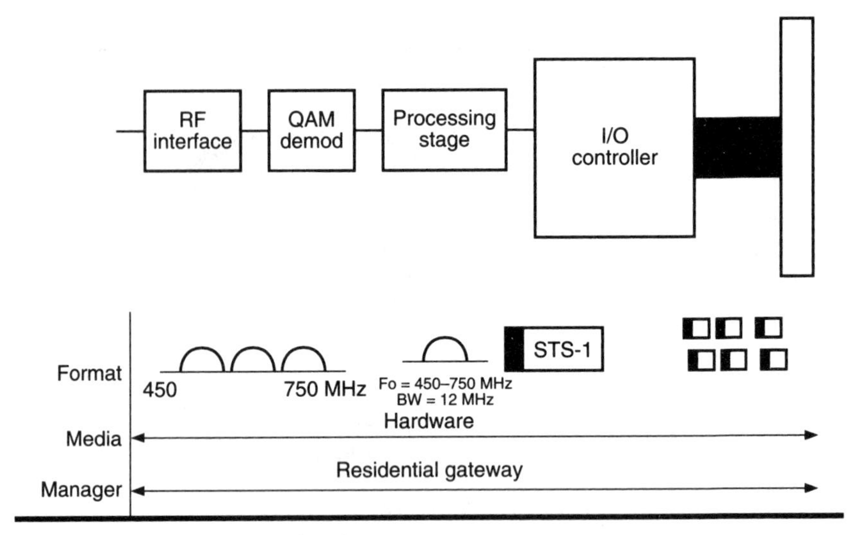

Figure 10.25 Gateway forward path.

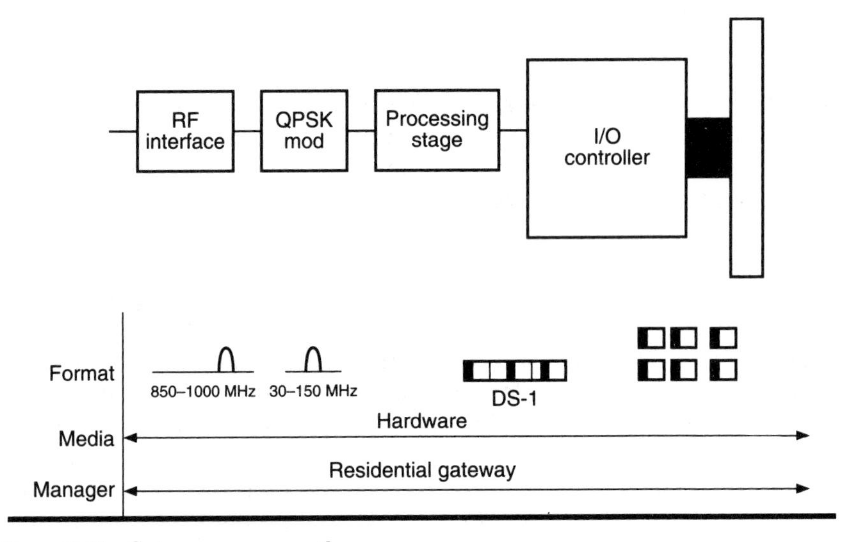

Figure 10.26 Gateway reverse path.

10.4.11 Setup converter

The setup converter (see Fig. 10.27) is one method of delivering video on demand and multimedia services. The setup converter consists of high-speed RF demodulation, video decompression, and reverse control circuitry. A variety of functionality can also be included in the design of a setup converter, including IR blaster, program guides, and game interfaces.

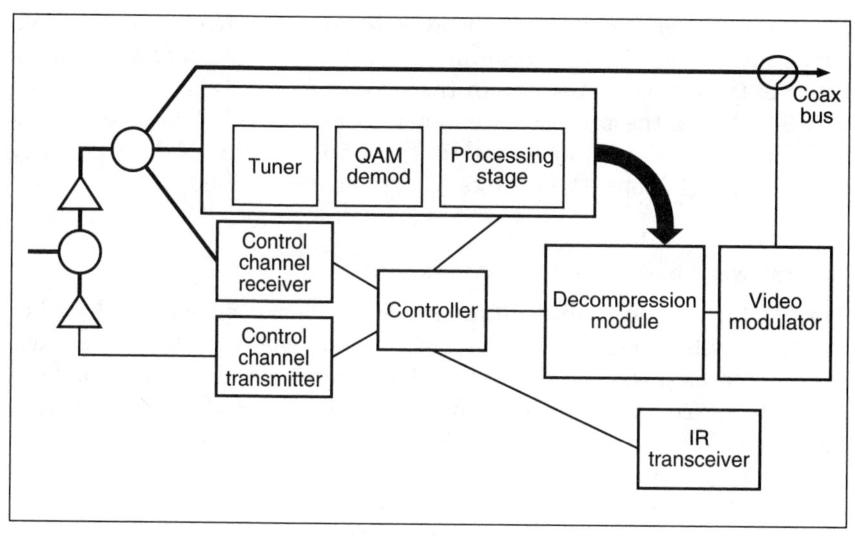

Figure 10.27 Setup converter.

The function of a setup converter is to receive the high-speed data from the neighborhood area node and to display this information on a television. The setup converter must be able to request programming and control the flow of the programming into the home via a reverse control channel.

The minimum capability for the setup box is:

Receive one 4-Mbps data stream from the neighborhood node

Transmit request and commands to the neighborhood node

The desired capability for the setup converter (beyond the minimum capability) is:

Receive one 4-Mbps data stream from neighborhood node

Transmit request and commands to the neighborhood node

Display program guide information on television screens

Control through IR blaster other devices such as TVs and VCRs

Interfaces to other devices that may need data such as video games and computer

10.4.12 Out-of-band control channels

The out-of-band control channel is the signaling path for all information distributed on the neighborhood network. When a subscriber requests bandwidth, the requests are received on the reverse control channel receiver. The neighborhood node manager then must acknowledge the request by the assignment of a path. The forward control channel transmitter operates in the 450- to 750-MHz frequency range at 1.544 Mbps; the protocol was not yet determined. The reverse control channel receiver operates in the 850- to 1000-MHz frequency range at 1.544 Mbps; the protocol has to be determined.

10.4.13 Neighborhood traffic management

This unit maintains a distributed relational database of the subscriber in the neighborhood. Each subscriber is assigned to a class of service that indicates the services a subscriber is authorized to use. This local class-of-service database is downloaded from the central metropolitan database of all citywide subscribers. When service is requested by the subscriber via the control channel, the class-of-service is used to validate the subscriber; this is the first line of network security.

Once authorized to use the network, the neighborhood traffic manager has the task of allocating the outbound digital channel that cells intended for the subscriber will appear on; it also has the task of assigning the inbound channel the subscriber's terminal should tune to.

For data traffic, the neighborhood traffic manager maintains retransmission queues so that if a packet check sum error is detected at the home terminal, the neighborhood manager responds to a retransmission request on the control channel, and retransmits the erroneous packet. This cycle is repeated until the packet is correctly received. A *go-back-N* algorithm is one possible strategy.

It should be possible to quickly replace a neighborhood traffic manager that has electrically failed; a *hot spares* option is desirable.

10.4.14 Home terminal operating system

The home terminal operating system must be downloadable from the server. A minimal ROM boot device communicates with the neighborhood traffic manager's control channel to request downloading. Examples of network bootable ROM devices are IBM's NetBIOS, and UNIX Reverse Address Resolution Protocol (RARP).

For application developers, the operating system must provide an interface to the network for interactive applications, an interface to graphics for virtual applications, and interfaces to other devices for special applications. A software tool kit must be provided to third-party application developments with appropriate licenses. A test suite of tools to ensure that network applications are always well behaved, i.e., always exit gracefully, is required. A style guide for a consistent look and feel of applications was planned to be made available from Time Warner Cable.

10.4.15 I/O bus

The I/O bus is a 16-, 32-, or 64-bit-wide path that allows memory-mapped devices to receive and transmit data to and from the I/O controller (see Figs. 10.28 and 10.29). The bus must be able to maintain constant bit-rate services for those services that require it, such as voice or video. It must be possible to plug in new devices to the I/O bus without interfering with other services on the bus and require only newly downloaded software to activate the new I/O device. The I/O bus must be an open, documented standard that allows third-party vendors, with the appropriate tool kit licenses, to develop plug-and-play applications for new services.

The I/O bus is a parallel bus interface. The devices that are connected to the bus respond to the I/O controller. Devices are connected to the bus service high-speed ATM cells that are memory mapped to a device that is connected through the media interface module.

The function of the I/O bus is to deliver bandwidth on demand to devices that are connected to the bus. The I/O bus achieves this by

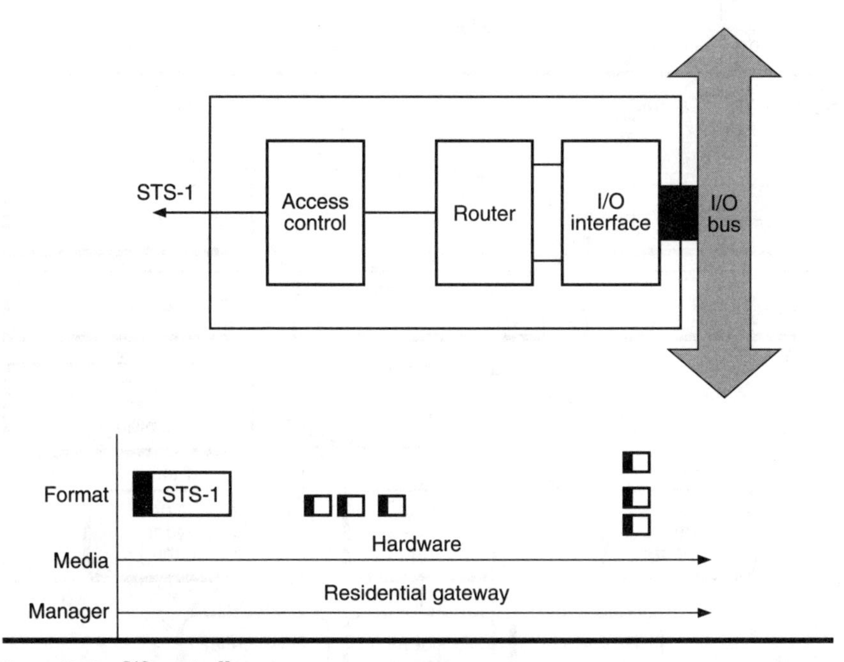

Figure 10.28 I/O controller.

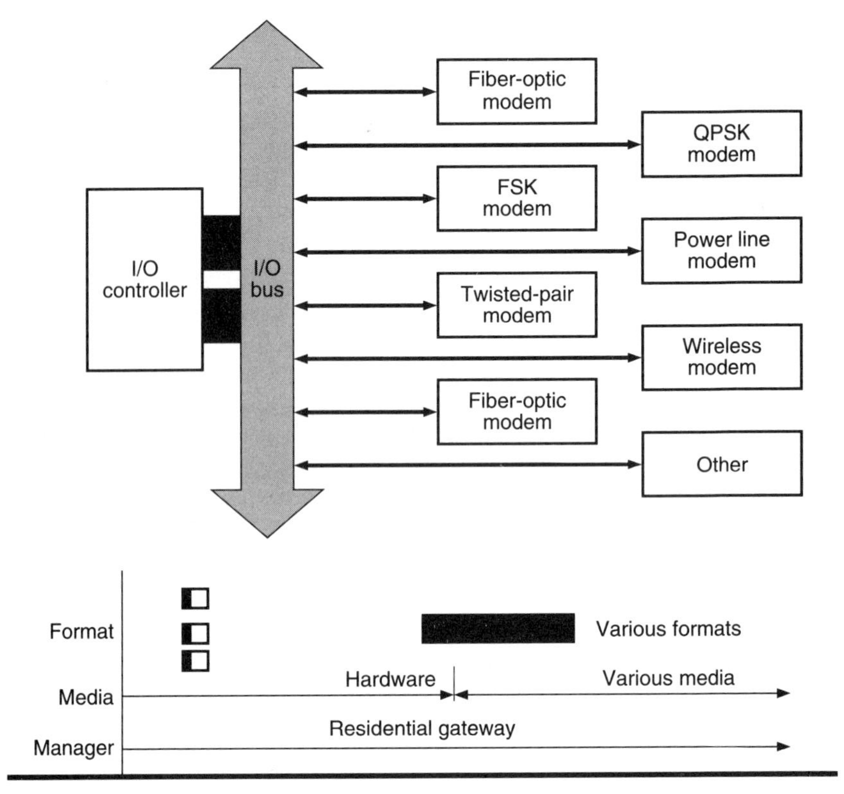

Figure 10.29 I/O bus.

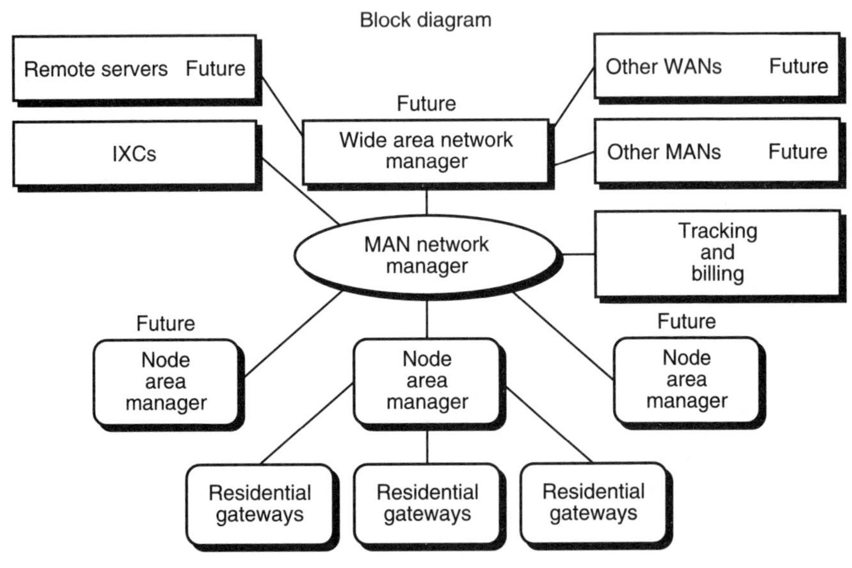

Figure 10.30 Time Warner Cable network control and administration.

routing ATM cells to I/O interfaces. The function of the I/O interfaces are to buffer ATM cells and to receive and transmit user data.

10.4.16 Operations and administration

Figure 10.30 depicts Time Warner Cable's view of network control and administration. Appropriate interfaces and capabilities must be supported.

References

1. G. Lawton, "Interoperating Cable Into The Great Opportunity," *Communications Technology,* November 1993, pp. 26 ff.
2. R. Pinkham, "Combining Apples and Oranges: The Modern Fiber/Coax Network," *Telephony,* February 7, 1994, pp. 28 ff.
3. Time Warner, "Development of a Full Service Network," *Time Warner RFP,* February 11, 1993.

Chapter

11

Terrestrial-Guided Distribution Systems: FTTC and FTTH

11.1 Introduction

Eventually, businesses and residences will be connected to the PSTN using fiber-optic cables from the central office to the curb (using FTTC technology); ultimately, they will be connected with fiber-optic cables end-to-end (using FTTH technology). The thrust of the national information infrastructure is, among other things, to enable such connectivity. The two key questions pertaining to this FTTC-FTTH goal are: How quickly will that happen, and how much will it cost? This chapter looks briefly at this topic from an implicit VDT-video distribution perspective. There is abundant literature on this topic going back at least five years (see as an example Ref. 1 and ancillary references). Only some highlights of the FTTC and FTTH technologies are covered here.

In the United States, fiber has penetrated the feeder segment of the PSTN, in conjunction with the deployment of *digital loop carrier* (DLC) systems; this approach was referred to as FITL in Chap. 4. Penetration of fiber-based DLCs, however, is still low, serving between 5 and 10 percent of all customers, depending on the RBOC and, so far, is only in support of traditional telephony services. About 50 percent of all *new* customers (new construction) are now provided with DLC systems, and most of these new DLC systems are supported by fiber from the CO.* The access network represents a major part of the investment of a telecommunication carrier (up to 50 percent of the total network cost),

---

* DLC can also be supported with metallic T1s.

and has a relatively long lifetime; hence, proponents argue that fiber optics is the only safe investment that will support the evolution from today's existing services to future ones.[2] Carriers are told by these proponents that "to remain viable, [they] must construct sophisticated optical access networks based on FTTH."[3]

FTTN architectures were implicitly covered in the previous chapter in the HFC discussion. The distinction between FTTN in support of an HFC arrangement and FTTC is one of proximity and scale. As noted in Chap. 1, the FTTN approach follows today's telephony model of using a specialized multiplexer at the end of the feeder plant, somewhere in the neighborhood; typically, from 100 to 500 (sometimes 2000) homes are supported. The FTTC approach puts the fiber termination closer to the consumer; typically, this arrangement supports 4, 8, or 16 homes (occasionally, as many as 128 or 256 homes may be supported). FTTC systems may use coaxial cable in the drops from the curb pedestal to the home; this could be 100 ft or so. Although there may be coaxial cable, the system does not fit the definition of HFC. In FTTC, the coaxial is passive (no amplifiers), serves only one home, and the length is small (100 to 200 ft). In an HFC system, the coaxial is active (it uses amplifiers), it serves multiple homes, and the length is long (say 1 to 2 miles).

Several providers have announced FTTC systems. Bell Atlantic was planning a 1995 rollout of such a system to 38,000 customers in Dover Township, N.J. SBC Communications and NYNEX are also using FTTC architectures. HFC systems can be upgraded, in time, to FTTC systems which offer more bandwidth in both the forward and return channels, cleaner signals, and increased service flexibility. SBC's trial in Richardson, Texas, will carry fiber to ONUs that serve, on the average, five homes each. From the ONU, a short coaxial cable carries the video signal and twisted-pair cable carries the telephony signal.

Because an FTTC architecture requires more equipment to be deployed (i.e., a piece of equipment for every 4 to 16 homes), the FTTN approach is likely to see a more immediate introduction for the support of video services. It is expected that FTTC will become more prevalent in two or three years, on the way to the ultimate FTTH solution. FTTH is further in the future. Therefore, this chapter focuses on a shorter term goal: FTTC architectures. FTTH is only briefly discussed at the end of the chapter.*

* Some also introduce the concepts of *Fiber To The Zone* (FTTZ), *Fiber To The Building* (FTTB), and *Fiber To The Floor* (FTTF). The term Fiber To The User (FTTU) is also used as a synonym of FTTH.

11.2 Approaches to FTTC

FTTC discussions in the industry focus on four areas:

- Technology (at the component level)
- Architecture
- Services supported
- Cost

In the cable TV industry, fiber-to-the-neighborhood systems have begun providing a more reliable and cost-effective alternative to long amplifier cascades. In the telephone industry, FTTC has reached cost parity with an all-copper network, to *support traditional telephony,* as measured in terms of installation costs—maintenance costs are typically less than those of traditional twisted-pair systems.[4]

By building an FITL infrastructure (FTTC being one manifestation) that shares fiber, optoelectronics, environmental enclosures, and powering equipment among more than one home, carriers can reduce the per-home cost of deploying traditional switched telephone services compared with nonfiber DLCs. The addition of video capabilities to the network, however, involves new design and economic considerations.[4]

Typically, in the cost comparisons that are often undertaken, designers look at FTTC networks delivering 16-Mbps video services, supporting three to four MPEG-2 video channels. FTTH networks are typically designed to deliver 622 Mbps of bandwidth to the user.

11.2.1 Technology

As noted, in FTTC, fiber is introduced as far as several tens to several hundred feet from the customer. For telephony-only services, twisted-pair wires are used to connect the individual users. For video services, coaxial drops may be used. ONUs typically are located at the curb. The fiber-based distribution network is known as the *optical distribution network* (ODN). The PON-PDN discussed in Chap. 4 is a specific example of an ODN (another type of ODN could use active optoelectronics). Multiplexing methods and supporting technologies are used to improve the cost per user by sharing elements of the ONU among several users. As discussed in Chap. 4, small, medium, and large ONUs can be employed, based on the specific application.

A number of technologies are available for FTTC systems. PDN technology is currently viewed as the most viable solution for deployment of fiber distribution services for narrowband telephony services (including ISDN). One- and two-fiber systems have been developed and deployed. The one-fiber PDN is, in theory, more cost-effective; however, since a large portion of the expense of connecting a home is the labor involved in the installation, the final differentiation is not clear: it costs nearly as much to install a one-fiber cable as a two-fiber cable.

Fiber operating at 1310 and 1550 nm is the technology of choice, although operation at 780 nm is also under consideration. CMOS LSI components running at 50 MHz are typical of the FTTC installations in the field. Optical amplification is also being introduced (see Sec. 11.5).

Providing power in a fiber system is still one of the most challenging issues in bringing fiber closer to the customers; a pedestal that feeds power to several ONUs is considered a practical solution for ensuring power-system reliability, maintainability, and economy.[5] Domicile-based ONUs can use commercial power along with a small secondary battery.

In an ODN network, an *optical line termination* (OLT) entity provides the network-side interface; the output side of the ONU provides the customer-side interface. The physical ODN can use a number of variations, as shown in Fig. 11.1. The passive star, which is known as PON, is the most promising technology for implementing a cost-competitive ODN that accommodates a small number of narrowband users. This is because in PON an OLT unit and a portion of the ODN can be shared by a number of ONUs (refer to Fig. 11.1). PON supports both distributive services as well as interactive services. To support interactive services, both logical and physical bidirectionality must be supported.

At the logical level, two-way services are supported with TDM. In the downstream direction, TDM signals from the OLT that contain all the subscriber information in predetermined time slots, are broadcast to all ONUs connected to the OLT; each ONU accesses only the assigned time slot.*[5,6] In the upstream direction, optical streams from each ONU are transmitted in such a manner as to avoid collision (e.g., random access or other techniques).

At the physical level, a number of optical bidirectional approaches can be used for PON, as seen in Fig. 11.2 and Table 11.1.[5]

11.2.2 Economics

Making fiber access networks cost-competitive with copper-based systems is the primary focus of ongoing fiber technology research. Fiber is already competitive in the feeder portion of the network, where it can be shared among many users. See Fig. 11.3. Given the greater number of distribution network links, it is important for fiber to achieve cost-

* Encryption may be needed to ensure privacy for VOD and other services.

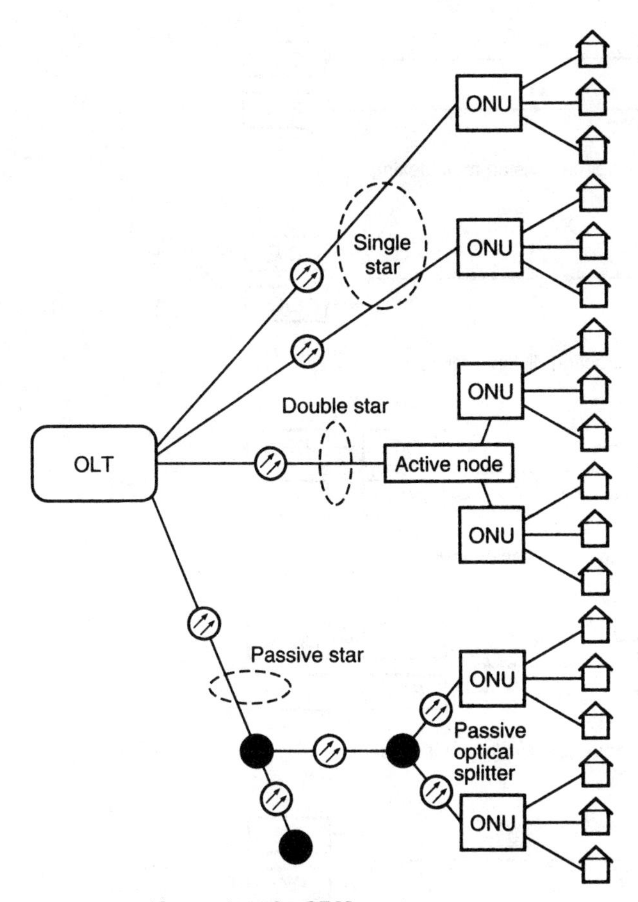

Figure 11.1 Alternatives for ODNs.

parity with twisted pair, *not only for telephony services, but more importantly, for broadband services.** Figure 11.4 depicts the cost of fibering a local loop, based on a number of situations and approaches. As can be seen from Fig. 11.4, as the number of homes sharing the ONU increases, the per-channel cost goes down; however, as the number of users increases, the bandwidth per user could go down (in some implementations). Because the cost of distribution networks is shared by relatively few customers, the selection of an optimal technology (among the set previously described) is critical to the ability of providing cost-competitive video services. Volume production is the key factor

* The reader should be careful in interpreting parity results reported in the literature; usually parity is described only in the context of delivering telephony services; to deliver video, the access facilities are likely to be more expensive.

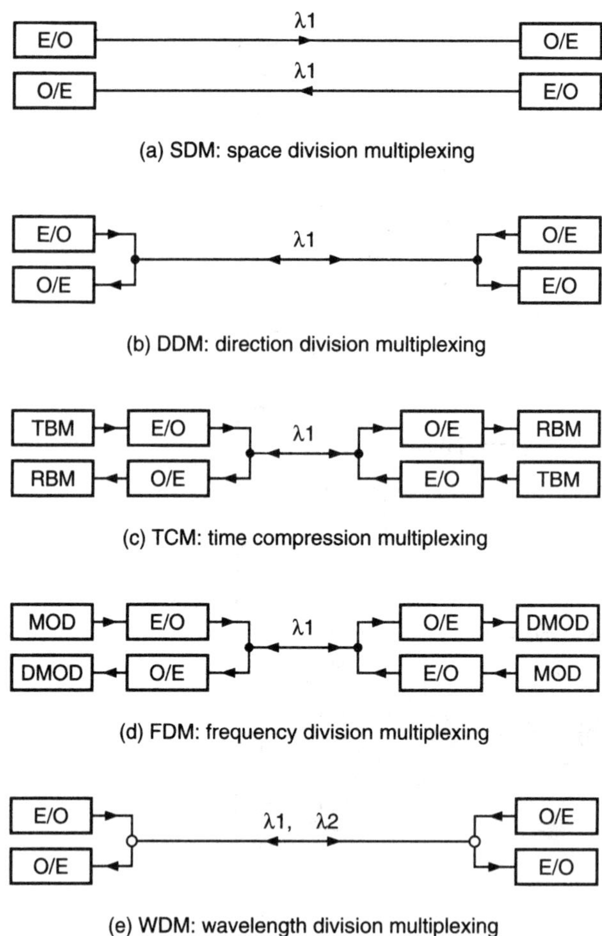

Figure 11.2 Optical bidirectional transmission configurations.

in driving the cost down. Figure 11.4 shows that narrowband FTTC networks can achieve initial cost-parity with a fiber-based DLC using twisted distribution, for large enough ONUs, for providing telephony to a *new-build* suburban area. As seen in this figure, today ONUs achieve parity only when they support 15 to 25 users; smaller ONUs with two to six lines will be economically viable at some future point. These same studies show the following press-time results (for a variety of underlying assumptions):

TABLE 11.1 PON Physical Bidirectional Techniques

Space division multiplexing	Uses two separate fibers for the downstream and upstream directions. Implementation is simple, but the total system cost is slightly higher than that of a single-fiber arrangement.
Direction division multiplexing (DDM)	Realizes bidirectional transmission over one fiber using an optical directional coupler at the same wavelength. Limited application because of technical consideration.
Time compression multiplexing (TCM)	Uses an optical burst-mode TDM technique (also called optical ping-pong compression), where information in each direction is time compressed in the transmit buffer memory and then transmitted over a single fiber. The direction of transmission alternates during successive transmission periods (only one signal is present on the fiber at any one time). The received information is then time decompressed.
Frequency division multiplexing	Uses one wavelength, separating the upstream and downstream signal in the electrical domain. Each direction of the signal transmission is modulated before the electrical-to-optical conversion so that the frequency spectrums of the two modulated signals are separated from one another. At the receiver side, electrical filtering in the demodulator suppresses signals from the other direction that are the result of optical reflections.
Wavelength division multiplexing	Uses two signals with different wavelengths to transmit the upstream and downstream components. Each direction of signal transmission is separated by a WDM coupler and filter.

- Cost of FTTH video and voice services per user: $1850 to $2350
- Cost of FTTC video and voice services per user: $1550 to $2050
- Cost of HFC video and voice services per user: $1350 to $1850

Table 11.2 depicts some combination of services that can be delivered over an FTTC system as compared with the relatively similar architecture of HFC discussed in the previous chapter.* In this table, narrowband services include traditional telephony, H.261-based video telephony, and low-bandwidth data applications (e.g., telecommuting applications). A carrier could choose to deploy either HFC, FTTC, or a mix of the two. Switched video services include VOD, NVOD, IVOD, IMVOD, etc. Alternative *a* provides all services over an FTTC system; this alternative does not use HFC systems. Alternative *b* employs a dual HFC and FTTC network where the HFC system is used for broadcast video and the FTTC system is used for VOD services. Alternative *c* employs a dual HFC and FTTC network where the HFC system is used to deliver broadcast services and VOD and the FTTC is used (ini-

* Comparisons with FTTH are discussed in Sec. 11.4.

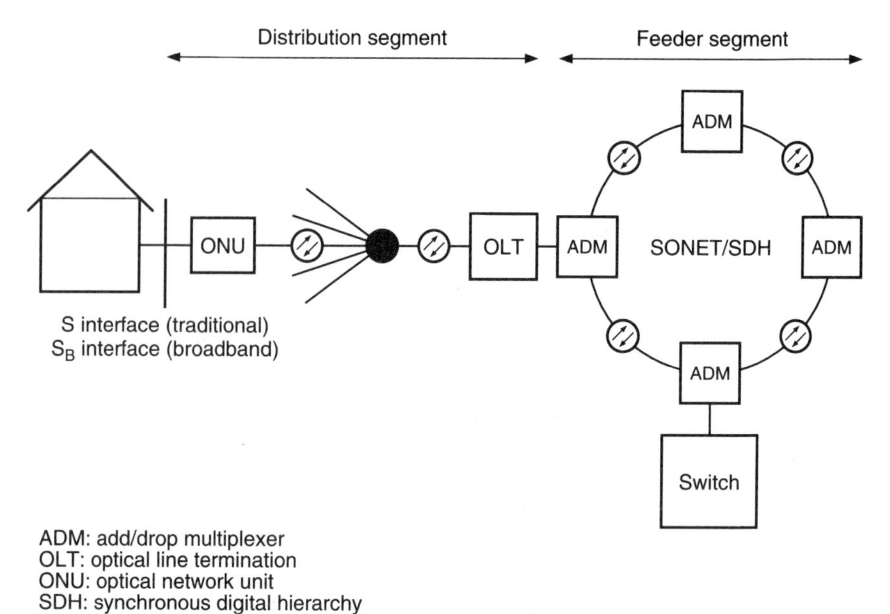

Figure 11.3 Feeder-distribution networks.

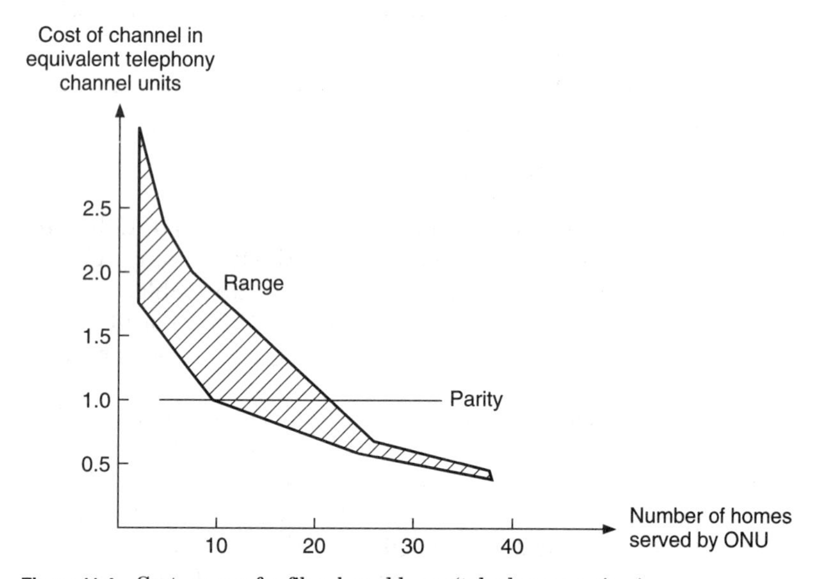

Figure 11.4 Cost ranges for fiber-based loops (telephony services).

TABLE 11.2 Service Mix for Two Potential RBOC Strategies

Alter-native	New plant or expansion of existing plant				Overlay (video only)
	a	*b*	*c*	*d*	*e*
HFC		Broadcast video services	Broadcast video services Switched video services	Narrowband services Broadcast video services Switched video services	Broadcast video services Switched video services
FTTC	Narrowband services Broadcast video services Switched video services	Narrowband services Switched video services	Narrowband services		

NOTE: Another option is ADSL, which supports integrated delivery of voice and video services.

tially) to carry narrowband services. Alternative *d* delivers all services over the HFC system (this architecture may be employed by a cable TV company). Alternative *e* involves no FTTC deployment (where, for example, a good twisted-pair plant is already in place and needs to fiber upgrade), but where an RBOC wishes to deliver video services; this entails the deployment of an overlay network using HFC.

Economic analysis of these and similar alternatives leads to the conclusion that no one architecture is optimal for all levels of service penetration, all types of video services, and all demographic-geographic situations. Some of the factors that come into play in an economic analysis include service factors, demand, technology costs, and other location-specific factors. Some observations that have been variously reported and/or are generally understood are as follows:*

1. In suburban environments, the construction of a new all-digital HFC system (see Chap. 10) using composite twisted-pair or coaxial cable and a (new) FTTC have similar deployment costs (i.e., within 20 percent), given similar service and demand factors. (For the incremental upgrade shown in Chap. 10, Sec. 10.2, about 35 percent of the cost is for the setup box, 37 percent is for the coaxial cable, 8 percent for the video ONU, 11 percent for the CO-to-node facilities, and 10 percent for CO expenses. The coaxial costs go down as the penetration increases, but the other costs remain approximately the same.)

* Detailed cost analyses require considerable documentation. It is not the purpose of this chapter to undertake such analyses.

2. Customer usage of PPV and equivalent services, as well as the cost of the setup box handling D-A conversion, are important parameters in system cost differentiation.* Planners are advised to apply conservative views in terms of PPV rates (as discussed in Chap. 2). Additionally, there is the factor of overall video service penetration (take rate). RBOCs could conservatively expect a 15 to 35 percent penetration, with an average of 25 percent. Taking this conservative view and assuming the use of composite drops for the HFC systems, it can be shown that an analog HFC system, an all-digital HFC system, and an FTTC system have similar deployment costs. Analysis shows that, as the penetration increases beyond the range just given, the mixed analog-digital HFC is cheaper than an FTTC architecture, given current costs. Along another dimension of these complicated cost analyses, as PPV rates increase, the all-digital alternatives are more cost-effective, even at low overall service take rates (at higher take rates, the costs are similar).

3. As noted in Table 11.1, FTTC can be used with TDM techniques (e.g., TCM), as well as with WDM techniques. Given costs at press time, FTTC-TDM is cheaper than FTTC-WDM for most ranges of applicable parameters (take rates, PPV rates, etc.). However, if an FITL system was already in place, then WDM might make some sense. Ultimately, it appears that it is better to deploy narrowband and broadband equipment simultaneously if possible.

4. For localities where the deployed twisted-pair plant is in good condition, the HFC can be installed as an overlay. For communities requiring plant upgrade or for new installations (e.g., a new development), the HFC can be deployed side-by-side with FITL to support video services without having to upgrade the narrowband FITL system. Cost-savings can be achieved by powering the narrowband FITL system over the coaxial cable used for video and by using composite coaxial/twisted-pair distribution plant drops.

5. The optimal number of homes per ONU for an all-digital HFC system is in the vicinity of 100, using optoelectronics costs for 1996; the optimal number for a hybrid analog-digital HFC system is in the vicinity of 130 homes.†

6. As discussed in Chap. 10, a per-subscriber cost increase is experienced when two coaxial cables are installed to each domicile in order to

* In analog HFC, the video information as well as the menuing information are already in analog form so that conversion equipment is not required; digital HFC systems and FTTC systems need such conversion.

† Higher costs (e.g., 1994 costs) increase the number of served homes; such numbers then become comparable to those used by the cable TV industry for the FSAs.

support a larger number of channels. In such a situation, FTTC systems compare more favorably.*

7. Based on the fact that fewer setup boxes are required in an all-analog HFC system, it has a lower activation cost than any of the other systems (all-digital HFC, a hybrid analog-digital HFC, an FTTC-TDM, and FTTC-WDM). This results in about 20 percent less expensive service activation.

8. An all-analog HFC system, it has a lower maintenance cost than any of the other systems (all-digital HFC, a hybrid analog-digital HFC, an FTTC-TDM, and FTTC-WDM). Further, the FTTC-TDM has a lower operation cost than FTTC-WDM.

9. The cost of laying cable is about $4.75 per meter for underground installation and $3.75 per meter for aerial installation; if the network is later rebuilt, the cost of aerial installation remains the same, while the cost of pulling underground cable increases to $20.75 per meter because of the high cost of rewiring an existing neighborhood.[4]

As seen in Table 11.2, a variety of integrated or overlay network options for delivering telephony and video services exist. In summary, in terms of telephony and video services for rehabilitated or new-build areas, integrated networks (such as HFC and FTTC) have a lower initial cost than the overlay video-only networks. However, an overlay network may be deployed for localities where the narrowband twisted-pair network is relatively new or still in good condition.

11.3 Examples of FTTC Systems

A number of vendors manufacture FTTC equipment, including ADC, Alcatel, AT&T, DSC, Northern Telecom, and BroadBand Technologies (BBT), particularly with an eye to video services. Figures 11.5 and 11.6 depict two typical commercially available configurations by ADC and BBT, respectively. (BBT's system is being used in Bell Atlantic VDT trial in New Jersey.)

Figure 11.7 depicts, for illustrative purposes, one example of an FTTC system manufactured by Alcatel.[2] The system uses analog-optical amplifiers, described in more detail in Sec. 11.4; Figure 11.8 depicts the amplifier technology employed. This FTTC system distributes TV and audio signals from the *headend* (HE) to the subscriber via two networks: the *transport network* (TN) (also known as the backbone net-

---

* As a side note, the cost of an HFC network is more sensitive to the number of broadcast analog channels that need to be carried than to the number of digital channels; as discussed in Chap. 4, this arises from the fact that numerous digital channels can be packed in one 6-MHz channel.

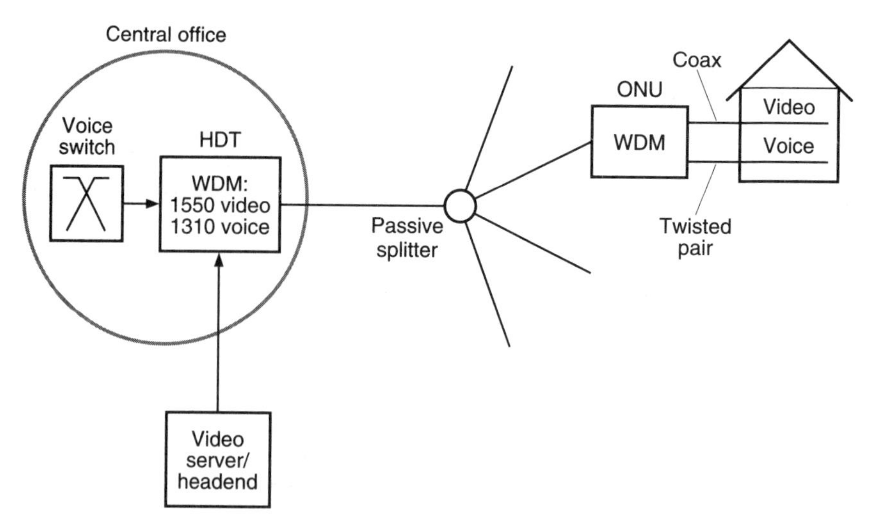

Figure 11.5 FTTC system by ADC (simplified view).

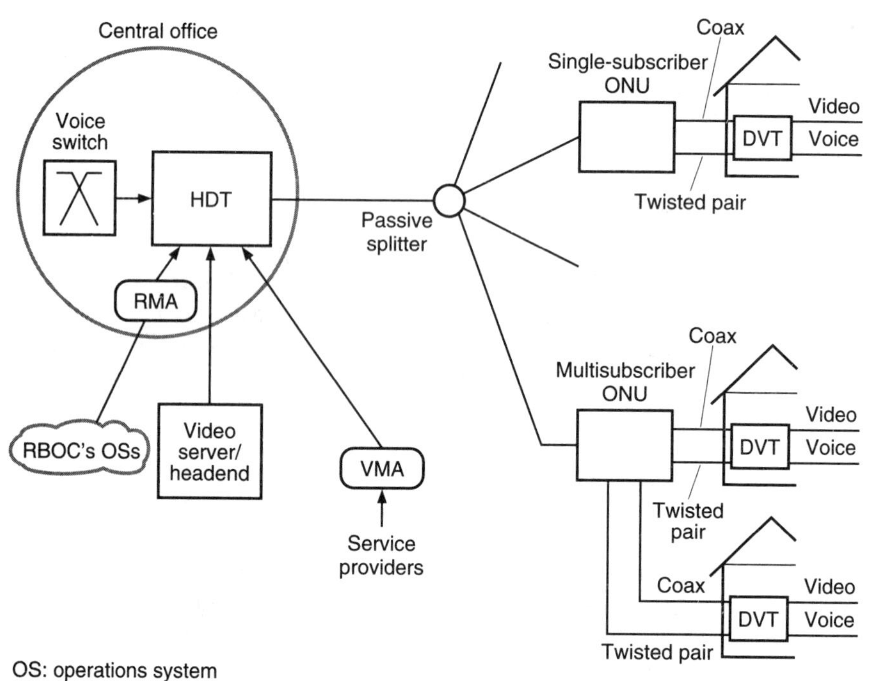

OS: operations system
RMA: remote administration module
VMA: video administration module
DVT: digital video terminal

Figure 11.6 FTTC system by BBT.

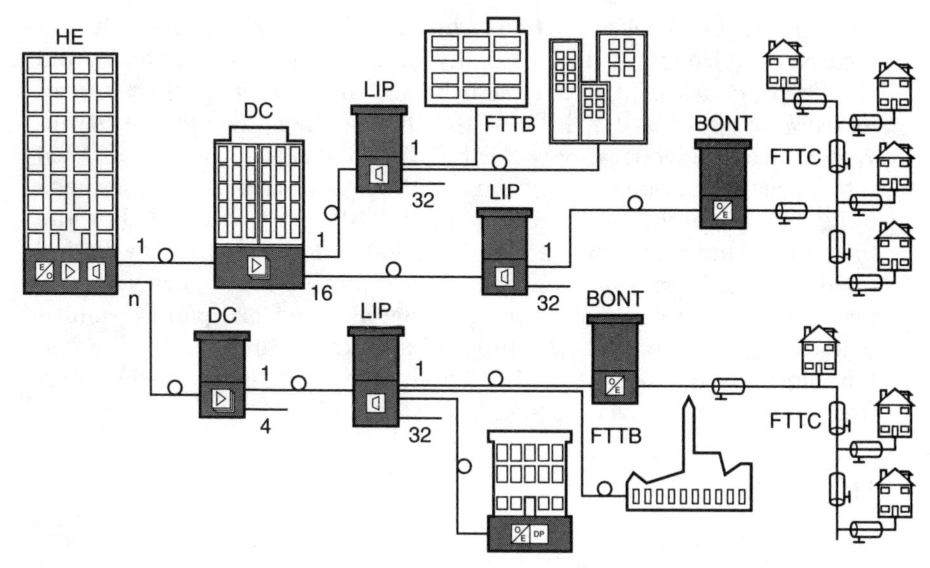

Figure 11.7 Example of FTTC system (Alcatel).

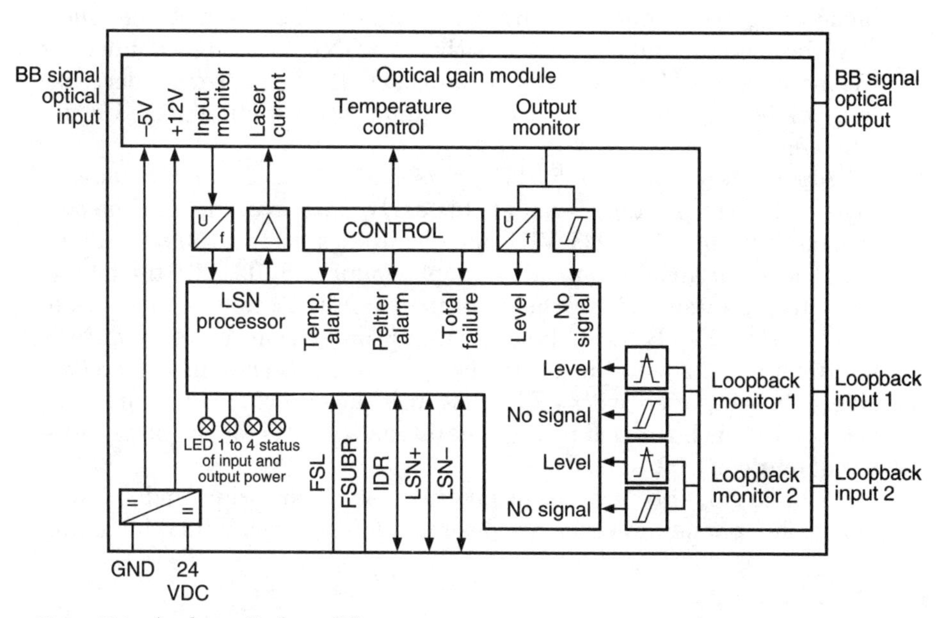

Figure 11.8 Analog optical amplifier.

work in this text) between the HE and the *distribution center* (DC) (in a carrier environment, this typically would be the central office), and the *access network* (AN) between the DC and the *broadband optical network termination* (BONT). At the BONT, the optical signal is converted into an electrical one. For FTTC applications, the signal is fed into a passive coaxial distribution and drop network to the subscriber's building where the *demarcation point* (DP) is established. For the fiber-to-the-building application, the DP is allocated with the BONT inside the subscription building. The typical fiber topology in the transport network is a point-to-point signal transport; point to-multipoint signal transport is the standard application in the AN. The system supports up to 50 AM-VSB TV channels in the 47- to 600-MHz frequency range (to be extended later to 860 MHz).

11.4 FTTH

FTTH is recognized as the ideal target fibered distribution network, since all homes have access to the kind of capacity needed to support the interactive broadband services now evolving. An FTTH architectural example, from Ref. 3, is shown in Fig. 11.9. Although many people at this time believe that FTTC is more cost-effective than FTTH, some argue that the opposite is true. Conservative analyses have shown that for an average 1.2 telephone lines per home, the deployment cost for an FTTH supporting narrowband* service is about $400 (about 30 percent) higher than the cost of an FTTC. Such studies show that the incremental cost of dedicating the ONU to a single domicile is the major cost differential, even if the cost of the ONU were to decrease due solely to volume (e.g., to the volume of one million ONUs per RBOC).

Proponents claim that FTTH offers cost and maintenance advantages over HFC systems even at this early time, even though the current generation of FTTH VIU optoelectronics costs $1500 to $2000. Proponents in favor of immediate deployment of FTTH cite the following reasons in support of their argument on the cost-effectiveness of FTTH[3]: with FTTH, ONU housing requirements can be relaxed; testing functions for metallic interfaces to customer terminals can be considerably simplified with FTTH because the interface lines are very short; FTTC is burdened by high maintenance costs for the large number of outdoor ONUs needed.

Other arguments are based on the cost of powering and maintenance. Telephone networks are designed for high reliability, including

* The cost of delivering broadband services would be higher.

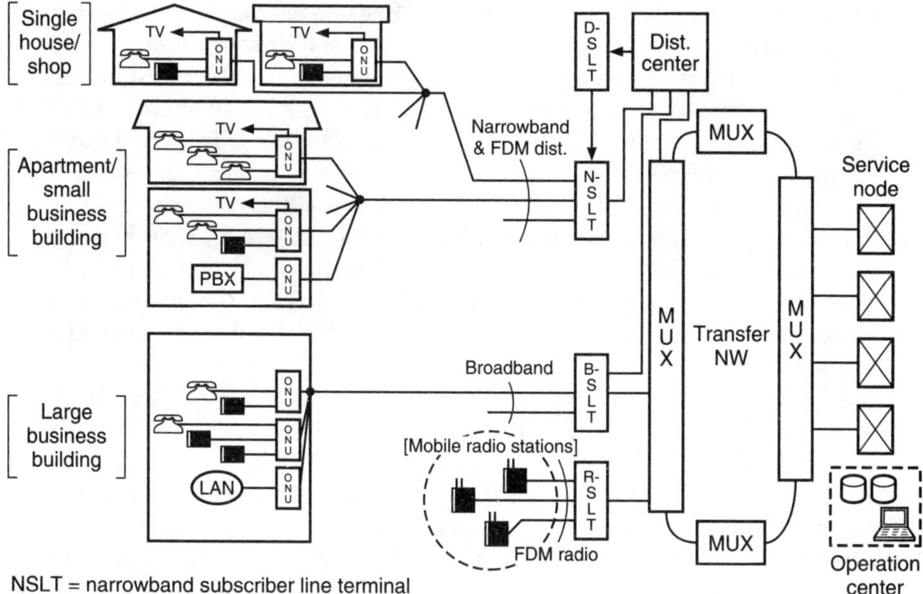

Figure 11.9 FTTH system being investigated by NTT.

separate powering. This reliability carries a cost of as much as $115 per watt over the life of the system (this includes power-conditioning equipment, batteries, and actual electrical costs).[4] The DLC continuously uses 1 to 4 watts per line; for a system with 96 lines, that could add $44,000 to the cost of the system. One way to deal with this cost is, therefore, to take the cable TV approach by providing consumers a service box which is locally powered.

Studies claim that FTTH architectures can save approximately 25 percent, or approximately $23 per subscriber per year in maintenance costs; proponents state that FTTH networks could cut maintenance costs even further.[4] It is well known that coaxial networks are more prone to failure than twisted-pair networks because of the cascade of amplifiers that is required. Hence, replacing the coaxial with fiber could, in theory, reduce maintenance costs. However, in the type of FTTC networks previously discussed, the coaxial may be only 100 ft long and have no amplification; therefore, the maintenance savings of FTTH compared with FTTC are questionable.

Proponents argue that the advantages of FTTH overcome the cost penalty of the higher number of ONUs and that FTTH is more desirable as the infrastructure for future services. For just a few telephone lines per node, given current prices, fiber-based loop systems are almost as cost-effective as metallic loop systems; this situation is typical of a suburban environment. In a high-density urban environment, FTTB may be cost-effective immediately, since a single fiber with a single passive splitter at the entrance point can be used. In mixed environments of high-density buildings and single or double-residence homes, fiber loop systems can still be more cost-effective than metallic loop systems, when the total cost for telephone lines to all customers in a fixed distribution area is considered. If many apartment buildings are included in one area, cost-savings gained from the apartments compensate for the higher costs incurred by the single houses in the same area.

As noted earlier, FTTH supports higher bandwidth (e.g., 622 Mbps) than FTTC systems (e.g., 16 Mbps). An interesting question is: What is the cost comparison between these two systems if they both carried the same per-user bandwidth, e.g., 16 Mbps? Conservative analyses using component costs at press time show that even when the FTTH system supports lower bandwidth it is still 25 to 35 percent more expensive than the FTTC network.

Early U.S. entrants in the FTTH arena include ADC Telecommunications, which has begun selling Homeworx that can be implemented as an FTTH network, and Optical Communications Corp. At press time Pacific Bell was launching an FTTH trial in a high-income residential development in Newport Beach, Calif., to study maintenance and provisioning issues; initially, the trial supported only telephony and data applications. (The capital costs are not optimized because the carrier is using electronics designed for an FTTC network.)

11.5 Optical Amplification Technology

As mentioned in Chap. 4, there is interest in optical amplifiers in the context of all fiber-based distribution systems (FTTN, FTTC, and FTTH). Two technologies are now available: *semiconductor optical amplifiers* (SOAs) and *erbium-doped fiber amplifiers* (EDFA). Optical amplification technologies are now seeing commercial introduction, although this technology is still expensive. EDFAs are penetrating some cable TV systems (headend applications) and in long-haul (e.g., underwater) telephone network applications. SOAs are still in the laboratory stage. Although they are cheaper than the EDFA ($7000 to $15,000 compared with $30,000 to $70,000, with the expectation that the cost will be down to $2000 by 1997), they are low-power devices with a typical output of 6 dBm. In addition, they introduce crosstalk and nonlinearities.

EDFAs operate at 1550 nm (note that most of the fiber now in place operates at the 1310-nm range). Optical amplifiers work by using a laser at one wavelength (say 880, 980, or 1060 nm) to energize the erbium-doped fiber, which emits energy in response to an incoming signal; when the energy from the pump laser is carried down the fiber, it can be used by the next amplifier to regenerate the signal.[7] Development of EDFAs have stimulated the interest of 1550-nm signal sources for analog cable TV transmission. Since in many geographical areas, standard single-mode fibers (Class 3A, nondispersion shifted) are already installed or planned for 1300-nm applications, these fibers also have to be used for 1550-nm transmission in the future; hence, the amplification equipment being developed must be compatible with the already installed single-mode fiber. The amplification process in the optical fiber (doped with rare earth Er ions) is stimulated by a pump laser that is coupled via the pump coupler together with the optical input signal to the erbium fiber. See Fig. 11.10.[8] Erbium-doped fiber amplifiers propagate both signal and pumping light simultaneously through an optical fiber to amplify the incoming signal light. When the Er atoms in the fiber are pumped to a high energy level by the excitation light, induced emission occurs, and the input signal light is amplified.[2]

Optical amplifiers, which are very linear and add only very low levels of noise, can be used in three ways:

1. *Power amplifier mode.* The amplifier is used to increase the strength of the laser signal at the source, enabling it to go for several hundred miles.
2. *Regenerator mode.* The amplifier is used to increase the optical signal after it has traveled a certain distance and it has become attenuated; this approach can also be used in conjunction with taps that go from one to many fibers.
3. *Preamplifier mode.* The amplifier is used to boost the incoming signal before it is handed off to the photodetector.

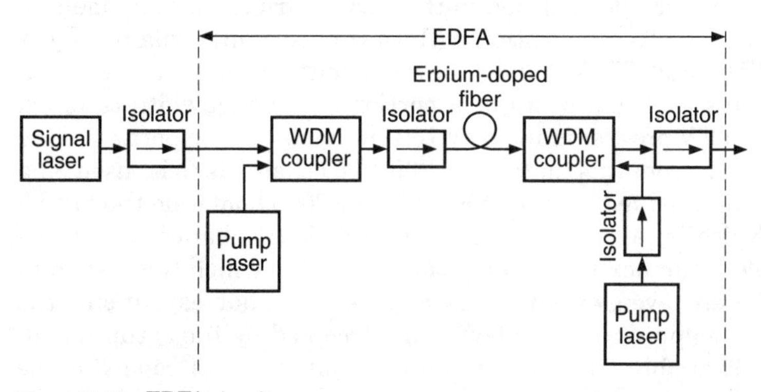

Figure 11.10 EDFA circuit.

Optical amplifiers have found early applications for rural cable operators where the cable has to go 10 to 20 miles to meet a customer. Using 50 or so cascaded RF amplifiers is undesirable, not only from a signal quality point of view, but also from a reliability point of view. A link with 50 cascaded RF amplifiers can have several failures a year. Optical amplifiers enable the cable operator to boost the signal (say to 40 mW) to reach the intended users, even across a tap of a dozen or more fibers. One can retain continuity through a local failure of power. This works by operating the amplifiers at saturation and thereby obtaining enough power to sustain a 5-dB drop through an amplifier that has lost external power. The amplifier without power maintains optical continuity and can use the pumping energy from the previous amplifier in the chain.

Externally modulated fiber-optic systems with optical amplifiers and required cooling capabilities can be expensive at the component level (laser can cost $20,000 to $70,000) but can be cheaper at the total system level when considering the elimination of the RF amplifiers on a coaxial system or optoelectronic regenerators on a fiber-optic system, and the improved availability. An optical amplifier has considerably fewer integrated circuits than an electro-optical regenerator. AT&T has a number of trials underway including underwater cable applications. AT&T's applications are long-haul; however, the tests are aimed at determining, among other things, how well it can amplify WDM signals. Some expect to see field deployment of the technology for 2.4-Gbps systems in 1995.

Proponents claim that for long runs (e.g., between headends) optical amplifiers are already cost-effective today. Others see the technology not being consistent with the move from broadcast video to narrowcast VOD services, since in this scenario, the signals are meant for fewer homes and do not need, ostensibly, to be split off at a tap point (optical amplifiers maximize the number of homes that can receive signal from a transmitting laser). If this were actually the case, the technology would not see much penetration in the local distribution component of VDT or other HFC applications. However, currently, many of the FFTN, FTTC, and FTTH (under development) systems use a multiplexed video signal that contains a portion of the bandwidth reserved for broadcast TV programming (say 100 channels), plus a portion of the bandwidth that contains up to (say) 200 VOD channels to be used on a first-come, first-served basis of 500, 1000, or 2000 homes on the logical cluster. Typically, all users receive all the VOD channels, but the addressable setup box selects only the channel to which the residence is entitled (here, even with ATM, all cells reach all homes, but only the cells with the appropriate VPI-VCI are accepted by the setup box in question). Therefore, the idea of splitting off identical copies of the same video signal still makes sense, since narrowcasting in these systems is implemented at the *logical* level, not at the *physical* level. (In a system such as ADSL, the signal reaching each user is individually and uniquely meant for that user, since here one has a traditional switching arrangement.)

Yet another argument against optical amplification in the local loop is that laser now puts out enough power (say 15 mW) to enable passive splitting of the signal among multiple fibers (for a system such as PON; see Chap. 4).

References

1. Special Issue on Fiber-Optic Subscriber Loops, *IEEE Communications Magazine,* vol. 32, no. 2, February 1994.
2. W. Schmid and U. Steigenberger, "A Fiber in the Loop Cable TV System Using Analog Optical Transmission with Optical Amplifiers," *Electrical Communication,* 3rd Quarter 1993, pp. 248 ff.
3. T. Miki, "Toward the Service-Rich Era," *IEEE Communications Magazine,* February 1994, pp. 34 ff.
4. G. Lawton, "Fiber's Reach: How Far, How Soon, How Much?" *Communications Technology,* December 1993, pp. 54 ff.
5. Y. Mochida, "Technologies for Local-Access Fibering," *IEEE Communications Magazine,* February 1994, pp. 64 ff.
6. A. A. de Albuquerque, et al., "Field Trials for Fiber Access in EC," *IEEE Communications Magazine,* February 1994, pp. 40 ff.
7. G. Lawton, "Optical Amplifiers: Ready For Prime Time?" *Communications Technology,* November 1993, pp. 28 ff.
8. C. E. Holborow and C. J. McGraph, "Promising CATV Technologies," *Communication Technology,* December 1991, pp. 22 ff.

Chapter

12

Other Distribution Systems

This chapter looks briefly at existing and evolving video distribution systems that represent competition to both cable TV and RBOCs' VDT systems. As a group, these technologies are called *wireless cable*. In general, these systems require a smaller capital outlay on the part of the service provider (about half of cable TV's expense of $800 to $1200 per subscriber[1]); in addition, these systems can be used by customers not served by cable. Their presence in a community has the effect of stimulating competition. However, wireless operators have had some difficulties in obtaining access to video programming sources, since the various distributors may already have exclusive rights in a locality with the entrenched cable TV provider. Also, there is the issue of limited bandwidth in the available spectrum, although video compression and digital modulation of the carrier using QAM or SVB techniques (discussed in Chap. 4) has the potential of relieving the bottleneck.* Usually, it is more difficult to support I-TV services unless a separate return path (e.g., a shared packet radio channel) is provided.

12.1 Direct Broadcast Satellite (DBS)

DBSs are newly launched satellites of sufficient power that the signal can be received using only a 1- to 2-ft antenna (the amount of received signal in a satellite system is a function of, among other things, the satellite transmit power and the area of the receiving antenna). DBS is a high-powered satellite operating with 200-W transponders, in the 12-GHz band (K_u-band). In traditional cable TV systems, the antenna

* This is the same reason that cellular telephone systems are migrating to a digital TDMA scheme.

that feeds the cable system is large (e.g., 30, 15, or 10 ft) and is, therefore, located in some antenna farm outside town, or in an out-of-sight location. About 2.8 million end users have installed their own 10-ft antennas on their property (e.g., backyards, roof). In a DBS system, the antenna can be conveniently located at the consumer's site, indoors or outdoors. Round or 12-in-plate antennas can be used. These devices can be placed on a windowsill, coffee table, etc., as long as there is a line of sight between the antenna and the satellite. See Fig. 12.1.

DBS video delivery now directly competes with terrestrial cable TV. The consumer can purchase the antenna, receiver, and descrambler equipment from an appropriate outlet for about $750 (expected to decrease to $500 by 1996). The programming is purchased by subscription

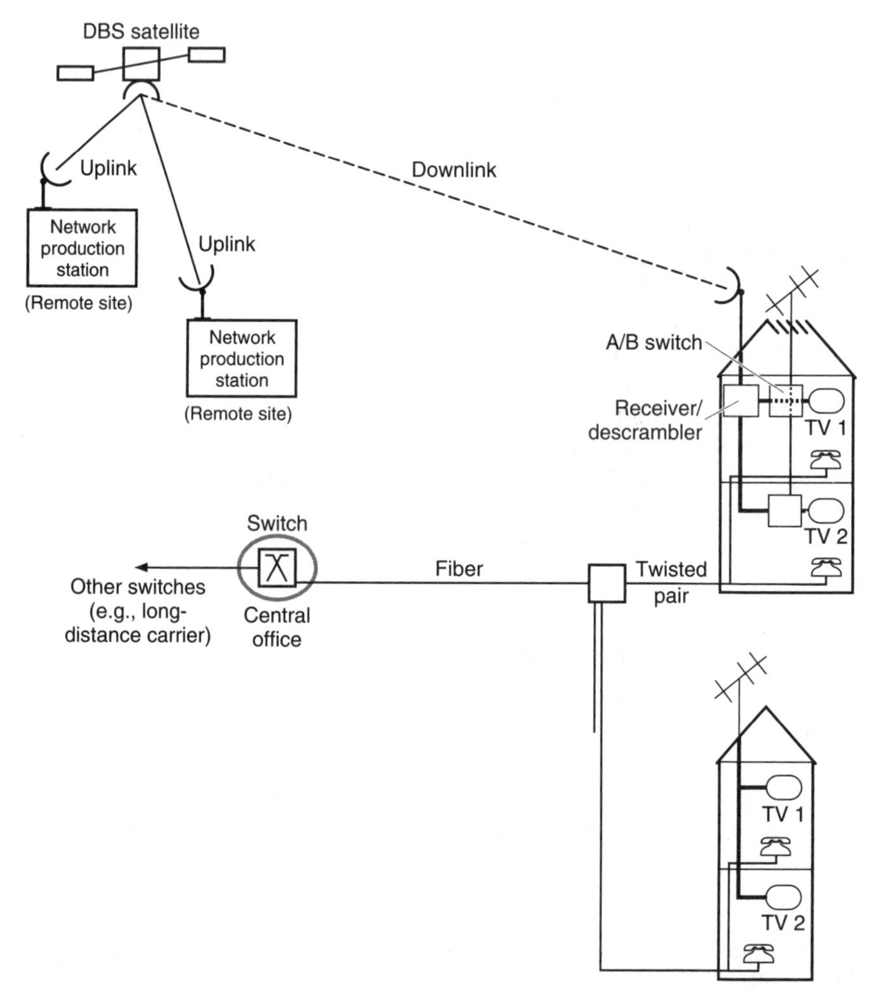

Figure 12.1 A traditional DBS system.

from a program provider. Typical monthly fees range from \$15 to \$50. Japan has used DBS for a number of years to deliver HDTV to homes and businesses. U.S. service started in 1994. The United States service will be available from two colocated Hughes satellites located at 101 degrees west. Nine DBS licenses have been awarded so far by the FCC. From 100 to 200 channels will be available, depending on the video compression technique used; PPV services are also supported. CLI is building the compression systems for DirecTV, which is one of the two services currently (planned) to be carried by DBS, based on MPEG-2.

Although DBS has a large initial capital requirement, the life expectancy of a satellite is 12 to 15 years. The operator does not have to capitalize on the receive equipment and the setup box because this is done by the consumer. Within its footprint, the reach of DBS is ubiquitous, except for extreme cases where the consumer does not have a clear southern exposure. Two factors cited as drawbacks of the service are that it does not (easily) support I-TV and the consumer does not receive local programming.

Early providers of DBS services included DirecTV (a unit of GM Hughes Electronics), United States Satellite Broadcasting (a division of Hubbard Broadcast), and PrimeStar (owned by a consortium that includes the five largest cable TV companies). Users of DBS have to use a telephone line to order a PPV service using their remote controls and electronic programming guides. In addition, it is not easy to hook up the service to multiple TV sets, and the number of distinct simultaneous channels that can be watched is limited to one or two. DBS equipment providers appear to have little appreciation for the fact that consumers have multiple TVs and that different members of a household may desire to watch different programs simultaneously.

According to some observers, DBS providers will become players in the interactive TV arena, but will remain small players for the foreseeable future.

DirecTV was planning to use about 70 of its 150 channels for PPV applications, with showings every 15 or 30 minutes. Subscribers could order from an on-screen menu, and the setup box uses the store-and-forward technique discussed in Chap. 13 to relay the order. DirecTV runs the movies off videotape (Sony Digital Betacam[2]). Consumers were initially able to buy the systems at Sears and Circuit City.

12.2 Multipoint Distribution Service (MDS)

MDS was created in the early 1960s as a local distribution method for closed-circuit business communications (e.g., video teleconferencing). Later, the FCC increased the bandwidth from 10 to 12 MHz. As a consequence of this decision, the FCC received many applications from providers of entertainment video. The first use of MDS for pay programming occurred in the early 1970s. MDS is regulated as a common carrier rather than as a broadcaster. As a result, any number of parties can lease transmit time from the licensees, and can purchase and resell programming. As can be inferred from Fig. 12.2, the setup cost for MDS is relatively low compared to that of cable TV because it only entails building an appropriate tower. This approach requires the customer to have a *line-of-sight* (LOS) view of the broadcasting tower.

12.3 Multichannel MDS (MMDS)

In 1983, the FCC relaxed the rules on spectrum allocated to ITFS (Instructional Television Fixed Service), allowing educational institu-

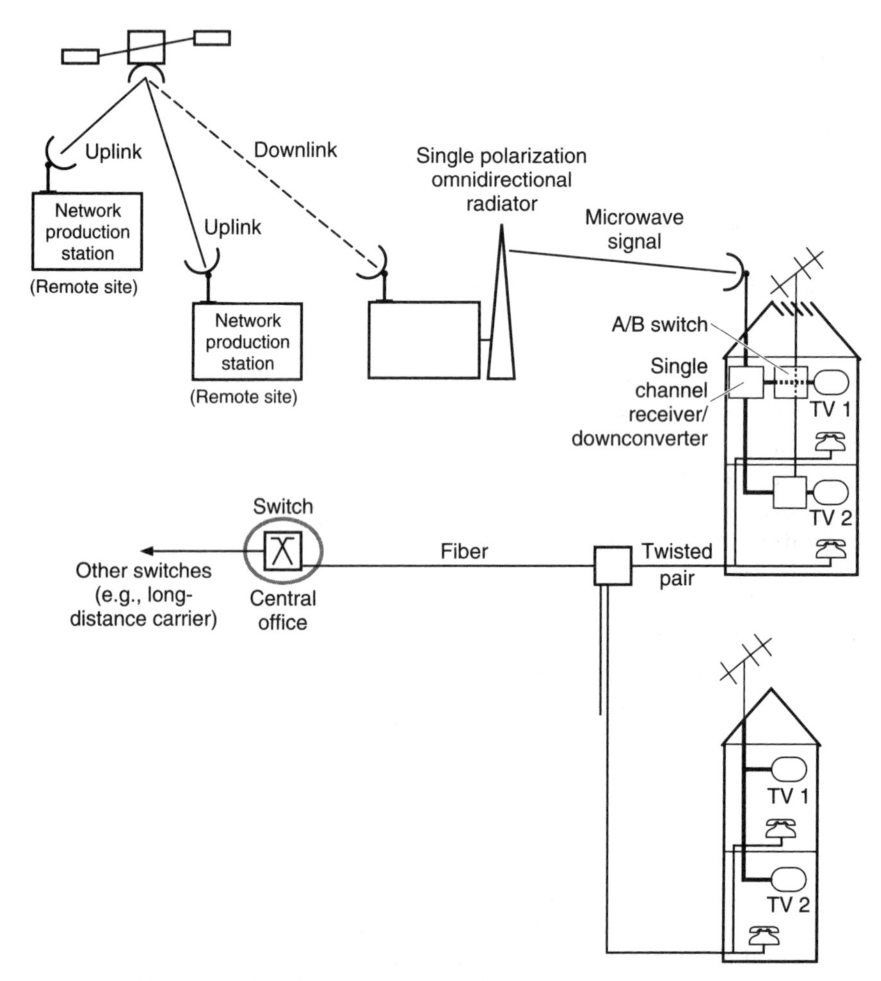

Figure 12.2 Multipoint distribution service.

tions to lease their excess capacity in the 2.5- to 2.7-GHz band. This enabled MMDS to come into existence. An MMDS operator can assemble up to 33 channels in a market through leasing channels from nonprofit institutions. With compression, they should be able to obtain 150 channels or more, including PPV. Because of its relatively low-cost structure, a wireless operator can achieve margins similar to that of the cable TV industry on 20 percent penetration.[1]

Approximately 17,000 applications to the FCC for MMDS have been received. Figure 12.3 depicts an MMDS system. The programming is received from satellite feeds and is broadcast to local subscribers from a central point, in a manner similar to that of ordinary broadcast television (with an omnidirectional antenna). No franchise licenses are

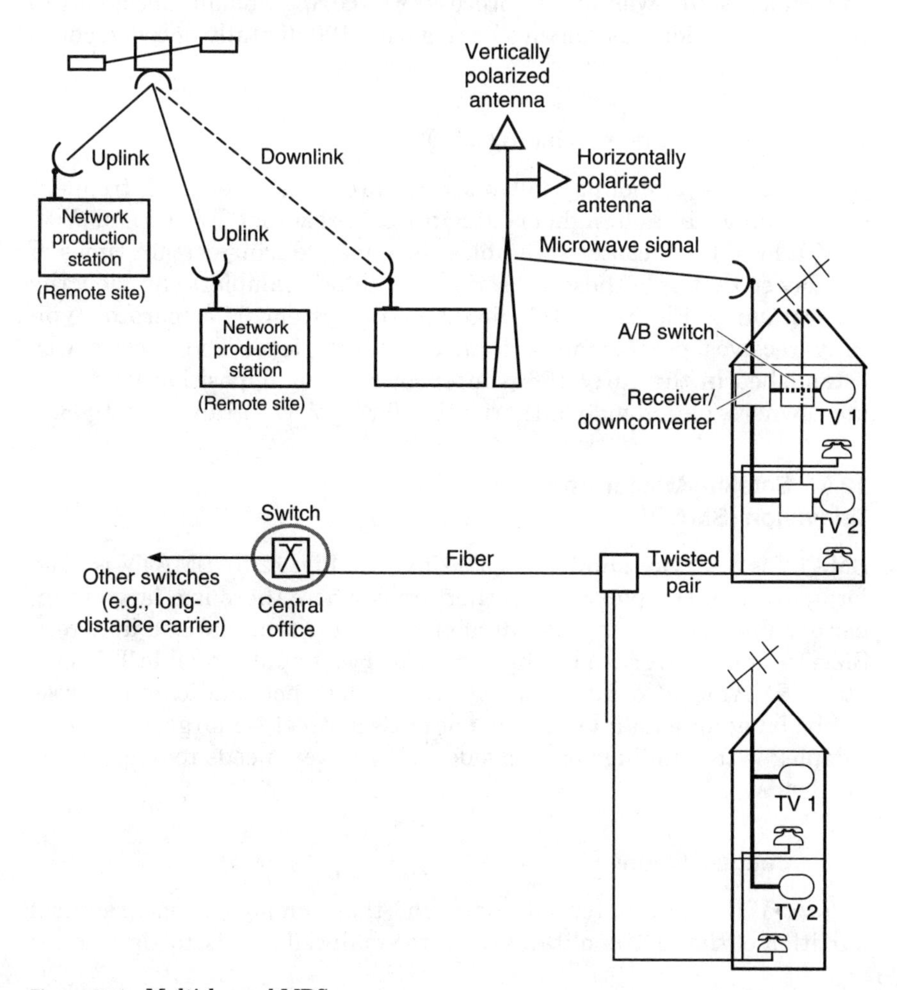

Figure 12.3 Multichannel MDS.

required to operate such a system. As for MDS, this approach requires the customers to have a LOS view of the broadcasting tower; occasionally, there are problems in seeing the tower because of terrain considerations. This service is usually cheaper than a terrestrial system, with rates in the $15 to $20 per month range, which is as much as $10 less than the cost from other providers.[3]

There are over 100 MMDS systems in the United States providing services to 500,000 customers. Although these systems typically support two dozen channels at this time, there are discussions about supporting 150 to 200 channels in the future using compression and QAM-VSB techniques. Typically, PPV service is initiated over a telephone line. As an example of such a system, Cross Country Cable of Riverside, Calif., was in the process of securing enough spectrum to reach four million Los Angeles homes with 100 digitally encoded channels by the spring of 1995.[3]

12.4 Subscription Television (STV)

STV is a one-channel pay video service broadcast over UHF frequencies (Channel 14 and higher) using a local broadcast TV station. With STV, a local broadcaster scrambles its signals during certain times of the day so that only those subscribers with descramblers can view the programming. Figure 12.4 depicts an STV system. The operator typically receives programming from a satellite feed. The service was established in the early 1960s after overcoming opposition from theater owners; nationwide authorization for STV was granted in 1968.

12.5 Satellite Master Antenna Television (SMATV)

SMATV is a service similar to cable TV (see Fig. 12.5). SMATV is used for apartment complexes and other areas where there is a large number of subscribers (in a small area of 200 to 400 tenants) and in hotels. SMATV is not governed by the same rules that apply to cable TV operators. SMATV (also called private cable) does not (and cannot) cross public land; for example, it does not cross a street—a large apartment complex with buildings on two sides of the street needs to employ two headends.

12.6 CellularVision

CellularVision is a patented approach* that employs a topographical partition of the metropolitan area into small cells of six to eight miles

* Patent held by CellularVision, Freehold, N.J.

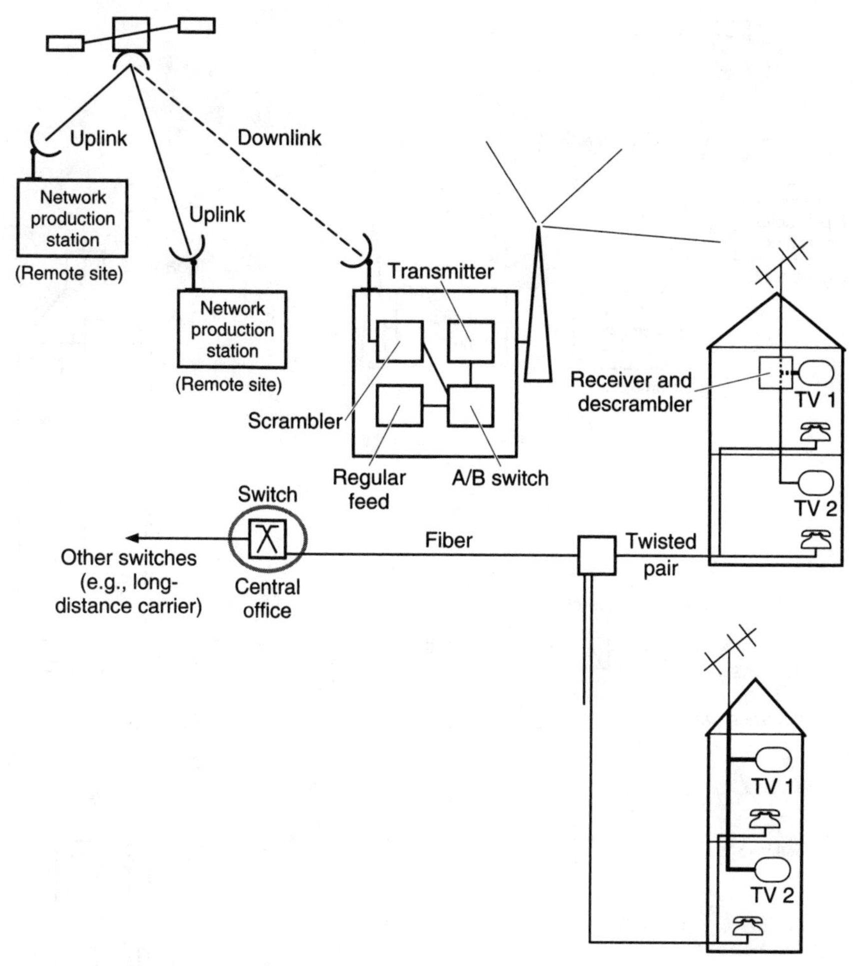

Figure 12.4 Subscription TV.

to achieve frequency reuse (as is the case in cellular telephony). Additionally, different polarizations can be used to increase the number of transmission options. Frequency reuse implies that more throughput can be derived than would otherwise be possible with the same set of frequencies. The transmission band is in the 27.5- to 29.5-GHz range (also known as K_a). Initially, a 1-GHz signal (from which one easily and cheaply obtains 1 Gbps) is employed. At the K_a frequencies, the signal has high directionality. The consumer employs a flat antenna of a few square inches. The service supports two-way communications. To provide interactive applications on CellularVision's network, however, consumers need two transceivers. Each channel consists of an FM signal occupying a 20-MHz band. Two-way communications can be

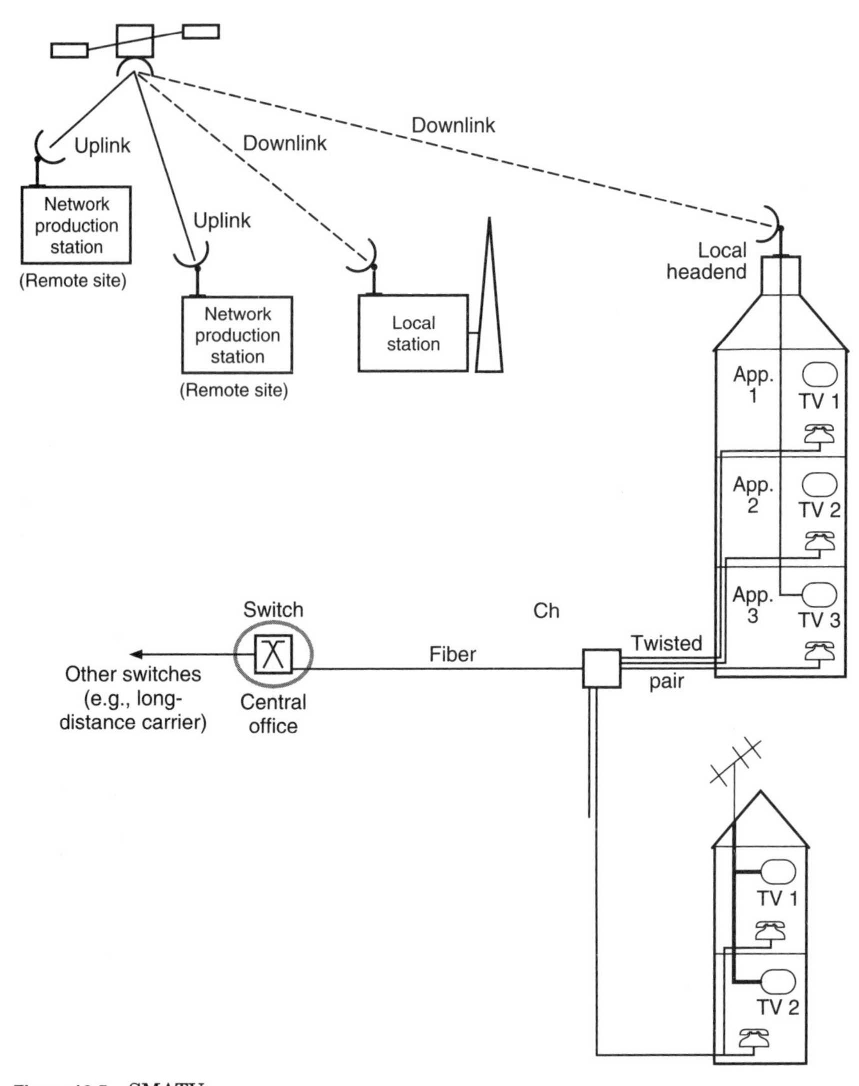

Figure 12.5 SMATV.

inserted between video channels for reverse transmission. The programming will be delivered to cell sites via fiber, coaxial, or a radio link at the same 28-GHz radio frequencies. The FCC has granted CellularVision a license to operate a national microwave cable network.[4]

A trial that started in 1992 was underway at press time in Brooklyn, N.Y. In the summer of 1994, the service was extended to Queens and Midtown Manhattan.[5] Bell Atlantic was also planning some CellularVision tests. The technology supports 49 channels of broadcast TV, data, and videoconferencing for users within three miles of the transmitter. In addition to Bell Atlantic, Ameritech, BellSouth, NYNEX, and U S WEST have all taken an interest in the potential of the 28-GHz radio technology (FCC rules on the use of this spectrum are expected for late 1995 or 1996).[6,7] At press time, however, the FCC had slowed down its approval process due to the fact that the 27.5- to 28.5-Hz band is also being considered for the global satellite-based wireless communication services supported by Iridium and Teledesic.

12.7 Upgrades to Traditional Broadcasting Systems

As discussed elsewhere in this text (e.g., Chap. 4), new digital compression techniques allow a broadcaster to use part of the spectrum to broadcast up to 13 MPEG-1 or three MPEG-2 channels, along with an HDTV channel that can be delivered to customers with addressable setup boxes using the STV approach. With digital technology and flexible use of the spectrum, broadcasters will be positioned to compete in tomorrow's marketplace.[8]

12.8 Interactive Network

As of press time, Interactive Network Inc. (IN) was the only operating wireless interactive television system for TV homes in the United States. It is a subscription-based patented entertainment system that allows TV viewers to interact directly with sports, game shows, prime-time, news-talk programs, and education programs.[9] IN initially served paying customers in the Sacramento and San Francisco Bay areas and in Chicago, and was planning to have service available in many top U.S. markets by the end of 1995.

References

1. D. H. Leibowitz, "New Agendas for Cable and Multichannel Media Taking Shape," *Cable TV and News Media,* February 1994, pp. 1 ff.
2. S. Scully, "From Videotape to Video Servers, Technology Drives PPV," *Broadcasting and Cable,* November 29, 1993, pp. 66 ff.
3. M. Stump, "Wireless Operator Takes Aim at L.A.," *Cable World,* February 7, 1994, pp. 1 ff.
4. J. Yoshida and L. Wirbel, "Cashing In on Convergence," *OEM Magazine,* June 1993, pp. 24 ff.
5. K. Johnson, "Acorns Sprout among the Oaks of the Telecommunications Field," *The New York Times,* July 5, 1994, pp. 1 ff.
6. A. Lindstrom, "The Business of Interactive Video," *Telephony,* November 1994, pp. 6 ff.
7. C. Wilson, "Wireless Video to the Home Still Stalled," *Inter@ctive Week,* January 30, 1995, pp. 32 ff.
8. The IEEE Institute, "Fear, Greed Breed New Technology Use, Keynoter Tells U.S. Broadcaster," vol. 18, no. 3, May/June 1994, pp. 1 ff.
9. "Cable Optics, Information Gatekeepers," Boston, January 1994, pp. 1 ff.

Chapter

13

The Setup Box and the New TV Set

13.1 Introduction

The setup box, also called set-top box or home communications terminal, will play a most critical role in the deployment of new digital-based video in general and VDT services in particular. It will be used to handle the communications (e.g., ATM), provide decompression (at least until such functions are built directly into the TV set), and support user control and signaling. This chapter examines some aspects of the evolving setup box technology.

Currently, only 15 percent of the households that have cable have setup boxes; in addition, an even smaller subset of consumers have addressable setup boxes. Hence, there is both a need and an opportunity for penetration of advanced, third-generation setup boxes*. However, cable operators are now spending $100 to $135 for new setup boxes to achieve a revenue of $30 per month; for operators to be willing to spend double or triple that amount on the latest generation of high-tech setup boxes (with multiple MPEG-2 decoders, AAL5 processors, ATM chips, and multilevel modems), they will need to substantially increase their income derived from the services.[1] Table 13.1 depicts some figures relating to TV equipment for illustrative purposes and to position the discussion in terms of demographics.[2]

* First-generation setup boxes were just converters; second-generation setup boxes were addressable devices; third-generation setup boxes, characterized as superhigh-tech by some, include the kind of setup boxes and features described in this chapter.

TABLE 13.1 TV Equipment Shipments in the United States

Article	1993 sales
Camcorders	2,968,861
Color TV	22,148,118
Color TV-VCR combination	1,861,847
Laserdisc players	203,993
Projection TV	440,273
Analog setup boxes	5,400,000*
VCR	11,940,470

* This figure is for worldwide sales (56% General Instrument, 15% Scientific Atlanta, 7.5% Pioneer, 7.5% Zenith, 14% all other).

13.2 Setup Equipment and Functions

13.2.1 Functionality

I-TV is so new that critical pieces of the hardware and software technology are still being developed.[3] The setup box is one example of such technology under development. The issue is not so much one of basic technology, since VLSI has progressed to the point where all the necessary functionality can be placed in one or two major chips, but one of costs and standardization. Figure 13.1 depicts an example of a setup box at the logical level. A third-generation setup box must support functions such as decompression, communication (including possibly ATM and AAL processing, e.g., see Ref. 4), digital-to-analog conversion, and signaling. Signaling is a particularly thorny issue, impacting openness. Signaling becomes even more complex in VDT, as contrasted to a single-cable system or headend arrangement, because of the multiplicity of parties involved (e.g., access network, backbone network, and multiple VIPs). The support of truly interactive services, involving not only quasi-symmetric, two-way bit-stream flows but also a rich interface to the human being behind the TV set, will prove to be a new challenge for the setup box manufacturers.

As is clear from this discussion, digital setup boxes for broadband I-TV are much more than simple tuners or descramblers, as they were in the past, yet their cost must be driven down to about $300 to $500 before broadband I-TV will be economically feasible. Specifically, a typical box might contain the following components[3]: a powerful CPU, such as an Intel 80486 or PowerPC; 1 to 3 MB of RAM; a high-speed graphics chip for screen overlays and video games; a display chip; a 1-GHz RF tuner; a demodulator; an error-correction chip; an MPEG-2 decoder; logic to strip the audio soundtrack from the incoming video; a Dolby decoder; two 16-bit audio D-A converters; a video RGB converter; an RF modulator; an infrared interface for remote control; ROM for the operating system; a security chip to prevent theft of service; and a switching power supply. Even under the best pricing scenarios, that would cost $1000 to $1500 at press time. Many companies and potential providers are planning their strategies on the assumption that the setup technology will be affordable, i.e., that it will be in the $300 to $500 range by 1995 or 1996.

The setup box must perform demultiplexing, whether in the ADSL, ATM, or HFC environment. It must also support telephony and other narrowband services. Additionally, it must support interaction with the consumer, not only in a walk-up mode but also in a remote-control mode. Hence, an infrared circuit must provide the means for the customer to signal the network. Where interdiction is used, an interdiction circuit in the setup box must receive signals from the network to control what users view. The RBOCs have the additional consideration of determining which portion of the setup box is on the network side of the user side of the user-to-network interface (the VIU-NI defined in earlier chapters). With the exception of the infrared control, all of the functionality of a setup box can be provided by a device physically located on the outside wall of a home (however, even here, the issue remains of where the VIU-NI interface is). Placing the unit on the outside requires raggedness and environmentally sound packaging.

Setup box
Function x
Function y
• • •
Function z

Communication functions	
Control plane	User plane
OSI application, presentation, session, transport, and network layer Session control for network signaling, network access signaling, and application signaling	OSI application, presentation, session, transport, and network layer
	MPEG-1, MPEG-2
Data link layer ATM adaptation layer: AAL5	Data link layer ATM adaptation layer: AAL5, AAL1, etc.
ATM	ATM
Physical layer: coax, ADSL, etc.	Physical layer: coax, ADSL, etc.

Figure 13.1 Logical capabilities of a (high-end) setup box.

Additionally, one must address the issue of powering, as discussed throughout this text, particularly for FTTH arrangements, and particularly when supporting telephony services in a one-wire architecture.

Given the aforementioned menu of functionality, some in the industry are questioning where cable TV companies and telephone carriers should concentrate the bulk of the processing power for the planned full-service systems: in the setup box or in the network? The goal of the service providers is to optimize the economics. Sophisticated setup boxes could cost thousands of dollars; since there are likely to be many replicas of these boxes, it may prove cost-effective to place common functions that can be remoted in the network (at the very least, the setup box will have to handle the MPEG-2 decoding). This would also enable future, technology-driven updates to be undertaken centrally without requiring the consumers to replace equipment. The question is: In the long run, which of these two approaches will be better? A press-time assessment by industry experts was to deploy a setup box with the equivalent capacity of an 80286 chip and a small hard drive. The amount of setup box sophistication will also depend on the envisioned services. A service such as VOD is less demanding than a service such as interactive shopping with advanced control.[5] Pacific Bell has reportedly found that the company can get a "decent" setup box in terms of memory capacity, FLASH-memory, MPEG-2 decoding, and interface support toward the user, for a reasonable price; they are focusing on a PowerPC processor with 2 MB of RAM.[6] Ameritech is looking at a PowerPC with 4 MB to support video games. The price of the setup box is clearly dependent on the amount of memory included.

Observers indicate that there has been little analysis of the demand on the network, the required technologies, and the cost of such technologies. Time Warner Cable's Orlando FSN trial, Viacom's Castro Valley, and U S WEST's Omaha trials used powerful setup boxes. However, some of these companies acknowledge publicly that they have not "identified the right combination of architecture and technologies to provide the services economically." Other operators have put less smarts in the home and more sophistication in the headend. For example, Bell Atlantic is on record saying that for its trials "80 percent of the intelligence is in the network."[5]

One major issue pertaining to setup boxes that may retard penetration of the services is the lack of compatibility between systems. An operator is now forced to purchase the headend equipment and the setup boxes from the same vendor. This becomes a problem as the operator expands, perhaps by acquisition of another cable system that may employ a different set of hardware. The issue, therefore, is whether the network-to-setup box (more specifically, headend-to-setup box) interface should be open. This becomes particularly important in VDT situations—more so than in a simple cable TV situations, because of the need to connect to the equipment of multiple VIPs. Some aspects of this issue were covered in Chap. 5; other aspects are covered later in this chapter.

An advisory committee of consumer electronics and cable representatives assembled at press time for an FCC compatibility rule making, the Cable-Consumer Electronics Compatibility Advisory Group, has been reported to "be contentious," as the two groups try to create a new definition of *cable-ready* for cable-consumer electronics coexistence.[7] The outcome is important because it can open up the cable in-house equipment business in a manner similar to opening up the purchase of telephone set equipment. Consumer groups are asking for the right to own the equipment in their homes, the right to choose where and what consumer electronics and cable interface is purchased, and the right of access for consumers to use their own equipment to access electronic services. Cable companies hope to ensure that no matter what new interactive or digital services they develop, access control and encryption/decryption will be retained. Other requirements include the ability to monitor the status of the system, control specified functions, and communicate with a central site, such as the headend. A *modular decoder* was planned to be available by December 31, 1996; this device offered by cable TV operators would provide access and other controls.

13.2.2 Typical equipment on the market

The setup box market is estimated at $500 million a year. There are about 20 manufacturers of this type of equipment.[8] Suppliers such as AT&T, Hewlett-Packard, DEC, IBM, and Sony are positioning themselves in this market by either supplying chips or selling complete devices.[9] Table 13.2 depicts some of the key suppliers of setup box equipment for advanced cable systems.

General Instrument is the nation's largest manufacturer of setup boxes. Through recent contract assignments (e.g., Bell Atlantic-AT&T cooperation), it appears to be emerging as a principal supplier of third-generation setup boxes.[10] General Instrument Corp. manufactures the DigiCable digital compression setup box that is being used by cable operators in areas of competition of channel capacity exhaustion. For example, in 1994, Adelphia Communications ordered 120,000 units (at $250 each) for delivery in 1995. Although in the past the MSO used Scientific Atlanta setup equipment, Adelphia was planning to use compression to increase the channel-carrying capacity in areas of competition (e.g., in Dover Township, N.J., where Bell Atlantic is establishing a VDT system).[11] Cox Cable announced that it was purchasing 200,000

TABLE 13.2 Suppliers of Setup Box Equipment for Advanced Cable Systems

3DO
Acorn Computer Group
Apple
AT&T
General Instrument*
Gold Star
GTE
Jerrold
IBM
Nintendo
Oak
Philips Electronics NV*
Pioneer
Scientific Atlanta* (also in cooperation with Toshiba)
Sega
Silicon Graphics
Sony
USA Video
Zenith

* Traditional setup box supplier.

DigiCable setup boxes. This brought the total orders for the DigiCable to 2,500,000.[12]

In 1993, Scientific Atlanta introduced a setup box that incorporated two Motorola 8-bit microcontrollers for handling features such as an electronic program guide, advanced messaging service, VCR controller, VOD, and downloadable software; a serial port may be added to the device to support connections to printers, PCs, or game-player consoles.[13] The company, in conjunction with Toshiba, developed a high-end setup box for Time Warner's 1994 VOD trial in Orlando, Fla.; this device incorporates a RISC processor and offers MPEG decompression. The 8600X can support a variety of interactive and related services. Digital compression capabilities can be added by a cable operator, as needed. Other features of the 8600X include[14] advanced graphics, electronic programming guides, expanded PPV for NVOD applications, VCR programming, downloadable software for new service upgrades, downloadable screen in different languages, additional programmability through insert cards for ROM and RAM, support for interactive games, and virtual channels that require no dedicated bandwidth for services such as weather reports and sports scores. Recently, Scientific Atlanta demonstrated capabilities of the interactive television system from Interactive Network Inc. on Scientific Atlanta's new 8600X setup box (as covered in Chap. 12, IN provides a wireless interactive television system).

United Video Satellite Corp. is developing an add-on module that was to provide interactive multimedia capabilities to the third-generation setup boxes now reaching the market. The module will be compatible with setup boxes from General Instruments, Scientific Atlanta, Zenith Cable Products, and others.[13]

Many setup boxes today, including Scientific Atlanta's, store a customer's PPV order and periodically call the operator's billing center to report activity. This technique is known as *store and forward.*[15] Zenith takes a different approach to ordering from the home: inside each setup box there is a small transmitter that can relay orders and information to the headend on the 5-to-30-MHz channel; this supports direct ordering and authorization from the headend.

3DO is considered by some to have a de facto standard (the interactive multiplayer) for TV-top multimedia boxes.[16] GoldStar and Samsung license the 3DO hardware. In 1994, 3DO and Creative Technologies developed a 3DO-compatible PC add-in card that enables a DOS-based PC to play software titles developed for the 3DO format. The interactive multiplayer merges a 32-bit RISC game player and a PhotoCD player into an "infotainment navigator"; dozens of software developers are writing applications for it. 3DO has partnered with AT&T, Matsushita, Time Warner, and MCA.

Among other suppliers, Toshiba, working on next-generation setup boxes, has taken a major equity stake in Time Warner, perhaps in light of the Full-Service Network discussed for illustrative purposes in Chap. 10[13]; however, Time Warner Cable was planning to use setup boxes supplied by Silicon Graphics.[17] In the recent past, Microsoft Corp. has worked with General Instrument and Intel Corp. to sell the idea of a $300 PC for the TV setup box.[13] Apple was expected to introduce a setup box built around a low-end Macintosh "with everything else that does not belong in a setup environment stripped out."[18] The setup box handles MPEG-encoded video. Some see the Apple-IBM-Motorola alliance on the PowerPC as having the potential of becoming the standard platform on which devices such as a setup box could be based.[13]

TCI was reportedly investing $200 million with General Instruments and AT&T for setup boxes capable of decompressing digitized video.[19] The driver is clear: as noted in Chap. 4, up to eight quality video channels can be placed in the bandwidth used by one analog TV channel. What has been holding up actual deployment on MPEG-based setup boxes, in addition to the finalization of the MPEG-2 standard, has been the fact that powerful enough chips to do the decompression (as well as headend real-time compression) are not yet available in the $50 range, leading to a finished setup box in the $200 to $300 range. Such setup boxes may appear in 1995 or 1996.

Microsoft has announced plans to develop an operating system for Sega's next-generation video game system, known as Saturn. The addition of a sophisticated operating system to a video game platform would make it similar in functionality to a PC, since, ultimately, both entities use a microchip at the core. There is an expectation that the move might position the machine as a powerful setup box that could run software and also act as an intelligent controller for network-provided interactive content, while at the same time positioning Microsoft as a potential provider in both the game and video entertainment area.[20]

Some (Gnostech of Allendale, Va., in particular) have proposed using a telephone keypad, rather than an expensive PC-driven setup box, to interact with their TVs. This approach places the intelligence and graphics capabilities in the network (RBOCs or headend) rather than in the user device. Users would be able to tune their TVs to a designated interactive channel and use their regular telephones to call in and establish an interactive link; they could use the buttons on the keypad (or a joystick generating the DTMF tones) to interact with the screen; the buttons act as a cursor control; for example, 6 means go east; 9 go southeast; 8 go south; 7 go southwest; etc. There is a realization that right now there is a preference in I-TV planning in favor of PC-based control and/or smart setup boxes control; this may work fine for trials of a few thousand customers, but for hundreds of thousands or even millions, the cost of the setup box becomes a critical factor.[21]

OEMs see the biggest new business opportunity in the setup PC, which they view as a derivative of the desktop and laptop computer.[13] This device, in the view of the OEMs, could become more than just a black box handling compressed and switched video signals between the home TV set and the video provider; for example, the setup box may also be an application as the home-based multimedia communication center*. There is now much activity surrounding what equipment is used in trials; in the opinion of observers, however, the real issue is what systems will be deployed (i.e., find market acceptance) by 1996 when digital video technology should begin to experience actual commercial introduction.[13] The installed base of digital setup boxes is expected to reach 1 million by 1997; the installed base of analog boxes will then be 35 million. The market's transition from analog to digital is expected to occur in two steps. First, there will be hybrid devices (some were already on the market in 1995); these allow users to access I-TV and NVOD, but will not include video decompression (e.g., MPEG-2). Eventually, there will be all-digital systems. By the year 2000, observers expect an installed base of 50 million units (43 percent analog, 43 percent hybrid, and 14 percent digital).

---

* This author, among others, has advocated for over a decade the creation of a new $400 to $700 home appliance to serve as a communication center (*comce*), which functions as a multiset home PBX with wireless phone sets (as many as there are people in the home, personalized to each person), answering machine, printer, fax, teletex device, database gateway, VCR, task organizer, electronic appointment book, memo, computer, calculator, energy manager, etc., to aggregate all the new communications functions into one easy-to-use device.

In summary, looking at the entire field of I-TV, some observers prognosticate that "it won't come cheap, it won't come easy, and it probably won't come as quickly as some people are predicting."[3] These assessments are based on the fact that new computers are needed for both ends of the I-TV network: clients (setup boxes) as well as servers (that is, high-throughput video servers).

13.3 Capabilities of Setup Boxes in the Context of Signaling

As discussed in previous chapters, there is a major need for standardization, or at least openness, in the industry to foster interworking and service expansion. This is particularly important as customers demand access to multiple VIPs. Some aspects of a digital television system have been standardized, for example, satellite transmission, HDTV, ATM, and MPEG-2. Other aspects remain to be standardized or opened. The VIU-NI is a particularly important area. The VIU-NI is not only ripe for standardization or openness at the technical level but is also one of the critical enabling factors, making this service commercially viable or infeasible. Where would the audio CD market be without standardization? Where would the VCR market be? Where would radio and television be without common standards? VIU-NI standardization must take place in the user plane supporting connectivity (e.g., ATM, ADSL), as well as in the control plane supporting signaling.

The material that follows focuses on signaling. The topic of VIU-NI signaling was already treated in Chap. 5, from a network service provider perspective. This section looks at the issue from a setup box perspective. This section contains, for illustrative and pedagogical purposes, a synopsis of two published Scientific Atlanta specifications for signaling in the context of I-TV.[22, 23]

The summary is included here for pedagogical reasons, to illustrate the concepts described in this chapter. Developers and other parties working on products of commercial value should refer directly to the original documentation, especially since only some key highlights are included. This discussion is not meant to imply that this approach is the best, the only one, or the recommended one. The discussion simply illustrates what such a protocol must encompass, how the various elements communicate, and so on.

Readers not further interested in signaling can skip to Sec. 13.4.

13.3.1 Overview

This specification,[22] describes the network interfaces required to interconnect to the signaling network that supports the *Interactive Video Services Network* (IVSN). Included are sections that describe the basic interfaces necessary to interoperate common telecommunications network devices on this broadband network. Several elements will be included that utilize specific variations of common telecommunications techniques and schemes.

Communications supported by IVSN conform to a passband architecture. Unlike a baseband architecture, a passband approach takes advantage of modulated RF carriers to combine several types of modulation schemes, contention techniques, and transport protocols over one shared medium. Critical to the understanding of this process is the concept of forward and reverse path signaling. In a traditional passband architecture, the RF spectrum is split into two separate and distinct directional communication pathways. The reverse path is the spectrum reversed for communications from users on the shared network to a centralized control area commonly referred to as a headend. Forward path signaling is the spectrum reserved for communication from the headend to the multiple users of the shared medium. The two communication pathways are separated by a guard band where no communication is carried, in order to eliminate interference.

This Scientific Atlanta specification is divided into distinct areas that cover the methodologies required to operate the signaling network. They include slotted ALOHA, time division multiple access (TDMA), description of the message cells, differentially encoded offset quadrature phase-shift keying (O-QPSK), 1.544-Mbps framing in the forward path and reverse path communication, and Ethernet*.

13.3.2 Slotted ALOHA signaling

Slotted ALOHA is used for managing contention of transmission over a signaling link. Scientific Atlanta uses slotted ALOHA as a technique for signaling between a subscriber's *set-top terminal* (STT) and the headend. Slotted ALOHA provides instant channel allocation for the STT.

The ALOHA technique is used for multiple subscribers who will have equal access to the signaling channel. It is probable that simultaneous transmission will occur. The ALOHA technique provides resolution of signaling throughput when simultaneous transmissions occur.

The signaling channel(s) for the STT and the headend use analog carrier for transmission; the modulation technique is O-QPSK. Two signaling channels are initially recommended, but additional channels can be added to accommodate high-traffic rates that will result in fre-

* Not all of these aspects are treated here; the interested reader is referred to the original documentation for a more complete discussion.

quent message packet collisions. The STT will randomly select one of the allotted channels for each packet transmission.

13.3.2.1 Slotting. The slotting technique allows the transmit start to be synchronized to a common clock source. Synchronizing the start times increases message throughput of this signaling channel since the message packets do not overlap during transmission. The period between sequential start times is identified as a slot. Each slot is a point in time when a message packet can be transmitted over the signaling link. Users are restricted to start transmission of packets only at slot boundaries. Clocking is received from a 1.544-Mbps signal generated at the headend and transmitted simultaneously to all STTs. Since all STTs reference the same clock source, the slot times are aligned for all STTs. Since there is propagation delay in any transmission network, the slot size must accommodate deviation of transmission times due to propagation delay. The slot duration time is 0.5 ms. Using this slot size at 1.544 Mbps provides a rate of 2000 slots per second.

13.3.2.2 Message packet. The message packet structure is 512 bits, as shown in Fig. 13.2. Note that the packet can encapsulate an entire ATM cell. The preamble is the header information needed to identify the start of the message cell. The preamble is a fixed pattern for providing a framing sequence to the STTs and is inserted at the first bit of the STT's slot interval. This preamble pattern is a repetitive 8-bit pattern: 11011000. The message cell is the actual message payload. An additional time of 168.4 μs (guard time) is allocated to accommodate the propagation delay of the broadband network. The propagation delay accounts for the time in which an STT receives its 1.544-Mbps clocking signal as well as the return transmission of the message packet to the IVSN.

13.3.2.3 Collisions. After transmission of a message packet, the STT will determine that the signal was successfully received. This is accomplished by having an echo message sent back to the sending STT. The echo message is the message cell only and does not include the preamble. If the echo is not detected by the STT, then signal collision is assumed. A collision occurs if two or more STTs attempt packet transmission during the same slot. A collision will be detected if an STT does

Preamble (88 bits = 11 octets)	Message cell (AAL5/ATM protocol data units) (424 bits = 53 octets)

Figure 13.2 Scientific Atlanta ALOHA-TDMA packet.

not receive its echo message. If a collision occurs, then the STT will retransmit its message packet.

13.3.2.4 Retransmission. Retransmission must occur if the preceding packet has collided with other transmitted packets and was not received by the IVSN. The STT determines that its packet signal has collided if the video is not detected within 200 message slots (100 ms). When this occurs, there are two parameters that influence the transmission and determine its probability for success. These are the retransmit time (the selected slot for retransmission) and the use of an additional signal channel.

The STT will randomly select a slot for retransmission at a time period beyond the 100-ms detection time. The time period for retransmission time is 0 to 50 ms. Within this period, up to 100 slots are available for retransmission, and the STT will randomly select a slot.

An alternative signaling channel(s) could be used to increase the probability for successful retransmission of message packets. Before retransmission of a message packet, the STT selects a signaling channel for retransmission. Since this is a random selection, the probability of success is increased.

13.3.3 Time division multiple access signaling

TDMA allows STT access onto a signaling channel. The TDMA technique is used by Scientific Atlanta for non-contention-based communication between the STTs and the IVSN, as well as the STTs and a VIP. TDMA is based on dividing access by multiple STTs onto a shared signaling channel. This technique provides a negotiated bandwidth allocation slot access method.

TDMA, as implemented by Scientific Atlanta, assigns time slots to individual STTs. The slot assignments are accomplished through messaging from the IVSN to the STT. Within the IVSN is a signaling controller that initiates, receives, and interprets the messages. This messaging is described in Ref. 23; some highlights are covered later.

The TDMA signaling channel serves a size of several hundred subscribers. Since this channel is used only for interactive session messaging, the number of users utilizing this link is limited to the number of functioning simultaneous interactive users. The signaling channel(s) from the STT to the IVSN use analog carrier for transmission. The modulation techniques are O-QPSK. One channel is used for communication, and additional channels can be allocated if additional message throughput is needed.

13.3.3.1 Slotting. The TDMA technique uses a slotting methodology that allows the transmit start times to be synchronized to a common clock source. Synchronizing the start times increases message throughput of the signaling channel since the message packets do not overlap during transmission. The periods between sequential start times are defined as slots. Each slot is a point in time when a message packet can be transmitted over the signaling link. Clocking is received from a 1.544-Mbps signal generated at the IVSN and transmitted simultaneously to all STTs. Since all STTs reference the same clock source, the slot times are aligned for all STTs. Since there is propagation delay in any transmission network, the slot size accommodates deviation of transmission due to propagation delay. Slot duration time is 0.5 ms. Using this slot size, 1.544 Mbps provides a rate of 2000 slots per second.

13.3.3.2 Message packet. The message packet is identical in format to the packet in the slotted ALOHA environment (see Fig. 13.2).

13.3.3.3 Slot assignment. Since the TDMA signaling link is used by STTs that are engaged in interactive sessions, the number of message slots on this channel is dependent on the number of simultaneous users. When messaging slots are not in use, an STT may be assigned multiple message slots for increased messaging throughput. Additional slot assignments are provided to the STT from the signaling controller in the IVSN through messaging packets described in Ref. 23, some highlights of which are covered later.

Using the 2000 slots per second as available capacity, multiple STTs are assigned available message slots. The signaling channel serves a node size of several hundred homes and of that node population, there is a subset of users that will simultaneously use interactive services.

13.3.4 Message cell format

Figure 13.3 depicts the message format used in the signaling from the STT to the headend. The message cell payload contains all or part of a message payload data unit, depending on the length of that message. If the message is shorter than 384 bits, it will fit one cell; otherwise, multiple cells are required. Figure 13.4 depicts the general message cell payload.

13.3.5 Control messages

The document in Ref. 23 describes the format for control messages that are exchanged between the service boundaries in the interactive video

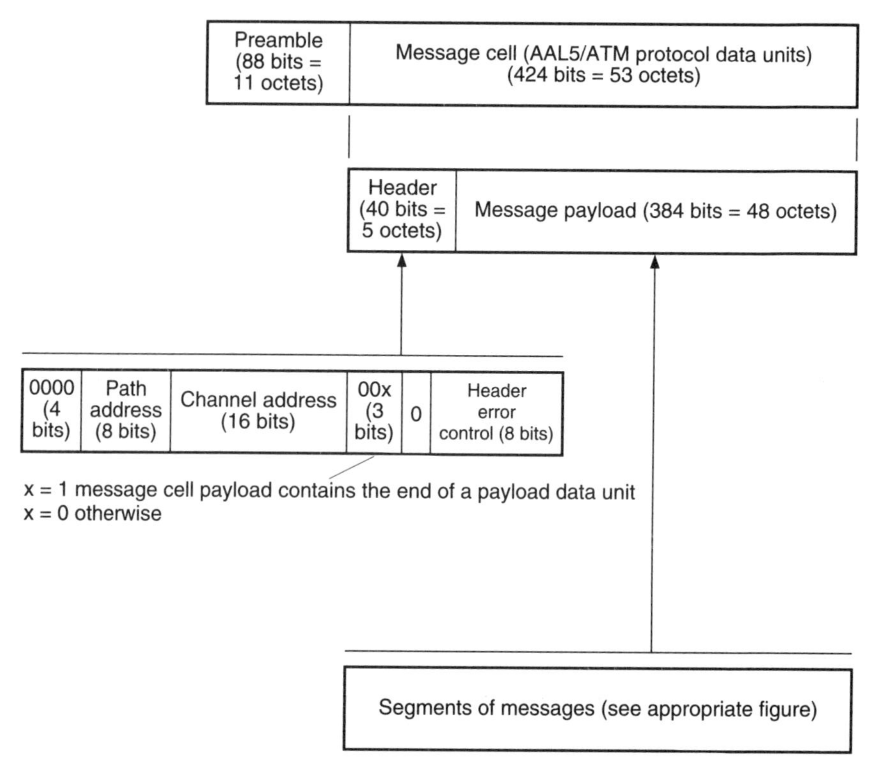

Figure 13.3 Format of message cell.

services system. This system enables an STT to request a service from a *service provider* (SP) through the use of control system messages that are defined in the specification. The requested service is transmitted from the SP over a portion of the bandwidth of a broadband transmission network that is connected to the STT.

The SP provides the service to the network through a UNISON-1 transport (see Chap. 8, Sec. 8.7.5). This service is then routed to the

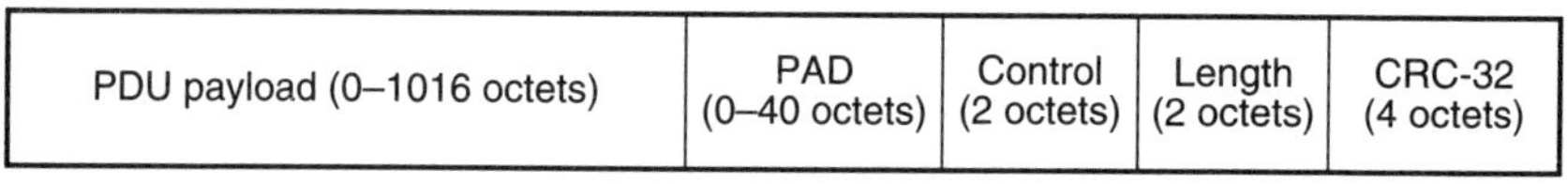

PDU payload (0–1016 octets)	PAD (0–40 octets)	Control (2 octets)	Length (2 octets)	CRC-32 (4 octets)

Notes:

PAD field is used to align PDU to a message cell boundary.

Control field is set to all ones.

Length field contains the number of octets in the PDU payload.

CRC-32 is calculated over entire cell PDU.

Figure 13.4 Message cell payload.

appropriate broadband channel and transmitted to the STT. UNISON-1 is an optical interface that is modeled after the standard SONET transport, except that the standard has been modified to facilitate the transmission of MPEG-2 transport streams.

The STT and SP communicate through an out-of-band signaling path that allows the STT to control a service interactively. One such application for this type of service is in the area of VOD services where the STT requests a movie to be transmitted from an SP. While the movie is being transmitted, the STT has the ability to send commands such as play, pause, and fast-forward to the SP. Other applications include on-line data services, shop-at-home services, and other services that require the use of an interactive connection.

In the specification, *Level 1 service* (L1) is the portion of the system known as the video dialtone. This portion of the system is responsible for setting up and maintaining the session between the STT and an SP. *Level 2 service* (L2) is the portion of the system responsible for providing the requested service to the L1 portion of the network (the SP) and for terminating the service at the user of the network (the STT).

The *connection management computer* (CMC) manages sessions between the STTs and the SPs. The CMC is responsible for the following functions:

- Provisioning the multiplexer-modulator
- Provisioning the set-top terminal
- Session management between the STT and the service provider

Either the STT or the SP may send a request for a service to the CMC. The CMC determines if there are resources available to transport the requested service and establishes a service from the SP to the STT. The CMC then sends the service information to both the STT and the SP to allow them to connect to the network and begin the requested service.

The following message flows between service levels L1 to L2 are described in the specification:

CMC ↔ STT

CMC ↔ SP

Level 2 ↔ Level 2 peer-to-peer messages

Figure 13.5 provides an overview of control and video data flows.

13.3.5.1 CMC message format. Signaling between the STT and the SP across the service boundaries is accomplished using the standard Internet Protocol (IP), as described in RFC 791. Each STT and SP has

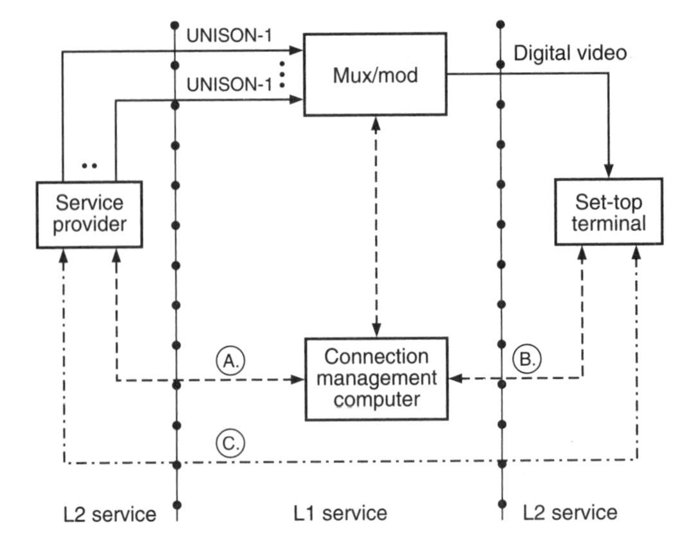

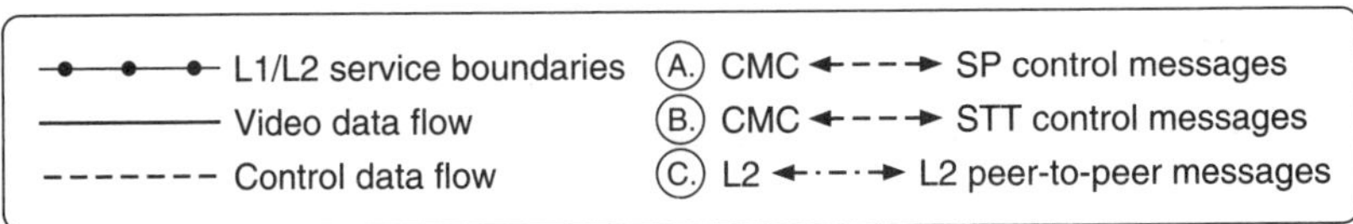

Figure 13.5 Overview of control and video data flow.

a unique address that is mapped to an IP address, which is assigned to the STT by the CMC. The physical address of the STT is a unique 4-octet number that is assigned to the STT at the time of manufacture. When the STT is powered up and connected to the network, the STT sends a provisioning request to the CMC to get its IP address. (Address resolution using ARP-RARP may be required but is not described here.) The IP address of the SP is provisioned into the STT by the CMC. There is a separate CMC for each SP on a network. An STT may communicate with only one SP. In addition to the IP address that identifies the STTs and the SP, communications packets also contain a User Datagram Protocol (UDP) header. The UDP header contains the specific port addresses for individual services within an STT.

Figure 13.6 defines the mapping of the IP messages into a message cell transport packet that is sent between the CMC and STT. The *payload data unit* (PDU) is reassembled at the STT or CMC into the original PDU. At press time, there were two possible STT PDU formats defined:

Format A: an 8-octet payload that uses a single message cell

Format B: a 56-octet payload that uses two message cells

After an STT has established a session to the SP, the STT communicates with the SP over a TDMA upstream application signaling link, as discussed in the previous subsection. Depending on the system load, there may be more than one TDMA data stream. The CMC instructs the STT where in the broadband spectrum it should tune for its TDMA channel and assigns the STT an allocation of slots in that data stream.

Table 13.3 depicts a summary of the messages as described in Ref. 23. Each of these messages is then further defined at the detailed parameter level in the specification.

13.4 Issues in Security

Security issues are often thought of in the context of the setup box. This section provides a very short treatment of the topic (some additional discussion was provided in Chap. 4). The ITU-R describes conditional access, an important form of security, as being defined by two key elements that comprise it. These are[24]:

- *Conditional access system:* Within a TV distribution system, it is the means to selectively provide TV programs to specific individual subscribers; the system includes means to track access for accounting purposes.
- *Scrambling:* The alteration of the characteristics of a broadband video-sound-data service to prevent unauthorized reception of the information in a clear form. The alteration is a specific process under the control of the conditional access system (sending end).

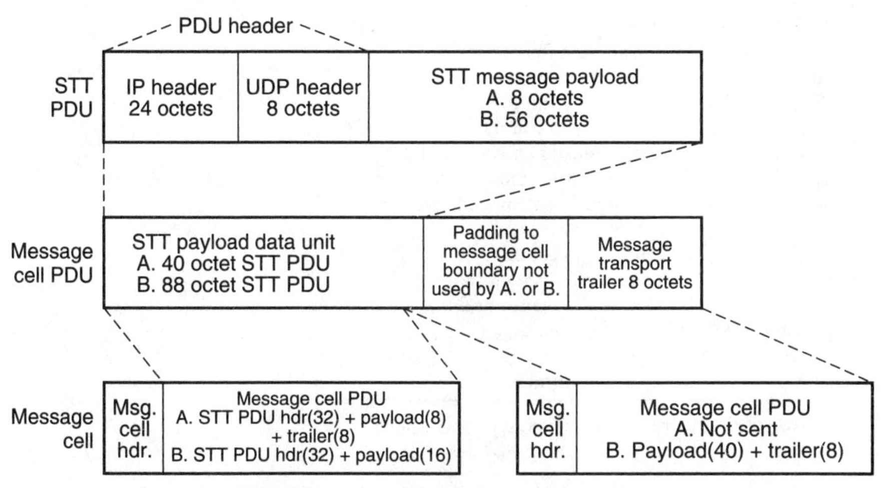

Figure 13.6 Mapping of STT PDU to message cell transport packets.

TABLE 13.3 Scientific Atlanta Message Set

CMC ↔ L2 command messages
01 STT session setup—0 × 20
02 STT session setup acknowledge—0 × 21
03 STT session connect request—0 × 22
04 STT session provision—0 × 23
05 STT session provision acknowledge—0 × 24
06 STT continuous session setup—0 × 25
07 STT session release—0 × 30
08 STT session release acknowledge—0 × 31
09 STT session teardown—0 × 32
10 STT session teardown acknowledge—0 × 33
11 STT session in progress—0 × 40
12 STT status request—0 × 41
13 STT status—0 × 42
14 STT reallocate TDMA slots—0 × 050
15 STT reallocate TDMA slots acknowledge—0 × 51
16 STT switch TDMA slot allocation—0 × 52
17 SP session setup—0 × 60
18 SP session setup acknowledge—0 × 61
19 SP session provision—0 × 62
20 SP session provision acknowledge—0 × 63
21 SP continuous session setup—0 × 64
22 SP session connect request—0 × 65
23 SP session connect acknowledge—0 × 66
24 SP session release—0 × 70
25 SP session release acknowledge—0 × 71
26 SP session teardown—0 × 72
27 SP session teardown acknowledge—0 × 73
28 SP session in progress—0 × 80
29 SP status request—0 × 81
30 SP status—0 × 82
31 STT peer-to-peer message—0 × 90
32 SP peer-to-peer message—0 × 91
33 STT broadcast message—0 × 92
34 SP broadcast message—0 × 93
35 STT procedure error—0 × 0a0
36 SP procedure error—0 × a1
CMC ↔ STT session management
01 CMC ↔ STT session setup commands
02 CMC ↔ STT session teardown commands
03 CMC ↔ STT session status commands
04 CMC ↔ STT TDMA reallocation commands
CMC ↔ SP session management
01 CMC ↔ SP session setup commands
02 CMC ↔ SP session teardown commands
03 CMC ↔ SP session status commands
peer-to-peer and broadcast messages
01 peer-to-peer message commands
02 broadcast message commands

NOTE: Refer to original specification for details.

- *Access control:* The function of the conditional access control at the sending end is to generate the scrambling control signals and to provide information to enable authorized users to descramble the program or service.

As cable TV systems and VDT systems begin to deliver digitally compressed signals end-to-end (headend-to-setup box or video server-to-setup box), the video distribution industry has the opportunity to apply fresh thinking to the requirements for conditional access. For about 15 years, access has been accomplished using addressable boxes used in conjunction with the analog signal. Industry experts are now advocating that conditional access move beyond this type of tiered addressability, as digital technologies become pervasive.[25] A new generation of setup boxes will be deployed, leading to the need, as well as the opportunity, for establishing new security mechanisms to protect the content of the ever increasing value of the new information-based services. There is an opportunity at this early time to do away with the proprietary approaches of the past and to introduce an industry standard for handling access and related controls. This comes against the following backdrop[25]:

- Widespread theft of satellite-delivered programming shows that scrambling or encryption of the voice channel alone is not an adequate protection against unauthorized use. Secure authentication, in the well-defined language of the ISO standards, as well as digital encryption, are needed. The cost of piracy must be made high in relation to the perceived value of the misappropriated programming and services.
- The set of capabilities of cable-addressable systems always has seemed to involve a struggle to catch up with market needs. There never seems to be enough tags or tiers to deal with the latest PPV offering.
- Addressable converter-descramblers afforded new ways to market cable services and to generate additional value. However, there have been limitations imposed by the lack of compatibility between setup boxes; lack of compatibility has occurred at every level, including security considerations.

Compatibility provisions of the Cable Act of 1992 challenge the analog addressing approach used by cable operators so far. For the past decade, proprietary scrambling approaches have dominated the setup box scene, ultimately impacting the cost of the equipment. As the industry enters a new era, there is the opportunity to eradicate these compatibility problems if appropriate planning is undertaken at an early time. One wants a commonality of approach that enables multiple vendors to build interchangeable equipment, while at the same time maximizing system-level security. Since some of the functions of a setup box, e.g., decompression, may migrate into the TV set, it is necessary to have open security standards, so that any TV set can be connected to the video distribution system.

At this juncture, there are significant departures in the offing from previous generations of technology; such changes invite the examination of opportunities to attain improvements over the current situation, in areas such as ease of interaction, interoperability, and strengthened security. It is clear that digital compression affords the possibility and promise of a quantum improvement in security. Encryption can by itself bring higher integrity compared with scrambling. It is unlikely that a hacker could actually break the encryption algorithm. However, there is the concern about compromise of keys through disclosure, theft, bulletin boards, cloning, and distributing unauthorized devices, etc. Often scrambling and encryption systems have been defeated by modifying the setup box-home receiver. Hence, the authentication system (which in turn enables authorization to partake of certain privileges) must be well designed. For entertainment distribution, security techniques were introduced in the early 1980s; since that time, there has been a steady increase in the adoption of encryption techniques. However, except for very high cost systems that could afford total digitization of the audio and video material, virtually all existing systems have techniques that use randomizing capabilities to deterministically scramble analog components and reorder or otherwise reassemble these components at the receiving end (examples are line shuffling, cut and rotation, and random inversion of video).[24]

The analog underpinning of the randomization processes used in the past makes this approach unacceptable as a long-term solution. Additionally, in most implementations, enough recognizable information remains to frustrate the goal of a robust conditional access system, which is to ascertain that the scrambled information contains no useful derivatives of the original signals and that reconstitution is not possible by examination of the scrambled waveform alone.

As a side note, many of the digital HFC designs include a pool of shared channels to support a service such as VOD. In the past, that information was broadcast into the fiber-coaxial system for all to nominally receive (but not necessarily decode); however, after the initial interaction with the specific recipient requesting the VOD movie, the information did not come down the channel with a label identifying the destination address of every block of information. This implies that the privacy of the individual was protected because, even if a consumer

TABLE 13.4 Security Attributes for Next-Generation Cable-VDT Systems

- Access must be reliable, not allowing compromise from one service into another.
- Conditional access must be recoverable from infractions.
- Conditional access must use strong authentication techniques.
- Encryption mechanism must be robust.
- Key management and distribution should be inexpensive.
- Open system security should not imply that knowledge of how the system works enables the intruder to gain access.
- Security must be implemented in an open manner, through a key mechanism.
- Subscriber hardware must be tamper-proof to prevent replacement of components (e.g., critical security components should not be accessible from the outside).

could break into the system, there was no identification of who was the recipient of the video information. With ATM, each cell has a distinguishable label so that, in theory, there is a potential loss of privacy, unless secure mechanisms are put in place.

Table 13.4 depicts some security attributes that need to be included in next-generation cable-VDT systems. Key functionality includes the following[25]:

- Large number and variety of channels or program choices, tiers, program packages, and other digital services must be controlled.
- Multiple operators and programmers must be accommodated.
- High-speed authentication, authorization, and deauthorization must be supported.
- The ability to securely deliver keys must be made available.

The establishment of an open security standard would be consistent with the current trend of developing standards for subsystems of the technology (e.g., compression, transmission, signaling), so that maximum benefits accrue from the interoperability of equipment as well from advances in other areas (e.g., development of ATM chips, developments in fiber technology).

Both the European industry (through the ITU-R) and the North American industry (through the Advanced Television Systems Committee) have investigated the standardization of conditional access.

13.5 Computer-Based Television Sets

There is an expectation of an impending merger between the computer and the TV set into a multimedia environment.[25] The Macintosh TV is discussed briefly here, for pedagogical reasons, as an early archetype of this metamorphosis. Although the device, which became available in 1994, did not meet the full expectations of all observers, it was considered an important step along the road to fully integrated multimedia systems.

The immediately visible feature of this PC is the fact that it, the keyboard, and all the peripherals are black, reminiscent of home electronic components. It includes a TV, a computer, and a CD player in one enclosure. The sleek-looking system uses up a small footprint; it includes a 14-in Sony monitor, built-in stereo speakers, ergonomically positioned volume and brightness control, and an infrared remote control.[26] The system has a 68030 microprocessor, a 32-bit data bus, 5 MB of RAM, a 160-MB hard drive, and supports 256-color video. The system has a pair of RCA audio jacks, an RCA video port for VCR and camcorder, and a coaxial port for connecting cable TV or an outside antenna. Inside the Mac is a cable-ready tuner (the tuner takes up the only available expansion slot). The quality of the display is not as good as one might imagine, because of the tradeoffs involved in designing a monitor for both NTSC that is interlaced (as noted in Chap. 4) and for a noninterlaced RGB signal.

One can switch between computer and TV mode using a key (currently it is not possible to have the TV program as a window while working at the PC). CD options likewise only work in the computer mode. Newly added system software lets the user control all TV and CD functions from the computer or from the remote (another feature not included is computer-based intelligence to program a VCR function or display on-line program listings). A new control panel is used to change TV settings and to establish a password to keep others from accessing the TV. Logical names can be assigned to channels. However, the use of both a setup box from a cable TV and a VCR becomes more complex. One is able to listen to a CD (or a TV channel) while working at the computer. Single-frame capture is possible, but movie clips consisting of two or more frames cannot be captured.

This is only a first example of the convergence of the PC and the TV that is expected in the relatively near future.[27]

References

1. *Communications Technology,* April 1993, pp. 18 ff.
2. *Video Technology News,* January 3, 1994, p. 6.
3. A. Reinhardt, "Building the Data Highway," *Byte,* March 1994, pp. 46 ff.
4. R. Woolnough, "Acorn Plants ARM Chip in Set-top Box," *Electronic Engineering Times,* July 11, 1994, pp. 1 ff.
5. "Smart Boxes or Networks? Simple Questions Prove Tough to Answer," *Video Technology News,* February 14, 1994, pp. 4 ff.
6. P. Bernier, "Carriers Sort Out Their Choices," *Telephony,* Nov. 7, 1994, pp. 18 ff.
7. "Cable, Consumer Electronics Search for Common Ground in Compatibility Tussle," *Video Technology News,* February 28, 1994, pp. 5 ff.
8. J. Carlton, "Macromedia and Microware to Unveil Technology to Boost Interactive TV," *Wall Street Journal,* August 1, 1994, p. B6.

9. G. McWilliams, "They Can't Wait to Serve You," *Business Week,* January 24, 1994.
10. E. L. Andrews, "AT&T Picked to Build Bell Atlantic's Network," *New York Times,* May 20, 1994, p. D1.
11. C. Weinschenk, "Jones, Adelphia Cutting GI Deals," *Cable World,* February 28, 1994, p. 8.
12. *Video Technology News,* January 17, 1994, p. 7.
13. J. Yoshida and L. Wirbel, "Cashing In on Convergence," *OEM Magazine,* June 1993, pp. 24 ff.
14. "Cable Optics, Information Gatekeepers," Boston, January 1994, pp. 1 ff.
15. S. Scully, "From Videotape to Video Servers, Technology Drives PPV," *Broadcasting and Cable,* November 29, 1993, pp. 66 ff.
16. *Multimedia Week,* March 14, 1994, p. 4.
17. G. Lawton, "Interoperating Cable into the Great Opportunity," *Communications Technology,* November 1993, pp. 26 ff.
18. "Apple Tunes in to Interactive TV," *Multimedia Week,* February 14, 1994, p. 1.
19. G. Lawton, "The Bits Are Coming: Deploying Digital for Video Services," *Computer Technology,* April 1993, pp. 38 ff.
20. *Telephony,* January 24, 1994, p. 16.
21. R. Karpinski, "Can Touch-Tone Signaling Drive Interactive TV," *Telephony,* January 10, 1994, p. 10.
22. Scientific Atlanta, *Signaling Requirements for Interactive Video Services,* no. 69L006Z, 1993.
23. Scientific Atlanta, *Interactive Video Services Control System Messages,* no. 69L008Z, rev. 5.7, December 1993.
24. T. Wechselberger, "Conditional Access and Encryption Options for Digital Systems," *Communications Technology,* November 1993, pp. 20 ff.
25. G. S. Stubbs, "Conditional Access for Compression Systems: Desirable Attributes and Selection Criteria," *Communications Technology,* November 1993, pp. 18 ff.
26. L. Sherman, "The TV-Computer: A Forced Marriage," *Newmedia,* March 1994, pp. 81 ff.
27. J. McGarvey, "Competition Heats Up Early Digital Set-Top Market," *Inter@ctive Week,* January 16, 1995, pp. 26 ff.

Chapter

14

The Players: Telephone Carriers Moving to Video

This chapter examines some of the RBOCs' activities in terms of competition, plans, technologies, trials, alliances, etc., in the VDT context. Although this information is time-dependent, it is useful in documenting the earnest efforts in the mid-1990s in the area of advanced video distribution systems. Many of the plans involving rollouts have a 3- to 7-year window of implementation; therefore, although in the future there will be other alliances, plans, and strategies, many of the activities listed in this chapter will continue to be of relevance for the foreseeable future. All information contained in this chapter is based on the open literature.

The reader may also want to review some of the market issues covered at the end of Chap. 2. As noted there, an interesting synthesis of what might happen in the next few years is provided by the city of Rochester, N.Y. In 1994, the Rochester area opened up the local loop to competition. Historically, Rochester Telephone provided local telephone service. Because of the new freedoms, Time Warner Cable was planning to provide telephone service over its cable system on a fully interconnected basis. MFS Communications Inc. also announced it was planning to start offering switched telephony services in Rochester. More generally, cable industry suppliers are building devices that adapt cable networks for multimedia use, supporting voice, video, data, and wireless service over cable. Furthermore, there is the inter-RBOC telephony competition through out-of-region cable TV systems. These market dynamics, if replicated on a national scale in the next three to five years would have profound repercussions on the local telecommunications scene in general and on the RBOCs in particular.

14.1 Competitive Strategies

At this juncture, RBOCs see transport of video over a VDT platform and video content as separate aspects of video services (this being clearly driven by the content restrictions discussed in Chap. 3). However, beyond that distinction, they have the market challenge of differentiating VDT services from traditional cable TV services. Several RBOCs have become "increasingly aggressive in their efforts to bolster stagnant telephone revenues by providing . . . VDT."[1] As their revenues in the local loop continue to flatten and competition grows, the RBOCs are looking to new revenue sources. As noted in Chap. 2, multimedia revenues are expected to grow at about 11 percent per year until 1998, while traditional telephone revenues are expected to grow at only 3 percent per year over the same period.[2] For the immediate future, there are several approaches to video services that the RBOCs may consider, as follows:

- Develop, deploy, and maintain a VDT platform; enlist (partner with) appropriate VIPs* to provide customers the type of programming they want to watch; and expand VDT offerings to include services such as teleshopping, games, and multimedia. For example, at the program content level, The Walt Disney Co. recently reached an understanding with Ameritech, BellSouth, and SBC Communications for the formation of a joint venture to develop, market, and deliver traditional as well as interactive programming in North America.
- Deliver some of the permitted content-delivery services (those with high interactivity); focus on entertainment first and then introduce other services of *convenience* for the residential and (later) the business market. According to researchers, video content, more than the elegance of the technical solution, is going to spell the difference between market acceptance of VDT and market rejection. Studies indicate that delivery of useful information is also a key driver, for example, distance learning, access to government data, access to Internet and VAN databases, etc. Also, in 1994, Creative Artists Agency, headed by Hollywood power broker M. Ovitz, and Bell Atlantic, NYNEX, and Pacific Telesis announced a large video entertainment venture to develop programming.
- Establish an out-of-region network, including content delivery and telephony.
- Target niche markets.

---

* These could be competitive cable TV providers in a given city, to leverage against entrenched providers, or multiple system operators, such as TCI and Time Warner Cable.

TABLE 14.1 Market Development Strategies for Interactive TV

- Develop compelling interactive content, with recreational, educational, and employment-usable information value
- Introduce standards that support an open platform for all hardware, service, and content providers
- Create an infrastructure that supports residential, business, and institutional clients
- Facilitate interconnection of regional interactive networks to achieve increased access to content resources
- Integrate services to support existing and evolving voice, data, and video applications

Table 14.1 depicts some key market development strategies that are applicable to both the traditional carriers and the cable TV companies.

For planning purposes, carriers can assume that a penetration of 15 to 35 percent, with a 25 percent average, can be achieved. They can also assume that viewers are willing to spend from $25 to $55 per month. Some carriers have actually used the high end of these estimates for planning purposes (i.e., 50 percent and $55[3]); however, these high-end points may be somewhat optimistic, at least initially.

As noted earlier, competition in general, and in the video distribution area in particular, will increase in the future. For example, there could be competition in one RBOC's region by an RBOC pursuing an out-of-region video content or distribution opportunity (e.g., SWBT operating in Bell Atlantic's region). On the other hand, there could be cooperation between an RBOC and a cable TV company (e.g., Bell Atlantic and Sammons*). Additionally, an RBOC could generate an opportunity by pairing up with a competitor cable TV company, against an entrenched cable TV provider (e.g., Bell Atlantic and FutureVision against Adelphia Cable in Toms River, N.J.). Since there are no franchise requirements to this type of new video delivery, the RBOC could show a competitive advantage.

Carriers should aim at deploying a VDT system that is scalable in terms of services and number of customers. The decision as to which technology to employ, from the set discussed in Chaps. 9 to 12, is based on both technical considerations and business considerations. The services that a provider wants to deploy during an initial phase and during a later phase also determine the architecture a provider needs to use. System requirements fall in four categories:

Service requirements

Subscriber requirements

---

* This plan to build a switched digital video network serving 12,000 customers in Floral Park, N.J., has now been abandoned.

Transport network requirements (backbone, switching, and local distribution)

VIP requirements

Service requirements. As discussed in this text, there are many applications of interest (VOD, NVOD, IVOD, IMVOD, PPV, MPGs, I-TV, interactive shopping, compressed video at different compression rates, interactive distance learning, telephony, video telephony, etc.). The VDT platform must be able to accommodate the desired subset of services and allow for new services to be added.

Subscriber requirements. User requirements include, among others, cost-effectiveness, availability, abundance of video program sources, picture and video quality, ease of use, navigability, and responsiveness (both in terms of network switching-signaling parameters and in terms of carrier-VIP support). As an average, a household needs to have access to from three to five simultaneous channels, dictating the type of delivering system that needs to be deployed.

Transport network requirements. The network must support an open network architecture, enabling access to multiple VIPs. It must support reliable, cost-effective, high-quality movement of information from the VIPs to the VIUs, including signaling information. In addition to video, the network should allow interconnection to other existing networks for voice, data, and information services. Typically the network needs to be a broadband network that supports two-way communication.

VIP requirements. The RBOC needs to deploy an infrastructure that supports equal access to the *set* of VIPs that wishes to enter the market. Sufficient transport and signaling capacity needs to be made available. The platform must support a plethora of services (as previously discussed), and accommodate the rapid introduction of new services when needed. The platform may also collect appropriate usage statistics.

Multimedia is experiencing major growth, and it may be appropriate for the RBOCs to pursue these types of services. The consumers have voted with their pocketbooks about their interest in multimedia programming. For example, multimedia computer sales grew 1833 percent in 1993 (worldwide), from 127,000 systems sold in 1992 to 2.46 million sold in 1993.[4]

Overall, the telephone company trials, such as Bell Atlantic's Northern Virginia trial and U S WEST's Omaha project, are following Time Warner Cable's Orlando model of "throwing as much money at the technology as necessary to do a serious market test."[5] Many U.S. firms are also making investments in foreign markets to acquire newly privatized telecommunications companies or to conduct technical trials of systems that these companies hope to replicate in the United States.[6] At this time, RBOCs wanting to provide commercial VDT service in the United States must file a "214-tariff" with the FCC.

In summary, telephone companies already have a service relationship with almost every consumer in the United States. However, they have limited experience in selling nonessential services to consumers; both the telephone companies and the cable TV companies will need to develop the capability to sell I-TV, VOD, MPG, interactive home shopping, educational services, and PCS to consumers. Another advantage of the telephone carriers is that they have national, state, and local political influence. They are regulated to a fair degree at the state rather than local level. They also have a tradition and/or capability for technical training programs for the staff (however, these programs must be enlarged to include video), and they generally have more capital resources. Brand-name recognition can also work in favor of the carriers. Finally, they operate, to a very large degree, in contiguous territory.

Cable TV companies' advantages include the fact that their staff is nonunionized. This staff tends to be cross-trained and is somewhat more prepared to deal directly with consumers on a variety of issues than the staff of a carrier. The management structure is less bureaucratic and more flexible. One of the challenges of the cable TV companies is dealing with multiple regulatory agencies, particularly at the local level. Cable TV companies are often spread out over noncontiguous territory, making the provision of voice services more difficult.

14.2 Broadband Trials and Activities

This section provides a short press-time summary of RBOC activities in the related field of ATM.

14.2.1 Ameritech fastpacket-ATM efforts

SMDS and frame-relay PVC were deployed in 1992. An RFP for ATM switches was issued in early 1993. AT&T was selected as the initial ATM switch supplier.

14.2.2 Bell Atlantic fastpacket-ATM efforts

SMDS was deployed in 1992; further SMDS and frame-relay PVC deployment occurred in 1993. The carrier was awarded ARPA contracts for a six-node experimental high-speed network in the Washington, D.C., area.

14.2.3 BellSouth fastpacket-ATM efforts

BellSouth deployed frame relay and SMDS in 1992. A trial that is underway, named VISTAnet, uses ATM switching in support of medical applications. ATM switches (Fujitsu's) have been deployed in North Carolina to support the North Carolina Information Highway for distance-learning applications. Further ATM deployments are planned to support frame relay and SMDS.

14.2.4 NYNEX fastpacket-ATM efforts

NYNEX deployed frame-relay PVC service in 1992. An ATM platform (Fujitsu-based) was announced for 1994 deployment. The carrier also announced Media Broadband Service, which is a technology trial using multimedia for medical and publishing applications.

14.2.5 Pacific Bell fastpacket-ATM efforts

SMDS was deployed in 1992 and frame relay in late 1993. ATM switches (Newbridge's initially; AT&T's later) were deployed at the end of 1993 in San Francisco and 1994 in Los Angeles. Cell-relay PVC market trials were initiated in early 1994, and tariffs were filed and approved. The carrier has established a CalREN program that involves Pacific Bell funding of application development using ATM services.

14.2.6 SBC Communications fastpacket-ATM efforts

Frame-relay service was introduced in 1992. An ATM technology trial known as Project Lindbergh was underway at press time. An RFI-RFP on ATM switches was issued in late 1993, with a decision expected by early 1995.

14.2.7 U S WEST fastpacket-ATM efforts

SMDS, frame relay, and transparent LAN interconnection service (TLS) were deployed in 1992. An ATM trial known as COMPASS was underway at press time, connecting three ATM switches. ATM services are being introduced in phases: In late 1993, Newbridge switches were selected as edge devices for a customer trial; two customer application trials of ATM service were undertaken in early 1994; other applications for leading-edge customers were planned for 1995.

14.3 Video and Video Dialtone Activities

This section reports about press-time RBOC activities for VDT services. Table 14.2 provides a snapshot of recent announcements and undertakings.

TABLE 14.2 Telephone Companies' Video Activities and Suppliers, Early 1995
Courtesy: *Telephony,* November 7, 1994.

	Ameritech
Deployment:	Commercial service to 1.2 million homes in suburban Chicago; Columbus, Ohio; Detroit; Indianapolis; Milwaukee. Pending approval.
Transmission:	ADC Telecommunications (fiber piece); Scientific Atlanta (coax piece).
Switch(es):	ATM switches from AT&T; maybe others.
Other:	Supertrunking equipment from ADC; headend component from Scientific Atlanta.
	Bell Atlantic
Deployment:	$11 billion regionwide full-service network using HFC, FTTC, and ADSL. 6 locations filed 214. Pending approval.
Setup box:	General Instrument (about 30%); others to be named.
Server(s):	Bell Atlantic Video Services' server will be supplied by nCube.
Software:	Oracle and nCube.
Transmission:	AT&T, BroadBand Technologies, and General Instrument.
Switch(es):	ATM switch from Alcatel and Siemens.
Deployment:	ADSL trial in Arlington, Va., in place. Awaiting FCC approval for expanded market of ADSL in the area.
Setup box:	Compression Labs Inc. is providing the decoder function only for the technical trial. IBM and Divicom will provide the setup boxes for the market trial.
Server(s):	For the technical trial, IBM. Bell Atlantic Video Services will use an nCube server and Sequent hardware for the market trial; other information providers will choose their own servers.
Software:	IBM is providing the software for the technical trial. Bell Atlantic Video Services will use software from Oracle for the market trial.
Transmission:	ADSL transmission equipment for market and technical trials is from Westell; DSC Communications Corp. is providing the cross-connect for both.
Switch(es):	DSC is providing the switch for the market trial of ADSL.
Other:	DSC server controller.
Deployment:	Fiber-To-The-Curb trials in Dover Township, N.J., and in 3 communities in Morris County, N.J. Approval received for Dover Township; awaiting approval for Morris County.
Setup box:	Philips Consumer Electronics and Compression Labs Inc. consortium.
Transmission:	BroadBand Technologies.
Switch(es):	No server will be used initially.
Other:	Cables from Alcatel and Siecor.
	BellSouth
Deployment:	Trial in Chamblee, Ga. Awaiting FCC approval.
Setup box:	Scientific Atlanta.
Server(s):	Hewlett-Packard.
Software:	Oracle Corp. (server).
Transmission:	Scientific Atlanta (fiber/coax)

TABLE 14.2 Telephone Companies' Video Activities and Suppliers, Early 1995 (*Continued*) Courtesy: *Telephony,* November 7, 1994.

	GTE
Deployment:	Manassas, Va., trial 214 application to FCC not yet filed. AT&T will provide all components, including setup boxes, a server, an ATM switch, HFC transmission equipment and digital encoding equipment as part of its end-to-end solution.
Deployment:	Cerritos, Calif. Started as a trial in 1990, operations continue under a court stay.
Setup box:	GTE Main Street and GTE ImagiTrek (both part of marketing test); and Apollo Cablevision.
Server(s):	None used (video stored on tapes in specially designed device).
Software:	GTE.
Transmission:	Scientific Atlanta (coax); AT&T, BroadBand Technologies, American Lightwave Systems (FTTH).
Switch(es):	GTE Labs broadband switch.
Deployment:	Commercial rollout of HFC network planned for Clearwater/St. Petersburg, Fla.; northern Virginia (including Manassas); Ventura County, Calif.; and Honolulu.
	NYNEX
Deployment:	Manhattan trial.
Setup box:	Zenith (digital).
Server(s):	Digital Equipment Corp.
Software:	On-Demand Technologies.
Transmission:	ADC Telecommunications (HFC); ADC's American Lightwave Systems (fiber loop access equipment).
Switch(es):	Dynair Electronics.
Other:	Stratus Computer Inc. is providing the fault-tolerant computer that is serving as the video system controller. Fiber modulators from Scientific Atlanta.
Deployment:	Warwick, R.I., $130 million deployment to 60,000 homes. Awaiting 214 approval.
Server(s):	Digital Equipment Corp.
Transmission:	Raynet FITL equipment.
Switch(es):	Fujitsu ATM switches.
	Pacific Telesis
Deployment:	Commercial service to 100,000 homes in San Diego, Orange County, Los Angeles, and Silicon Valley.
Server(s):	Hewlett-Packard (Level 2 server).
Software:	Sybase, Oracle Corp., Microsoft Transmission: AT&T.
Other:	AT&T and Andersen Consulting working with Pacific Bell to develop billing system.
	Rochester Telephone
Deployment:	HFC test of video on demand in Brighton, N.Y., in place. ADSL may be tested later.
Setup box:	USA Video.
Server(s):	USA Video.
Software:	USA Video.
Transmission:	ADC Telecommunications (HFC).

TABLE 14.2 Telephone Companies' Video Activities and Suppliers, Early 1995 (*Continued*) Courtesy: *Telephony,* November 7, 1994.

	SNET
Deployment:	$4.5 billion investment to roll out commercial interactive services over integrated voice, video, data network.
Transmission:	AT&T (HFC).
Deployment:	West Hartford, Conn., trial of video on demand over HFC in place.
	SBC Communications
Deployment:	FTTC trial in Richardson, Texas, pending FCC approval.
Software:	Microsoft.
Transmission:	AT&T for telephony; BBT, a subcontractor of AT&T, switched digital FTTC.
	Sprint
Deployment:	Wake Forest, N.C., trial using HFC, awaiting 214 approval.
Transmission:	Philips, Scientific Atlanta, C-Cor, Texscan, Jerrold/General Instrument (amplifiers); Commscope, Times Fiber, Trilogy (main line cable and drop cable); Scientific Atlanta, Jerrold/GI, Regal (taps); Scientific Atlanta, Jerrold/GI, Regal (splitter); Gilber, PPC, LRC, Raychem, Pyramid (connectors).
Other:	Interdiction equipment will be used.
	U S WEST
Deployment:	Omaha, Neb., interactive network trial, in place.
Setup box:	Scientific Atlanta.
Server(s):	Digital Equipment Corp.
Software:	3DO (setup box chips and hardware).
Transmission:	Scientific Atlanta (video); AT&T (telephony components).
Other:	Century Communications is providing subscriber-management software; Alpha Technologies is providing power equipment; interdiction equipment from Scientific Atlanta.
Deployment:	Broadband deployment of $750 million in Denver; Twin Cities; Portland, Ore.; Boise, Idaho, planned. AT&T and DSC will provide telephony components.

14.3.1 Ameritech

In 1993, Ameritech filed complaints with U.S. district courts in Illinois and Michigan challenging the restrictions of the 1984 Cable Act. In January 1994, Ameritech announced that it was planning to spend $4.4 billion to bring video services to 6 million customers by 1999. In February 1994, the RBOC filed applications with the FCC to deliver VDT services to more than 1.26 million customers in its region in 134 communities, starting in 1994.[7] The initial customer base included 501,000 customers in the suburbs of Chicago, 262,000 in Columbus, 232,000 in Detroit, 115,000 in Indianapolis, and 146,000 in Milwaukee (the filing requested approval for 134 communities). Customers included residential consumers, schools, hospitals, libraries, and other institutions; education was given emphasis in the various announcements. The plans are to add 1 million customers per year to reach the 6 million total by 1999. Ameritech also announced the intention of making a total investment of $29 billion by 2009.[7]

The 750-MHz HFC system planned for initial use enables the delivery of up to 390 analog and digital channels (70 analog broadcast, 240 digital broadcast, and 80 digital on demand; the actual number of digital channels will depend on the compression standard chosen by the VIP). The switched channels will be supported using ATM switches from AT&T (at this time, the ATM network supporting residential video is separate from the network supporting business data applications). At press time, Ameritech was in the process of awarding contracts to video equipment including video servers and intelligent setup boxes with remote control.[8] The company was expected to select Digital Equipment Corporation as the supplier of video servers.[9] VIPs were also being identified; as of the time of this writing, no teaming or partnerships had been announced. One VIP was expected to act as the manager of the common block of the broadcast channels, in turn selecting programming on a nondiscriminatory basis from multiple providers (as noted in Chap. 3, until cross-ownership is lifted, an RBOC is precluded from the selection, pricing, and packaging of programming). Services include VOD, interactive education, MPGs, entertainment, and information services. Additional information on this HFC system was provided in Chap. 10.

14.3.2 Bell Atlantic

Bell Atlantic was the first Bell Operating Company receiving FCC permission in July 1994 to offer interactive television services to its customer base.*[10] Bell Atlantic created the *Bell Atlantic Video Service* (BVS) subsidiary to deploy interactive broadband networks. The carrier is viewed by observers as an early adopter of technology.[11] Bell Atlantic is following several VDT avenues, with the goal of deploying a regionwide, feature-rich system. After challenging the constitutionality of the restrictions of the Cable Act of 1994, they initiated a number of trials.

Much press was made on October 13, 1993, when Bell Atlantic and TCI announced the intention to merge. The now defunct deal was val-

---

* In early 1995, Bell Atlantic (along with GTE) was also given permission to produce and distribute their own programming by a U.S. District Court in Virginia. This decision opens the discussion as to whether Bell Atlantic must negotiate franchise agreements with the city of Arlington. The FCC has placed “strict safeguards” on Bell Atlantic “to prevent its telephone ratepayers from cross-subsidizing the new video services and to ensure fair competition among providers of multichannel video delivery services.”

ued at $33 billion and would have given Bell Atlantic access to about 25 percent of the cable subscribers. Reasons offered for the termination of the deal include mandated FCC rate cuts, uncertain regulatory landscape, diverging corporate culture, and stock market reaction.

From 1994 to the year 2010, Bell Atlantic plans to spend $15 billion to replace the twisted-pair plant with a fiber plant, with the intention of reaching 1.2 million households in 1995 and 9 million households in 2000 in the top 25 markets.[12] The carrier was targeting to enter six major markets in 1995.

The network access technology (local loop transport) for their Stage 1 Video Dialtone Service is ADSL at 1.544 Mbps for the downstream channel and a 16-kbps two-way channel for signaling at the VIU-NI. Concentration of the signaling channels from discrete VIUs is undertaken in the network; the combined stream is delivered to the VIPs at 56 kbps. The user plane network technology is digital crossconnect-based circuit switching.[13] This system was discussed in Chap. 5.

The Alexandria, Virginia, trial uses ADSL (trial started in 1994), delivering MPEG-1 compressed movies on demand at 1.544 Mbps.[14, 15] The information is sent through the switched network to the decoder in the TV's setup box, which regenerates the NTSC format for the traditional TV. This trial started out with Bell Atlantic employees and evolved to a market trial of 2000 consumers by April 1994; the goal of the $160 million life cycle investment is to reach 300,000 homes and generate a yearly cash flow of $115 million by 2003.

Bell Atlantic has also built a two-way FTTC system in Morris County, N.J., and was planning to lease the system to the entrenched cable operator, Sammons[14] (the decision was later rescinded). It is also planning to deploy a similar system in Dover Township, N.J., and lease it to Futurevision, to compete against the entrenched cable operator, Adelphia Cable. Key applications being investigated for the VDT effort are VOD, interactive shopping and banking, games, education and health programs, and tailored advertisement. The agreement is to provide VDT service to approximately 38,000 homes in Dover Township, HFC technology, in competition with the local cable operator.

Bell Atlantic's New Jersey trial initially employed HFC equipment from BroadBand Technologies (BBT, Durham, N.C.). The HDT combines telephony signals from the voice switch and digital video signals from a digital headend (i.e., a VIP) and sends them over a traditional digital (SONET) fiber. (A device at the cable headend converts the analog TV signal to a compressed digital signal.) In this FTTN architecture, the node close to the domicile segregates the video and telephony into a coaxial cable for video delivery to a digital setup box and into a twisted pair for regular (POTS) telephony to a standard telephone (this architecture was illustrated in Chap. 11). BBT's setup box was developed in cooperation with Philips Consumer Electronics and Compression Labs.[6, 16] The system is digital end-to-end. In Phase 1 equipment, the headend provides for analog-to-digital conversion (to DS3) and for multiplexing of 64 video channels. The HDT provides a circuit-switching function at the DS3 level. The home-based digital video terminal terminates the coaxial line supporting three DS3s, provides the digital-to-analog conversion (from DS3 to NTSC), and supports signaling. In Phase 2 equipment, the headend becomes an ATM multiplexer; it provides for analog-to-digital conversion and compression (to MPEG-1 and/or MPEG-2) and for multiplexing of over 1500 video channels. The HDT provides a circuit-switching function at the DS3 level. The home-based digital video terminal terminates the coaxial line, undertakes cell extraction and decoding, provides the digital-to-analog conversion (from DS3 to MPEG-1 and/or MPEG-2), and supports signaling.

As with other HFC architectures, this approach entails installing fiber relatively close to the served domiciles to avoid long coaxial runs, with ensuing amplification and noise problems. Carriers, as well as cable companies, are already in the process of installing such fiber, and unlike other systems that use multiple fiber, analog-modulated fiber, or both, this approach uses a single digitally modulated fiber. The trial employs a star rather than bus (coaxial) topology and delivers video to the user via switching methods. Customers have access to a bidirectional information stream, although the bandwidth is asymmetric.

Bell Atlantic also announced a partnership with CellularVision to build, market, and operate a wireless cable system in New York City. The technology discussed in Chap. 12 uses 28-GHz radio to deliver up to 49 TV channels using architectures similar to cellular telephony.

In another venture, Bell Atlantic Video Services and Knight-Ridder, Inc., announced a plan to jointly develop news, information, and ad services, which the Bell Atlantic's Stargazer interactive multimedia system will deliver. The trial was set for Washington, D.C., in 1994, with regionwide rollout in 1995.[17] In addition, Bell Atlantic is working with retailers such as JCPenney, Land's End, and Nordstrom to develop video catalogs to facilitate the introduction of teleshopping services. Bell Atlantic has also announced a strategic alliance with Oracle to support multimedia services through Oracle's software and hardware from nCube. Under the agreement, Oracle will supply Bell Atlantic with video servers that run on hardware designed by nCube (nCube is owned almost entirely by Oracle's CEO). The deal is valued at $25 million.[18, 19] Supercomputer sales, affected by the decreases in military spending, are looking for a new outlet. With respect to content, Bell Atlantic has established video production studios in Virginia (see footnote, p. 450). The company is also focusing on building relationships with a number of content providers.

In mid-1994, Bell Atlantic announced that it chose AT&T Corp. and General Instrument Corp. to equip its $11 billion multimedia-VDT network.[11] General Instrument is to supply one million setup boxes in the next five years, with the deal estimated at $1 billion.[20] The setup boxes support VOD, I-TV, games, and other services; encryption is also provided. As part of this announcement, Broadband Technologies was picked to provide an HFC optic transmission system for networks serving as many as 50,000 homes in northern New Jersey.[12] Bell Atlantic aims at rewiring its major metropolitan service areas: northern New Jersey, Philadelphia, Pittsburgh, Baltimore, Washington, D.C., and Virginia Beach. The HFC-FTTN system serves 500 homes per node. One million customers were planned to be connected by the end of 1995.[3]

As noted earlier, Bell Atlantic was the first Bell Operating Company receiving FCC permission in July 1994 to offer interactive television services to its customer base. At that time, the FCC approved an application by Bell Atlantic to offer the service for an initial set of 38,000 homes. As reported in *The New York Times,*[10]

> This is the first time that the FCC has cleared the way for a telephone company to offer commercial cable television, and it suggests similar action on a backlog of 21 similar requests from other regional Bell companies that would cover millions of homes nationwide. Dozens of applications have been pending before the commission, some of them for more than one year. Though the agency has in principle been enthusiastic about competition between telephone and cable television, officials have been hesitant thus far to approve anything more than technical trials. The cable TV industry, meanwhile, has filed . . . objections in an attempt to stall their well-financed would-be rivals. Bell Atlantic executives said they intended to build an advanced fiber optic network within the next year capable of transmitting 384 channels of television in Toms River, N.J. The company said it would immediately undercut existing cable television prices by 20 percent, and executives predicted that competition would eventually drive prices even lower. . . . Initially, the new system will offer 60 to 80 channels of television and a modest array of interactive features. Among them will be the ability to subscribe to premium channels or pay-per-view movies and sporting events just by clicking a remote control device at on-screen menus. Local businesses will be able to send electronic messages to individual homes, and customers will be able to request more information in response to advertisements. In its decision the FCC said Bell Atlantic's plan would "promote the public interest" by stimulating investment in new technology, increasing competition in video services and promote lower prices for consumers. "A new day is dawning," FCC Commissioner Susan Ness said. "No longer will telephone companies simply provide telephone services and Cable companies merely provide video programming services." . . . But the commission also ordered Bell Atlantic to build a network that could accommodate independent programmers,

> separate from Futurevision, its designated partner. Bell Atlantic had proposed that its system would provide less than 100 channels initially, and all but a tiny fraction of that capacity would have been allocated to Futurevision. Under the terms . . . Bell Atlantic will have to start transmission with a system that delivers 384 channels. The National Cable Television Association, which has opposed the telephone companies' efforts to enter the cable business, immediately vowed to fight the decision in court. . . .

14.3.3 BellSouth

BellSouth has been involved in a trial with Viacom and AT&T that could lead to a nationwide service. In late 1993, BellSouth acquired over 20 percent ownership in Prime Management, which is a cable system operator based in Texas serving over 500,000 subscribers in Houston, Chicago, and Las Vegas. The deal was worth $250 million. BellSouth invested $1.5 in QVC as part of the Paramount acquisition proceedings ($1 billion in QVC stock, $0.5 billion in Liberty Media stock) to research, develop, and deploy I-TV services for domestic and international markets. It also owns Hospitality Network, which provides interactive TV services in hotels.

By early 1994, BellSouth had FTTC applications in one Kentucky location, one South Carolina location, one Georgia location, one Tennessee location, and one Florida location; it had FTTH application in one Tennessee location, two North Carolina locations, two south Carolina locations, one Georgia location, and three Florida locations.[1] BellSouth will use Sybase Inc. for interactive television software; initially the company had chosen Oracle. In the BellSouth test, Sybase's Intermedia software will be used to run such a network for 12,000 homes in Chamblee, Ga. At press time, BellSouth was waiting for regulatory approval to start the test by the middle of 1995.

14.3.4 NYNEX

NYNEX is known to be vigorously developing a VDT architecture for trial and deployment.[9] It has teamed up with Viacom International, investing $1.2 billion for video-related services.

Industry contacts report that in late 1993 NYNEX initiated efforts to develop and implement over the next few years a broadband network infrastructure to support multimedia, video, and telephony. Services such as video dialtone, fastpacket data services, and enterprise services will share this common infrastructure, as will video entertainment services (e.g., VOD), individual communication services (e.g., PCS, home education), and business network services (e.g., messaging, media communication, and video information access). The common infrastructure consists of a broadband network infrastructure and an information management infrastructure. The broadband infrastructure encompasses switching, transport, access, video servers, multimedia servers, and operation support capabilities. The information management infrastructure includes service and information management systems, as well as subscriber and provider systems. More specifically, this includes service management, resource management, session management, gateway, information management, and transaction management systems.

One can envision this as a three-layer architecture: the network infrastructure is at the lowest logical layer; the software management capabilities, which might be thought of as an information management operating system, are in a middle layer; and all the application segments sit at the highest layer. Whether VIPs will be involved or whether NYNEX plans to provide the video and multimedia servers directly remains unclear; one scenario is to provide the servers as part of the infrastructure but have the VIPs provide the actual content.

In early 1994, NYNEX announced that it will deploy a broadband fiber-optic/coaxial cable network reaching between 1.5 and 2 million customers by the end of 1996. The first 60,000 lines were planned for Warwick, R.I., for the middle of 1995. There was no indication at the time where the other lines were to be deployed; the company indicated that the final rollout plan depended on the regulatory climate in the states where NYNEX does business (Rhode Island was selected first because of the investment climate in that state).[21] Initially, NYNEX will deploy the broadband network as an overlay (i.e., the broadband network carries video, while telephony services remain on the existing network), since integrated architectures are not yet commercially available. By late 1995, NYNEX is planning to deploy integrated HFC systems operating at 750 MHz that carry telephony, broadcast video, and I-TV services.

After 1996, NYNEX is publicly committed to deploy the HFC network at the rate of about 1 million lines per year (NYNEX has 15.5 million lines). The point has been made publicly that, as yet, NYNEX is not committing to universal broadband service because "the technology does not appear to be economic to deploy in some of our rural areas." In those other areas, NYNEX is tentatively considering wireless communications and/or ADSL-1 or ADSL-3.[21]

NYNEX was expected to select Digital Equipment Corporation as the supplier of video servers as it ventured into VOD trials in New York City, Portland, Maine, and eastern Massachusetts, pending regulatory approval.[9, 22, 23]

NYNEX is also approaching the Japanese market, where only 3 percent of the households now have cable. NYNEX is working on a pilot project to deliver cable-telephony services through the Yokahama cable system.[24]

14.3.5 Pacific Bell

Pacific Bell announced the "California First" initiative. The $16 billion plan aims at establishing VDT and other advanced services to 5 million homes by 2000 (1.3 million homes by early 1997).[12] In early 1994, the Pacific Telesis Group asked the FCC for permission to provide video services over the telephone network to 1.3 million homes in California, the first step in the $16 billion plan to build an advanced network.[1] AT&T HFC equipment will be used, and AT&T is acting as the system integrator (Hewlett-Packard will supply the video servers in a deal valued at $20 million; four such HP 9000 servers were purchased in 1994). Additional information on this HFC system was provided in Chap. 10. At the same time, they filed a suit against the 1984 Cable Act.

In early 1994, Pacific Telesis Video Services (an unregulated subsidiary of Pacific Telesis) announced an early VDT trial in Malpitas, Calif., serving 1000 customers by the end of 1994. The AT&T video equipment used for the trial, including the GCNS-2000 ATM switch, allows customers to view video in multiple windows on their TV screens, preview programming by watching video clips, and navigate and select programming offerings.[25] AT&T will also provide some programming from New York City using DS3 facilities.

As part of the plan just described, Pacific Telesis was planning to spend at least $178 million between 1994 and 1996 to build a fiber-coaxial network to deliver video in Los Angeles, San Francisco, San Diego, and Orange County. Although several RBOCs have asked regulatory approval to provide video services on a trial basis, as of the time of this writing, only Bell Atlantic and Pacific Telesis had specifically asked to provide VDT services on a commercial basis. Such approval was expected to be obtained, with initial service to some homes by early 1995.[1]

Pacific Bell is planning initial services to 100,000 residences in San Diego, Orange County, Los Angeles, and communities in Silicon Valley through an HFC system employing nodes that support 450 homes each.

14.3.6 SBC Communications Inc.

After aggressively pursuing cable TV and broadband ventures outside its territory in 1993, SBC Communications (formerly Southwestern Bell) made its first in-region video move by announcing a video service trial in Richardson, Texas, in early 1994.[8] SBC and a newly formed subsidiary, Southwestern Bell Video Services, filed suit in U.S. district court seeking to overturn the Cable Communications Act of 1984, which, as noted in Chap. 3, prevents RBOCs from holding companies offering video services within their operating region. Unlike some other RBOCs, SBC reportedly decided to wait for a court ruling to provide full video services in its region (VDT transport platform as well as programming).

In the meantime, the infrastructure to serve 2000 homes in Richardson by 1995 was being put in place. The trial, to last one year, is both a technology and market trial. The transport architecture is an HFC system that carries both voice and video to between 100 and 400 homes per node. Video services include both existing cable programming and PPV and VOD. The company is also investigating interactive games, shopping, and education.[8] Southwestern Bell Video Services will acquire and provide all programming for the trial (programmers were still being selected at press time).

The architecture was expected to employ ATM switches to support services on demand; the video servers were to be placed at the central office as well as in vaults within the neighborhoods. The issue of where to decompress the video signal (e.g., at a network interface device just outside the customer's home or at the setup box) was under investigation. Other areas that were being studied included the issue of different billing options, for example, billing customers separately for video and voice services or providing a single itemized bill. Willingness to pay is another area of study.

AT&T is assisting SBC Communications to establish a VOD and I-TV trial reaching as many as 47,000 households by the end of 1996.

There also have been the out-of-region activities, including purchasing two cable systems from the former Hauser Communications in Washington, D.C., and the now defunct Cox deal. The $650 million Hauser deal enabled SBC to reach 225,000 subscribers in 400,000 passed homes. The Cox Cable deal entailed a $1.6 billion investment for 40 percent of the ownership, with an option to increase. The arrangement would have provided 21 out-of-region cable systems (excluding Cox's systems in the SBC region and in the Hauser systems), reaching 1.6 million subscribers. In mid-1994, SBC announced that it was going to offer telephony services in Bell Atlantic's territory over the cable system it acquired earlier in the year from Hauser Communications. A $100 million upgrade to the cable system is needed to add switching capabilities. This is an early example of inter-RBOC competition.[26]

14.3.7 U S WEST

In 1993, U S WEST issued an RFP to supplies to submit bids for the components of a broadband telecommunications network capable of providing video, voice, and data communications. The network is aimed at both residential and business customers, and, in the spirit of VDT, "the network will ultimately make it as easy to dial up video programs as it is to place a phone call today . . ."[27]

In 1993, U S WEST and Time Warner announced an alliance to develop what they labeled *electronic superhighways.* U S WEST spent $2.5 billion to acquire about 26 percent of Time Warner Entertainment. U S WEST and Time Warner have forged a strategic alliance to build the Full-Service Network discussed in Chap. 10. In the deal, U S WEST obtains access to franchise areas outside their own region as well as Time Warner's library of TV shows and movies. Time Warner secured cash and access to telephony technology and subscriber management systems.

A VDT technical-market trial was planned for Omaha at press time, involving as many as 60,000 subscribers.[28] The system being tested is an HFC technology developed by suppliers such as 3DO, AT&T, Digital Equipment Corporation, DSC, Scientific Atlanta, and Oracle. Services include VOD, PPV, games, and interactive shopping.

U S WEST has made announcements about VDT services to be extended (by 1995) to 330,000 subscribers in Denver, 290,000 subscribers in Minneapolis-St. Paul, and 130,000 subscribers in Portland. Expenditures on the project during 1994 to 1995 were put at three-quarters of a billion dollars. Beyond that there are plans for at least 15 other areas in the next five years (at an expense of about $2.5 billion). U S WEST is working with retailers such as JCPenney, Land's End, and Nordstrom to develop video catalogs to support interactive shopping.

U S WEST has also deployed a 10-site distance interactive learning system in Glendale, Arizona, that uses approximately 90 miles of fiber that employs WDM techniques to transport analog video on some wavelengths and digital information on other wavelengths. American Lightwave Systems, Inc., equipment is employed, supporting 16 channels. Each classroom is equipped with three cameras, four receive monitors, one transmit monitor, and one audio system. Each school transmits on one analog channel and receives all on the remaining nine analog channels. To receive a channel, a site tunes in to the channel in question.

At press time, U S WEST was in the process of acquiring, at the expense of $1.2 billion, two cable TV companies that serve contiguous regions in the Atlanta metropolitan area (Wometco and Georgia Cable Television). Combined, the two cable companies serve about a half-million subscribers. Contiguous networks are important if the company is to offer out-of-region telephony services to this area served by BellSouth; such an arrangement enables service interworking and achieves economies of scale in operating personnel. U S WEST was planning to spend about $750 per household between 1994 and 1998 to upgrade the system to support a bandwidth of 750 MHz.

In addition to the well-publicized 25.5 percent purchase relationship with Time Warner, U S WEST is working with TCI in the U.K.

Observers believe that these arrangements position U S WEST as a formidable force in the emerging interactive video market.

U S WEST is also approaching the Japanese market, working on a virtual reality project that could lead to services over a cable system owned by Japan's Tokyu Group, a major cable operator in Japan.[24]

14.3.8 SNET

At the end of 1993, federal regulators approved Southern New England Telephone's (SNET) Section 214 application to build a hybrid fiber-coaxial network in West Hartford, Conn., to offer video dialtone services. Set up as a trial, SNET offers video on demand, 18 channels of pay-per-view, in addition to broadcast TV and most cable TV networks currently offered by cable companies, including HBO, Showtime, and Nickelodeon.[11, 29] The trial network passes 1500 homes. SNET considers its trial in West Hartford a marketing trial (as opposed to a technology trial) and will test customer demand for an increased number of channels and advanced services.

SNET uses 110 Super-VHS cassette recorders to deliver on-demand video. Despite the lack of digital technology in its headend, SNET has implemented a reverse path in its plant so that customers can access over 1000 titles in its library via their set-top box remote control. As of July 1994, about 400 households had subscribed to the service—which competes directly with TCI in West Hartford—making it the largest trial of its kind to date. The company selected AT&T as the system integrator.[11]

In January 1994, SNET applied to the FCC to expand the 1500 West Hartford VDT trial to 150,000 homes in Hartford and Fairfield Counties. At press time, the company was awaiting FCC approval for such a commercial service; the service was planned for early 1995. In addition to movies on demand, SNET hopes to offer telephony, local news, interactive education and healthcare, and telecommuting to its portfolio of services. SNET hopes to add on-line services and Internet access and does not rule out the possibility of supporting personal computers in addition to TVs.

About the same time, SNET announced a 15-year plan to build a statewide broadband network, pledging to spend $4.5 billion. The plan was dubbed I-SNET. In April 1994, SNET selected AT&T Network Systems as strategic supplier for an initial 3-year, $1 billion Phase One construction plan designed to bring broadband connectivity to 500,000 customers. The AT&T package includes *host digital terminals* (HDTs), optical lasers and receivers, and network interface units. The deal represents a large commitment to hybrid fiber-coaxial architecture.

Recently, the governor of the state of Connecticut signed a bill opening every telephone market in the state to competition; by press time, 63 potential service providers applied for a license.[30] In response, SNET has undertaken the construction of the I-SNET broadband fiber-optic network that is planned to connect every customer in the state by 2009.

14.3.9 GTE

In early 1994, GTE Telephone Operations announced a plan for an interactive video trial, to include PPV and broadcasting services over an HFC system using AT&T equipment. The trial will involve 1000 residents in Manassas, Va. The company chose the site because a court ruling allowing Bell Atlantic to own and operate cable TV programming services within its operating territory in Virginia also applies to GTE, which has telephone operations in the same area.[25]

GTE was planning to provide VDT services to 550,000 homes by the end of 1995 through a cable TV-VIP partner, with an investment of $250 million.[31] Initial areas of coverage include northern Virginia, St. Petersburg-Clearwater, Fla., Thousand Oaks, Calif., and Honolulu, Hawaii. By 2003, the company anticipates offering broadband services, broadcast and VOD, and I-TV to 2 million homes, with the number to grow ultimately to 7 million in 66 markets. GTE will be using AT&T as the major supplier of video products. GTE's in-region service will operate over a separate network so that customers will get a second wire in their homes. The HFC-FTTN system is similar to the one announced by Pacific Bell and Bell Atlantic.

14.3.10 Rochester Telephone

Rochester Telephone favored deregulating the local loop. As a consequence, Time Warner was planning to offer telephony services on a fully interconnected basis. Rochester was expected to be the first market where telephone companies and cable companies compete in providing local telephone service to residential and business customers.[32] The plan received initial approval from the New York State regulators, and full approval was expected soon thereafter.

Reciprocally, Rochester Telephone has secured an agreement with USA Video to test market video and information services. The telephone companies provide transport and marketing coordination. USA Video provides in-home equipment, video servers, and a library of movies as well as music videos and recordings.

Interestingly, Rochester Telephone recently completed a study that suggests once again that there may not be enough demand for VOD services. For six months the company offered 100 movies, ranging from "how-to" titles to new releases from Universal, Warner Brothers, MGM,

Paramount, and HBO, to 52 households in a suburb of Rochester; prices varied from $0.49 to $3.99 per viewing. The take rate was low. Consequently, Rochester Telephone "has no current plans to offer the services to its regular customers . . . it cannot justify putting computers, setup boxes, and other equipment in place just to deliver VOD . . . . In the final analysis, VOD alone is not enough to sustain an economically viable service: it would not be sane to just jump in. To succeed, a commercial venture would have to have more types of services, such as online news, video telephony, shopping, and banking."[34] The test did indicate, however, that the technology to deliver VOD does work.

On other fronts, IBM, Rochester Telephone, and Cincinnati Bell Information Services have formed a team to develop a fiber-optic ring to link cable TV networks in Pennsylvania in competition to Bell Atlantic.[33]

14.3.11 Canadian companies

Two major Canadian cable operators (Rogers Cablesystems, Canada's largest MSO, and Shaw Cablesystems, the third largest Canadian operator) have advocated that cable operators and telephone companies jointly build and run multiservice fiber networks to avoid building side-by-side duplicated and unconnected networks. The telephone companies already have 34,000 miles of fiber, while the two largest Canadian cable companies, Rogers Cablesystems and Maclean Hunter (serving 35 percent of the 7 million subscribers), have 1200 miles of fiber. The Canadian telephone companies do seek an open communications market, and in late 1993 Stentor asked the Canadian Radio-TV and Telecommunications Commission to begin opening up the video business to telephone companies, which have a revenue stream of Canadian $13.7 billion, in exchange for opening up the local loop to competition from the cable companies, which have a revenue stream of Canadian $1.6 billion.[35] Bell Canada was already testing VOD for educational applications at two universities in 1993, with a hard-drive-based server that can support up to 30 simultaneous users.[36]

14.3.12 United Kingdom

The statistics of Table 14.3 give at least a static view of the United Kingdom telephone and cable market.[37] British Telecom was planning for a mid-1995 VDT trial to about 2500 households in the east of the United Kingdom. The services were to be provided free. BT has been prohibited to commercially transmit TV and video over its network until 2001, to allow cable TV companies to establish themselves. As an interesting note, home shopping does not appear to be as popular and profitable in the United Kingdom as it is in the United States.

TABLE 14.3 A Static View of the U.K. Telephone-Cable Market

Cable TV households passed	3.1 million
Cable TV household marketed	2.8 million
1993 cable TV households passed	1 million
1993 cable TV households passed	1.6 million
Cable TV subscribers	0.6 million
Cable TV penetration rate, %	22
Cable TV employees (permanent)	6,300
Cable TV employees (contract)	2,800
Residential telephone lines on cable	0.27 million
Business telephone lines on cable	0.04 million

In a related observation, European concerns have started to fund R&D in VOD and related technologies to prevent U.S. and foreign companies from controlling the European market. Filmnet, a 700,000-subscriber operator serving 13 countries, has taken the lead in this effort.

References

1. "Cable Optics, Information Gatekeepers," Boston, January 1994, pp. 1 ff.
2. "The Insight Research Corp. (Livingston, N.J.), Telephone Revenue Sources," *Cable World,* February 7, 1994, p. 25.
3. E. L. Andrews, "AT&T Picked to Build Bell Atlantic's Network," *The New York Times,* May 20, 1994, pp. D1 ff.
4. *Multimedia Week,* February 21, 1994, p. 3.
5. "Smart Boxes or Networks? Simple Questions Prove Tough to Answer," *Video Technology News,* February 14, 1994, pp. 4 ff.
6. A. Reinhardt, "Building the Data Highway," *Byte,* March 1994, pp. 46 ff.
7. C. Weinschenk, "Ameritech Asks FCC to OK Plans for VDT," *Cable World,* February 7, 1994, p. 7.
8. P. Bernier, "Southwestern Bell Gears up for Broadband Video Trial," *Telephony,* February 7, 1994, pp. 6 ff.
9. "NYNEX, Ameritech Are Said to Choose Digital Equipment Corporation to Supply Video-Server Systems," *Wall Street Journal,* February 25, 1994, p. B8.
10. E. L. Andrews, "FCC Allows Bell Atlantic to Offer Cable TV," *New York Times,* July 7, 1994, pp. D1 ff.
11. S. McCartney, "DSC Communications Doesn't Get Bell Atlantic Call," *The Wall Street Journal,* May 19, 1994.
12. J. J. Keller, "Bell Atlantic Throws Multimedia Dice, Kicking off $11 Billion Network Plan," *The Wall Street Journal,* May 20, 1994.
13. Bell Atlantic, *Signaling Specification for Video Dial Tone,* TR-72540, iss. 1, rel. 1, August 1993.
14. G. Lawton, "The Bits Are Coming: Deploying Digital For Video Services," *Computer Technology,* April 1993, pp. 38 ff.
15. R. Karpinski, "ADSL: Alive and (Seemingly) Well," *Telephony,* March 14, 1994, pp. 26 ff.
16. J. Yoshida and L. Wirbel, "Cashing In on Convergence," *OEM Magazine,* June 1993, pp. 24 ff.
17. C. Weinschenk, "Bell Atlantic, Knight-Ridder Team Up in Stargazer Effort," *Cable World,* February 7, 1994, p. 7.
18. "Bell Atlantic Selects Oracle Video Servers," *TE&M,* February 1, 1994, pp. 22 ff.
19. G. McWilliams, "They Can't Wait to Serve You," *Business Week,* January 24, 1994.

20. J. J. Keller, "Bell Atlantic Taps General Instrument, AT&T to Equip $11 Billion Network," *The Wall Street Journal,* May 19, 1994.
21. *The Cable-Telco Report,* February 28, 1994, p. 16.
22. E. Schroeder, "Video-on-demand," *PC Week,* February 21, 1994, p. 18.
23. "DEC Finds More Success With Video Server," *Network Week,* March 4, 1994, p. 1.
24. D. Ingelbrecht, "Opening Up Japan," *Cable World,* February 28, 1994, p. 2.
25. P. Bernier, "GTE, PacTel Video Choose AT&T Gear for Interactive Trials," *Telephony,* January 24, 1994, pp. 6 ff.
26. A. Bryant, "Southwestern Bell To Invade Bell Atlantic Phone Market," *The New York Times,* May 23, 1994, p. D1.
27. *Communications Technology,* April 1993, pp. 15 ff.
28. K. Patch, "Video Servers," *PC Week,* January 24, 1994, p. 111.
29. *Communications Technology,* December 1993, p. 12.
30. K. Johnson, "Acorns Sprout Among the Oaks of the Telecommunications Field," *The New York Times,* July 5, 1994, pp. 1 ff.
31. J. J. Keller, "GTE Corp. Plans to Offer Video Services by End of 1995, Still Seeks Cable Partner," *The Wall Street Journal,* May 25, 1994, p. B3.
32. G. Naik, "MFS to Offer Local Calling in Rochester," *The Wall Street Journal,* May 19, 1994.
33. "Cable Optics, Information Gatekeepers," Boston, January 1994, pp. 1 ff.
34. T. Steinert-Threlkeld, "How Much On-Demand Demand Is There?" *Inter@ctive Week,* February 13, 1995, p. 15.
35. J. Careless, "In Canada, Operators Gird for Telco Entry," *Cable World,* February 7, 1994, p. 19.
36. *Personal visit,* Ottawa, March 17, 18, 1994.
37. *Cable World,* February 7, 1994, p. 21.

Chapter

15

The Players: Cable TV Providers Moving to Telephony; Other Providers

There is a lot of public sentiment favoring competition for cable TV services. This chapter examines some of the dynamics of the cable industry in terms of competition, plans, technologies, trials, alliances, etc. Although this information is time-dependent, it is useful in documenting the earnest activities in the mid-1990s in the area of advanced video distribution systems. Many of the plans involving rollouts have a three- to seven-year window of implementation; therefore, although in the future there will be other alliances, plans, and strategies, many of the activities listed in this chapter will continue to be of relevance for the foreseeable future.

15.1 The Cable TV Industry

Since the 1950s, the cable TV industry has gone through three stages:

1. Retransmission of programming
2. Creation and delivery of programming specifically developed for cable
3. Interactive (digital) television services and other services, including voice and data

According to some, the two biggest changes since cable emerged 40 years ago were the development of addressable setup boxes (i.e., setup boxes with individual IDs that can accept messages broadcast through the cable system) and the introduction of analog video modulation over fiber media.[1] However, digital TV, compression, and ATM will represent an even more radical change.

In the recent past, there has been a trend for mergers and acquisitions, both within the cable industry as well as from the outside (e.g., RBOCs). The assessment in the early 1990s was that most cable TV companies could eventually be owned by an RBOC. This put pressure on both the cable TV companies to find a suitably acceptable buyer and on the RBOCs to find appropriate cable operators to take over. In particular, the proposed TCI-Bell Atlantic merger ignited a wave of expectations about additional consolidation and market activity. Clearly, the merger would have set a new stage for competition even among and between the RBOCs, because of TCI's national presence. Such acquisition trends slowed down after the reregulation of the cable TV industry via the Cable Act of 1992 and the following period of regulatory interpretation. Observers expect this trend to pick up again in 1995 to 1996. In particular, companies such as TCI and Cox are expected to be aggressive buyers of smaller cable TV outfits.[2]

Some feel that the RBOCs are orchestrating a preventive strike by deploying broadband ATM technology: effectively, the RBOCs are forcing the cable companies to come up with money, which they can generally ill afford, to upgrade their networks to match the technical improvements being made by the carriers. NTCA estimates that the cable operators need to raise $20 billion in the next few years to upgrade systems to compete with RBOC networks.[3]

Cable TV operators in the United States offer from 6 to 150 channels; the average is 36 channels. Thirty percent of the subscribers have cable TV systems supporting 54 or more channels; 63 percent of the subscribers have systems supporting from 30 to 54 channels; the other 7 percent of the subscribers have systems with 29 or fewer channels. Services include basic service, premium service, and PPV. Typical programs include broadcast and superstation TV, movies, music, sports, weather, shopping, special events, and some information services. Table 15.1 provides a forecast of cable TV penetration statistics for 1995.

Most of the 11,250 cable TV systems in the United States are one-way systems, although about a third of the operators polled indicated they were planning to deploy interactive systems in the near future. Fifteen percent of U.S. cable systems were two-way systems at press time, and of those, most suffer noise ingress that makes realization of even a single 6-MHz upstream channel nearly impossible.[1] The approaches to interactivity include two-way cable, telephone link, or a radio link established by the setup box. In general, cable TV companies have been slower in exploring, trialing, and introducing VOD services than the RBOCs.

TABLE 15.1 Estimated Cable TV Penetration Statistics (Projected 1995), Top MSOs

Cable company	Subscribers	Percentage of subscribers with fiber in the trunk
Tele Communications Inc. (TCI)	10,566,300	10
ATC/Time Warner Cable	7,120,300	42
Comcast Cable Communications	3,064,600	45(*)
Continental Cablevision	2,940,300	34
Cox Cable Communications	1,832,100	98
Cablevision Systems	1,830,900	7
Storer Cable	1,796,300	25
Jones Intercable	1,642,300	40
Newhouse Broadcasting	1,416,800	45
Total	32,209,900	
Percentage of total subscribers in United States	57	18†

NOTE: Forecast by author based on 1992 NCTA data.
* Estimated.
† This represents 10.2 million subscribers.

As noted in Chap. 2, video rentals are worth tens of billions of dollars per year, and cable TV operators hope to harness technology to obtain a portion of that market.[4] By combining traditional technology such as videotape recorders and automated tape libraries, new technology such as digital video storage, convenient navigation systems, and compression, programmers hope to gain market share in the near future. *Navigation systems* are computer programs that categorize and present large volumes of information in such a way that the viewer can rapidly and easily get around a database (e.g., a database of program listings) without complicated commands.

Some prognosticate that, while holding on to their video delivery business, cable companies could provide stiff competition to the RBOCs by providing local access to long distance carriers.[1] There also appears to be a major opportunity for cable carriers in the interconnection of several smaller, nonclustered networks into a single system, sharing capabilities such as mass storage, ad insertion, switching, and billing.[5] The key infrastructure need of the cable companies is to deploy fiber closer to the final delivery point, in the proximity of the domiciles. Prior to 1989, there was virtually no fiber in cable TV systems. In 1992 (the last year for which data was available at press time), the American market for fiber grew at a rate of 5 million km; the cable industry alone grew 100 percent in 1992 to 9 percent of the American market or 450,000 km. The 1993 deployment of fiber by cable companies was expected to grow by another 50 percent in 1993.[6] The cable companies are introducing fiber capable of supporting bidirectional communication. Some claim that a cable company's ability to deliver more channels (via fiber) will help them keep a franchise at renewal time.[7]

In August 1994, six of the major cable companies (TCI, Time Warner Cable, Continental Cablevision, Comcast, Cox Communications, and Viacom International), representing about 26 million subscribers, announced a $2 billion plan to offer a variety of local telephone services.[8] The project is to be overseen by Cable Television Laboratories Inc., the cable industry research consortium in Louisville, Colorado. The investment seeks to deploy network interface cards capable of supporting telephone and video connections at the consumer's domicile, as well as business and operations support systems that allow network management and billing for voice, video, and data services. Observers note that the cable TV companies have a "tremendous amount of work ahead of them" in this arena.[9]

Some observers note that, at the technical level, cable systems are moving in the same general direction as the networks of the RBOCs, which is why partnerships seem logical.[1] However, the defunct TCI and Cox deals show that the RBOCs may be wary of getting into other arenas than their own, that are highly regulated. In the context of the 1993 to 1994 reregulation of cable, the RBOCs may perceive cable TV to be too strictly regulated and/or unprofitable.

A cable TV company entering the local telephone market as a full-service provider (as contrasted to being a specialized or niche provider) faces at least two challenges: to obtain regulatory approval and to secure a satisfactory agreement with the existing LEC. The latter is dictated by the fact that subscribers expect to be able to call anyone who has a telephone, regardless of the carrier servicing the subscriber. This presents both engineering problems as well as financial problems. Ultimately, the competing carriers need to compensate each other for originating and terminating calls.

In addition to I-TV, cable TV companies are pursuing a number of service-business directions, including information services, AAP and other alliances, telephony, PCS, and data services.

Information services. Cable companies are starting to deliver information services over cable (see discussion on this topic at the end of Chap. 2). For example, over 20 million subscribers have access to the Web Sports Wire. Several hundred systems (about 800) support X*PRESS, which provides national and international news, financial news, sports information, weather, and educational information (the service is used by about 10,000 schools). Another service available to about 2.5 million subscribers is Tribune Media Services, which provides business, general news, and sporting information. Some cable operators (e.g., TCI with Digital Equipment Corporation's equipment) are adding a feature for the support of Ethernet services over cable TV for telecommuting applications (10 Mbps are supported up to a distance of about 75 miles). Another product, Zenith Electronics' Channel-Mizer, is a cable TV modem that features an "Ethernet" port for data transmission via a direct coaxial cable attachment (it allows four 0.5-Mbps subnetworks for a distance of 100 miles). Using this product, Zenith was working with Spry (Seattle, Wash.) to stimulate cable-based Internet access via Spry's Air Navigator, an inexpensive Windows-based software package.

AAP and other alliances. To strengthen their position, cable companies are pursuing a number of business alliances. For example, cable TV companies own, control, or have alliances with AAPs. At press time, there were more than 133 fiber-optic AAP networks in 72 U.S. cities; more than 36 other networks are under construction.[10] Relationships include Century Cable/Electric Lightwave, Continental Cablevision/Hyperion, Cox/Teleport, Jones Intercable/Jones Lightwave, TCI/Teleport, and Time Warner/Fibernet. The top five cable firms (TCI, Time Warner Cable, Continental Cablevision, Cox Cable Communications, and Comcast) jointly own Teleport Communications Group (TCG), which is the largest AAP (TCG connects local corporate users to interexchange-carrier POPs).[1] Much has been written in the press about the RBOCs-cable TV partnerships. These partnerships include:

- TCI and U S WEST (VOD trial in Denver using AT&T equipment)
- Cablevision of Loudon and Bell Atlantic (trial in Virginia)
- Liberty Cable and NYNEX
- Sammons (Morris County, N.J.) and Bell Atlantic (later rescinded)
- FutureVision (Toms River, N.J.) and Bell Atlantic

Table 15.2 depicts (on a permuted basis) some of the video services alliances that have been announced in the months preceding the publication of this text, to illustrate the extent of the interest on the part of cable operators, RBOCs, and equipment suppliers.

Telephone service. Consistent with the new competitive environment discussed in Chap. 3, cable TV companies are adding various types of telephony services, including fully digital, switched services over the coaxial cable carrying video, data, and voice. One example is Time Warner in Rochester, N.Y. Initially, customers could call only other customers on-net that had been accepted to use the service, but eventually, through colocation of equipment, there will be universal internetworking. Colocation is now beginning to happen. Time Warner also has received permission to provide limited local service in New York City and is planning to enter 24 other markets by the end of 1995. In early 1995, MFS Intelenet reached an interconnection agreement

TABLE 15.2 Some Alliances Pertaining to Video Distribution (Permuted)*

3DO, Digital Equipment Corporation, U S WEST
Apple, nCube, Oracle
AT&T, General Instruments, TCI
AT&T, Pacific Telesis
Bell Atlantic, CellularVision
Bell Atlantic, Compression Labs
Bell Atlantic, Oracle
BroadBand Technologies, Philips Electronics, Compression Labs
Cablevision Systems, Digital Equipment Corporation
CellularVision, Bell Atlantic
Cincinnati Bell Information Services, IBM, Rochester Telephone
Comcast, Continental Cablevision, Cox Cable Communications, TCI, Time Warner
Compression Labs, Bell Atlantic
Compression Labs, BroadBand Technologies, Philips Electronics
Continental Cablevision, Cox Cable Communications, TCI, Time Warner, Comcast
Cox Cable Communications, TCI, Time Warner, Comcast, Continental Cablevision
Digital Equipment Corporation, Cablevision Systems
Digital Equipment Corporation, General Instrument Corp.
Digital Equipment Corporation, Philips Electronics
Digital Equipment Corporation, Scientific Atlanta
Digital Equipment Corporation, U S WEST, 3DO
Digital Equipment Corporation, Zenith
General Instrument Corp., Digital Equipment Corporation
General Instruments, Intel, Microsoft
General Instruments, TCI, AT&T
IBM, Rochester Telephone, Cincinnati Bell Information Services
Intel, Microsoft, General Instruments
Interactive Network, TCI, Sega, Zing
Kaleida, Motorola, Scientific Atlanta
Microsoft, General Instruments, Intel
Microsoft, TCI
Microsoft, Time Warner, TCI
Motorola, Kaleida, Scientific Atlanta
nCube, Oracle, Apple
nCube, U S WEST, Oracle
Oracle, Bell Atlantic
Oracle, nCube, Apple
Oracle, nCube, U S WEST
Pacific Telesis, AT&T
Philips Electronics, Digital Equipment Corporation
Philips Electronics, Compression Labs, BroadBand Technologies
Rochester Telephone, Cincinnati Bell Information Services, IBM
Scientific Atlanta, Digital Equipment Corporation
Scientific Atlanta, Kaleida, Motorola
Scientific Atlanta, Toshiba, Time Warner, Silicon Graphics
Sega, Zing, Interactive Network, TCI
Silicon Graphics, Scientific Atlanta, Toshiba, Time Warner
TCI, AT&T, General Instruments
TCI, Microsoft
TCI, Microsoft, Time Warner
TCI, Sega, Zing, Interactive Network
TCI, Time Warner, Comcast, Continental Cablevision, Cox Cable Communications
Time Warner, Comcast, Continental Cablevision, Cox Cable Communications, TCI
Time Warner, Silicon Graphics, Scientific Atlanta, Toshiba
Time Warner, TCI, Microsoft
Toshiba, Time Warner, Silicon Graphics, Scientific Atlanta
U S WEST, 3DO, Digital Equipment Corporation
U S WEST, Oracle, nCube
Zenith, Digital Equipment Corporation
Zing, Interactive Network, TCI, Sega

* Although this information is time-dependent, it illustrates the widespread interest in video services on the part of suppliers, providers, cable companies, and RBOCs.

with NYNEX; under this agreement, NYNEX will pay MFS for completing calls from NYNEX customers to MFS customers, and vice versa. Similar deals are likely in the future between other AAPs and other RBOCs. In July 1994, Cablevision Systems announced that it was starting to carry local calls for two major businesses in Long Island.[10]

PCS. In the PCS arena, the cable companies have acquired many experimental licenses, while waiting for FCC allocation of spectrum (see Table 15.3). The cable facilities can be used for trunk backbones to connect cell sites and other mobile service areas. As an example, Cox Cable Communications sees itself as a leader in the development of PCS, as illustrated by the FCC's award of one of only three Pioneers' Preferences; Cox has indicated publicly the desire to "continue to be a catalyst for broad cable participation in PCS."[5] Related to this, the FCC accumulated $216 million in July 1994 on its auction of small segments of the radio spectrum to be used for interactive television services such as home-shopping, advertising, and information services. The monetary figure achieved through the auction, which greatly exceeded expectations, indicates, according to observers, that companies believe interactive services will generate substantial revenue within a few years.[11]

Data services. As alluded to previously, a number of cable TV companies now support data services. Examples include Continental Cablevision, TCI/Hybrid Networks, and TCI/Zenith.[14] Some are planning to use these services in support of telecommuting. In some situations, the data transfer is only unidirectional, but, for example, telecommuters could use these services to download information from a headquarters building that has, itself, a bidirectional link to the Internet. Alternatively, the return path could be via a dial-up path, as one product on the market facilitates.

The delivery of program guide information represents another opportunity. The delivery of as many as 500 channels (or access through VDT to thousands of switched channels) raises the question of how a viewer can determine what is available. A market opportunity exists for the provision of user-friendly mechanisms to navigate through programming information for a small fee. For example, TCI and *TV Guide* are collaborating to provide an electronic program guide service called *TV Guide on Screen.* Additional players, such as United Video, Zenith/Starsight Telecast, and others may enter the market.

15.2 Activities Toward New Video Delivery Systems by Key Operators

This section looks at some activities of key MSOs. The next section looks at activities of some other providers and parties.

TABLE 15.3 Experimental PCS Licenses Accumulated by Cable TV Companies as of Early 1994

Alhambra, Calif.	Cencom Cable Associates
Baltimore, Md.	Comsat
Baton Rouge, La.	United Artists Cable (TCI)
Billings, Mont.	Satcom
Boston, Mass.	Cablevision Systems, Continental Cablevision
Brunswick, Maine	Susquehanna Cable
Buffalo, N.Y.	Adelphia Cable Communications
Chicago, Ill.	Cablevision Systems
Cincinnati, Ohio	Time Warner
Cleveland, Ohio	Cablevision Systems, Pertel
Columbus, Ohio	Time Warner
Concord, California	Concord TV Cable
Dayton, Ohio	Viacom
Denver, Colo.	United Artists Cable (TCI)
Fairfax, Va.	Media General Cable
Fuitondale, Ala.	Cencom Cable Associates
Grand Island, Neb.	Cable USA
Hastings, Neb.	Cable USA
Hermet, Calif.	Inland Valley Cablevision
Jacksonville, Fla.	Continental Cablevision
Kahului, Hawaii	Chronicle Cablevision Hawaii
Kearney, Neb.	Cable USA
Las Cruces, N.M.	La Cruces TV Cable
Liburn, Ga.	Wometco Cable TV
Miami, Fla.	Adelphia Cable Communications
Milwaukee, Wis.	Viacom
Missoula, Mont.	Satcom
Monterey, Calif.	Monterey Peninsula Cable
Nashville, Tenn.	Viacom
New York City, N.Y.	Cablevision Systems, Cox Enterprises, Time Warner
Oakland, N.J.	United Artists Cable (TCI)
Olivette, Mo.	Cencom Cable Associates
Omaha, Neb.	Cable USA
Pearl, Mass.	Susquehanna Cable
Philadelphia, Pa.	Comsat, Pertel
Pittsburgh, Pa.	Adelphia Cable Communications, Pertel
Providence, R.I.	Susquehanna Cable
Riverside, Calif.	Cencom Cable Associates
San Diego, Calif.	Cox Enterprises
San Francisco, Calif.	Viacom, Western Cable
Seattle, Wash.	Viacom
Spokane, Wash.	Satcom
St. Petersburg, Fla.	Time Warner
Stockton, Calif.	Continental Cablevision
Tequesta, Fla.	Adelphia Cable Communications
Toledo, Ohio	Buckeye Cablevision
Trenton, N.J.	Comsat
Tulsa, Okla.	United Artists Cable (TCI)
Union, N.J.	Suburban Cablevision
Ventura, Calif.	Ventura County Cablevision
West Palm Beach, Fla.	Comsat
Westchester, N.Y.	United Artists Cable (TCI)
York, Pa.	Susquehanna Cable

15.2.1 TCI

TCI is the largest MSO; it has presence in 49 states and operates in 60 of the top 100 markets (with a projected 10.6* million subscribers in 1995). In 1993, TCI announced that by 1996, 80 percent of its subscribers will be served by a fiber-based backbone system; it also announced that it would spend $2 billion over the next few years to upgrade its system with fiber nodes and support for video compression.[7] The initial goal was to install 7000 miles of fiber in 100 cities. Additional expenditures in the years following 1993 (to $2 billion) have been aimed at establishing a broadband infrastructure in 400 cities by 1996, which TCI offered up as being the information superhighway. Services to be provided include VOD, simultaneous delivery of multiple VOD movies per home, interactive teleshopping, interactive educational programming, and data and voice communication capabilities. TCI has been an early and consistent supporter of MPEG-2 digital compression standards.[13] TCI plans have enough capacity for 500 channels for a variety of business and consumer services, including traditional broadcast, PPV, VOD, videoconferencing, voice telephony, and on-line access.[1]

TCI is implementing a scheme that uses two coaxial cables to each feeder (one of the cables will be "dark" until sometime in the future). The primary cable will be configured with a downstream channel in the frequency range from 50 to 750 MHz or higher, and an upstream channel in the lower frequency band. Services are optimized in favor of information delivery. The upstream bandwidth available on the first cable supports telephony, two-way data, PCS, and H.261 video telephony. TCI's second cable is midsplit with the "free" portions of 500 MHz of bandwidth allocated in each direction. When activated, this cable would allow TCI to provide a range of services.[1] This approach is used in conjunction with HFC systems where six fibers support a node serving 600 homes. To the user, this means an upgrade from 450 to 550 MHz, or even to 750 MHz.

TCI has a majority control in United Artists and owns half of Showtime. It has shares in Discovery Communications (including the Discovery Channel), Heritage Media, Interactive Network, Liberty Media, Reiss Media, Teleport Communications Group, and Turner Broadcasting (which includes TBS, TNT, and CNN). Ownership ranges from a low of 15 percent to a high of 80 percent. TCI also owns 30 percent of the new Sprint-cable telephone company joint venture; it also has investments in TeleWest Communications (U.K.), and in several affiliated cable systems, interactive services, and television programming outfits.

---

* 11.7 million after the acquisition of TeleCable in 1995. TCI also reaches over 9 million additional households through affiliates.

Much press was made on October 13, 1993, when Bell Atlantic and TCI announced the intention to merge. The now defunct deal was valued at $33 billion and would have given Bell Atlantic access to about 25 percent of the U.S. cable subscribers. Many viewed the merger as being anticompetitive. The deal reportedly fell through because of the perceived diminution of the value of this (or any) cable TV company as a result of the Cable Act of 1992 and subsequent interpretation and rate reduction.

In early 1994, an alliance with Microsoft was announced for a trial and service rollout in Seattle in 1994 to 1995 that entails transmission (TCI) and software (Microsoft) facilities for I-TV services.[14] TCI was working with Hybrid Networks in a trial in California to download Internet data at 10 Mbps over cable*; the company also had another trial in Utah, using Zenith HomeWorks, which lets home users communicate with office-based computers over the cable network at 0.5 Mbps.[12] (HomeWorks, which sold at press time for $495, consists of an external cable modem with Ethernet output and a PC bus card.)

TCI, Home-Shopping Network, and Japan's Sumitomo Corp. announced plans in 1994 to set up an interactive home-shopping service in Tokyo, Japan, through Suginami Cable TV.[15]

15.2.2 Time Warner Cable

In early 1993, Time Warner announced the *Electronic Superhighway* initiative. Time Warner served approximately 7.2 million subscribers in 34 states in 1994. It plans to deliver entertainment, VOD, games, home shopping, telephony services, videoconferencing, and education. The company operates in 36 states (including 85,000 customers in U S WEST territory). Some see Time Warner's as a bold move, likely to make it a formidable competitor to the RBOCs. The publicly stated posture of Time Warner is to acquire "account control," achieved by providing a full-service network and selling the customer as many as possible of the services just listed. Time Warner owns Warner Brothers, HBO and Cinemax as programming entities. In July 1994, the company signed a five-year contract with AT&T for a fiber upgrade to its backbone network using 5ESS switches and SONET equipment. The upgrade will enable the company to offer voice and data services in Rochester, N.Y., at an early time, and by late 1995, in 25 other cities across the country.† [8]

---

* The Hybrid Networks product uses RF modem technology for 10-Mbps downstream data over subsplit frequencies and sends return data over conventional dial-up lines at 9.6 to 19.2 Kbps.

† A 22-mile fiber-optic ring was under construction in Manhattan, N.Y., at press time.

By the end of the century, Time Warner plans to have upgraded its plant in the major markets with an HFC architecture, with nodes serving 500 households. With this new infrastructure, the company can add about 25 more channels per household to the current base of 50 channels. Also, there will be an increased deployment of addressable setup boxes, to enable consumers to order PPV services using the remote control rather than a separate telephone line.

In late 1994 to 1995, Time Warner was running an interactive services trial of the Full-Service Network, discussed in Chap. 10, in Orlando, Fla. This trial uses ATM end-to-end, from the large video server to setup boxes in subscriber homes.[1] After two years in planning, Time Warner began testing its Full-Service Network in late 1994. Time Warner and Silicon Graphics employees were the first customers. Silicon Graphics designed the network's operating system and provided the hardware (that is, the video servers), the setup boxes' chips, and the storage vaults for the trial. This architecture also supports telephony services. The company sees major revenue opportunities in this larger telecommunications market. The challenge is related to obtaining permission to enter that market, but the company was reported to be ready to seek such permission on a market-by-market, state-by-state basis, if needed.

As noted in Chap. 4, Time Warner plans to extend cable technology for HFC to support 1 GHz. Time Warner's Quantum Cable TV system is being tested in Queens, N.Y., using First Pacific Network's technology for telephone service over a cable system. It employs a fiber-to-the-feeder (that is, FTTN) approach with a 1-GHz analog system. The system supports two-way video, data, and voice (it provides connection to MCI for interexchange service). Time Warner has applied to the FCC for a PCS license.

Time Warner has applied to become a full-service telephone carrier in Ohio, serving both residential and business customers. Services are provided over existing and new coaxial cable systems, as well as over HFC systems. Time Warner plans to introduce state-of-the-art digital switching equipment to support the service. As of press time, the company already operated as a competitive access provider in 15 markets, including such Ohio cities as Cincinnati, Columbus, and Lima.

Time Warner also announced that it plans to create an interactive news-on-demand service where customers are able to choose and control the content, length, and order of their news programming. The service was planned to be offered at Time Warner's services trial in Florida. The service includes local, national, and international news, business and financial news, sports, reviews, health care, and weather (the information can also be directed to a printer).[5]

Time Warner Cable's full-service network test in Orlando, Fla., will also deploy *TV Guide*'s On-Screen Interactive electronic program guide. The 4000 customers in the trial will use the navigator to access On-Screen Interactive, which enables the customer to select program listings by time, channel, and category.[16] (Time Warner Cable has more than 500,000 subscribers in Florida.[17])

15.2.3 Comcast Cable Communications

Comcast provides cable services to about 3.1 million subscribers and has a revenue stream of $1.3 billion. The company has been an advocate for liberalization of the regulation that prevents cable companies from offering switched telephony services. Comcast was in the news at press time for foiling the proposed CBS-QVC merger. The operator is pursuing a number of I-TV efforts, including a 20,000 subscriber trial in conjunction with Your Choice TV (a subsidiary of Discovery Communications). In addition, Comcast is a member of the Teleport alliance, along with TCI, Cox Cable, Continental Cablevision, and Time Warner Cable, owning 20 percent interest. More recently, Comcast and Viacom announced they would test delivery of Prodigy and America Online over cable systems in Castro Valley, Calif., and elsewhere, using Hybrid Networks' equipment.[1]

Comcast ordered 150,000 digital setup boxes from General Instrument for delivery in 1994, enabling it to offer VOD services. In late 1994, Comcast Cable was planning to test a high-speed cable modem developed with General Instrument and Intel; the 64-QAM technology, along with compression, can greatly increase the information-carrying capacity of its cable systems. Storer Cable Communications of Monmouth County and Storer Cable Communications of Ocean County, two New Jersey subsidiaries of Comcast Cable, are linked with two other cable operators (Adelphia Cablevision and Monmouth Cablevision) in a bicountywide, high-capacity, fiber-optic interconnection arrangement, offering video and data communications to about 80 communities in New Jersey. This advanced technology project may be replicated in other localities in the future. As noted earlier, these regional groupings may be advantageous in that they facilitate sharing of resources, such as video servers and gateways to the Internet. At press time, Comcast also had a telephony trial in the planning stage, the trial was targeted for late 1995 or early 1996.[8]

15.2.4 Continental Cablevision

Continental Cablevision owns coaxial cable and fiber-optic networks that support up to 50 channels of programming. The operator plans to continue to install several thousands of miles of fiber a year. In addition to basic and premium cable service offerings, Continental Cablevision offers PPV, locally produced programming, Cable in the Classroom, and Internet access. It has revenues in excess of $1 billion, serving about 3 million subscribers in 650 localities. In late 1993, Continental Cablevision and Time Warner Cable became the first cable TV companies to interconnect their networks to support end-to-end telephony services using PCS technology.

Continental Cablevision, along with Comcast Corp., Cox Cable, TCI, and Time Warner Entertainment, announced in early 1994 that they had agreed in principle to form a joint venture to develop new telecommunications services using digital video, fiber optics, and wireless technologies.[5] The goal was to speed up the availability of competitive access telephone service, as well as develop new advanced services such as video telephony. The cable companies involved in this arrangement are already in the process of upgrading their existing fiber-coaxial networks with digital video transmission and two-way communications capabilities.

In late 1993, Continental Cablevision announced that it was offering a 10-Mbps link over cable into the Internet for Cambridge customers. Continental has teamed up with Performance Systems International (PSI), an Internet access provider. The monthly cost of the service was estimated at $70 to $100, and the customer can access the network using a $150 Ethernet adapter.[12]

15.2.5 Cox Cable

Cox Cable has 24 systems in 17 states serving approximately 1.8 million customers. Its networks have a high concentration of fiber transmission. The company has been an advocate of the use of cable as a landline connection medium for PCS microcells. Cox Communications, Cox Cable's corporate parent, has used acquisitions and alliances aggressively over the years. The cable operator offers basic programming, premium programming, satellite programming, VOD (Your Choice TV service being given a trial by The Discovery Channel), access to Prodigy, and classified advertising and information (trial with BellSouth). In early 1994, Cox Cable announced an I-TV trial in Omaha, Neb., by mid-1994. The trial included VOD and other interactive services. The trial was planned for expansion to several thousand customers by late 1994, supporting services such as VOD, games, interactive multimedia guides to local dining and entertainment, electronic classified advertising, education, interactive catalogs, and other home-shopping services.[5] Cox Cable was planning to use technology from ICTV, which employs IBM digital servers with optical jukeboxes capable of storing thousands of movies, catalogs, and games.

15.2.6 Paramount

Paramount was in the news in 1993 and 1994 because of the vigorous battle for ownership. Viacom (along with Blockbuster Video and NYNEX) bid against QVC (along with Liberty Media Corp., TCI, Cox Enterprises, and advance Publications) and won. The Paramount-Viacom combination includes entertainment (TV, movies, theaters), TV networks (cable, pay TV, regional sports channels), and broadcasting.

15.2.7 Time Warner Entertainment

Time Warner Entertainment is a leading entertainment and media organization comprising a number of publishing (e.g., *Time, Sports Illustrated, Fortune, Money*), music, and film concerns. Time Warner Entertainment consists of:

1. Time Warner Cable, which is estimated to have 7.1 million subscribers in 1995 in 36 states (discussed earlier)
2. Warner Brothers Inc.
3. *Home Box Office* (HBO)
4. Part ownership in seven alternative access providers

As discussed in Chap. 10 and earlier in this chapter, the company is a leader in the industry in replacing coaxial systems with fiber-optic systems. ATM is also being pursued. U S WEST invested $2.5 billion in Time Warner Entertainment; however, under the Cable Act of 1984, Time Warner will need to divest itself of interests in cable companies in the U S WEST region.

15.3 Other Players

15.3.1 Computer hardware and software developers

The computer industry is eager to enter the video arena. Many computer vendors are partnering with the local and interexchange carriers and with the cable TV companies to be part of the new video delivery systems being contemplated for introduction in the next few years. Obviously, if 85 percent of the 60 million subscribers are going to be needing setup boxes at $333 each, this represents a $17 billion market to be penetrated. It is no surprise that some computer suppliers claim that there is now pressure to get into the market and to find a way to differentiate the products.

Many companies have already developed video at the desktop for CD-ROM-based multimedia applications (there were over 4500 CD-

ROM titles at press time). The move to delivery of video is a logical extension of these activities. Also, the computer companies need to find applications for both idled mainframes and 100-MIPS workstations; video is one such application enabling these manufacturers to continue selling their products. Hardware companies such as Apple, Digital Equipment Corporation, Hewlett-Packard, IBM, and Silicon Graphics, and software companies such as Microsoft and Oracle have made various entries into the video distribution arena.

For example, as noted in Chap. 7, Digital Equipment Corporation manufactures video servers used by several RBOCs (NYNEX, U S WEST, Ameritech, Rochester Telephone, Stentor, and other providers). The company is also supplying the video server based on the Alpha AXP RISC processor, for a 1995 Swedish I-TV trial supporting 500 households. They also manufacture a product to provide Ethernet LAN connectivity over cable TV facilities, in support of work-at-home applications (they are building equipment that permits interconnection of Ethernet LANs over cable systems, supporting 10-Mbps speeds at a distance of up to 70 miles). Although observers claim that the contracts on video servers may not greatly impact the revenue stream of Digital Equipment Corporation in the immediate future, the contracts should provide positive publicity for the company's RISC processor (Alpha APX).[18] Digital's architecture has been noted for the scalable design and the way it handles the storage of video, enabling the operator to start small, but grow as requirements increase. The system has a tiered storage architecture that enables the most popular movies to be held in instantly accessible memory, with movies that are less in demand stored on hard disk arrays, in addition to a larger collection of "golden oldie" films in a tape library. The company is reportedly becoming involved beyond just selling hardware, through a media production center that compresses video into digital format for delivery over broadband networks. The company was also in the process of developing a new version of the Alpha processor, specifically targeted to video; video servers based on this technology are expected to handle up to 100,000 concurrent sessions. Digital Equipment Corporation expects 10 percent of its total business in 2000 to come from video servers.[19] (Silicon Graphics has also been quoted as indicating that 15 percent of its revenues will come from video servers.[9]) DEC has partnered with TCI, Continental Cablevision, Times Mirror Cable, and others.[1]

Digital Equipment Corporation has announced the availability of a series of "home shopping environments" jointly created with MicroMall. The on-line shopping malls started out with Land's End catalogs, but was expected to add "stores" in the future. The offering provides retailers with the ability to create interactive storefront programs where consumers can navigate with ease. The shopping environments utilize Digital's video servers. Digital has established a Digital Media Center in Tarrytown, N.Y., to support program creation, conversion, compression, and distribution. Digital Equipment is participating in U S WEST's Omaha, Neb., trial, in the NYNEX trial in Rhode Island, in an Australian trial with Telstra/Telecom, and in the Svenska Cable trial in Sweden.

With industrywide mainframe sales falling 10 percent in 1993, mainframe manufacturers in general and IBM in particular see a bonanza in video server opportunities.[19] IBM announced in the press that it was investing $100 million to seed the developments in video distribution for both business and residential users. IBM plans to use its mainframes as video servers and either ATM or a proprietary technology called *packet transfer mode* (PTM)* for the communication technology (packets range from 12 to 128 octets). The server, ES/9000, is a scalable system that can be grown as needed. An early application identified was a partnership with NBC to deliver news to the corporate PC. Another partnership involved ICTV and New Century Communication to provide on-demand services over a traditional coaxial cable system. Their stated direction is to deemphasize the business of selling pure hardware but to partner as needed to become a force in the multimedia arena and generate revenues on the movement of information. For example, IBM, Rochester Telephone, and Cincinnati Bell Information Services have formed a team to study the possibility of using a fiber-optic ring to link cable TV networks in Pennsylvania into a single system called FiberSpan Pennsylvania, by 2000, as a strong competitor to Bell Atlantic. (Cable operators involved in the project include Time Warner Cable, Cablevision Systems, Lenfest Group, and TCI.)[5] In early 1994, IBM sold the ES/9000 mainframe-based video server to the California Polytechnic Institute. The Institute was planning to develop a networked instructional environmental system that serves as a *virtual university* to provide students with on-demand access to full-motion video lectures, demonstrations and self-paced tutorials, electronic mail, and conferencing system. Ameritech, Cox Communications, and Bell Atlantic have also expressed interest in the server.[19] IBM conducted video trials with Bell Atlantic in 1993 to 1994 using multiple Power Series RS/6000s.[19]

IBM and ICTV of Los Gatos, Calif., have announced plans to work together on VOD and I-TV technologies, to develop new products to be sold to the cable TV companies and to the RBOCs. The two companies

---

* Specifically, the *Packetized Advanced Router Integrated System* (PARIS) supports 1.2 Gbps in a packet format on a 1-GHz cable serving 1000 homes in a star-bus arrangement. IBM proposed 550 MHz for analog channels and 450 MHz for digitized channels; an upstream channel of 25 MHz (100 Mbps) is also supported.

were already working together on the 1995 Cox Cable trial in Omaha, Neb. The trial includes VOD, games, musical videos on demand, interactive shopping, and interactive education. In the Omaha trial, IBM is supplying network communication equipment, software, and RS/6000 video servers, while ICTV is providing interactive TV hardware and the software for the headend system.

At press time, Motorola announced a relatively low-cost system called *CableComm,* to enable TV cable systems to support telephone, data, and wireless services. By attaching a control unit at a cable headend and a small box outside a customer's home, the Motorola system will reportedly support wireless or conventional telephone service, video telephony (up to 384 kbps), videoconferencing, and data transmission over the cable.[20] Reportedly, Time Warner is looking at this technology as a first step toward an integrated full-service network. CableComm was to be tested in 1995 market trials; shipments were to start in late 1995.

Intel Corp. and General Instrument announced an arrangement with TCI and Rogers Cablesystems (Canada) to enable the two operators to deliver high-speed data to PCs.[21] Services planned include access to Prodigy and America Online; shopping, entertainment, and other data services will also be supported. A PC adapter board is to be delivered in 1995. The system is for those cable systems that have upgraded the networks to HFC, which have special amplifiers that support two-way communication. The cable operator will need to deploy new networking devices and servers in the headend to make the service a reality.

Intel has enlisted a number of software developers in support of its CablePort project, which aims at delivering high-speed, multimedia interactive services to home-based PCs over a cable TV coaxial cable. The system is being trialed in Castro Valley, Calif., by Viacom, and at another (undisclosed) location by Comcast. Intel was hoping to deliver commercial products by early 1995 and to reach 10 markets by the end of the year and 20 markets by 1996.

Microsoft aims to be a major player in video delivery. It has defined the user interface for interactive services, whether delivered over cable or wireless, onto desktop PCs or TVs. Its first-generation product, Modular Windows, used the same API as Windows and was promoted to developers as an easy way to leverage multimedia PC development onto TV-like devices.[1] Microsoft is now pursuing non-Windows-based technology, which is expected in a number of pilots and trials. Deutsche Bundenpost Telekon is working with Microsoft Corporation to develop a range of multimedia, VDT, and I-TV technologies, in support of services such as teleshopping and video games. The basic GUI platform is based on Windows 95. In the United States, TCI and Microsoft were slated to conduct extensive tests of interactive services to PCs over coaxial cable (TCI was planning to invest $125 million in the Microsoft Network).

One of the main barriers to the rapid introduction of this technology is the difficulty in modifying multimedia PC programs to run on interactive television networks. A number of software companies, including Macromedia, Microware Systems, Oracle, and Microsoft, are addressing this problem.[22] In particular, Macromedia and Microware announced a technology to automate the conversion of PC programs to run on such networks. Currently, a developer of a multimedia program such as an electronic encyclopedia or a restaurant guide may have to spend as much as a half million dollars and a year to rewrite the program for interactive TV. Because the interactive TV market is nascent, as discussed throughout this text, software developers are reluctant to make such investments. Companies planning to test this I-TV concept in about a dozen major trials underway or planned at press time are starved for content other than movies and home shopping.[22]

15.3.2 AT&T

AT&T seems to have taken the position of developing the underlying technology for next-generation video distribution systems so as to be in a position to sell hardware regardless of the outcome of the competitive proceedings (alliances, partnerships, mergers, fights, etc.). The overarching goal (as implied by AT&T's advertisement campaigns) is to deploy an interactive long distance video carriage network for the distribution of video information to local distributors in the United States (e.g., local cable TV companies, RBOCs) and around the world.

By press time, AT&T had won all three video-related contracts that were actually awarded by the regional telephone companies: first Pacific Bell, then Southern New England Telephone, and then Bell Atlantic.[23] GTE will also use AT&T's products.[24]

In the recent past, AT&T demonstrated equipment or announced plans to develop interactive cable capabilities, for example, a SONET-based turnkey system to support video, telephony, and interactive multimedia, called Cable Services Integrated Network (CSIN)*. CSIN is being employed by Pacific Bell with some modifications. AT&T is involved with Ameritech, U S WEST, Pacific Bell, GTE, Time Warner,

---

* This ATM-SONET architecture supports 550 MHz of analog bandwidth, 1.5 Gbps of digital bandwidth, and upstream channels. Two fibers are used to the node to serve 1000 homes. Each of 30 10-MHz slots (for a total bandwidth of 300 MHz) carries 52.84-Mbps SONET streams using 64 QAM; in turn, each of these channels carries 12 digital video streams at about 4 Mbps each.

Viacom, and BellSouth trials. They are the system integrators for Pacific Telesis for the eventual delivery of video to 5 million homes. The AT&T-Viacom-BellSouth trial is a technology and market trial. At the technology level, AT&T is testing its video server technology. Over the 18 months of the trial, the 4000 subscribers will have access to PPV, VOD, MPGs, interactive home shopping, interactive advertisements, information services, interactive educational programming, telephony, and video telephony. AT&T has also announced alliances with providers of video games (e.g., Sega and 3DO) to develop MPGs among households over cable networks. AT&T, TCI, and U S WEST undertook a cooperative market trial for VOD and PPV services. The trial, which was completed in 1994, revealed that the usage rate for the two services offered, Hits at Home and Take One, were in fact about one order of magnitude higher than the buy rates previously assumed from preliminary market studies by Paul Kagan Associates.[25]

AT&T provides an end-to-end integrated video system that includes[26,27,28,29]:

Video server

Fiber-optic transmission equipment (SLC-2000)

Fiber-coaxial distribution system

ATM switch (GCNS-2000)

Video manager

Setup boxes

OS and operations, administration, management, and provisioning equipment

Authoring and digitization tools

In 1994, AT&T was working on two ATM switches in the SN-2000 ATM integrated services platform family, to complement the core GCNS-2000 switch.

On a related issue, AT&T recently purchased McCaw, making it a formidable cellular and PCS provider with 700 wireless licenses and 105 local cellular operations.

15.3.3 MCI

MCI has announced a $20 billion plan to enter new services over the next few years, including the local exchange market, PCS, residential video distribution, and I-TV. This is part of the NetworkMCI blueprint. The initial investment of a few billion dollars is aimed at establishing alternate access networks in the top 20 U.S. markets to save on RBOC-collected access charges. In early 1994, MCI stated that they were going to salvage the early investments they made in SMDS service, and they were going to bring out that service in lieu of ATM, at least in the short term, ostensibly to curtail the expenditures required to bring out ATM. Note that SMDS does not handle video.[30]

In 1993, an alliance with BT was announced, generating over $4 billion for MCI. MCI officers announced that the money would be used to finance acquisitions in the cable television, multimedia, and cellular industries. The entry in the wireless voice-data arena may be facilitated by a partnership announced in 1994 with Nextel, Comcast, Motorola, and NTT, aiming at an aggressive rollout for 95 percent coverage of the United States by 1996. (The partnership with Nextel was later rescinded.) MCI has established an alliance with Jones Intercable to provide telephony services over coaxial cable in Virginia and in Illinois.

15.3.4 Sprint

Sprint Corp. announced in 1994 the formation of a joint venture with three of the nation's largest cable operators, TCI, Cox, and Comcast, to attack the local telephone market. Other cable companies have been invited to join the venture. It has been estimated that $4 billion has been invested in the venture. Sprint will own 40 percent of the venture, TCI 30 percent, Comcast 15 percent, and Cox 15 percent. The 290,000 miles of cable which the partners already have in place could serve about a third of U.S. homes. The cable partners jointly own Teleport Communications. Sprint's long-haul network, the cable operator's fiber and coaxial cable network, and the firms' existing wireless licenses form a formidable backbone for nationwide video, voice, and PCS services.

References

1. A. Reinhardt, "Building the Data Highway," *Byte,* March 1994, pp. 46 ff.
2. *USA Today,* April 26, 1994, p. B1.
3. V. Pasdeloup, "Telecoms Reform Advanced by Hollings' New Bill," *Cable World,* February 7, 1994, p. 2.
4. S. Scully, "From Videotape to Video Servers, Technology Drives PPV," *Broadcasting and Cable,* November 29, 1993, pp. 66 ff.
5. "Cable Optics, Information Gatekeepers," Boston, January 1994, pp. 1 ff.
6. *Communications Technology,* April 1993, pp. 15 ff.
7. *CNYCA Newsletter,* February 1994, pp. 4 ff.
8. B. Burch, "Cable Kings Target Telcos," *Network World,* August 8, 1994, pp. 1 ff.
9. "SGI Ready Opposing Video-server Architectures: Video On Demand Battle," *Electronic Engineering Times,* October 4, 1993, p. 1.
10. B. Santo, "RBOCs Make Case to Scrap Restrictions," *Electronic Engineering Times,* July 11, 1994, p. 10.
11. M. L. Carnevale, "FCC's Take of $617 Million for Licenses Shows Demand for New Paging Services," *The Wall Street Journal,* August 1, 1994, p. A3.

12. G. Lawton, "Interoperating Cable into the Great Opportunity," *Communications Technology,* November 1993, pp. 26 ff.
13. R. E. Chalfant, "MPEG-II with B-Frames: Video Compression Standard for the Decades to Come," *Communications Technology,* April 1993, pp. 42 ff.
14. "TCI/Microsoft Hookup," *Multimedia Week,* March 14, 1994, p. 4.
15. D. Ingelbrecht, "Opening Up Japan," *Cable World,* February 28, 1994, p. 2.
16. "Newswire," *Cable World,* February 28, 1994, p. 3.
17. G. Lawton, "The Bits Are Coming: Deploying Digital For Video Services," *Computer Technology,* April 1993, pp. 38 ff.
18. "DEC Finds More Success with Video Server," *Network Week,* March 4, 1994, p. 1.
19. G. McWilliams, "They Can't Wait to Serve You," *Business Week,* January 24, 1994.
20. G. C. Hill, "Motorola Launching System to Convert Cable Networks for Multimedia Use," *The Wall Street Journal,* May 20, 1994.
21. D. Clark, "Intel to Announce Deals with Firms on Delivery of Data to PCs Via Cable," *The Wall Street Journal,* May 18, 1994.
22. J. Carlton, "Macromedia and Microware to Unveil Technology to Boost Interactive TV," *Wall Street Journal,* August 1, 1994, p. B6.
23. S. McCartney, "DSC Communications Doesn't Get Bell Atlantic Call," *The Wall Street Journal,* May 19, 1994.
24. J. J. Keller, "GTE Corp. Plans to Offer Video Services by End of 1995, Still Seeks Cable Partner," *The Wall Street Journal,* May 25, 1994, p. B3.
25. "AT&T Completes A Video Trial," *Electronic Engineering Times,* July 11, 1994, p. 18.
26. P. Bernier, "GTE, PacTel Video Choose AT&T Gear for Interactive Trials," *Telephony,* January 24, 1994, pp. 6 ff.
27. D. Minoli, "Video Dialtone: Overview," *Datapro Report,* April 1995.
28. D. Minoli, "AT&T Vendor Profile," *Datapro Report on the Convergence of Cable and Telephony,* April 1995.
29. D. Minoli, "Video Dialtone Strategies and Opportunities," *Datapro Report on the Convergence of Cable and Telephony,* April 1995.
30. R. Karpinski, "MCI Pulls SMDS Trigger, Explains ATM Caution," *Telephony,* January 24, 1994, pp. 7 ff.

Index

ABOUT THE AUTHOR

Daniel Minoli is principal consultant with DVI Communications, a leading systems engineering consulting firm specializing in voice, video, and data communications and information processing. He is also a lecturer at Stevens Institute of Technology and an adjunct professor at New York University's Information Technology Institute. Mr. Minoli previously worked for Bellcore, where he was actively involved in developing ATM-based VDT technology and services. He has published numerous books in the areas of broadband technology and basic telecommunications, including *ATM and Cell Relay Service for Corporate Environments,* available from McGraw-Hill.